Lecture Notes in Networks and Systems

Volume 511

The series "Lecture Notes in Networks and Systems" publishes the latest developments in Networks and Systems—quickly, informally and with high quality. Original research reported in proceedings and post-proceedings represents the core of LNNS.

Volumes published in LNNS embrace all aspects and subfields of, as well as new challenges in, Networks and Systems.

The series contains proceedings and edited volumes in systems and networks, spanning the areas of Cyber-Physical Systems, Autonomous Systems, Sensor Networks, Control Systems, Energy Systems, Automotive Systems, Biological Systems, Vehicular Networking and Connected Vehicles, Aerospace Systems, Automation, Manufacturing, Smart Grids, Nonlinear Systems, Power Systems, Robotics, Social Systems, Economic Systems and other. Of particular value to both the contributors and the readership are the short publication timeframe and the world-wide distribution and exposure which enable both a wide and rapid dissemination of research output.

The series covers the theory, applications, and perspectives on the state of the art and future developments relevant to systems and networks, decision making, control, complex processes and related areas, as embedded in the fields of interdisciplinary and applied sciences, engineering, computer science, physics, economics, social, and life sciences, as well as the paradigms and methodologies behind them.

Indexed by SCOPUS, INSPEC, WTI Frankfurt eG, zbMATH, SCImago.

All books published in the series are submitted for consideration in Web of Science.

For proposals from Asia please contact Aninda Bose (aninda.bose@springer.com).

More information about this series at https://link.springer.com/bookseries/15179

Marcelo Zambrano Vizuete ·
Miguel Botto-Tobar · Angela Diaz Cadena ·
Benjamin Durakovic
Editors

Innovation and Research - A Driving Force for Socio-Econo-Technological Development

Proceedings of the CI3 2021

Springer

Editors
Marcelo Zambrano Vizuete
Universidad Técnica del Norte
Ibarra, Ecuador

Angela Diaz Cadena
Universitat de Valencia
Valencia, Spain

Miguel Botto-Tobar
Eindhoven University of Technology
Eindhoven, The Netherlands

Benjamin Durakovic
International University of Sarajevo
Sarajevo, Bosnia and Herzegovina

ISSN 2367-3370 ISSN 2367-3389 (electronic)
Lecture Notes in Networks and Systems
ISBN 978-3-031-11437-3 ISBN 978-3-031-11438-0 (eBook)
https://doi.org/10.1007/978-3-031-11438-0

This Springer imprint is published by the registered company Springer Nature Switzerland AG
The registered company address is: Gewerbestrasse 11, 6330 Cham, Switzerland

Preface

The Second International Conference on Research and Innovation-CI3 2021 was held virtually on September 1–3, 2021. It aimed to disseminate the research project results that are being carried out in different higher education institutions, research centers, and the business sector. CI3 2021 was jointly supported and co-organized by the most relevant Ecuadorian Institutes: Instituto Superior Tecnológico Rumiñahui, Instituto Superior Universitario Bolivariano de Tecnología, Instituto Superior Universitario Central Técnico, Instituto Superior Universitario Espíritu Santo, Instituto Superior Tecnológico José Chiriboga Grijalva, Instituto Superior Universitario ISMAC, Instituto Superior Universitario Policía Nacional del Ecuador and Instituto Superior Universitario Vida Nueva; and sponsored by the Universidad Nacional Mayor de San Marcos (Peru), Universidade Federal de Goiás (Brazil) and City University of New York (USA), and GDEON. The content of this volume is related to the following subjects:

- Communications
- ICT in education
- Electronics and automation
- Energy and environment
- Security
- Software
- Technology trends

CI3 2021 received 162 submissions written in English by 596 authors coming from 18 different countries. All these papers were peer reviewed by the CI3 2021 Program Committee consisting of 55 high-quality researchers. To assure a high-quality and thoughtful review process, we assigned each paper at least three reviewers. Based on the peer reviews, 50 full papers were accepted, resulting in an 31% acceptance rate, which was within our goal of less than 40%.

We would like to express our sincere gratitude to the invited speakers for their inspirational talks, to the authors for submitting their work to this conference, and to the reviewers for sharing their experience during the selection process.

September 2021

Marcelo Zambrano V.
Miguel Botto-Tobar
Ángela Díaz Cadena

Organization

General Chair

Marcelo Zambrano V.	Universidad Técnica del Norte, Ecuador

Editorial Committee

Marcelo Zambrano V.	Universidad Técnica del Norte, Ecuador
Miguel Botto-Tobar	Eindhoven University of Technology, The Netherlands
Ángela Díaz Cadena	Universitat de Valencia, Spain

Publication Chairs

Marcelo Zambrano V.	Universidad Técnica del Norte, Ecuador
Miguel Botto-Tobar	Eindhoven University of Technology, The Netherlands

Organizing Committee

Wladimir Paredes	Instituto Tecnológico Universitario Rumiñahui, Ecuador
Marcelo Zambrano V.	Universidad Técnica del Norte, Ecuador

Steering Committee

Ángel Ernesto Huerta	Instituto Tecnológico Universitario Rumiñahui, Ecuador
Dra. Carmita Suárez	Instituto Tecnológico Universitario Rumiñahui, Ecuador

Program Committee

Daniel Ripalda	Universidad Nacional de la Plata, Argentina
Francesc Wilhelmi	Centre Tecnològic de Telecomunicacions de Catalunya, Spain
Santiago Vidal	Universidad Nacional del Centro de la Provincia de Buenos Aires, Argentina
Jhenny Cayambe	Pontificia Universidad Católica del Ecuador Sede Ibarra, Ecuador
Santiago Ushiña	Likatelec, Ecuador
Ricardo Rosero	Instituto Superior Tecnológico Sucre, Ecuador
Leandro Bezerra De Lima	Universidad Federal de Mato Grosso Do Sul, Brasil
Andrea Carolina Flores	Embraer Defensa y Seguridad, Brasil
Elmer Levano Huamaccto	Universidad Federal Do Parana, Brasil
Ari Lazzarotti	Universidad Federal de Goiás, Brasil
Ellen Synthia Fernandes de Oliveira	Universidad Federal de Goiás, Brasil
Fernanda Cruvinel Pimentel	Universidad Federal de Goiás, Brasil
Pablo Minango	Universidad Estatal de Campinas, Brasil
Danny De La Cruz	Universidad de las Fuerzas Armadas EPE, Ecuador
Johana Tobar Quevedo	Universidad de las Fuerzas Armadas EPE, Ecuador
Angel Jaramillo	Universidad de las Américas, Ecuador
Holger Aníbal Capa Santos	CACES, Ecuador
Alonso Estrada Cuzcano	Universidad Nacional Mayor de San Marcos, Ecuador
Mariana Lima Bandeira	Universidad Andina Simón Bolivar, Ecuador
Norma Molina	Universidad Israel, Ecuador
Elfio Pérez	Universidad Indoamérica, Ecuador
Alfredo Rodriguez Guzmán	Cedepro Ecuador, Ecuador
Diana Portelles Cobas	Cedepro Ecuador, Ecuador
Cristian Tasiguano	Instituto Tecnológico Universitario Rumiñahui, Ecuador
Lizbeth Suarez	Instituto Superior Vida Nueva, Ecuador
Carlos Ruiz	Instituto Superior Vida Nueva, Ecuador
Wilfrido Robalino	Instituto Superior Vida Nueva, Ecuador
Dario Morales	Instituto Superior Vida Nueva, Ecuador
Pamela Villareal	Instituto Superior Vida Nueva, Ecuador
Darwin Tituaña	Instituto Superior Vida Nueva, Ecuador
Ismenia Araujo	Instituto José Chiriboga Grijalva, Ecuador
Luz Marina Rodríguez	Instituto José Chiriboga Grijalva, Ecuador
Isabel Cristina Meléndez	Instituto Superior Tecnológico Libertad, Ecuador
Luis Alzate	Instituto Superior Bolivariano, Ecuador
Noemi Delgado	Instituto Superior Bolivariano, Ecuador
Marítza Salazar	Instituto Superior Bolivariano, Ecuador

Martha Fernandez	Instituto Superior Bolivariano, Ecuador
Néstor Xavier Maya	Instituto Superior Bolivariano, Ecuador
Michael Enrique Carrión	Instituto Superior Bolivariano, Ecuador
Lenin Daniel Valdivieso	Instituto Superior Bolivariano, Ecuador
Miguel Granja	Universidad Internacional del Ecuador, Ecuador
Marcelo Flores	Universidad Politécnica Salesiana, Ecuador
Diego Vizuete	Instituto Superior Sucre, Ecuador

Organizing Institutions

Sponsors Institutions

Contents

Communications

Optimal and Quasi-optimal Relaxation Parameter for Massive MIMO Detector Based on SOR Method

Juan Minango(✉) and Cristian Tasiguano Pozo

Instituto Superior Universitário Rumiñahui, Sangolqui, Ecuador
juancarlos.minango@ister.edu.ec

Abstract. The linear Minimum Mean Square Error (MMSE) detector uses the inversion matrix to detect in massive Multiple-Input Multiple-Output (MIMO) systems, which involves high cubic complexity in reference to the number of transmitting antennas. Thus, the low-complexity detector using the successive over-relaxation (SOR) method was employed in order to avoid the MIMO matrix inversion. In the SOR method, its relaxation parameter was achieved by intensive computational simulations and only for a given massive MIMO system configuration. Thus, in this paper, we develop and propose an approach to obtain the optimal and the quasi-optimal relaxation parameter of the SOR-based detector by using the massive MIMO channel properties such as near orthogonality and large number theory. It is shown that the quasi-optimal relaxation parameter has been obtained as a relation of the number of receiving and the number of transmitting antennas of the MIMO system where the computation of this quasi-optimal relaxation parameter is feasible.

Keywords: Massive MIMO · Successive over-relaxation · Matrix inversion · MMSE · BER

1 Introduction

Masisve multiple-input multiple-output (MIMO) are an essential technology for 5G [1,2]. Basically, massive MIMO uses a big number of receiving antennas in order to communicate with tens of users. Theoretical results shown that linear zero-forcing (ZF) and minimum-mean-square-error (MMSE) achieve near-optimal performance due to suitable propagation in massive MIMO [3–5]. Besides, ZF and MMSE use unfavorable matrix inversion, which results in high complexity for massive MIMO where the number of transmitting user grows [6].

Recently, some research endeavored to further reduce the complexity of MMSE detector [6,7]. Thus, Neumann series (NS) proposed in [7] to convert the matrix inversion to matrix multiplications. However, marginal complexity reduction was achieved. To not compute the inverse of the channel matrix, massive MIMO detection based on successive over-relaxation (SOR) iterative method

M. Zambrano Vizuete et al. (Eds.): CI3 2021, LNNS 511, pp. 3–10, 2022.
https://doi.org/10.1007/978-3-031-11438-0_1

was suggested in [6], which reduces the complexity and achieves a near-optimal performance of linear MMSE detector. Besides, a problem of SOR-based detector is the selection of the relaxation parameter ω, which controls the detector convergence. The authors of [6] obtained ω by means of simulations and to a given massive MIMO system configuration. Thus, we propose an approach to compute the optimal and the quasi-optimal ω for SOR-based detector in any massive MIMO system.

The remainder of this paper is organized as follows. Section 2 introduces the massive MIMO channel model and low-complexity SOR detector. Section 3 develops an approach to obtain the analytical optimal relaxation parameter for SOR-based detector. Section 4 determines the quasi-optimal relaxation parameter for SOR-based detector, which is obtained as a relation between the number of receiving antennas at base station and the number of transmitting users. Simulations are shown in Sect. 5. Finally, the conclusions are presented in Sect. 6.

2 Massive MIMO System

Following [6], massive MIMO systems employ M receiving base station antennas to communicate with K users is considered. Massive MIMO system follows the condition $M \gg K$. At the base station antennas, the received vector $\mathbf{y}$ with dimension $M \times 1$ is:

$$\mathbf{r} = \mathbf{H}\mathbf{x} + \mathbf{n}, \tag{1}$$

with $\mathbf{H}$ being a matrix with dimension $M \times K$, which follows a Rayleigh distribution, whose entries are considered independently and identically distributed (i.i.d.) complex Gaussian random variables with zero mean and unit variance [4], $\mathbf{x}$ represents the transmitted vector with dimension $K \times 1$ and $\mathbf{n}$ denotes the additive white Gaussian noise (AWGN) vector with dimension $M \times 1$ and composed by i.i.d complex Gaussian random variables following the distribution $\mathcal{CN}(0, \sigma_n^2)$.

From (1), the detection of $\mathbf{x}$ by MMSE is achieved as [3]:

$$\hat{\mathbf{x}} = \left(\mathbf{H}^H\mathbf{H} + \sigma^2\mathbf{I}_K\right)^{-1}\mathbf{H}^H\mathbf{r} = \mathbf{F}^{-1}\tilde{\mathbf{r}}, \tag{2}$$

where $\tilde{\mathbf{r}} = \mathbf{H}^H\mathbf{r}$ is the $K \times 1$ matched-filter vector of $\mathbf{r}$ and $\mathbf{F}$ denotes MMSE filter matrix given by:

$$\mathbf{F} = \left(\mathbf{H}^H\mathbf{H} + \sigma^2\mathbf{I}_K\right). \tag{3}$$

The computation of $\mathbf{F}^{-1}$ demands a cubic complexity in reference with K, that is $O\left(K^3\right)$.

2.1 Successive Over-Relaxation Method

Once $\mathbf{F}$ is an Hermitian positive definite (HPD) matrix, the authors of [6] have employed the SOR to iteratively resolve (2), decreasing the complexity from $O\left(K^3\right)$ to $O\left(K^2\right)$. Besides, an important topic of SOR-based detector is the selection of ω, which affects SOR convergence.

By decomposing the matrix $\mathbf{F}$ as:

$$\mathbf{F} = \mathbf{Di} - \mathbf{T} - \mathbf{T}^H, \tag{4}$$

where $\mathbf{Di}$, $\mathbf{T}$ and $\mathbf{T}^H$ are the diagonal, the strictly lower and upper triangular matrices of $\mathbf{F}$. Thus, SOR-based detector detects $\mathbf{x}$ without matrix inversion as follows [6]:

$$\hat{\mathbf{x}}^{(i+1)} = (\mathbf{Di} - \omega \mathbf{T})^{-1} \left[\left((1-\omega)\, \mathbf{Di} + \omega \mathbf{T}^H \right) \hat{\mathbf{x}}^{(i)} + \omega \tilde{\mathbf{r}} \right], \tag{5}$$

where $\hat{\mathbf{x}}^{(i)}$ and ω represent the solution of the i-th iteration and the relaxation parameter. Thus, (5) is given by:

$$\hat{\mathbf{x}}^{(i+1)} = \mathbf{M}_{\mathrm{SOR}} \hat{\mathbf{x}}^{(i)} + \mathbf{Di}, \tag{6}$$

where $\mathbf{M}_{\mathrm{SOR}}$ represents the iteration matrix given by:

$$\mathbf{M}_{\mathrm{SOR}} = (\mathbf{Di} - \omega \mathbf{T})^{-1} \left((1-\omega)\, \mathbf{Di} + \omega \mathbf{T}^H \right), \tag{7}$$

and $\mathbf{d}$ is an iteration vector given by:

$$\mathbf{d} = \left(\frac{1}{\omega} \mathbf{Di} - \mathbf{T} \right)^{-1} \tilde{\mathbf{r}}. \tag{8}$$

Defining the error vector as $\mathbf{e}^{(i)} = \hat{\mathbf{x}}^{(i)} - \hat{\mathbf{x}}$, where $\hat{\mathbf{x}} = \mathbf{F}^{-1}\tilde{\mathbf{r}}$ is the true solution, it follows from (6)–(8) that:

$$\mathbf{e}^{(i)} = \mathbf{M}_{\mathrm{SOR}} \mathbf{e}^{(i-1)}. \tag{9}$$

For the convergence of SOR-based detector, i.e., for the condition:

$$\lim_{i \to \infty} \mathbf{e}^{(i)} = \mathbf{0} \tag{10}$$

to hold for any arbitrary initial error vector $\mathbf{e}^{(0)}$, it is necessary for all the eigenvalues of $\mathbf{M}_{\mathrm{SOR}}$ to lie within the unit circle [8, Theorem 7.2.2].

Note that when ω is equal to the unity, the iterative procedure of SOR-based detector is similar to Gauss-Seidel (GS) based detector [9], which means that GS is a special case of SOR detector. The purpose of using a value of ω other than unity is to reduce the spectral radius of the matrix $\mathbf{M}_{\mathrm{SOR}}$. We denote this spectral radius by $\lambda_{\max}$. Ideally one would choose the best or the optimal relaxation parameter ω_{opt} which minimizes $\lambda_{\max}$. Thus, the convergence rate of SOR-based detector is minimum.

3 Optimal Relaxation Parameter

In the general case, it is difficult to estimate ω_{opt} satisfactorily. For this reason, the authors of [6] obtained an optimal relaxation parameter by means simulations, which is not practical in a real massive MIMO configuration. However, if we can impose certain restrictions on the form of matrix $\mathbf{F}$ given by (3), then, an optimal relaxation parameter can be obtained.

By assuming $\mathbf{F}$ follows Young's Property "A" [10] (see Appendix), that is, for any positive scalar q exists a diagonal matrix $\mathbf{G}_p$ such that:

$$q^{1/2}\mathbf{T} + q^{-1/2}\mathbf{U} = \mathbf{G}_q\left(\mathbf{T} + \mathbf{U}\right)\mathbf{G}_q^{-1}. \tag{11}$$

Then, it is possible to shown that SOR-based detector converges when ω is in the range $0 < \omega < 2$ [6, Lemma 2]. Now, let λ_n be the n-th eigenvalue of the matrix $\mathbf{M}_{\mathrm{SOR}}$ given by (7), then its corresponding $K \times 1$ eigenvector $\mathbf{r}_n$, which is non-zero complex-valued vector, satisfies the following equation [8]:

$$\mathbf{M}_{\mathrm{SOR}}\mathbf{r}_n = \left(\mathbf{Di} - \omega\mathbf{T}\right)^{-1}\left(\left(1-\omega\right)\mathbf{Di} + \omega\mathbf{T}^H\right)\mathbf{r}_n = \lambda_n\mathbf{r}_n, \tag{12}$$

Rewriting (12), we have that: $\left(\left(1-\omega\right)\mathbf{Di} + \omega\mathbf{T}^H\right)\mathbf{r}_n = \lambda_n\left(\mathbf{Di} - \omega\mathbf{T}\right)\mathbf{r}_n$
$\omega\left(\mathbf{T}^H + \lambda_n\mathbf{T}\right)\mathbf{r}_n = \left(\lambda_n + \omega - 1\right)\mathbf{Di}\mathbf{r}_n$
$\omega\lambda_n^{1/2}\left(\lambda_n^{-1/2}\mathbf{T}^H + \lambda_n^{1/2}\mathbf{T}\right)\mathbf{r}_n = \left(\lambda_n + \omega - 1\right)\mathbf{Di}\mathbf{r}_n$, provided $\lambda_n \neq 0$. Using (11) and rearranging (3), we have that:

$$\mathbf{Di}^{-1}\left(\mathbf{T} + \mathbf{T}^H\right)\mathbf{G}_{\lambda_n}^{-1}\mathbf{r}_n = \frac{\lambda_n + \omega - 1}{\omega\lambda^{1/2}}\mathbf{G}_{\lambda_n}^{-1}\mathbf{r}_n. \tag{13}$$

From (13), we can define

$$\mathbf{Di}^{-1}\left(\mathbf{T} + \mathbf{T}^H\right) \tag{14}$$

as the Jacobi iteration matrix and $\mathbf{G}_{\lambda_n}^{-1}\mathbf{r}_n$ as it n-th eigenvector. If the corresponding n-th eigenvalue is μ_n, we find out the following relation:

$$\left(\lambda_n + \omega - 1\right)^2 = \lambda_n\omega^2\mu_n^2. \tag{15}$$

For know values of μ_n and ω this quadratic equation gives the eigenvalues λ_n. Both roots are always eigenvalues. On the other hand, from (15), we have that ω_{opt} must satisfy the relation:

$$\omega_{\mathrm{opt}}^2\mu_{\max}^2 - 4\left(\omega_{\mathrm{opt}} - 1\right) = 0, \tag{16}$$

where $\mu_{\max}$ is the spectral radius given by (14). Thus, from (16), the optimal ω_{opt} for SOR detector can be computed as:

$$\omega_{\mathrm{opt}} = \frac{2}{\sqrt{1-\mu_{\max}^2}}. \tag{17}$$

4 Quasi-optimal Relaxation Parameter

From (17), to compute ω_{opt}, it is necessary to know previously $\mu_{\max}$, which in a real massive MIMO scenario is not feasible. Also, if $\mathbf{H}$ varies fastly, $\mathbf{F}$ changes too. Thus, we need to recompute ω_{opt} for each detection, which demands an incremental complexity. However, it is possible to determine a quasi-optimal relaxation parameter $\hat{\omega}_{\mathrm{opt}}$ as we will show.

Considering the Jacobi iteration matrix given by (14), we can rewrite it as:

$$\mathbf{Di}^{-1}\left(\mathbf{T}+\mathbf{T}^{H}\right)=\mathbf{Di}^{-1}\left(\mathbf{F}-\mathbf{Di}\right)=\mathbf{Di}^{-1}\mathbf{F}-\mathbf{I}_{K}. \tag{18}$$

Besides, employing the law of large numbers [11], once $M \gg K$, the matrix $\mathbf{Di}^{-1}$ can be computed by $\frac{1}{M}\mathbf{I}_K$. Then, (18) can be approximated by:

$$\mathbf{Di}^{-1}\left(\mathbf{T}+\mathbf{T}^{H}\right)\approx\frac{1}{M}\mathbf{F}-\mathbf{I}_{K}. \tag{19}$$

As our goal is find the spectral radius $\mu_{\max}$ of he approximated Jacobi iteration matrix given by (19), we have that:

$$\mu_{\max}\left(\frac{1}{M}\mathbf{F}-\mathbf{I}_{K}\right)=\frac{1}{M}\mu_{\max}\left(\mathbf{F}\right)-1. \tag{20}$$

On the other hand, since $\mathbf{F}$ given by (3) follows complex central Wishart distribution [3], we have that the matrix $\mathbf{F}$ has a spectral radius given by:

$$\mu_{\max}\left(\mathbf{F}\right)\cong M\left(1+\sqrt{\frac{1}{\alpha}}\right)^{2}, \tag{21}$$

with $\alpha = M/K$ denotes the loading factor. By replacing (20) and (21) in (17), and performing some mathematical manipulation, the quasi-optimal relaxation parameter for SOR-based detector is given by:

$$\hat{\omega}_{\mathrm{opt}}\cong\frac{2}{1+\sqrt{1-\frac{1}{\alpha}\left(2+\sqrt{\frac{1}{\alpha}}\right)^{2}}}. \tag{22}$$

From (22), we can observe that $\hat{\omega}_{\mathrm{opt}}$ depends only α which is deterministic once it is previously known M and K. Thus, once $\mathbf{H}$ and $\mathbf{F}$ change, $\hat{\omega}_{\mathrm{opt}}$ is not compute again.

5 Simulation Results

Figure 1 presents the effects of the relaxation parameter ω in the range of $0 < \omega < 2$, on the spectral radius $\lambda_{\max}$ for different loading factors α. From Fig. 1, we can observe that the curves have a vertical asymptote from the left at $\omega = \breve{\omega}_{\mathrm{opt}}$ and a linear behavior for $\omega > \breve{\omega}_{\mathrm{opt}}$, where $\breve{\omega}_{\mathrm{opt}}$ denotes the simulated optimal relaxation parameter.

In Table 1, we proceed to compare the simulated $\breve{\omega}_{\mathrm{opt}}$ given in Fig. 1, the analytical ω_{opt} given by (17) and the approximated $\hat{\omega}_{\mathrm{opt}}$ given by (22) against different loading factors α. It is evident from the Table 1 that the simulated $\breve{\omega}_{\mathrm{opt}}$

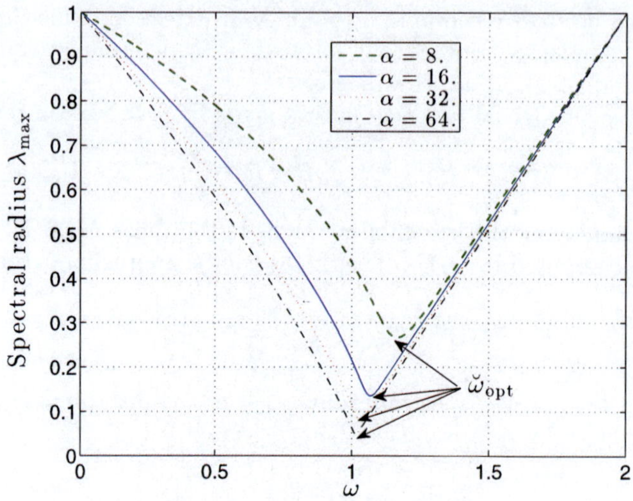

Fig. 1. Spectral radius $\lambda_{\max}$ versus ω for a variable loading factors $\alpha = M/K$.

Table 1. Optimal relaxation parameter comparisons for different loading factors $\alpha = M/K$.

$\alpha = \frac{M}{K}$	$\breve{\omega}_{\mathrm{opt}}$ given in Fig. 1	ω_{opt} given by (17)	$\hat{\omega}_{\mathrm{opt}}$ given by (22)
8	1.146	1.142	1.286
16	1.069	1.067	1.094
32	1.034	1.036	1.040
64	1.016	1.017	1.020

and the analytical ω_{opt} converge to the same results, especially when α increases, which shows that the optimal relaxation parameter can be obtained through the analytical expression given by (17) contrary to intensive simulations as [6]. On the other hand, $\hat{\omega}_{\mathrm{opt}}$ is quite close to ω_{opt}, which shows the efficiency of $\hat{\omega}_{\mathrm{opt}}$ and its easy computation for different loading factors α in real massive MIMO scenarios.

Finally, Fig. 2 compares the ω_{opt} and $\hat{\omega}_{\mathrm{opt}}$ given by (17) and (22), respectively, against the loading factor α. We can see a gap between ω_{opt} and $\hat{\omega}_{\mathrm{opt}}$ which is despicable for a high α.

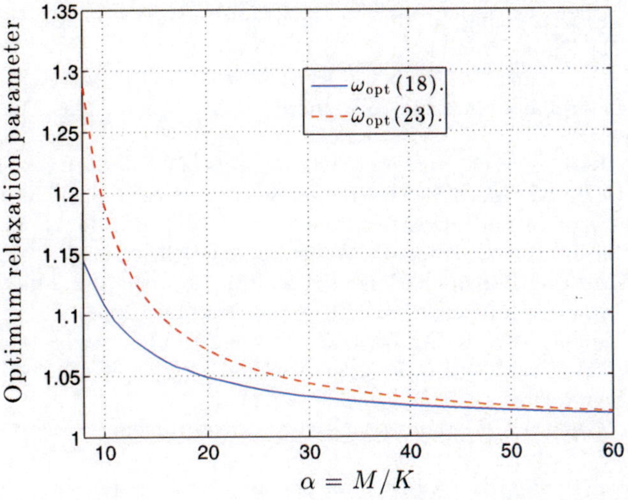

Fig. 2. Comparison between ω_{opt}, and $\hat{\omega}_{\text{opt}}$ versus α.

6 Conclusion

In this paper, we have obtained the analytical optimal relaxation parameter of SOR detector used in massive MIMO. Further, the orthogonal channel matrix asymptotic properties allow to propose a not complicated method to compute the quasi-optimal relaxation parameter, which depends only on the number of receiving and the number of transmitting antennas. This quasi-optimal relaxation parameter results in an easy way to compute and ideal for practical massive MIMO system configuration.

A Property "A"

A given matrix **A** has Property "A" if there exist two disjoint sub-sets S and T of V, the set of the first N integers, such that $S \cup T = V$ and if $a_{ij} \neq 0$ then either $i = j$ or $i \in S$ and $j \in T$, or $i \in T$ and $j \in S$. An equivalent definition is as follows: the matrix **A** has Property "A" if there exists a vector $\mathbf{q} = (q_1, q_2, \ldots, q_N)$ with integral components such that if $a_{ij} \neq 0$ and $i \neq j$ then $|q_i - q_j| = 1$.

References

1. Al-Falahy, N., Alani, O.Y.: Technologies for 5G networks: challenges and opportunities. IT Prof. **19**(1), 12–20 (2017)
2. Marzetta, T.L.: Noncooperative cellular wireless with unlimited numbers of base station antennas. IEEE Trans. Wirel. Commun. **9**(11), 3590–3600 (2010)

3. Rusek, F., et al.: Scaling up MIMO: opportunities and challenges with very large arrays. IEEE Signal Process. Mag. **30**(1), 40–60 (2013)
4. Hoydis, J., ten Brink, S., Debbah, M.: Massive MIMO in the UL/DL of cellular networks: how many antennas do we need? IEEE J. Sel. Areas Commun. **31**(2), 160–171 (2013)
5. Ngo, H.Q., Larsson, E.G., Marzetta, T.L.: Aspects of favorable propagation in massive MIMO. In: European Signal Processing Conference (EUSIPCO). Proceedings of the European Signal Processing Conference, EURASIP, pp. 76–80 (2014)
6. Gao, X., Dai, L., Hu, Y., Wang, Z., Wang, Z.: Matrix inversion-less signal detection using SOR method for uplink large-scale MIMO systems. In: 2014 IEEE Global Communications Conference, pp. 3291–3295, December 2014
7. Wu, M., Yin, B., Wang, G., Dick, C., Cavallaro, J.R., Studer, C.: Large-scale MIMO detection for 3GPP LTE: algorithms and FPGA implementations. IEEE J. Sel. Top. Signal Process. **8**(5), 916–929 (2014)
8. Bjorck, A.: Numerical Methods for Least Squares Problems. Society for Industrial and Applied Mathematics (1996)
9. Greenbaum, A.: Iterative Methods for Solving Linear Systems. Society for Industrial and Applied Mathematics, Philadelphia (1997)
10. Young, D.: Iterative methods for solving partial difference equations of elliptic type. Trans. Am. Math. Soc. **76**(1), 92–111 (1954)
11. Voiculescu, D.: Limit laws for random matrices and free products. Invent. Math. **104**(1), 201–220 (1991). http://eudml.org/doc/143880

Co-channel Interference Mitigation Using Convolutional Enconder via Joint Decoding Viterbi Algorithm

Juan Minango[1(✉)], Marcelo Zambrano[1,2], and Jorge Caraguay[2]

[1] Instituto Superior Universitário Rumiñahui, Sangolqui, Ecuador
juancarlos.minango@ister.edu.ec

[2] Universidad Técnica del Norte, Ibarra, Ecuador

Abstract. This paper proposes and develops a modification in the traditional Viterbi algorithm for a co-channel interference scenario in cellular networks where the desired and interferer users employ convolutional encoder. Thus, a decoding that explicitly take into account the error correction encoded and modulation format of both desired and interferer user, in order to decode jointly and combat co-channel interference is proposed. Furthermore, it is demonstrated by simulation that the joint decoding Viterbi algorithm proposed outperforms the traditional Viterbi algorithm, which assumes that interference co-channel is Gaussian. Further the proposed joint decoding Viterbi algorithm does not present bit error rates floors even in presence of strong co-channel interference user. The simulation results show that the joint decoding Viterbi algorithm presents a few dB of loss when compared to a system without co-channel interference. Thus, the results presented in this paper can be useful to related applications such as spatial multiplexing and Non-orthogonal multiple access NOMA systems.

Keywords: Co-channel interference · Convolutional codes · Viterbi algorithm

1 Introduction

The rapid growth of wireless networks is expected to continue into the coming years i.e. 5G and 6G. Thus, one of the main problems is the interfering co-channel which is an issue for increase the cell throughput [1].

Co-channel interference is treated in some papers as part of thermal noise, even it is considered as a Gaussian stochastic process [2–4]. Thus, a no correct of co-channel interferes can lead for an analysis far from reality. Interference randomization techniques used to use the interference more noise-like were proposed in [5–7]. In these paper the presence of one dominant co-channel due the co-cells is considered.

A co-channel interfering is different from Gaussian noise in its statistical characteristics. We are going to suppose that co-channel interference signal is

M. Zambrano Vizuete et al. (Eds.): CI3 2021, LNNS 511, pp. 11–21, 2022.
https://doi.org/10.1007/978-3-031-11438-0_2

similar to the target signal, that is, both employ error correction encoded and hereafter modulated by a constellation.

In the 5G and 6G a proper treatment of interfering co-channel is desired to performance improvement [8,9]. In this way, new receiver designs with advanced interfere management are essential.

Thus, these new receivers detect jointly the signals from the desired and co-channel interference users. Thus, new receivers bring performance gains over ordinary detection techniques. Thus, we propose and develop a modification to the Viterbi algorithm to combat the interfering co-channel user in a cellular system.

Section 2 presents the channel model, taking into consideration the presence of a strong interfering user. Section 3 describes the Viterbi algorithm and the modified Viterbi algorithm which takes in consideration co-channel interference. Section 4 presents the performance analysis for both algorithms. In Sect. 5 numerical simulations and comparisons between both algorithms are presented. Finally, Sect. 6 shows the conclusions.

2 System Model

Figure 1 shows a digital transmission system composed by two users: a desired and a co-channel interferer. In the system being considered, the desired and interferer users, generate sequences of information bits denoted by $\mathbf{b_0}$ and $\mathbf{b_1}$, respectively.

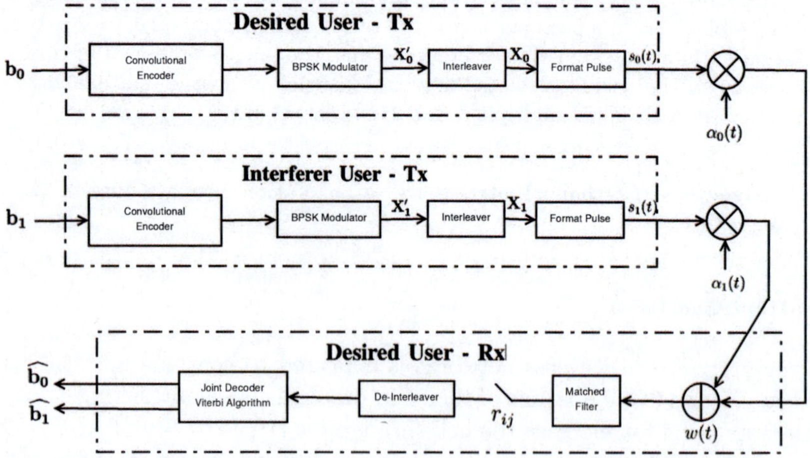

Fig. 1. Digital transmission system with one dominant co-channel interfere.

It is considered $\mathbf{b_0}$ and $\mathbf{b_1}$ that the probabilities of a 0 or 1 bits ocurring are the same in each sequence. Considering only the desired user, their information bits sequence $\mathbf{b_0}$ is encoded by a convolutional encoder with rate $\mathrm{R}_0 = k_0/n_0$ and with memory elements v_0, where k_0 and n_0 are the number of inputs and outputs, respectively.

The convolutional encoder output is multiplexed, and each one of the encoded bits are modulated by a BPSK modulator, generating the BPSK symbols sequences $\mathbf{X}'$. The BPSK symbols sequence $\mathbf{X}'$ enter into an interleaving with infinity depth, producing the sequence $\mathbf{X}$. Finally, the sequence of BPSK symbols passes by the pulse $g(t)$ and then it is transmitted. Thus, the transmitted signal by the desired user can be writeen by:

$$s_0(t) = \sum_{i=-\infty}^{\infty} Ax_{0_{ij}} g(t - iT_b)\cos(2\pi f_c t), \tag{1}$$

where A is the amplitude, $x_{0_{ij}}$ denotes coded BPSK related to the i-th symbol of the j-th branch in the trellis generated by the convolutional encoder of the desired user, T_b is the bit period, $g(t - iT_b)$ is the pulse format that satisfies the Nyquist criterion for zero inter-symbol-interference (ISI) and f_c is the carrier frequency. The pulse energy of the pulse is normalized such that $\int_0^{T_b} |g(t - iT_b)|^2 \, dt = 1$.

Likewise, the co-channel interferer user whose transmitter performs the same procedure as the desired user, thus, the transmitted signal by co-channel interferer user is given by:

$$s_1(t) = \sum_{i=-\infty}^{\infty} \beta A x_{1_{ij'}} g(t - iT_b)\cos(2\pi f_c t), \tag{2}$$

where β is an amplitude factor to vary the signal-to-interference relation, and $x_{1_{ij'}}$ denotes the i-th coded BPSK symbol of j'-th branch in the trellis generated by the convolutional encoder of the interferer user.

Both signals $s_0(t)$ and $s_1(t)$ are transmitted at the same time, in other words both signals are synchronous, which represents the worst case [8], thus, the received signal at the base station is:

$$r(t) = \alpha_0(t)s_0(t)\cos(2\pi f_c t + \phi_0) + \beta\alpha_1(t)s_1(t)\cos(2\pi f_c t + \phi_1) + w(t), \tag{3}$$

where $\alpha_0(t)$ and $\alpha_1(t)$ are the fading amplitude of the desired and interferer user in a slow non-selective Rayleigh fading channel, respectively, $w(t)$ is the additive white gaussian noise (AWGN), with variance $\sigma^2 = (N_0 R_b \mathrm{R}_0)/4$, where N_0 is the unilateral noise power spectral density.

The received signal given in (3) is passed through a filter matched to the pulse $g(t)$ that we are going to suppose rectangular and sampled every T_b seconds. Thus, each sample r_i can be written as:

$$r_i = \frac{\alpha_{0_i} A x_{0_i}}{2} + \frac{\beta\alpha_{1_{i'}} A x_{1_{i'}} \cos(\phi_0 - \phi_1)}{2} + w_i. \tag{4}$$

We have also considered a perfect channel estimator, which knows the fading amplitude at each bit time interval in order to be used at the Viterbi decoding process. Using (4), the signal-to-interference relation (SIR) is written as:

$$\frac{S}{I} = \frac{1}{\beta^2}, \tag{5}$$

where $\overline{\alpha_0^2} = \overline{\alpha_1^2}$ and $\overline{s_0^2(t)} = \overline{s_1^2(t)}$.

Thus, the amplitude parameter β depends on the SIR as:

$$\beta = \sqrt{\frac{1}{S/I}}. \tag{6}$$

3 Viterbi Algorithm

In this section, we first describe the ordinary Viterbi algorithm. It ignores the interference, that is treated as part of the AWGN noise. Then, a modification to the ordinary Viterbi algorithm is proposed, which we call as Joint Decoding Viterbi Algorithm (JDVA). This modification allows it to obtain estimates of the information bit sequence for both the desired and interferer user.

3.1 Ordinary Viterbi Algorithm

Ordinary Viterbi algorithm is a maximum-likelihood sequence estimator, which chooses the path with the lowest metric in a trellis. Thus, $n_b = 2^{k_0+v_0}$ metrics in each trellis section must be calculated, where n_b is the number of branches in each trellis section, k_0 is the number of inputs to the encoder and v_0 represents the number of memories associated to each input of the encoder.

The Viterbi algorithm uses the add-compare-select operation for the state metric calculation, considering all branches that reach a given state. Thus, the metric of each branch is calculated by:

$$m = \sum_{j=1}^{n_b} \left| r_{ij} - \alpha_{0_i} \widehat{x_{0_{ij}}} \right|^2, \tag{7}$$

where r_{ij} is the received sample amplitude corresponding to the i-th BPSK symbol given by (4) of the j-th branch in the desired user trellis.

In (7), the co-channel interferer user are ignored. In other words, the interference term $\beta\alpha_{1_{i'}} Ax_{1_{i'}}$ in (4) is lumped together with the additive white gaussian noise w_i.

3.2 Joint Decoding Viterbi Algorithm (JDVA)

In the proposed modified Viterbi algorithm, the convolutional encoder and BPSK constellation of the co-channel interferer is taken into consideration. Thus, the path with the lowest metric in a new trellis, to which we call concatenated or combined trellis, is chosen. The lowest metric path contains the sequence of bit information estimates, $\widehat{\mathbf{b_0}}$ and $\widehat{\mathbf{b_1}}$ of the desired and interferer, respectively.

In order to generate the new trellis for modified Viterbi algorithm, we must taken into consideration both the convolutional encoder used by the desired and interferer. Thus, each section of the new trellis, concatenate or combine the trellis sections of the desired and interferer. Thus, we need to calculate $n_b = 2^{(k_0+v_0+k_1+v_1)}$ metrics in each concatenated trellis section. Therefore, the metric of each branch in the concatenated trellis is calculated by:

$$m_{(JDVA)} = \sum_{j=1}^{n_0} \sum_{j'=1}^{n_1} \left| r_{ij} - \alpha_{0_i} \widehat{x_{0_{ij}}} - \beta \alpha_{1_i} \widehat{x_{1_{ij'}}} \right|^2, \tag{8}$$

where $\widehat{x_{1_{ij'}}}$ is the i-th BPSK symbol result from the convolutional coding process and n_1 is the number of encoded bits of the interferer user convolutional encoder.

On the other hand, for the JDVA, the number of states resulting from the concatenated trellis is the product of the number of states of the desired and interferer trellis, which is given by:

$$N_{\text{States}} = 2^{v_0} 2^{v_1}, \tag{9}$$

where v_1 denotes the number of memory elements in the convolutional coder of the interferer user, respectively.

4 Performance Analysis

The performance in terms of bit error rate (BER) is analysed for both Viterbi algorithms in a Rayleigh fading channel with AWGN in this section.

4.1 Ordinary Viterbi Algorithm

The BER for the ordinary Viterbi algorithm in the presence of one interferer was analysed in [10]. Thus, the BER upper bound for binary convolutional codes and BPSK modulation under Rayleigh fading channel is given by:

$$P_b \leq \frac{1}{k_0} \sum_{d=d_{free}}^{\infty} B_d \int_0^{\infty} \int_0^{\infty} \left[\frac{1}{2} Q \left((\alpha_0 - \beta \alpha_1) \sqrt{\frac{2d\text{R}E_b}{N_0}} \right) \right] f(\alpha_0) f(\alpha_1) d_{\alpha_0} d_{\alpha_1}, \tag{10}$$

with d_{free} being the free distance of both convolutional code, B_d is the total number of non-zero information bits on all paths of distance d, R is the code rate, $f(\alpha_0)$ and $f(\alpha_1)$ are the Rayeligh distribution of the fading amplitude of the desired and interferer user, respectively.

4.2 Joint Decoding Viterbi Algorithm (JDVA)

The upper bound of BER for a convolutional encoder with concatenate trellis decoding can be written as:

$$P_b \leq \frac{1}{k} \sum_{d=d_{free}}^{\infty} B_{d_{(J.D)}} P_d(\mathbf{X}, \hat{\mathbf{X}}), \tag{11}$$

where k is the number of inputs for the concatenated convolutional code given by $k = k_0 + k_1$, $P_d(\mathbf{X}, \hat{\mathbf{X}})$ is the pairwise error probability of the concatenated trellis when the Hamming distance between the concatenated transmitted sequence $\hat{\mathbf{X}} = \{\mathbf{x_1}, \mathbf{x_2}, \cdots, \mathbf{x_i}, \cdots\}$ composed by coded BPSK symbols from de desired and interferer and the concatenated estimated sequence from the concatenated trellis $\hat{\mathbf{X}} = \{\widehat{\mathbf{x_1}}, \widehat{\mathbf{x_2}}, \cdots, \widehat{\mathbf{x_i}}, \cdots\}$ is d.

Note that each element of $\mathbf{X}$, i.e. $\mathbf{x_i}$, is a vector composed by the concatenation of BPSK symbols transmitted by the desired and interferer user, that is $\mathbf{x_i} = \left[\alpha_{0_i} x_{0_{ij}}, \alpha_{1_i} x_{1_{ij'}}\right]$.

Considering that the concatenated transmitted sequence $\mathbf{X}$ has finite length P, that is $\mathbf{X} = \{\mathbf{x_1}, \mathbf{x_2}, \cdots, \mathbf{x}_P\}$, the total metric from (8) is given by:

$$\mathbf{m}(\mathbf{r}, \mathbf{X}) = \sum_{p=1}^{P} \sum_{j=1}^{n_0} \sum_{j'=1}^{n_1} \mid r_{p_{ij}} - \alpha_{0p_i} \widehat{x_{0p_{ij}}} \beta \alpha_{1p_i} \widehat{x_{1p_{ij'}}} \mid^2 . \tag{12}$$

A pairwise error occurs when the sequence $\hat{\mathbf{X}}$ is chosen instead of $\mathbf{X}$ whenever $\mathbf{m}(\mathbf{r}, \hat{\mathbf{X}}) \geq \mathbf{m}(\mathbf{r}, \mathbf{X})$. Thus, the pairwise error is given by:

$$P_d(\mathbf{X}, \hat{\mathbf{X}}) = P\left\{\mathbf{m}(\mathbf{r}, \hat{\mathbf{X}}) \geq \mathbf{m}(\mathbf{r}, \mathbf{X}) \mid \mathbf{X}\right\}, \tag{13}$$

where conditioning on $\mathbf{X}$ in (13) indicates the components of observation $\mathbf{r}$ have to be computed assuming that $\mathbf{X}$ was transmitted. Using the same procedure as [11], the pairwise error probability for the concatenate convolutional code can be evaluate as:

$$P_d(\mathbf{X}, \hat{\mathbf{X}}) = Q\left(\sqrt{\frac{2RE_b}{N_0} \sum_{\mathbf{n}\in\eta} |\mathbf{x_n} - \hat{\mathbf{x_n}}|^2}\right) \tag{14}$$

where $R = k/v$ with $v = v_1 + v_1$ is the code rate for the concatenated trellis and η is the set of all $\mathbf{n}$ for which $\mathbf{x_n} \neq \widehat{\mathbf{x_n}}$.

Note that implicitly $\mathbf{x_n}$ and $\widehat{\mathbf{x_n}}$ are conditioned on the fading amplitudes since $\mathbf{x_n} = \left[\alpha_{0n_i} x_{0n_{ij}}, \alpha_{1n_i} x_{1n_{ij'}}\right]$. Therefore, the (14) must be evaluated numerically as (10).

5 Simulation Results

In this section, the performance in terms of BER for the modified Viterbi algorithm is evaluated by means of Monte Carlo simulation versus different signal-to-noise ratio $\overline{\gamma_b}$. Performance results of the proposed decoder are compared with ordinary Viterbi algorithm.

Table 1. Convolutional parameter I.

Convolutional encoder I	
Generator matrix	$G = [7, 5]_8$
Encoder rate	$R = 1/2$
Number of memory elements	$v = 2$

Figure 2 presents BER as a function of $\overline{\gamma_b} = \overline{\alpha_0} E_b/N_0$ in dB, where the desired and interferer use convolutional encoder with same parameters that are presented in Table 1, for $S/I = 3$ dB. In Fig. 2 clearly notice that the ordinary Viterbi algorithm has a very poor performance in terms of BER. For instance, BER is approximately $5 \cdot 10^{-2}$ for $\overline{\gamma_b} = 10$ dB. On the other hand, the joint decoding Viterbi algorithm presents a BER that falls as $\overline{\gamma_b}$ increases. Thus, for $\overline{\gamma_b} = 10$ dB, the BER is lower than 10^{-3}. However, the joint decoding Viterbi algorithm performance for BER $= 10^{-3}$ is worse only in approximately 1.5 dB when compared with a system without the presence of co-channel interferer.

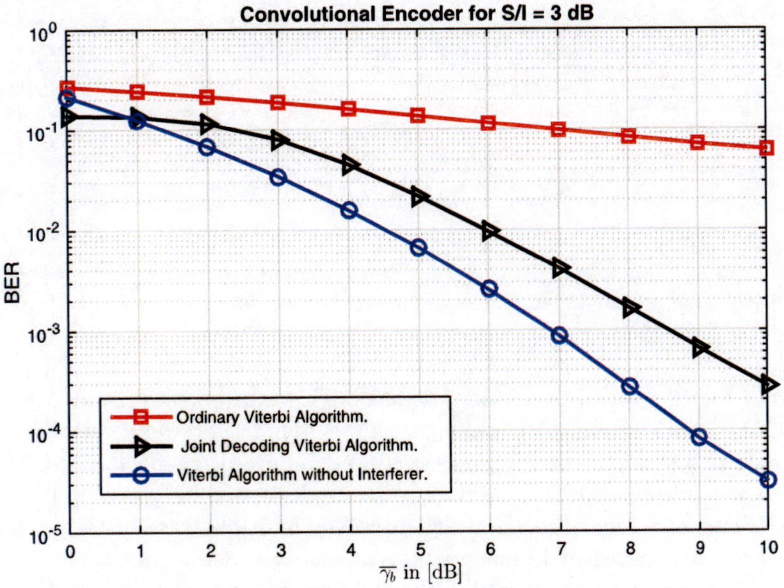

Fig. 2. BER as a function of $\overline{\gamma_b}$, where the desired and interferer user, use convolutional codes with same parameter in Rayleigh fading channel, for a $S/I = 3$ dB.

Figure 3 shows BER verus $\overline{\gamma_b}$ for $S/I = 0$ dB where we notice that the ordinary Viterbi algorithm presents a BER floor of approximately 1/4, which represents a poor performance. On the other hand, the joint decoding Viterbi algorithm does not exhibit an error floor in the BER. Its performance improves

10 dB in $\overline{\gamma_b}$ per decade of BER, making this decoder an optimal tool to combat the co-channel interference effects, especially in situations where there are a small ratio of S/I. In other words, in situations where the desired and interferer user transmit with same power. However, note that the joint decoding Viterbi algorithm has the performance loss of 1.5 dB compared to a system without interference.

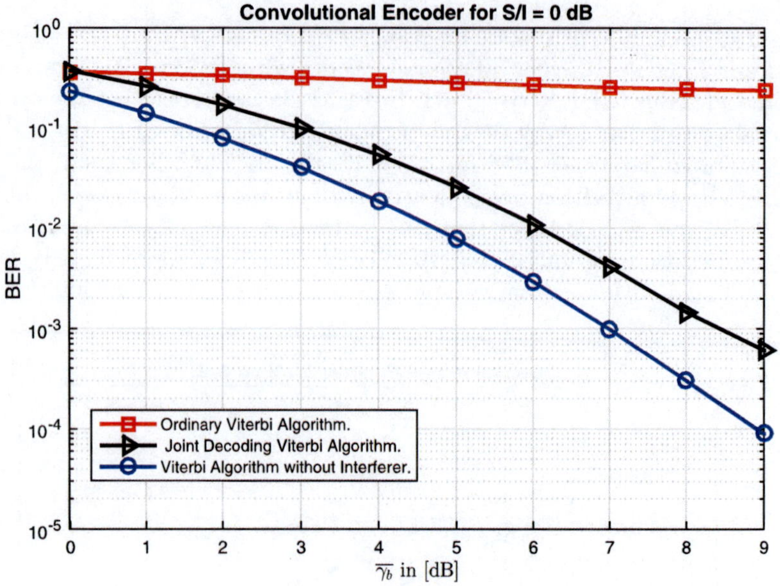

Fig. 3. BER as a function of $\overline{\gamma_b}$, where the desired and interferer user, use convolutional codes with same parameter in Rayleigh fading channel, for a $S/I = 0$ dB.

From Fig. 3, we can conclude that although both users transmit with same power, the joint decoding Viterbi algorithm is able to obtain an estimate of the information bits sequence of the desired and interferer users. For this reason, our decoder is an excellent candidate to be implemented in spatial multiplexing MIMO systems, which are systems that employ multiple transmitting antennas, where each one transmit independent symbols, in order to boost the bit rate. Thus, JDVA decoder can be used in the absence of interference. Thus, JDVA can be translated to the ability of decoding spatial transmit sequence of the desired user.

Considerer the transmission scheme of Fig. 1 from the point of view of spatial multiplexing. Assume that now both information bits of desired and interferer user belong to a spatial multiplexing MIMO with two transmitting antennas.

Thus, in Fig. 4 the performance of a spatial multiplexing MIMO system with two transmitting $N_t = 2$ and one receiving antenna $N_r = 1$ is presented, which uses the joint decoder or concatenated trellis in order to decode their information bits. Furthermore, this system is compared to a spatial multiplexing MIMO hard decision system, that is, BPSK symbols of each transmitting antenna are detected and demodulated through a maximum likelihood detector (MLD) and then are decoded independently by two conventional Viterbi algorithms. In order to maintain the total transmitting power constant, the power transmitted by each antenna is half of the total transmitting power.

In Fig. 4, we observe clearly that JDVA outperforms notably in performance to spatial multiplexing MIMO hard decision system.

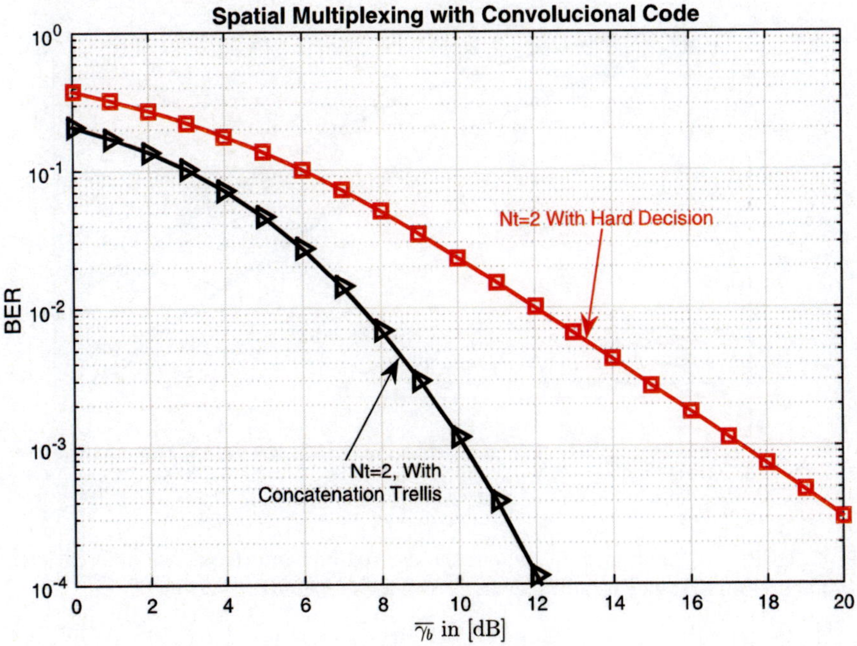

Fig. 4. BER as a function of $\overline{\gamma_b}$ for spatial multiplexing MIMO system with two transmitting and one receiving antenna with BPSK modulation and convolutional code in Rayleigh fading channel.

Finally, we consider that the interferer uses a different convolutional encoder given in Table 2.

Figure 5 show BER for a $S/I = 3$ dB. We observe that ordinary Viterbi algorithm exhibit an error floor in the BER, while JDVA does not. However, JDVA is worse by 2 dB than a system without interferer. Note that JDVA performs well even when the desired and interferer make use of different convolutional encoders.

Table 2. Convolutional parameter II.

Convolutional encoder II	
Generator matrix	$G = [64, 74]_8$
Encoder rate	$R = 1/2$
Number of memory elements	$v_1 = 3$

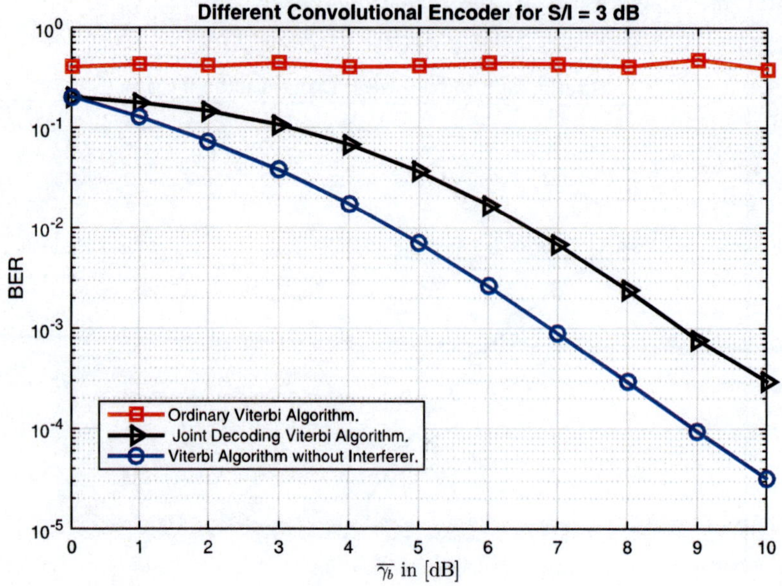

Fig. 5. BER as a function of $\overline{\gamma_b}$, where the desired and interferer, use different convolutional codes in Rayleigh fading channel, for a $S/I = 3$ dB.

6 Conclusions

This paper proposed a modification in the Viterbi algorithm in order to decoding jointly the information bits sequence of desired and interferer user. Such modification was built by concatenating the trellis desired and interferer users. It is shown by simulation that the proposed Joint Decoding Viterbi Algorithm outperform the conventional Viterbi algorithm and does not exhibit BER floors. Thus, the absence of BER floor for the Joint Decoding Viterbi Algorithm makes its use attractive by the fact that makes an interference limited channel into a fading limited channel. Therefore, although advanced decoding schemes require more information about interferer and are more complex to implement, the improved performance can make their use attractive to combat co-channel. Further, it was shown that Joint Decoding Viterbi Algorithm works even when the desired and interferer make use of different convolutional encoders, and that in addition

Joint Decoding Viterbi Algorithm is an optimal candidate to be used in spatial multiplexing MIMO systems.

References

1. Leinonen, M.E., Jokinen, M., Tervo, N., Kursu, O., Pärssinen, A.: Radio interoperability in 5G and 6G multiradio base station. In: 2020 IEEE 92nd Vehicular Technology Conference (VTC2020-Fall), pp. 1–5 (2020)
2. Lopez-Perez, D., Guvenc, I., de la Roche, G., Kountouris, M., Quek, T.Q., Zhang, J.: Enhanced intercell interference coordination challenges in heterogeneous networks. IEEE Wirel. Commun. **18**(3), 22–30 (2011)
3. Orooji, M., Soltanmohammadi, E., Naraghi-Pour, M.: Evaluating the effects of co-channel interference in wireless networks. In: 2013 IEEE International Conference on Acoustics, Speech and Signal Processing, pp. 4693–4697 (2013)
4. Huo, Y., Dong, X., Xu, W.: 5G cellular user equipment: from theory to practical hardware design. IEEE Access **5**, 13992–14010 (2017)
5. Etkin, R.H., Tse, D.N.C., Wang, H.: Gaussian interference channel capacity to within one bit. IEEE Trans. Inf. Theory **54**(12), 5534–5562 (2008)
6. Gulati, K., Chopra, A., Evans, B.L., Tinsley, K.R.: Statistical modeling of co-channel interference. In: GLOBECOM 2009 - 2009 IEEE Global Telecommunications Conference, pp. 1–6 (2009)
7. Liu, Y., Qin, Z., Elkashlan, M., Ding, Z., Nallanathan, A., Hanzo, L.: Nonorthogonal multiple access for 5G and beyond. Proc. IEEE **105**(12), 2347–2381 (2017)
8. Ma, W., Zhao, H., Liu, Y., Shao, S., Pan, W.: A co-channel interference rejection method for 5G ultra dense heterogeneous networks. In: 2018 IEEE International Conference on Communications Workshops (ICC Workshops), pp. 1–5 (2018)
9. Gulati, K., Evans, B.L., Tinsley, K.R.: Statistical modeling of co-channel interference in a field of poisson distributed interferers. In: 2010 IEEE International Conference on Acoustics, Speech and Signal Processing, pp. 3490–3493 (2010)
10. Viterbi, A.: Convolutional codes and their performance in communication systems. IEEE Trans. Commun. Technol. **19**(5), 751–772 (1971)
11. Bhuyan, N.H.M., Ahmed, M., Faisal, M., Salauddin, M.: Bit error rate performance analysis of convolutional coded orthogonal frequency-division multiplexing system. In: 2016 57th International Scientific Conference on Power and Electrical Engineering of Riga Technical University (RTUCON), pp. 1–6 (2016)

Direction Finding with Angle Estimation in Antenna Array Using Radiation Pattern

Ruben León[1,3], Alexis Tinoco[1,3], Santiago Andrade[1], Paul Machado[1(✉)], Manolo Paredes[1], and Fernando Lara[1,2]

[1] Cicte - Centro de Investigación de Aplicaciones Militares, Universidad de las Fuerzas Armadas ESPE, 171103 Sanqolquí, Ecuador
{rdleon,pfmachado}@espe.edu.ec

[2] Maestría de Investigación en Electronica mención Telecomunicaciones, Universidad de las Fuerzas Armadas ESPE, 171103 Sanqolquí, Ecuador

[3] Department of Electrical, Electronics and Telecommunications, Universidad de Las Fuerzas Armadas ESPE, 171103 Sanqolquí, Ecuador

Abstract. Directional antennas concentrate the radiated energy in a localized manner, increasing the transmission or reception gain of a radio signal. One of the purposes of electronic warfare is the detection of arrival signals with their angle of arrival. In this article it is proposed to find the radiation pattern of an element of the antenna array, to determine the received power distribution in the 360° with an interval of 1°. The radiation pattern allows integrating a circular antenna array in order to detect the angle of arrival of a radio signal in the 80 MHz to 2 GHz band, also, a coefficient is proposed to estimate the angle of arrival using the two elements that receive the highest power from the circular antenna array.

Keywords: Circular antenna array · Direction finding · Radiation pattern · Digital receiver

1 Introduction

A fundamental element in communication systems are antennas, since, allows to transform the electrical signal into electromagnetic waves or vice versa. These can transport the information to a geographically distant place to other, exploiting the benefits of the radio electric spectrum, allowing the fulfillment of the objectives for which a telecommunications system is created [8, 9]. The antennas are characterized based on parameters: polarity, gain, directivity, bandwidth and frequency [3]; and within these parameters knowing the directivity of the antenna is a priority, to determine the angle of arrival of a signal.

The most used tool to visualize some of the characteristics of an antenna are radiation pattern, since, with this it is displayed: radiated power in elevation, azimuth, azimuthal aperture and elevation aperture, also, determine the type of antenna by its directivity [5–7].

M. Zambrano Vizuete et al. (Eds.): CI3 2021, LNNS 511, pp. 22–32, 2022.
https://doi.org/10.1007/978-3-031-11438-0_3

Determining the direction of arrival of a radio signal, contemplates a complex process, accuracy in estimating the angle of arrival of a signal decreases as the intensity of the radio signal decreases, attenuation is due to various physical factors: geographical environment, frequency, distance, power, antenna characteristics, among others. Therefore, knowing the radiation pattern of the antenna improves the estimation [4, 10].

In [1], the use of direction finding is proposed in an array of 4 antennas, to estimate the angle of approach of a small unmanned aerial vehicles (sUAV). To determine the angle, the use of a deep neural network is proposed. In [11], direction finding of environments with multiple paths is presented. To compensate for the multiple paths, the use of machine learning is proposed with a mean error of 11°.

The objective of this article is to determine an estimated arrival angle of the transmitter signal, for which, determine the radiation pattern of an element of circular antenna array, from this process, find the relationship between the signal received by the central and adjacent antenna, and with this coefficient determine the estimated angle.

The rest of this paper is organized as follows. Section 2 presents the design of the positioning system and the methodology for data collection. The Sect. 3 presents the results obtained in the device tests and finally, in the Sect. 4, the conclusions are presented.

2 Radiation Pattern

The present investigation focuses on determining the radiation pattern of an element, to combine the capabilities of several similar elements, and constitute an array of antennas and determine the angle of arrival of a radio signal.

To measure the reception power, the directional antenna (reception) is placed in the opposite direction to the reference transmitting antenna (transmission). The results of the radiation diagram show the highest reception power is at the position of 180° . The Fig. 1a presented a photography of element of circular antenna array with vertical polarization, and the Fig. 1b shows horizontal polarization.

The representation of the signal level, in polar coordinates, allows to visualize the characteristics of the antenna in each direction. The Fig. 2b shows the polar diagram of antenna with horizontal polarization and, it shows that the greatest amount of energy is radiated at 180°; that is, when the antenna is face to face with the received signal. In the Fig. 2b two lateral (posterior) lobes in the antenna are presented. With the principal lobe antenna beam width is calculate, the half power beam width is 88° (see Fig. 2b).

The Fig. 2a shows the polar diagram of single antenna with vertical polarization, this diagram presented one principal lobe and one lateral lobe, the beam width of the vertically polarized antenna is 66°, which is obtained by measuring the angle between the points where the radiated power is greater than or equal to half (−3 dB) its maximum value.

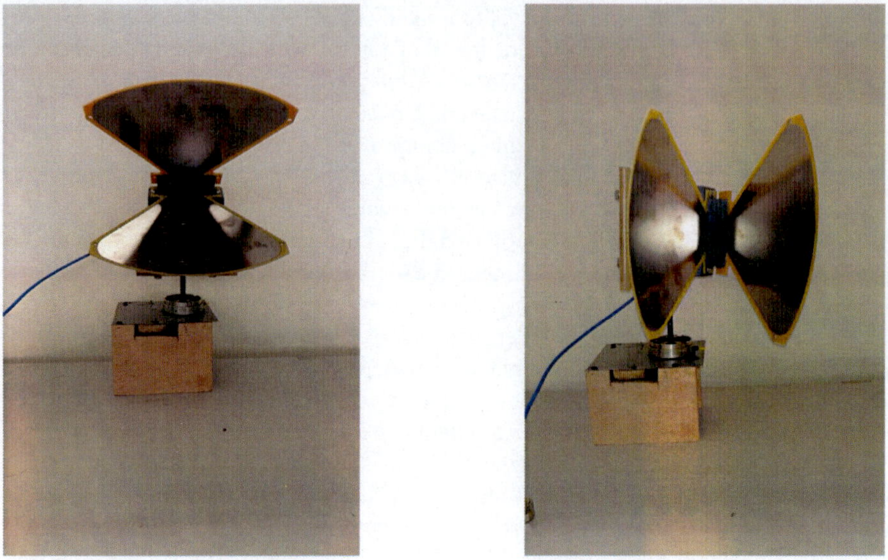

(a) Vertical Polarization (b) Horizontal Polarization

Fig. 1. Element of circular antenna array

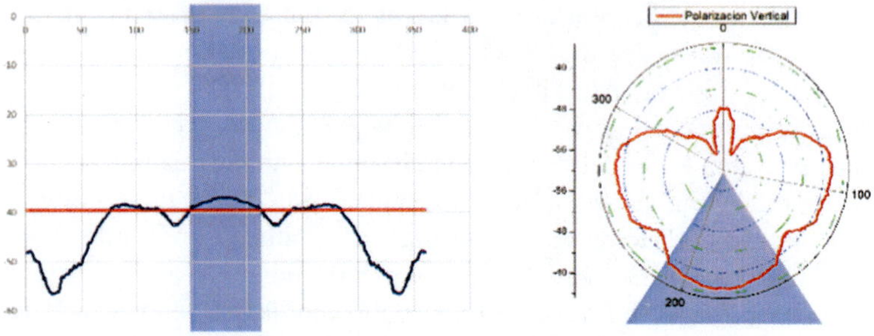

(a) Vertical Polarization

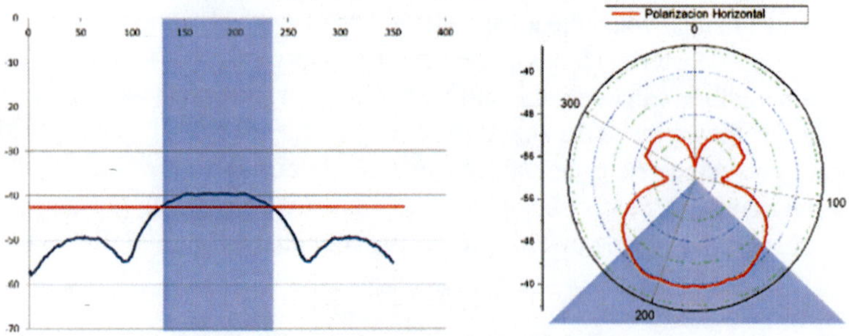

(b) Horizontal Polarization

Fig. 2. Element of circular antenna array

2.1 Antenna Simulation

Simulation is done in HFSS software [2], the Fig. 3 shows the geometry of the antenna in the simulator.

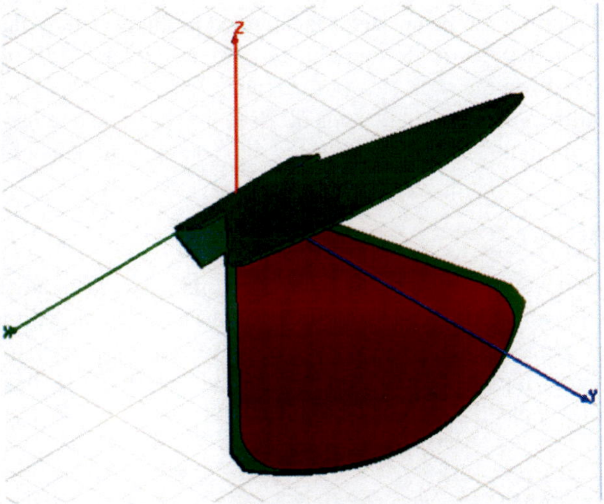

Fig. 3. Antenna model simulated in the HFSS program

From the simulated model, the dispersion parameter $|S_{11}|$ is obtained (see Fig. 4), this shows that the antenna operates between the frequencies of 1 GHz to 3 GHz.

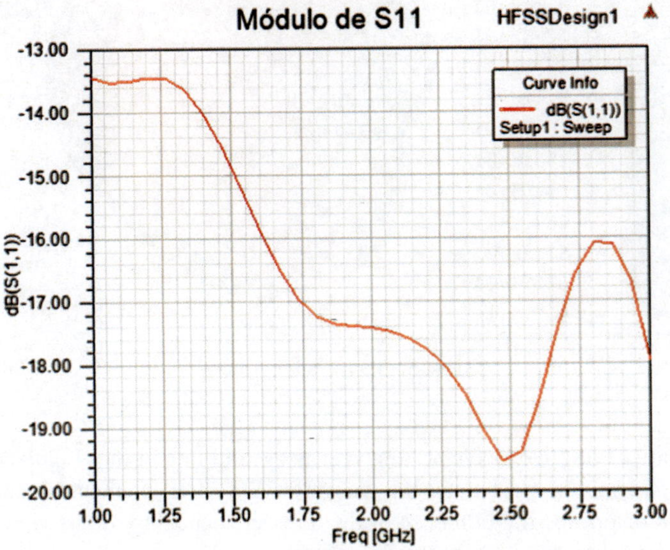

Fig. 4. Reflection coefficient of simulated antenna

The Fig. 5 shows the three-dimensional radiation pattern, for a frequency of 1 GHz, in this figure aiming direction, primary lobe, secondary lobes, 3 dB angles, primary lobe to secondary lobe (SLL) ratio, and front-to-back ratio (FBR) are shown.

The polar diagrams or principal planes for horizontal and vertical polarization were obtained in the simulator, and compared with those measured.

With the simulation, the beam width is obtained in the principal planes. The beam width in the horizontal plane is 88° (see Fig. 6b), because it has –3 dB of gain at 46° on the left side and 134° on the right side of the main lobe.

The beam width in the vertical plane is 60° (see Fig. 6a), since, it has a value greater than −3 dB of gain in 60° on the left side and 120° on the right side of the main lobe.

2.2 Circle Antenna Array

The circular array antenna TCI 641 [12] is used in this work. The antenna array uses a switching system in an approximate time of 478.83 ms between the 8 directive antennas, during this time the reception power is measured for the frequency set in the digital receiver. The digital receiver used is Signal Hound BB60C [13], this receiver have 27 MHz of instantaneous bandwidth.

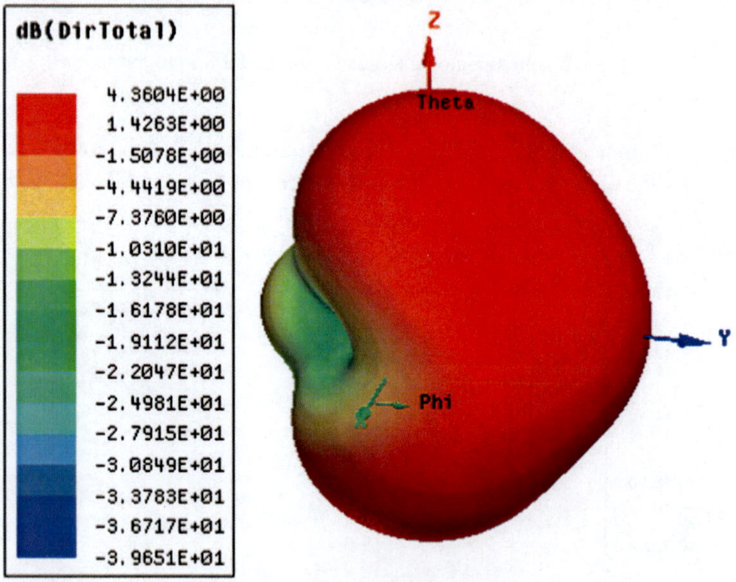

Fig. 5. 3D radiation pattern

The Anritsu MS2036A Vector Analyzer is used to obtain the S-parameters of the directional antenna in field tests, the parameter S_{11} of an element in the array was measured to determine the influence of the other elements. Two tests were performed with the dome cover and without the dome cover.

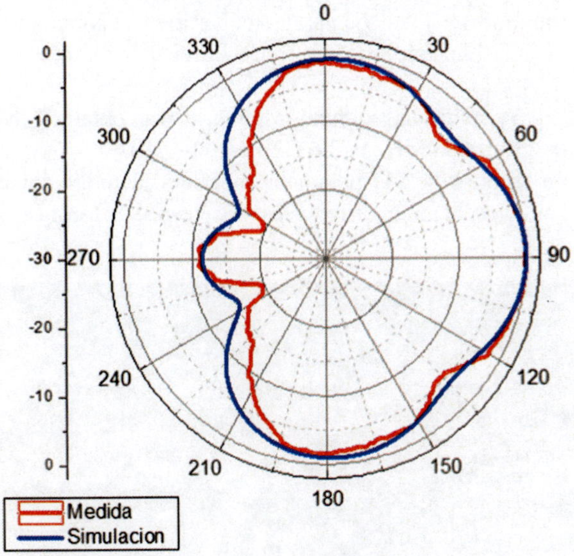

(a) Vertical Polarization

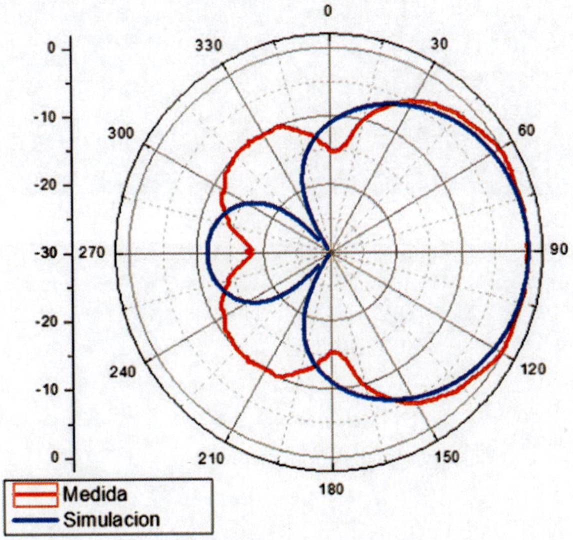

(b) Horizontal Polarization

Fig. 6. Simulated and measured polar diagram

According the authors [14], the S_{11} parameter has to meet condition $|S_{11}| < -10$ dB, in the case of a transmission coupling, in the case of reception can be expressed $|S_{11}| < -3$ dB.

After measuring S_{11} with and without a dome, it was obtained that the array can operate in the range from 600 MHz to 3.6 GHz (see Fig. 7).

In the Fig. 7b presented the S11 parameter measured with the dome of the antenna arrangement does not present major differences, with respect to the measurement without

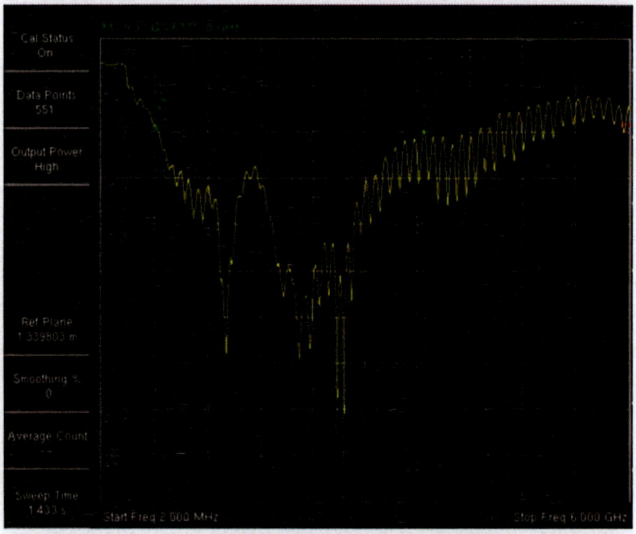

(a) With dome

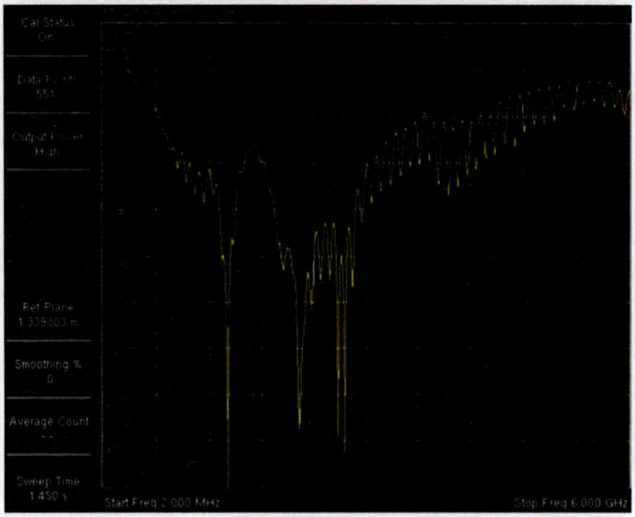

(b) Without dome

Fig. 7. Measurement of parameter S_{11}

a dome (see Fig. 7a), but around of the band 1400 MHz and 2200 MHz better reception is seen in relation to what it presents without the dome located.

3 Algorithm and Results

From the radiation pattern and the condition of higher power of an antenna when being face to face, antenna Tx and Rx, it is proposed to find the coefficient between the central antenna, the antenna that is face to face with the transmitter, and the neighboring antenna with higher energy either left or right.

As it is a switched circular array, only one antenna is receiving in an instant of time, and finding the coefficient between the two antennas with the highest energy allows estimating the angle of arrival of the transmitter signal, in the Fig. 8 this scenario is presented.

Each element of the array has a coverage of 40°, and the algorithm intends to determine the estimated angle of arrival within the coverage area of each element. The first condition is to divide the coverage area into two 20° left and 20° right. By determining the adjacent antenna (left or right) with the highest power, the sub-region in which the transmitter is located is determined.

Once the central antenna and the adjacent antenna have been determined, the relationship between the received powers in each antenna is determined to estimate the angle of arrival. To find the relationship of the angle of arrival and the coefficient proposed in the test scenario, the location of the transmitter is known a priori.

In the Table 1, a practical example is presented, where the power in the central antenna at 170° is 39.50 dBm, while the received power of the adjacent right antenna is

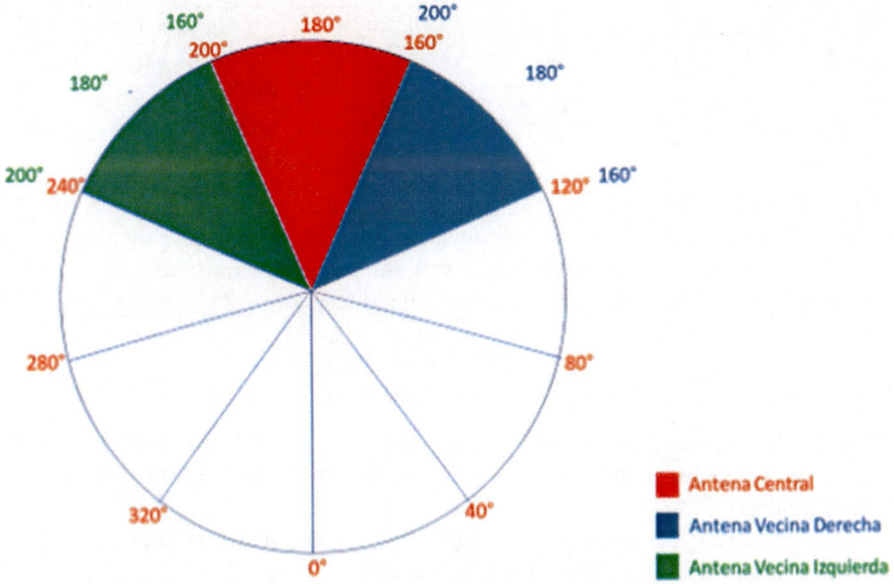

Fig. 8. Relationship between the central antenna and its neighboring antennas

Table 1. Ratio of the antenna with higher radiation and its right neighbor antenna

Cental Ant.	Central Ant.	Rigth Ant.	Rigth Ant.	Relationship
	dBm		dBm	Rigth/Center
160	−39,60	200	−39,60	1,000000000
161	−39,59	201	−39,61	0,995405417
162	−39,58	202	−39,62	0,990831945
163	−39,57	203	−39,63	0,986279486
164	−39,56	204	−39,64	0,981747943
165	−39,55	205	−39,65	0,977237221
166	−39,54	206	−39,66	0,972747224
167	−39,53	207	−39,67	0,968277856
168	−39,52	208	−39,68	0,963829024
169	−39,51	209	−39,69	0,959400632
170	−39,50	210	−39,70	0,954992586
171	−39,49	211	−39,83	0,924698174
172	−39,48	212	−39,96	0,895364766
173	−39,47	213	−40,09	0,866961876
174	−39,46	214	−40,22	0,839459987
175	−39,45	215	−40,35	0,812830516
176	−39,44	216	−40,48	0,787045790
177	−39,43	217	−40,61	0,762079010
178	−39,42	218	−49,88	0,737904230
179	−39,41	219	−40,87	0,714496326
180	−39,40	220	−41,00	0,691830971

210° is 39.70 dBm. To determine the coefficient, the units are transformed from decibels [dB] to milliwatts [*mw*], with the next equation the coefficient is estimate, and in the Table 1, the coefficient is presented by moving the position of the transmitter between 160° and 180°, with the right antenna as the adjacent antenna with higher power.

$$C_r = \frac{Power_{neighbor}[mw]}{Power_{central}[mw]}$$

In the Table 2, an example is presented where the adjacent antenna is on the left, placing the transmitter between 140 and 160 of the array (this area is covered by the central element).

After obtaining the tables with the ratio coefficient, and repeating the tests to determine the error, a mean error of ±2° was found.

Table 2. Ratio of the antenna with higher radiation and its left neighbor antenna

Cental Ant.	Central Ant. dBm	Left Ant.	Left Ant. dBm	Relationship Left/Center
140	−41,00	180	−39,40	0,691830971
141	−40,87	181	−39,41	0,714496326
142	−40,74	182	−39,42	0,737904230
143	−40,61	183	−39,43	0,762079010
144	−40,48	184	−39,44	0,787045790
145	−40,35	185	−39,45	0,812830516
146	−40,22	186	−39,46	0,839459987
147	−40,09	187	−39,47	0,866961876
148	−39,96	188	−39,48	0,895364766
149	−39,83	189	−39,49	0,924698174
150	−39,70	190	−39,50	0,954992586
151	−39,69	191	−39,51	0,959400632
152	−39,68	192	−39,52	0,963829024
153	−39,67	193	−39,53	0,968277856
154	−39,66	194	−39,54	0,972747224
155	−39,65	195	−39,55	0,977237221
156	−39,64	196	−39,56	0,981747943
157	−39,63	197	−39,57	0,986279486
158	−39,62	198	−39,58	0,990831945
159	−39,61	199	−39,59	0,995405417
160	−39,60	200	−39,60	1,000000000

4 Conclusions

The measurements in the field, with respect to the simulation using the HFSS software, show very similar diagrams, generating reliability of the results. Small differences are due to measurement errors and small differences in antenna construction.

The beamwidth of the irradiation patterns of the directive antenna, both in horizontal polarization and in vertical polarization, ensures that using an array of 8 directive antennas distributed equally in a circular way, estimates of the angle of arrival can be made. with 360° coverage.

The coefficient between the power received by two antennas at the frequency of interest allows estimating the angle of arrival within a sub-region of the central antenna with an error of ±2°.

Acknowledgment. This work was supported by Centro de Investigación de Aplicaciones Militares CICTE (Universidad de las Fuerzas Armadas ESPE), and Agrupamiento de Comunicaciones y Guerra Electrónica (AGRUCOMGE - Ecuador).

The authors gratefully acknowledge the contribution of Universidad de las Fuerzas Armadas ESPE for the economical support in the development of this project under Research Grants 2019-PIV-001-INV and 2019-PIV-002-INV.

References

1. Abeywickrama, S., Jayasinghe, L., Fu, H., Nissanka, S., Yuen, C.: Rf-based direction finding of uavs using dnn. In: 2018 IEEE International Conference on Communication Systems (ICCS), pp. 157–161. IEEE (2018)
2. Ansoft: HFSS High Frequency Structure Simulator user' guide. Pittsburgh (2005. [Online].). URL http://anlage.umd.edu/HFSSv10UserGuide.pdf
3. Balanis, C.: Antenna theory: analysis and design. John wiley & sons (2015)
4. Bazzi, A., Slock, D., Meilhac, L.: Efficient maximum likelihood joint estimation of angles and times of arrival of multiple paths. In: 2015 IEEE Globecom Workshops (GC Wkshps), pp. 1–7. IEEE (2015)
5. Boustani, B., Baghdad, A., Sahel, A., Badri, A.: Performance analysis of direction of arrival algorithms for smart antenna. Int. J. Electri. Comp. Eng. **9**(6), 4873 (2019)
6. Chen, Z., Gokeda, G., Yu, Y.: Introduction to Direction-of-arrival Estimation. Artech House (2010)
7. Elhag, N., Osman, I., Yassin, A., Ahmed, T.: Angle of arrival estimation in smart antenna using music method for wideband wireless communication. In: 2013 International Conference on Computing, Electrical and Electronic Engineering (ICCEEE), pp. 69–73. IEEE (2013)
8. Godara, L.: Handbook of antennas in wireless communications. CRC press (2018)
9. Haykin, S.: Digital communications. Wiley, New York (1988)
10. Khan, A., Wang, S., Zhu, Z.: Angle-of-arrival estimation using an adaptive machine learning framework. IEEE Commun. Lett. **23**(2), 294–297 (2018)
11. Mukherjee, T., et al.: Rssi-based supervised learning for uncooperative direction-finding. In: Joint European Conference on Machine Learning and Knowl edge Discovery in Databases, pp. 216–227. Springer (2017)
12. Sharp, E., Hsi, S.: Broadband fan cone direction finding antenna and array. US Patent 6,198,454 (2001)
13. SignalHound: Signal Hound BB60C User Manual. Battle Ground, WA (2020). [Online]. URL https://signalhound.com/sigdownloads/BB60C/BB60C-User-Manual.pdf
14. Suárez, C., Carroll, J., Puerto, G.: Antena compacta para aplicaciones en satélites de reducido tamaño. Información tecnológica **27**(3), 189–198 (2016)

Analysis of BPSK Modulation Using the NI ELVIS III Communications Module

Flores Viera Dayana[1(✉)] and Valdés Córdova Ernesto[2]

[1] Instituto Superior Universitario Central Técnico, Quito, Ecuador
floresd829@gmail.com
[2] Instituto Superior Universitario Bolivariano, Guayaquil, Ecuador
evaldes@itb.edu.ec

Abstract. The present project was developed to perform an analysis of BPSK modulation, recovery of original, data and degradation caused by noise.

A practice was implemented with the NI ELVIS III communications board where a BPSK signal was generated, the noise (AWGN) was introduced, and the demodulation process was performed to observe the behavior of the data at the input of the receiver. Besides, a simulation was performed in MATLAB covering BPSK modulation and demodulation and compared to actual data for analysis of results.

Methodologies such as scientific and experimental research were addressed, Where the information necessary for proper implementation and simulation of BPSK was obtained on communications cards. The results obtained confirm both in the theory and practice: the digital input signal, when changing the logical level, from 0 to 1, the phase of the output signal changes, by 180° or 0°. In this way, a BPSK signal is generated.

Keywords: BPSK · Phase · Modulation · Demodulation · Noise

1 Introduction

Modulation is the process where the parameter of a signal is varied proportionally to a second signal that carries the information [1]. The benefit of the modulation in a communications channel at the time of transmitting data is to obtain a better performance against possible noises and interferences.

According to Carlson and Crilly [2], in base of their study they affirm, "The main objective of modulation in a communications system is to generate a proper modulated signal to the characteristics of the transmission channel".

The research project analyzed the process of modulation and demodulation of the BPSK digital signal to collect information from the NI ELVIS III communications card that allows the analysis of the data at the input and output of the communications channel.

M. Zambrano Vizuete et al. (Eds.): CI3 2021, LNNS 511, pp. 33–49, 2022.
https://doi.org/10.1007/978-3-031-11438-0_4

1.1 BPSK Modulation

BPSK binary phase shift modulation uses two symbols, capable of analyzing each bit of an input binary information stream that is characterized in a form of angular modulation and which consists of varying the carrier phase, the phase variation of 0° when it is 1 logical and 180° for the logical 0. "BPSK modulations occupy a higher bandwidth, have a limited speed and generally these modulations are used for Bluetooth, RFDI and wireless LAN communications" [3].

BPSK has results of possible two output phases for a single frequency and amplitude. The first output phase represents logical level 1 and the other logical level 0, as the digital signal changes phase, the output carrier changes, between two different angles of 180°, as shown in (1):

$$S_i = A\cos\left[\omega_{RF} t + \varphi_i(t)\right] \quad (1)$$

where,

Si = BPSK modulated signal.

A = Carrier amplitude.

ω_{RF} = Angular frequency of the carrier rad/s.

t = Time.

φ_i = Initial phase.

In addition, you can represent the digital symbols that are received in the time period using the constellation diagram as shown in Fig. 1.

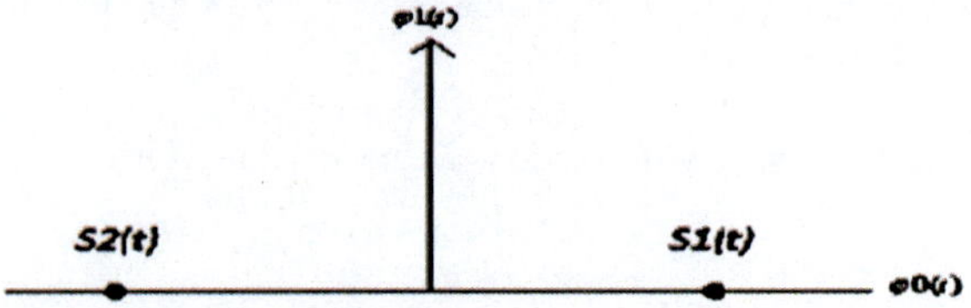

Fig. 1. BPSK constellation diagram [4].

BPSK modulation is angular as shown in the Fig. 2, that is, it consists of varying the phase of the carrier between several discrete values, the modulated signal with its phase discontinuities appears at the beginning and end of each interval T (period), in transitions from 0 to 1, and vice versa [5].

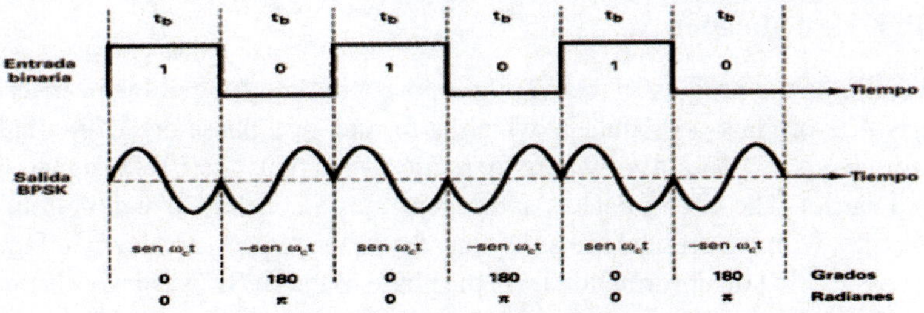

Fig. 2. BPSK wave time relation [5].

1.2 BPSK Modulator

The modulator is a simple circuit that performs a multiplication of the input signal, that is complemented by a bandpass filter which eliminates non-significant harmonics, obtaining a lower percentage of distortion (Fig. 3). According to Vélez [6], "BPSK modulator allows two output phases through only one carrier, which is between 0 and 1 logical, if there is a variation in the input signal, the angle of the carrier phase offset will be 180°".

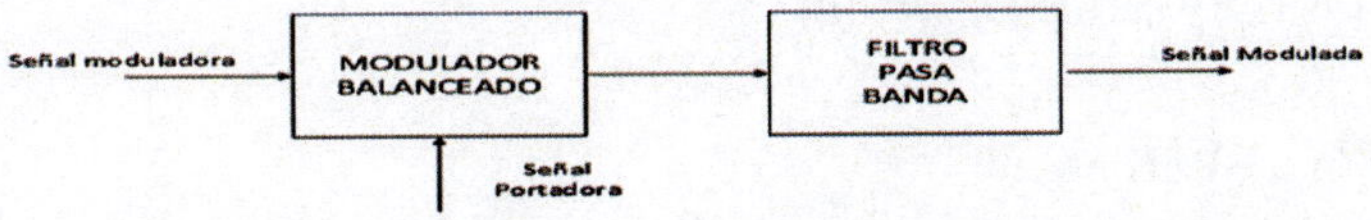

Fig. 3. Modulator BPSK.

1.3 BPSK Transmitter and Receiver

The balanced modulator functions as a phase inversion switch depending on the logical condition of the digital input, the carrier is transferred to the output, by 180° phase shift with reference to the direct carrier (Fig. 4).

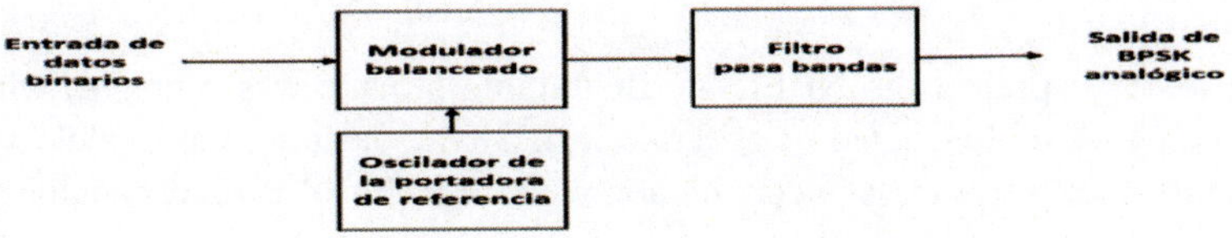

Fig. 4. BPSK transmitter

The study made by Castañeda [7] with the implementation of a QPSK digital modulator-demodulator based on an FPGA determines that "to increase the rate of data transmission there has to be an increment in the availability of bandwidth and SNR".

1.4 BPSK Demodulation

Demodulation is the process to recover the information transmitted by a carrier wave, it consists of a product demodulator which is formed of a phase detector, a balanced modulator, and a low pass filter where there are two signals the BPSK signal and the recovered carrier. The circuit detects and regenerates a coherent signal, both in phase and frequency, with the original transmission carrier, this process is shown in Fig. 5 and it is done by filtering the information from the phase of the BPSK wave whose argument is shown in Eq. (1).

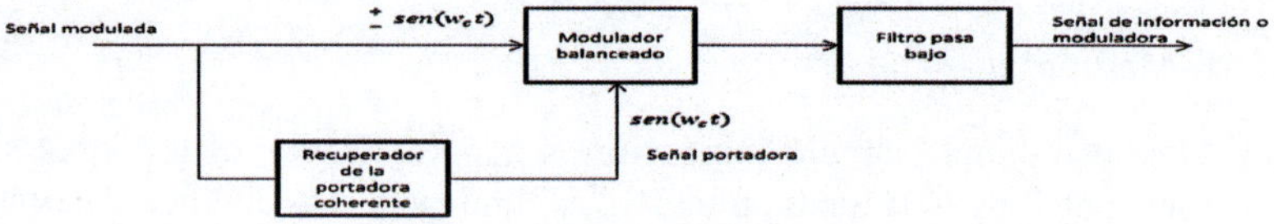

Fig. 5. BPSK demodulator [8].

For demodulation of a BPSK signal, it is necessary to recover the carrier of the same received signal. "The carrier retriever is a circuit which extracts from the received PSK signal the necessary carrier signal" [9].

1.5 Noise in a BPSK Signal

BPSK modulated signal, when is transmitted through a channel, must be considered that it modifies the signal that travels through it, it means, which interferences or variations are generated that are transmitted simultaneously, thus, a signal of an unwanted random nature. A basic noise model is additive Gaussian White Noise or AWGN where the signal distorted by a white noise has a constant level of frequency, and at the desired SNR.

The SNR represents the ratio of the signal-to-noise in the decibel (dB) considering the higher the value in dB, the better the communication. Because BPSK is only working in two states the existence of noise affects little.

2 Materials and Methods

For the research project the NI ELVIS III training module was acquired, where all the materials were recognized to generate a BPSK modulation, the module was the main material for the process in conjunction with the EMONA card, for analog /digital baseband communications.

2.1 Instruments Used

- Time base oscilloscope
- Oscilloscope with the Fourier transform (FFT)
- Function generator

To use the instruments, it was necessary to have communication to the National Instruments platform where interactive learning was obtained through simulations and virtual instrumentation to go from the theory to the real behavior of the circuits.

2.1.1 Hardware: EMONA Communications Card

- Four BNC connectors to 2mm banana cables
- Several 4 mm banana-type connection cables

2.2 Hardware and Software Installation

The power supply and USB cable were connected to the station, then the compatible EMONA card was inserted securing it effectively and correctly.

The NI ELVIS module was turned on where the status of the LEDs was first taken into account and the connection with the computer was continued, the NI DAQmx drivers were installed which is a support for data acquisition cards along with the Measurements Live support software.

The Measurements Live software is opened, it goes to the browser where the NI ELVIS III is detected. Another method that was used to access the virtual instruments was where the IP address of the training module was placed.

2.3 EMONA Communications Card Modules

The electronic instruments that were used are suitable for communications represented by electrical signals that through the process could be interpreted by using virtual tools such as oscilloscopes and multimeters, the modules of the EMONA card used in practice are detailed in Table 1.

Table 1. EMONA card parameters

Modules	Parameters	
Adder	Max. frequency 1 MHz	Signal output with inverted polarity
Noise Generator	No input or signal required	Noise level min. 0 dB and max. 22 dB
Sequence Generator	Digital clock frequencies between 1 Hz to 1 MHz	Analog clock frequencies 500 Hz to 10 kHz Supports bipolar signals
Low pass filter tunable	Rejection band attenuation > 50 dB and approx. 0.5 dB	Max. voltage within the values ± 5 v

To generate the BPSK signal, the configuration in Fig. 6 was performed, the black pins of the oscilloscope cables were inserted into the ground points (GND).

From the module "Master signals" the carrier wave is obtained at a frequency of 100 kHz, the information signal with a frequency of 8.33 kHz that enters the sequence generator is measured in the oscilloscope of channel 1. The BPSK signal is generated at a frequency of 100 kHz, which is obtained at the output of a multiplier.

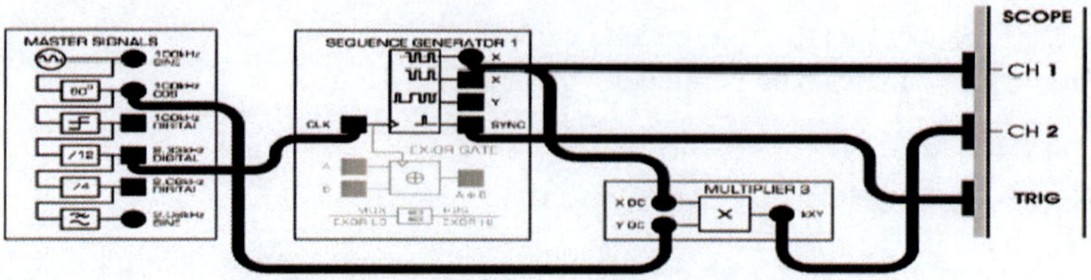

Fig. 6. Connection for BPSK generation [10].

We proceeded to modify the connection of the module "Master signals" to better appreciate the behavior of the BPSK signal due to the change of frequency of the digital input signal to 100 kHz, the configuration of Fig. 7 was made.

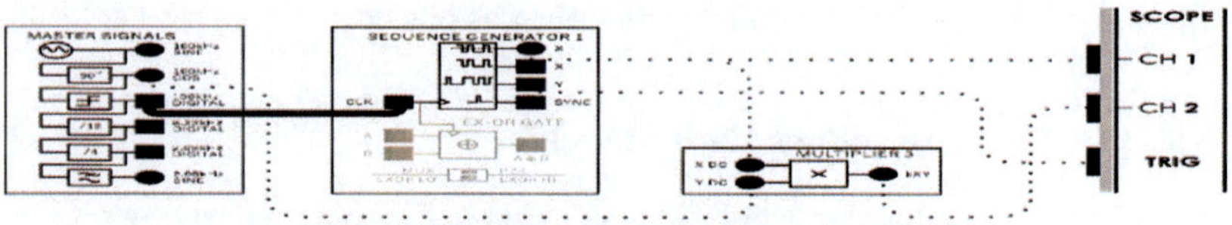

Fig. 7. Connection for modified BPSK [10].

For the demodulation was configured according to Fig. 8 where it is coupled in the LPF module and the GAIN control was adjusted to half.

It was adjusted in half because the LPF module has a function to let pass-low signals, where the amplitude of the output signal varies, a second multiplier was used, and the filter module low-pass tunable to implement a product detector to retrieve the digital data of the BPSK signal.

The digital information signal for the demodulation configuration Works at a frequency of 8.33 kHz.

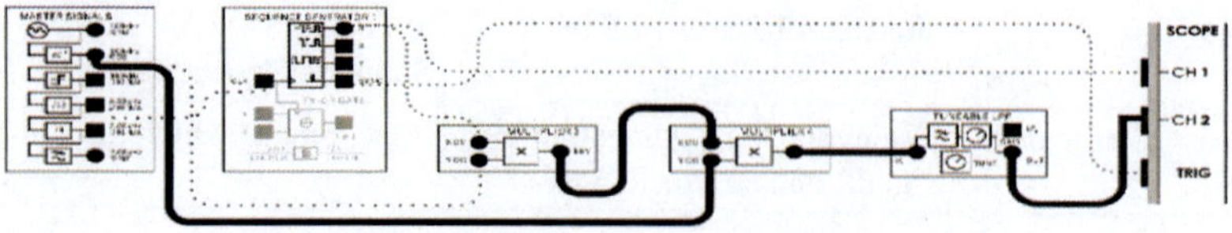

Fig. 8. Connection for product demodulation [10].

The previous connection was modified to use a comparator to restore the data as shown in Fig. 9.

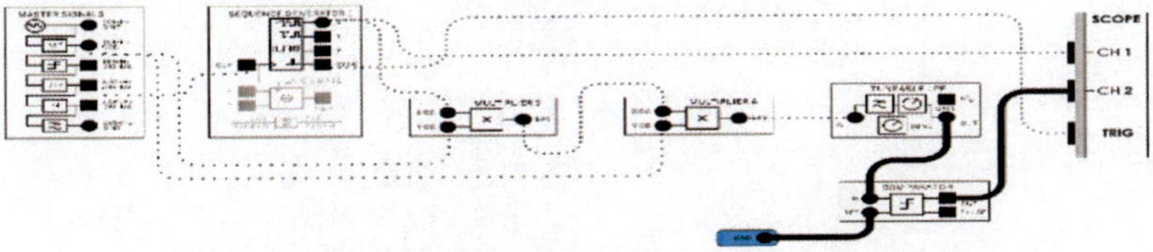

Fig. 9. Connection for data recovery [10].

The configuration in Fig. 10 was connected while retaining the previous connection to generate the behavior of a noisy channel where the noise source comes from the function generator module.

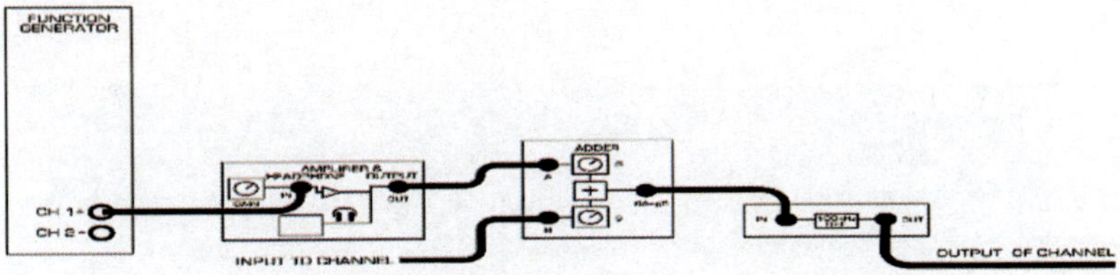

Fig. 10. Connection up the noisy bandlimited channel [10].

The configuration of the implemented BPSK modulation circuits is observed in Annex 1 section, so each of the installations is referenced to the real equipment.

An external program was used to show the reader comparisons of the digital signal, carrier, and modulated signal. MATLAB is a programming environment used to analyze, design learning systems, signal processing. It was used as a part of that research project to simulate BPSK modulation and demodulation and achieve a comparison with the signal generated by the NI ELVIS III training device.

3 Results

Once the learning objectives have been achieved through all the information collected, the following activities were understood:

- A BPSK signal was generated.
- The noise was introduced into the transmission channel
- Data retrieved in BPSK modulation and demodulation in MATLAB.
- The values obtained at each stage of the practice were analyzed.

3.1 Section 1: BPSK Modulation

3.1.1 Generating a BPSK Signal

Most of the results are generated by the EMONA communications card and the National Instruments platform, we proceed to the first condition in generating a BPSK signal (blue signal), with the help of the block diagram in Fig. 11.

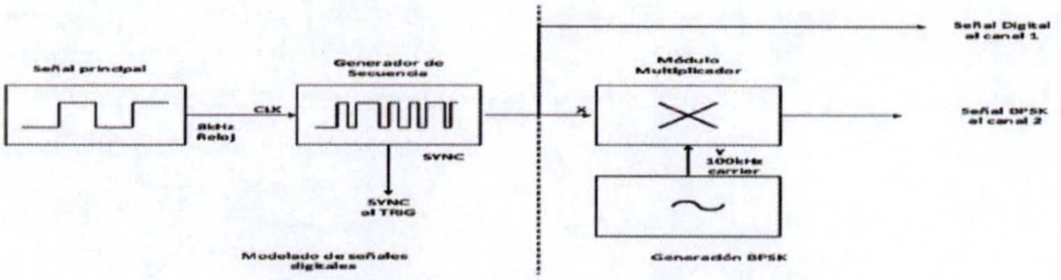

Fig. 11. Block diagram for BPSK generation.

The digital signal input (yellow signal) presented in this first configuration in the company of the modulated signal (blue signal), the output of the sequence generator module changes the logical level (Fig. 12).

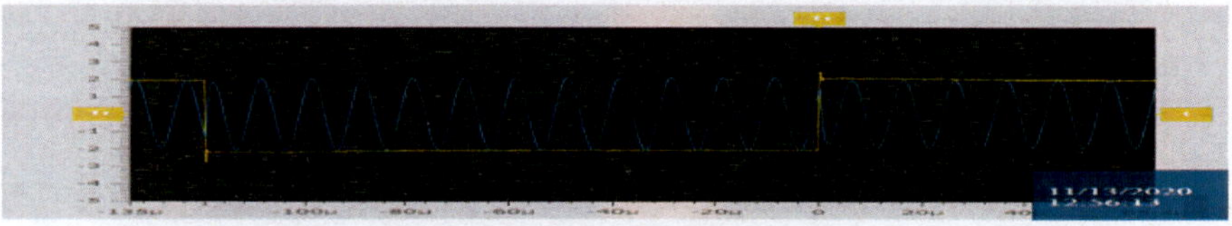

Fig. 12. Digital signal and BPSK modulated signal output.

A modification was made to the sequence generator module, the 8 kHz digital information signal (yellow signal), and the 100 kHz carrier signal where these two signals enter to a multiplier to generate a BPSK modulated signal (blue signal) and the logical transitions of the data flow were observed.

It can be seen in Fig. 13 that, the message signal or the digital signal, when changing the logical level, that is from 0 to 1 or 1 to 0, the phase of the BPSK signal changes, in such a way that the signal reverses the phase by 180° (Table 2).

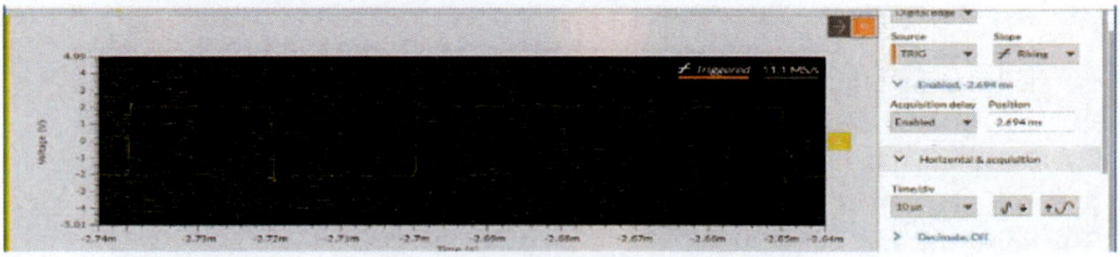

Fig. 13. Digital signal and BPSK signal with 10 us/div scale.

Table 2. Measurement data was obtained by generating a BPSK signal.

Channel	VPP	RMS	Frequency	Period
Ch 1	4.960 V	2,026 V	25 kHz	40 μs
Ch 2	8.854 V	3,073 V	100 kHz	10 μs

Note: The inputs of channel 1 (digital information signal) and channel 2 (BPSK modulated signal) of the oscilloscope are based on the sequence generator module and the BPSK signal.

3.1.2 Change in Carrier Frequency

Figure 14 shows the digital input signals (yellow signal) and BPSK (blue signal) with better stability due to a change that was made in the "Master Signals" module of the carrier source from100 kHz Cos to 100 kHz Sine (Table 3).

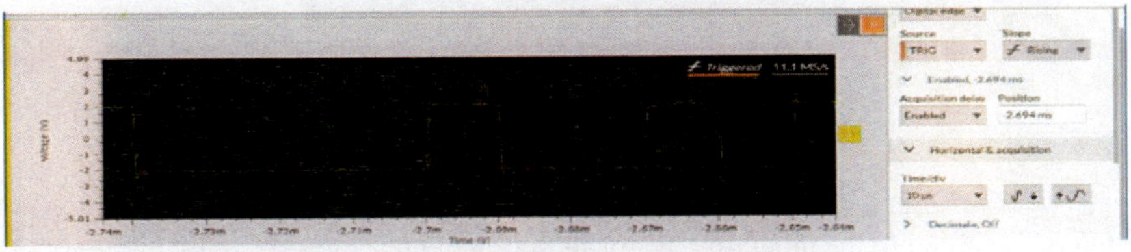

Fig. 14. Modulated signal and 100 kHz signal.

Table 3. Measurement data was obtained by switching from 100 kHz Cos to 100 kHz Sine.

Channel	VPP	RMS	Frequency	Period
Ch 1	4.975 V	2,026 V	20 kHz	50 μs
Ch 2	8.128 V	2.71 V	66.76 kHz	14.98 μs

Note: Channel 1 represents the original digital signal and the values measured on channel 2 represent the BPSK modulated signal.

3.2 Section 2: Demodulation of a BPSK Signal Employing a Product Detector

The following is the block diagram of the demodulator Fig. 15.

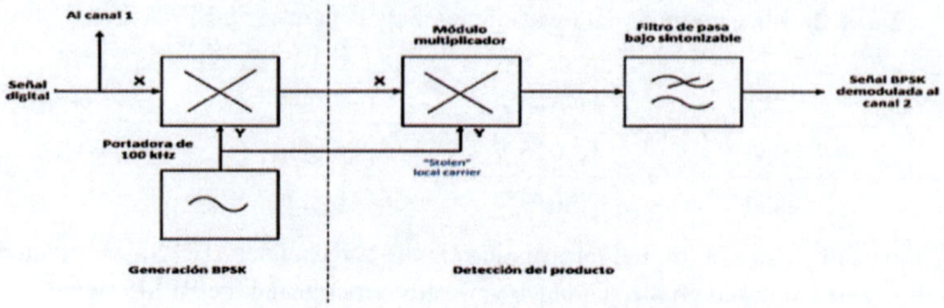

Fig. 15. Block diagram for product demodulation.

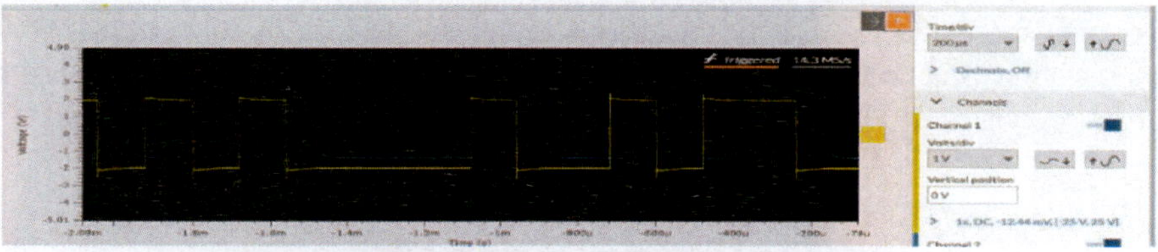

Fig. 16. Demodulated signal after the product detector.

It could be observed that the demodulated signal (blue signal) in Fig. 16 the waveform obtained is similar to the input digital signal (yellow signal) this is because after passing through the LPF (low pass filter) allows the passage of low frequencies and thanks to the obstruction of high frequencies we obtain the demodulated signal with values similar to the digital signal, and this is shown in Table 4.

Table 4. Measurement data obtained from BPSK demodulation.

Channel	VPP	RMS	Frequency	Period
Ch 1	4,997 V	2,002 V	4,167 kHz	240 μs
Ch 2	5,594 V	2,427 V	4,168 kHz	239,9 μs

Note: In the demodulation of a BPSK signal the data generated on channel 1 (digital signal) and channel 2 (demodulated signal) tend to have similar frequencies of 4,167 kHz because their waveforms were obtained are similar.

3.2.1 Restoring the Recovered Data with a Comparator

The use of a comparator to restore distorted digital signals, clean a demodulated BPSK signal is the action that was performed from the block diagram of Fig. 17 where results of the demodulation (blue signal) were obtained with data in the two channels, the frequency which the waves work is the same because it compares the amplitude of the signal detecting a logical 1 and a logical 0 if it is greater or less than a base level respectively, the demodulated signal that is obtained is from the digital signal (yellow signal). Measurements were collected to better understand demodulation, as shown in Table 5 (Fig. 18).

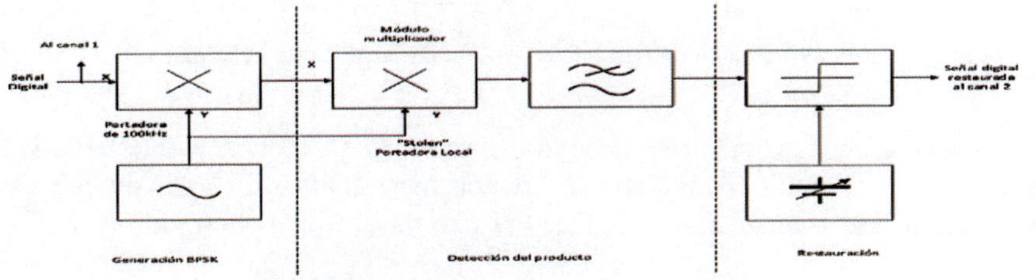

Fig. 17. Block diagram for data recovery.

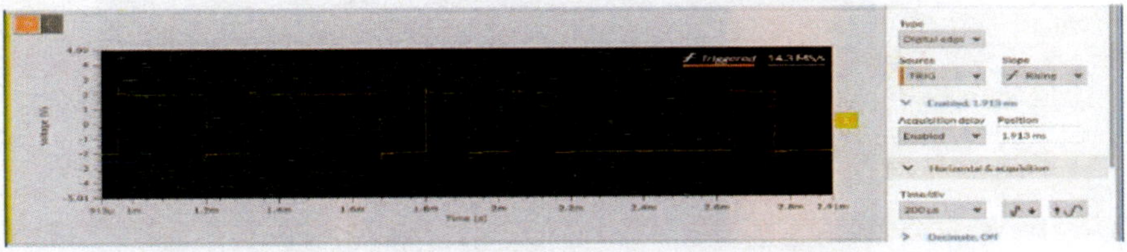

Fig. 18. Signal recovered to the comparator output.

Table 5. Measurement data obtained from the restoration of recovered data.

Channel	VPP	RMS	Frequency	Period
Ch 1	5,01 V	2,001 V	2,778 kHz	360 μs
Ch 2	5,421 V	3,428 V	2,778 kHz	360 μs

Note: Channel 1 is values obtained from the measurement of the digital signal. Channel 2 shows values of the recovered signal checking that there is a similarity between frequencies and voltages.

3.3 Section 3: Introducing the Noisy Channel

3.3.1 Modeling of the channel limited by the band AWGN (Additive White Gaussian Noise) with noise

In this section, it's added an amplifier module which works at its minimum value, to continue with the implementation is presented the block diagram of Fig. 19, where it is essential to follow the order and not disconnect any of the cables of the previous configuration, only to add to determine the behavior of a real channel adding noise to the BPSK signals.

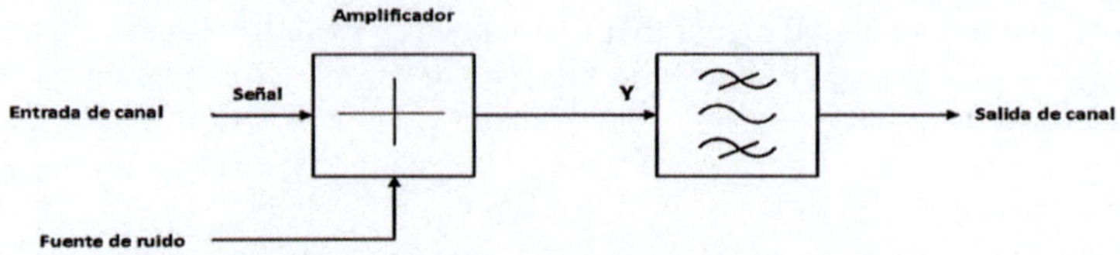

Fig. 19. Block diagram for noisy AWGN band limited channel.

The results obtained in this section through an AWG noise channel (blue signal) are shown in Table 6, in this configuration a function generator was used that produces and delivers signals of triangular, rectangular, and sine waveform as shown in Fig. 20.

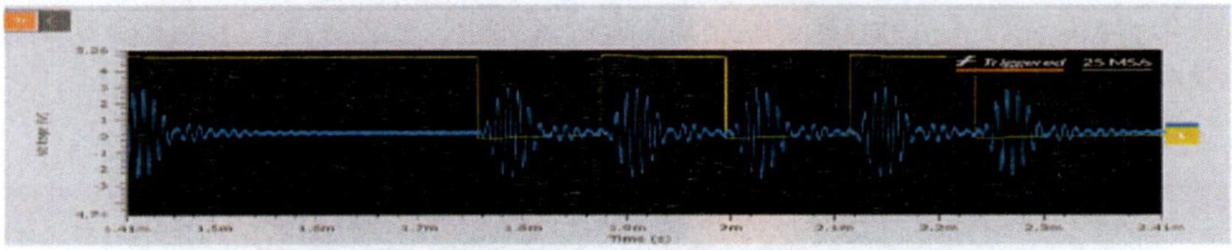

Fig. 20. Signal display when adding the noisy channel.

Table 6. Measurement data was obtained by adding the AWG channel.

Channel	VPP	RMS	Frequency	Period
Ch 1	5,154 V	3,275 V	4,167 kHz	240 μs
Ch 2	283,4 V	36,87 V	99,69 kHz	10,03 μs

Note: Channel 2 is the AWG channel which, according to the measured values is higher than the values of the carrier wave.

ECB_120k-noise.csv is a file that was added to the function generator with a refresh rate of 500 kS/s. which generated a noise signal in a frequency range of 120 kHz.

When charging, it was made certain modification to the circuit which allowed the entry of noise, obtaining a noisy BPSK signal (blue signal) by varying the gain of the amplifier module for different noise levels and this can be observed in the amplitude of the noisy signal (yellow signal) being superior, less than or equal to the size of the signal. Among the gain values used were −20 dB and 0 dB.

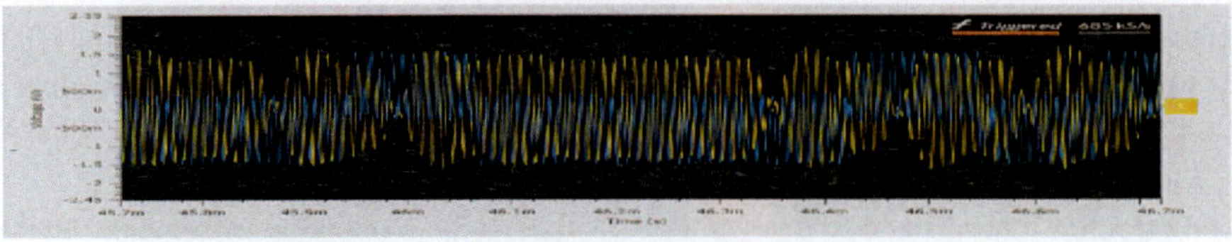

Fig. 21. BPSK signal with added noise 0 dB.

With the AWG channel a comparison was made together with the noisy BPSK signal (blue signal) as can be seen in Fig. 21, it was appreciated that the noise is higher than the BPSK wave and this can be seen in the measured values shown in Table 7.

Table 7. Measurement data was obtained by adding noise to the BPSK signal.

Channel	VPP	RMS	Frequency	Period
Ch 1	3,483 V	969,7 mV	15,36 kHz	65,1 μs
Ch 2	6,907 V	2,14 V	1,418 kHz	705 μs

Note: The gain values used at 0 dB have as a higher wave than the noise wave (yellow signal) and everything is presented in the values in Table 7 with a high degree of comparison of channel 1 frequency.

The noise gain that was used for the demonstration was −20 dB and this is verified in the frequency of channel 1 of the oscilloscope where a lower frequency than channel 2 of the BPSK signal was used demonstrated in Table 8, together with Fig. 22 that shows the wave types obtained by having the gain in −20 dB. As seen in the noisy signal (yellow signal) compared to the BPSK signal (blue signal).

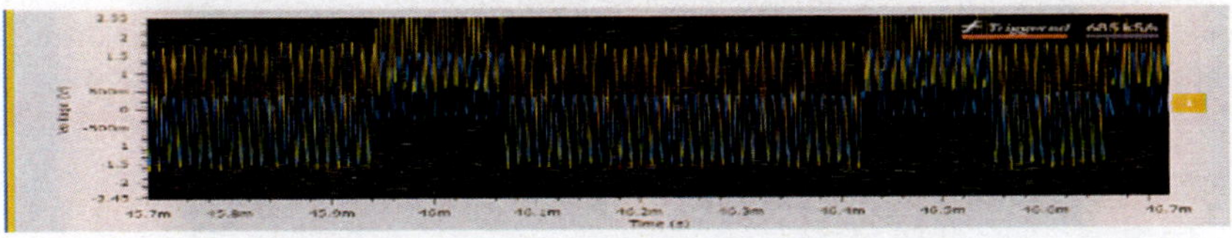

Fig. 22. BPSK signal with added noise −20 dB.

Table 8. Measurement data was obtained by adding noise to the BPSK signal.

Channel	VPP	RMS	Frequency	Period
Ch 1	6,879 V	1,975 V	1,042 kHz	959,8 μs
Ch 2	6,901 V	2,14 V	1,418 kHz	705,1 μs

Note: When generating noise of −20 dB the wave size is similar, because it has no variation in frequency, the values generated by channel 1 and channel 2 are almost similar.

The function of the BPF module of the card is to allow the passage of a certain frequency within a specific bandwidth, in this case it is 100 kHz attenuating the frequencies that are outside.

4 Discussion

4.1 BPSK Modulation

4.1.1 Generating a BPSK Signal

The procedure that was generated throughout the performed tests obtained changes that were necessary to obtain the signals, for example, time base, digital edge, trigger source, channel in DC and the inputs to the oscilloscope that are arranged to display output from the sequence generator module and the BPSK signal coming out from the multiplier module.

The sequence generator module is used to model a digital signal and the multiplier module to generate the BPSK signal by implementing its mathematical model. The Phase shift is achieved when there is a variation in the bit value of the digital signal that produces 180° shifts.

Changes were made in the carrier frequency, in the beginning, we worked with 100 kHz in sine function; it means, that the wave will move by 0°, when changing to 100 kHz in cosine function, it is observed in the graphs that there is an advance of the wave of 90° concerning sine and this can be observed in each bit change where the offset of 180° of the wave must exist.

4.1.2 Investigation of the BPSK Spectrum

In the study, it is possible to visualize the power spectral density of the BPSK signal, which works at a bandwidth resolution of 444 Hz in which the center or carrier frequency is 100 kHz with a span of 129 kHz, this occurs when moving the baseband spectrum to the frequency of the carrier Fig. 23.

The upper and lower side frequencies are separated from the carrier frequency of 100 kHz, but due to errors in the instruments the frequency that was obtained was 98kHz, with an amplitude of 1 dB can be seen in Fig. 23.

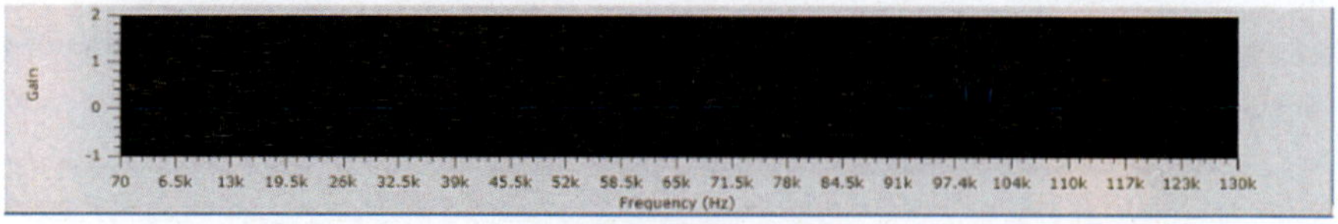

Fig. 23. BPSK spectrum.

4.2 Demodulation of a BPSK Signal Using a Product Detector

An important fact at this point is the use of a tunable pass-low filter module in the EMONA card, to only let the low signals pass and prevent the high ones because in the demodulation it is a matter of recovering information carried by a carrier wave.

It was determined that the recovered digital signal is not identical to the original digital signal, this is because demodulating a BPSK signal requires an envelope detector, the reason is that it tends to have peaked because the envelope of a BPSK signal is not purely square.

4.2.1 Investigation of the BPSK Spectrum During Demodulation

The use of the Fourier transform consists of multiplying two signals in the time domain, to obtain signals in the frequency domain. Therefore, in BPSK the baseband symbols are the spectrum centered around the frequency of the carrier, this is given if the data sequence is a string of zeros and ones alternately. This spectrum contains powers in odd multiples of the fundamental frequency.

The spectrum that occurs around the frequency from 0 Hz to 200 kHz is noted that there is a large amount of random vibration that increases the background noise in the spectrum.

In Fig. 24, the spectrum of BPSK demodulation its frequency is not very defined, this is by the perturbations presented in the channel, it can be observed that in the frequency of 0 Hz the gain is less than the transmitted one demonstrating with this that the is higher the noise.

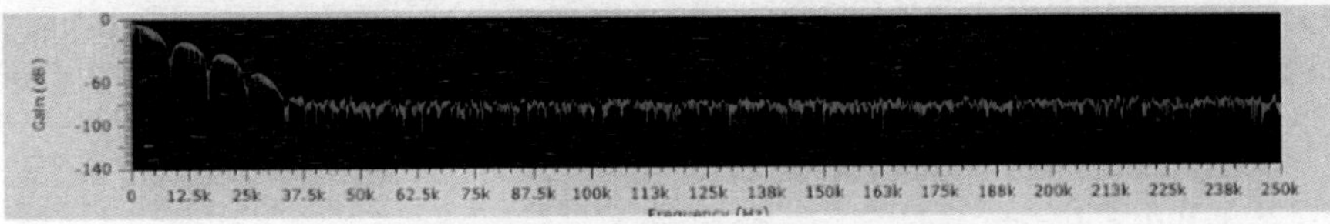

Fig. 24. BPSK demodulation spectrum.

MATLAB was used to simulate the modulation, demodulation and noise in BPSK by generating a code with the parameters obtained from the practice, the development of the code is observed in the section of Annex 2, it was seen in Fig. 25 that by properly formulating the BPSK base equations together with the necessary commands that when executing it was achieved obtained visualization product of generating a binary signal next to a sine wave that represents the cover and as this gives the result of a BPSK wave when passing through the mathematical equations entered.

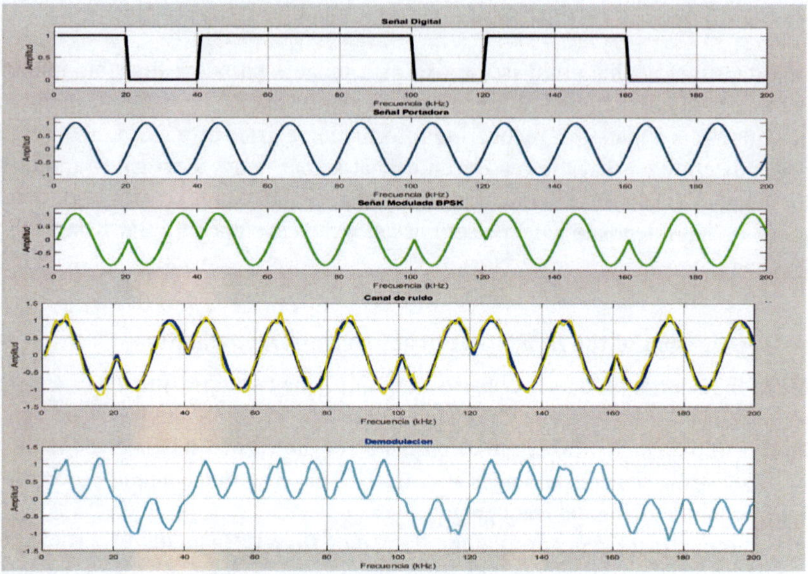

Fig. 25. MATLAB visualizations: binary signal, carrier signal, BPSK signal, Noise channel, demodulation.

5 Conclusions

As a result of the analysis of the BPSK modulation, the following points were determined:

The BPSK demodulation by product detector, the wave obtained are presented certain similarities to the input binary signal, with the help of an LPF for the obstruction of high signals, but with some peaks determining that there is some kind of distortion, and that the signal obtained is not square. Unlike the use of a comparator which is a circuit that restores distorted digital signals.

In the BPSK spectrum, the frequency obtained was 98 kHz this value was the result of working with a frequency of 100 kHz being lower due to the bandwidth resolution of a value of 444 Hz and the existing errors in the instruments.

A noise signal with a bandwidth of 120 kHz was generated which was added to the function generator and incorporated into the BPSK signal, where the different noise levels between 0 dB and −20 dB were observed through the use of an adder.

The AWGN channel is chosen as a model for the analysis of communication systems for its simplicity, for this reason noise is considered in many tools and methods of study since it is a physical phenomenon.

With MATLAB programming was tested the BPSK digital modulation by a vector of binary bits and the input of a desired frequency of work, the results of the visualizations obtained when analyzing them were a satisfactory conclusion that the diagrams of the waves are similar to the visualizations that were obtained from the virtual oscilloscope provided by the platform of the NI ELVIS III module.

References

1. Guerrero Condo, L.M.: Diseño y desarrollo de prácticas de laboratorio para comunicaciones analógicas basadas en modulación de amplitud con el uso de los equipos TIMS (2016)
2. Carlson, A.B., Crilly, P.B.: Communcation Systems: An Introduction to Signals and Noise in Electrical Communication. McGraw-Hill, New York, NY, USA (2010)
3. Nieva Suarez, D.A., Sanabria Lemos, J.R.: Sistema de Comunicación SDR Para Modulación PSK Y QAM. https://red.uao.edu.co/bitstream/10614/10601/5/T08268.pdf (2018)
4. Barco Vargas, H.F.: Implementación de escenarios de simulación en sistemas de comunicación por satélite utilizando técnicas de modulación PSK. http://repositorio.ucsg.edu.ec/bitstream/3317/10621/1/T-UCSG-POS-MTEL-96.pdf (2018)
5. Tomasi, W.: Sistemas de Comunicaciones Electrónicas. PEARSON EDUCACIÓN, México (2003)
6. Vélez Agua, J.A.: Análisis Cuantitativo del Ruido en Sistemas de Modulación Digital ASK, FSK Y BPSK. http://repositorio.ug.edu.ec/bitstream/redug/36310/1/2 - TESIS - ANÁLISIS CUANTITATIVO DEL RUIDO EN SISTEMAS DE MODULACIÓN DIGITAL ASK%2C FSK Y BPSK.pdf (2018)
7. Castañeda Alcalá, D.O.: Implementación de un Modulador - Demodulador Digital QPSK en Base a un FPGA como Prototipo para un Microsatélite. http://132.248.52.100:8080/xmlui/handle/132.248.52.100/8653 (2015)
8. Herrera Pajaro, L.E., Torres Viloria, A.J.: Diseño e Implementacion de un Modulador y Demodulador Bpsk. https://biblioteca.utb.edu.co/notas/tesis/0062410.pdf (2011)
9. Zúñiga Eraso, C., Cruz López, W.A.: Análisis del desempeño de un sistema de comunicaciones con modulaciones BPSK/QPSK/OQPSK/8PSK basado en hardware reconfigurable. http://repositorio.unicauca.edu.co:8080/bitstream/handle/123456789/1912/Anexos.pdf?sequence=2&isAllowed=y (2014)
10. EMONA Instruments Pty Ltd.: National Instruments corp.: Binary Phase Shift Keying (BPSK). https://education.ni.com/teach/resources/1197/binary-phase-shift-keying-bpsk

Optimization of Radiation Parameters for Antennas Used in Cell-Based Wireless Networks

Marcelo Zambrano[1,2](✉), Ana Zambrano[3], Juan Minango[2], Edgar Maya[1], and Mauricio Dominguez[1]

[1] Universidad Técnica del Norte, Av. 17 de Julio 5-21, CP 100105 Ibarra, Ecuador
omzambrano@utn.edu.ec

[2] Instituto Tecnológico Universitario Rumiñahui, Av. Atahualpa 1701 y 8 de Febrero, CP 171103 Sangolquí, Ecuador

[3] Escuela Politécnica Nacional, Ladrón de Guevara E11-253, CP 170525 Quito, Ecuador

Abstract. Dead zones and interference are two of the biggest problems in cell-based wireless networks. In practice, these drawbacks are often caused, on many occasions, by the imperfect radiation parameters of the antennas that make up the base station (s). This paper describes an experimental method to optimize, or better adapt to the design, the antenna radiation parameters used in cellular wireless networks. It is proposed to install reflective shells or shields, with customized geometries, that allow redirecting the radiation lobes power of the antennas that are causing interference or that are not delivering the necessary power in any area or sector.

Keywords: Antennas · Radiation parameters · Radiation lobes · Wireless cellular networks

1 Introduction

The advantage of wireless networks in relation to their fixed or wired pairs lies in the use of Electromagnetic Waves (EMW) as a transmission medium, allowing them to provide service to mobile users or in places where fixed networks do not have coverage [1, 2]. One of the most important types of wireless networks with the greatest market penetration are cellular networks (e.g., WiFi [3] or Advanced Mobile Service (SMA) [4, 5]), which are characterized by dividing the target geographic area in smaller sectors (cells), through the implementation of a communications backbone made up of multiple transmitters/receivers (base stations) of lower power, size and, obviously, scope.

Within this same context, and in addition to the aforementioned advantages, cellular networks, also known as mobile networks [4], have made it possible to reduce cabling, visual pollution and the costs associated with communication networks, with a quality of service that increasingly resembles that of its fixed competitors [6]. However, in terms of coverage, it is technically not possible to cover a wide area or territory with a single radio station, due to the high radiation power that would be required in the antennas

M. Zambrano Vizuete et al. (Eds.): CI3 2021, LNNS 511, pp. 50–62, 2022.
https://doi.org/10.1007/978-3-031-11438-0_5

and the possible health risks and costs involved under this alternative. Additionally, if an architecture of this type exists (single cell or point-multipoint), there would be many dead zones that would be generated (zones without service within a theoretically covered area) [5], considering, in a general way, the obstacles that interrupt the line of sight (LOS) between transmitters and receivers, due to civil structures and, in this case, the orography of Andean countries such as Ecuador. For these reasons, in practice, mobile networks divide their coverage area into multiple elementary units or cells, each one covered by a base station [7]. Under this model, it is possible to reduce both the radiation power of the antennas and the possible dead zones [8]. Furthermore, this architecture allows the frequencies reuse and considerably improves the network spectral efficiency, increasing the number of serviceable users and the quality of the services offered [9].

Cellular networks offer several advantages compared to other wireless communication systems: great capacity for data transmission, optimization of the power required for transmission/reception, great coverage, and of course, the integration of services such as video, audio and data. [4, 6]. The transmission/reception technology used in this type of network demands that the base stations be widely integrated into the civil infrastructure of the areas that want to be served, especially for the latest generation technologies such as 5G [10]. This is because the signal must be able to reach the user wherever they is, so that, if necessary, cells are usually subdivided into smaller units such as microcells or sectors [1], according to the required quality parameters and network design.

Given the frequencies at which this type of network works, it is essential that there be a line of sight (LOS) between the transmitters and receivers antennas (Fig. 1).

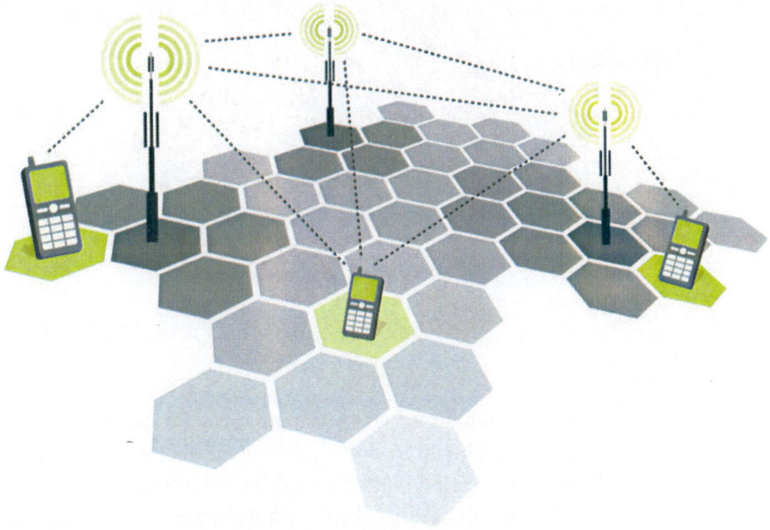

Fig. 1. Cellular network [6].

An antenna is a device designed to radiate or receive OEMs within a given frequency range; it is characterized by its radiation parameters, which are usually represented graphically by means of a radiation diagram [12, 13].

Radiation diagrams make it possible to obtain a geometric representation (three-dimensional or two-dimensional) of the radiation parameters; they are generally represented in spherical coordinates (radius (r), polar angle (θ) and azimuthal angle (Ø)) [14] and as a function of the electric and magnetic field and/or the radiated power density. The most important parameters represented in a radiation diagram are (Fig. 2):

- Aiming direction: maximum radiation from the antenna, defined based on its directivity and gain. Directivity is a measure of the concentration of radiated power in a particular direction. The gain refers to how much energy is transmitted in the direction of maximum radiation [15].
- Main lobe: closed region where most of the radiation power is concentrated. There may be more than one [13].
- Secondary lobes: closed regions where radiation also exists, but with less power than that existing in the main lobe [13].
- Beamwidth: It is the angle formed by the two imaginary axes of connection of the antenna with the points where the gain has decreased by 3 dB with respect to the maximum gain, having the maximum gain axis as a bisector. According to this parameter, the radiation zone is defined where the signal pickup is adequate (between 70% and 100% of the maximum gain).

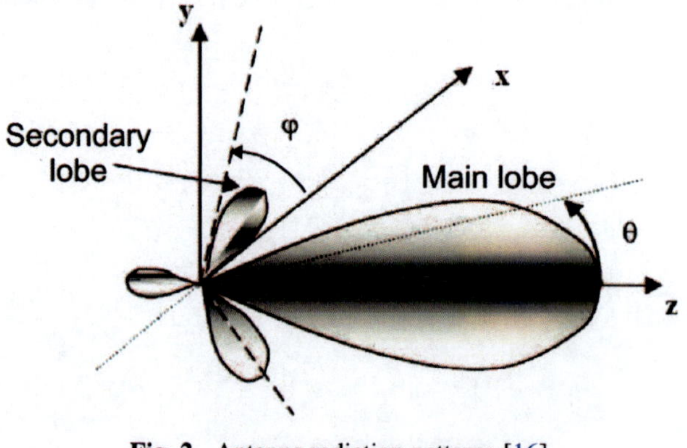

Fig. 2. Antenna radiation patterns [16].

According to the shape of the radiation pattern, antennas can be classified as isotropic, omnidirectional or directional, as shown in Fig. 3. An ideal antenna that has the same radiation power in all directions is called Isotropic and, in the In practice, it is used as a comparison parameter for real antennas. Omnidirectional antennas are characterized by a wide and short-range radiation beam, theoretically radiated over 360° of the horizontal plane. Directional antennas are those whose power is concentrated within a narrow and long-range beam in the desired direction (Fig. 4) [17].

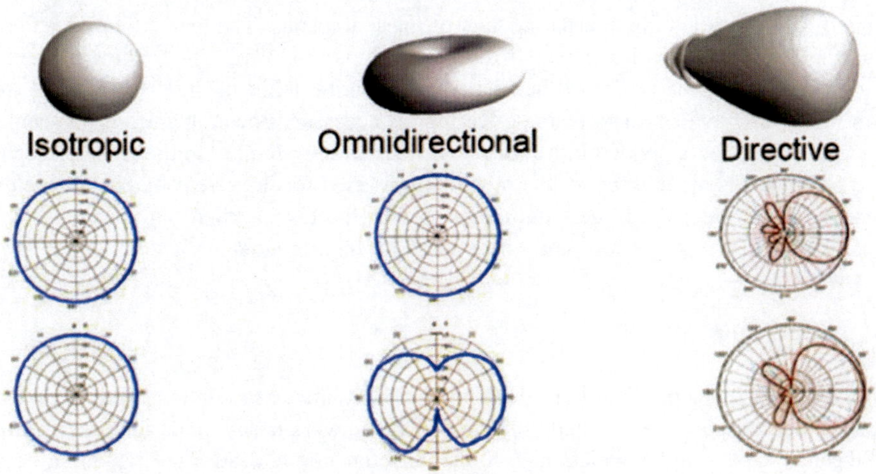

Fig. 3. Antenna types [17].

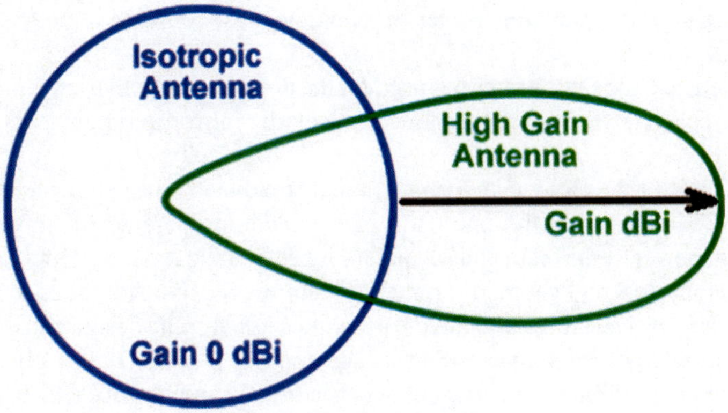

Fig. 4. Gain antenna [18].

Two of the most important problems in this type of network are dead zones and interference. Dead zones are generated in those sectors where the signal power is poor or non-existent, due to obstacles that are interposed between the transmitter and the receiver (s), limiting the line of sight. On the other hand, the interferences arise in those areas where the secondary radiation lobes (lateral and posterior) [1, 10] of the antennas that serve adjacent cells or sectors, have a greater power and/or range than desired. In summary, the aforementioned issues are usually caused by the imperfect radiation parameters of the antennas that make up the communications backbone of those same networks.

In this paper describes the preliminary results of an experimental method for optimizing the antennas radiation parameters used in cell-based wireless networks. The research main objective is to mitigate interference and reduce dead zones caused by imperfections in the radiation parameters of these antennas. The installation of reflective shells or shields with a personalized geometry is proposed, which allow reducing and/or redirecting the power of the secondary radiation lobes of the antennas that are causing interference or that are not delivering the necessary power in any area or sector.

The document is divided into four parts: first, an introduction to the research topic and objectives is made; second, the methodology used for the electromagnetic shields design is described; third, the simulations results and tests carried out are presented; finally, fourth, the conclusions and future work for this research are listed.

2 Methodology

As mentioned in the previous section, in practice, the antennas radiation parameters of a cellular network are hardly exactly matched to the network design or civil infrastructure of the area to be covered, which leads to in interferences and dead zones which, in turn, make that coverage and quality of the services offered not as desired. It is proposed to modify the antennas radiation parameters that make up the communications backbone of a cellular wireless network, using reflective shields or shells, which allow optimizing and/or redirecting the radiation power of said antennas, coupling them to the desired parameters.

As a reference, the same principles used for the design of reflective antennas [19] have been taken, characterized using a conductive-metallic mirror to optimize its directivity and gain [20].

The features of the electric and magnetic fields radiated by an antenna that includes a reflector will depend on the one hand of the radiation parameters of said antenna (beamwidth, power of the main and secondary lobes, polarization, etc.) and, on the other hand, of the physical and geometric reflector features (focal length, diameter, material, etc.) [21]. The physical and mathematical principles used for the design and construction of reflective antennas are based on the laws of geometric optics (GO) [22], physical optics (PO) [23] and the diffraction theories Geometric (GDT) and Physics (PDT) [24, 25].

The theoretical validation of the proposal described in this paper was carried out through multiple simulations using the Ansys HFSS [26] software. In these simulations, the radiation parameters of an microstrip omnidirectional antenna [27] were modified, using reflective shields, with different geometries and distances on the back of it, since it was assumed that the radiation to be modified was that of the posterior lobe. The results are shown in the next section.

3 Discussion and Results

The first thing that was carried out was the design and simulation of the microstrip antenna, with characteristics such as those shown in Fig. 5.

a) Front view.

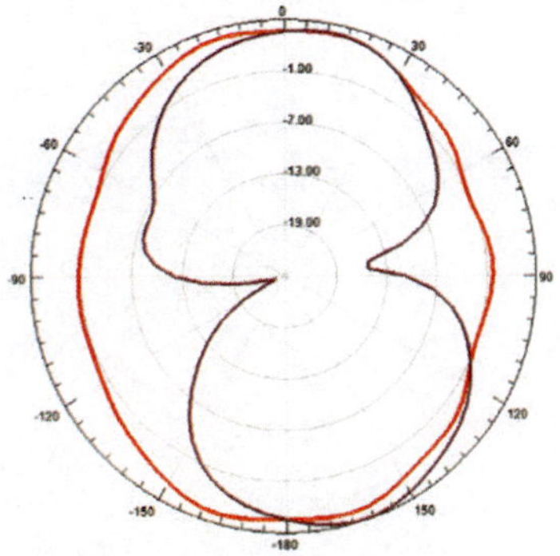

b) Horizontal radiation diagram.

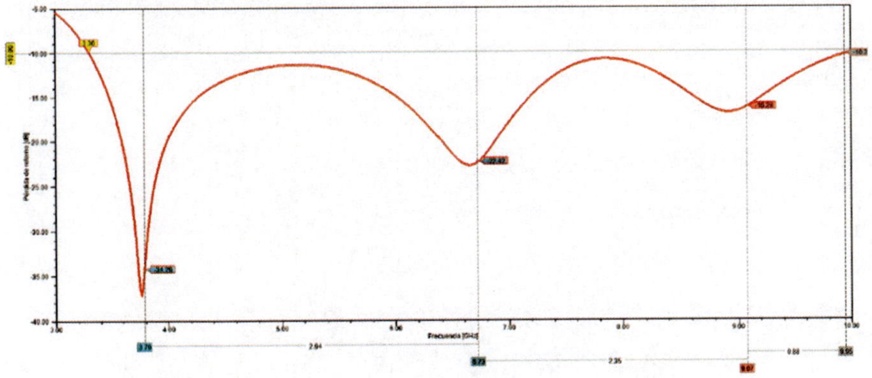

c) Bandwidth and resonance frequency (3 GHz to 10 GHz).

Fig. 5. Microstrip antenna.

Both the antenna and the reflective shields were simulated using the aluminum as the material. Below are some of the antenna simulations with different types of electromagnetic shields (Fig. 6).

3.1 Antenna with a Reflective Shield of 1 mm Thick and an Antenna-Shield Distance of 10 mm

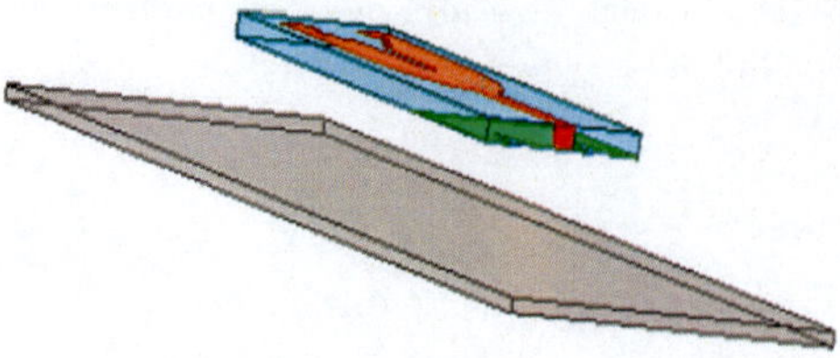

a) Lateral view.

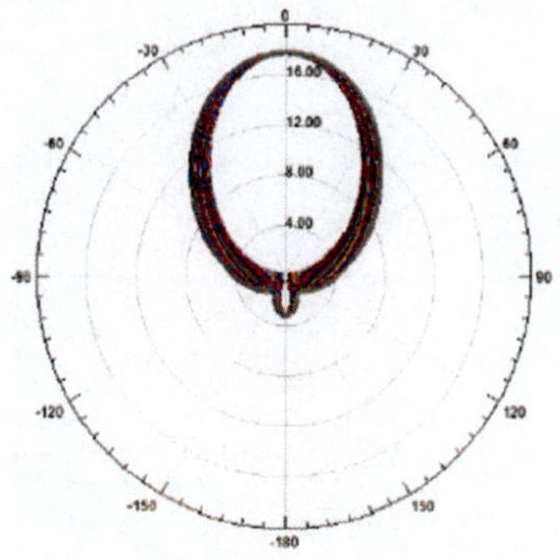

b) Horizontal radiation diagram.

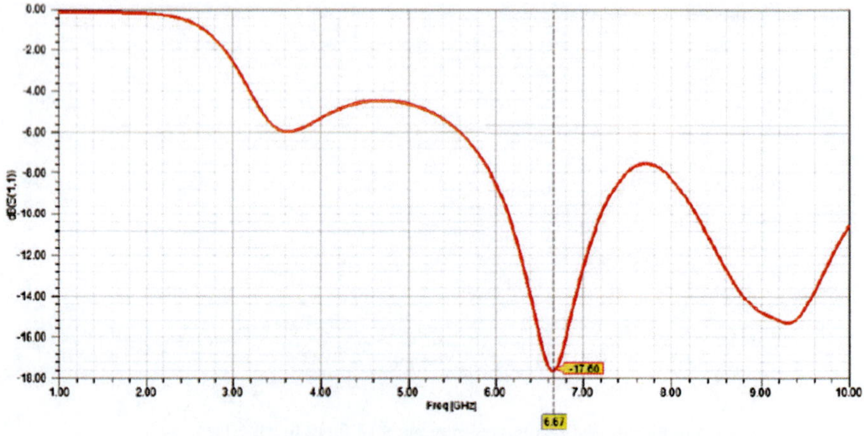

c) Bandwidth and resonance frequency.

Fig. 6. Antenna with a reflective shield of 1 mm thick and an antenna-shield distance of 10 mm.

3.2 Antenna with a Reflective Shield of 1 mm Thick and an Antenna-Shield Distance of 50 mm

In Fig. 7, it can be seen that as the antenna-shield distance increases the directivity decreases, forming lateral radiation lobes with significant power. It is important to emphasize that the posterior radiation lobe is minimal; in addition, the gain and frequency parameters also undergo changes.

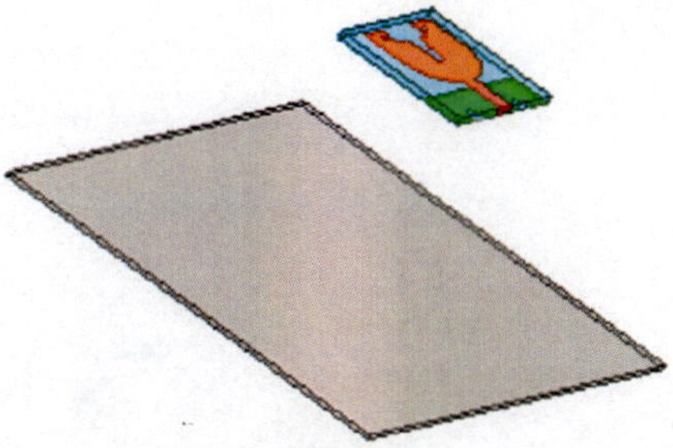

a) Lateral view.

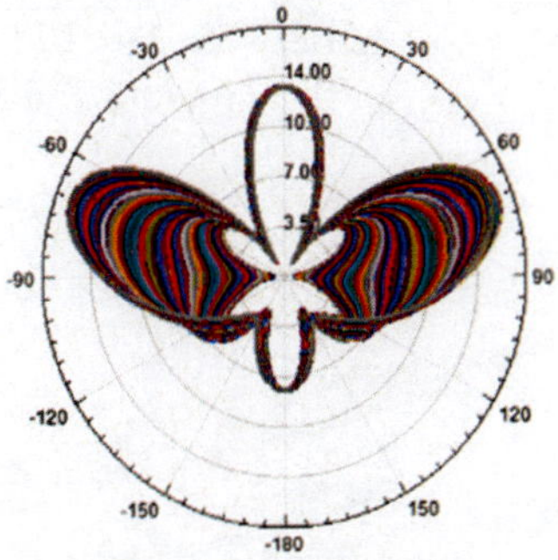

b) Horizontal radiation diagram.

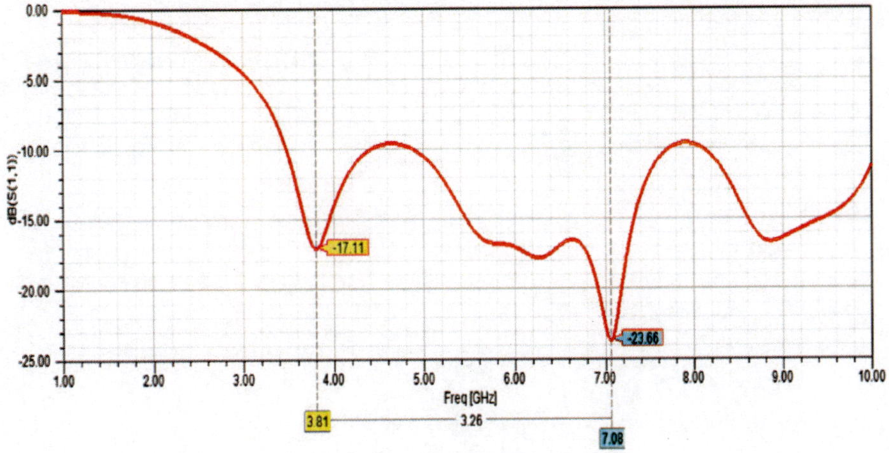

c) Bandwidth and resonance frequency.

Fig. 7. Antenna with a reflective shield of 1 mm thick and an antenna-shield distance of 50 mm.

3.3 Antenna Between Two Reflective Shields of 1 mm Thick; Separation Distance Between the Antenna and Shields: 10 mm

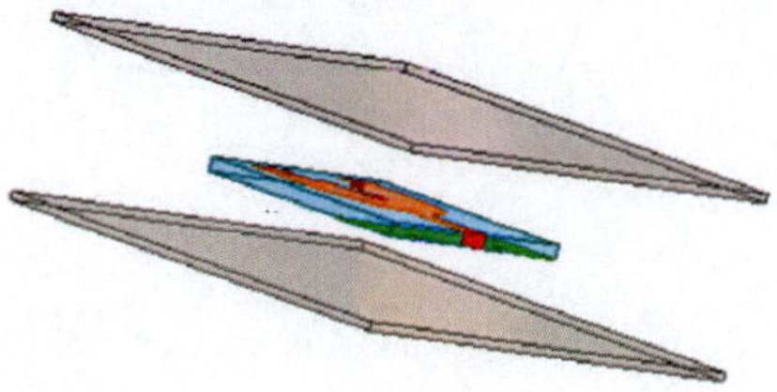

a) Lateral view.

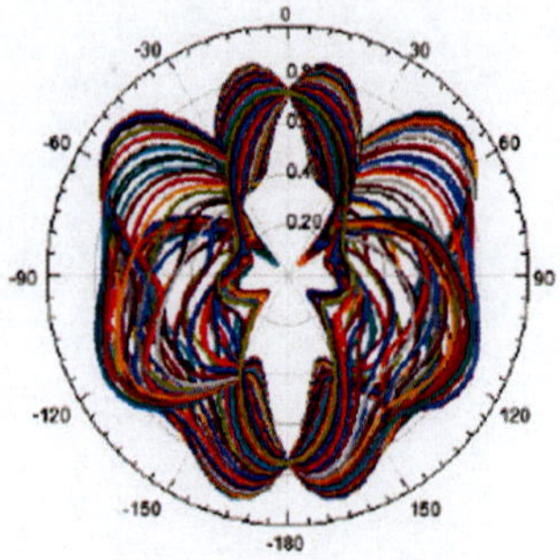

b) Horizontal radiation diagram.

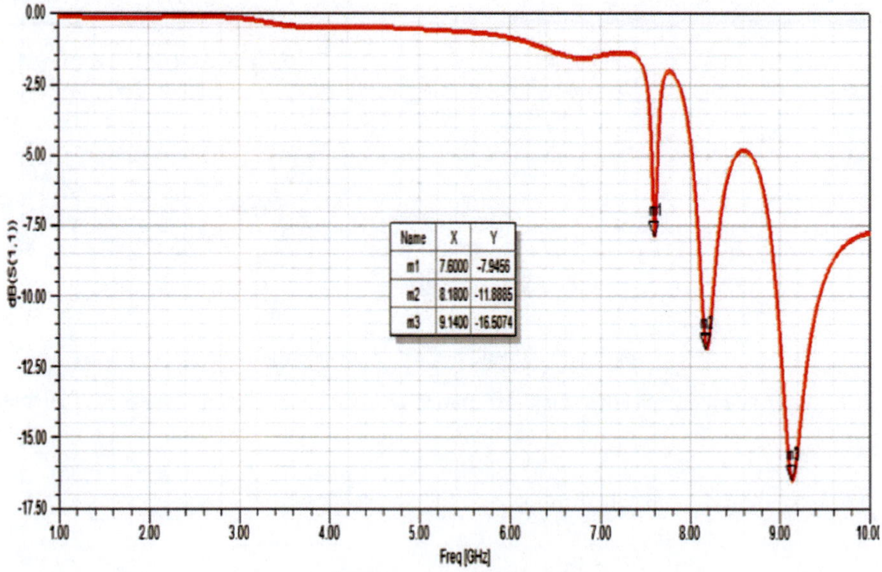

c) Bandwidth and resonance frequency.

Fig. 8. Antenna between two reflective shields of 1 mm. thick; separation distance between the antenna and shields: 10 mm.

3.4 Antenna Between Two Reflective Shields of 1 mm Thick; Separation Distance Between the Antenna and Shields: 50 mm

Comparing Figs. 8 and 9, it can be clearly seen that the distance between the reflective shields and the antenna is a fundamental parameter that directly affects the radiation pattern (Fig. 10).

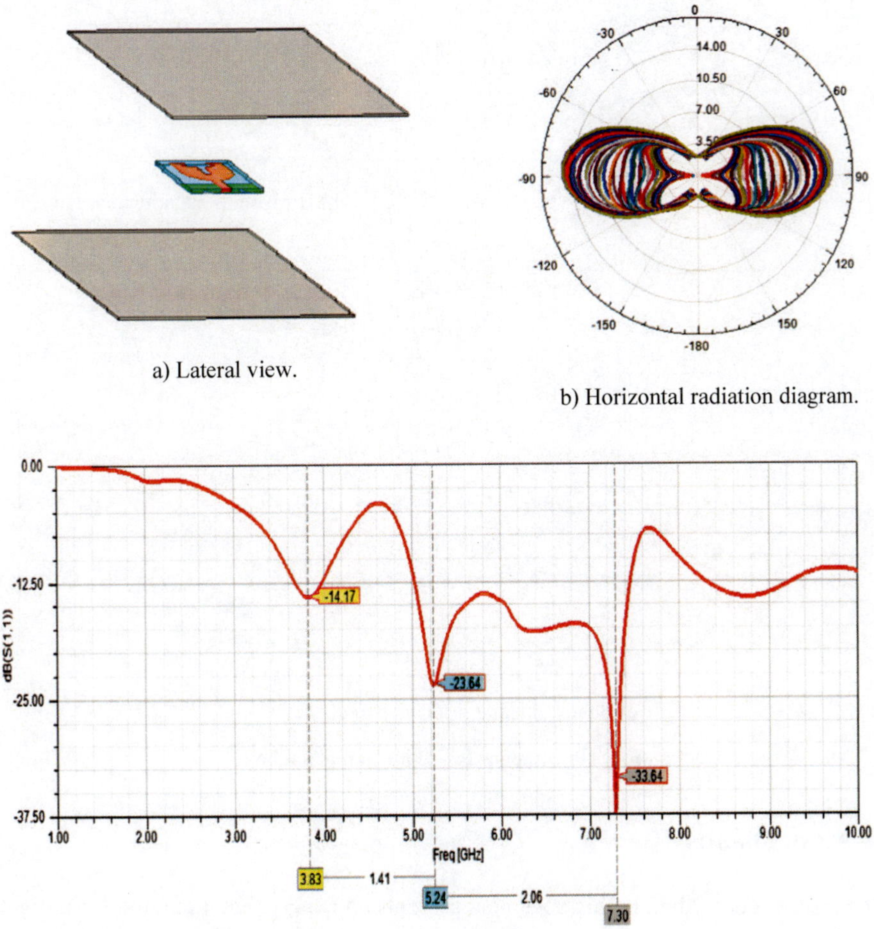

c) Bandwidth and resonance frequency.

Fig. 9. Antenna between two reflective shields of 1 mm. thick; separation distance between the antenna and shields: 50 mm.

3.5 Antenna Inside a 1 mm Metal Box

a) Front view.

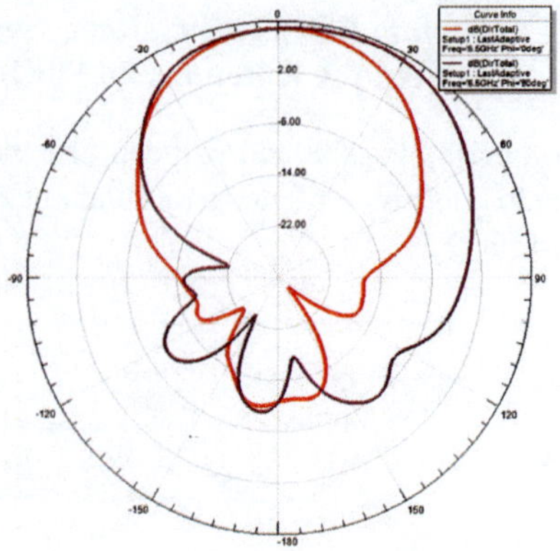

b) Horizontal radiation diagram.

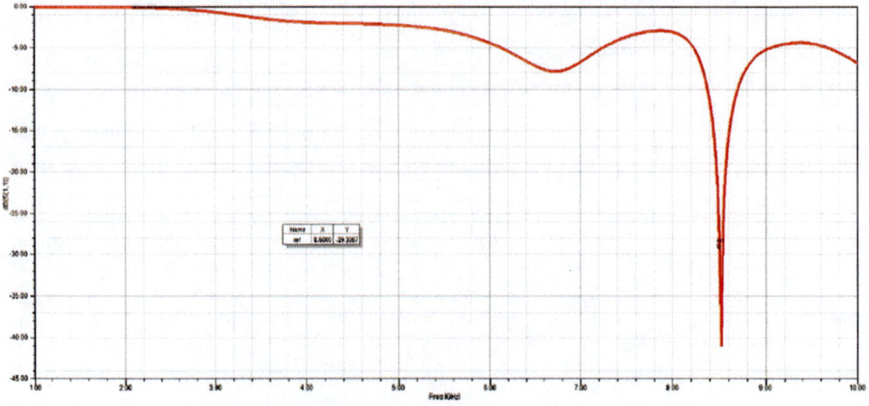

c) Bandwidth and resonance frequency.

Fig. 10. Antenna inside a 1 mm metal box.

4 Conclusions

The research described in this document presents an experimental method for mitigating dead zones and interference in cellular wireless networks. The implementation of metallic reflective shields in the antennas that make up the base stations of the cellular network is proposed, with the aim of optimizing and/or adjusting to the design the radiation parameters of the cellular network.

After the simulations and tests carried out, it can be concluded that it is not only feasible to reduce or mitigate the radiation power of the unwanted or interfering secondary lobes, but that power can also be redirected to the main lobe, increase or improving network coverage.

It is important to emphasize that, with the use of reflective shields, is not only possible to modify the directivity and gain of the antennas, but also parameters such as bandwidth and resonance frequency undergo changes.

As future work, we will continue to experiment with different forms of shields, this time with concave and convex geometries that manage to direct the electromagnetic fields laterally and at will. Equations that mathematically describe the relationship between the electromagnetic field and the geometry of the shields must also be established.

References

1. Sauter, M.: From GSM to LTE-Advanced Pro and 5G: An Introduction to Mobile Networks and Mobile Broadband. John Wiley & Sons, Chichester, UK (2017)
2. Rappaport, T.S.: Wireless Communications: Principles and Practice, vol. 2. Prentice Hall, New Jersey (1996)
3. Gast, M.: 802.11 Wireless Networks: The Definitive Guide. O'Reilly Media, Inc. (2005)
4. ARCOTEL, «Servicio Móvil Avanzado SMA,» Quito, 2018
5. Youssef, L., Ruichek, Y., Touahni, R.: Multi-hop communications inside cellular networks: a survey and analysis. Int. J. Electr. Electron. Eng. Telecommun. (IJEETC) **8**(6), 297–306 (2019)
6. Agiwal, N.: Next generation 5G wireless networks: a comprehensive survey. IEEE Commun. Surv. Tutorials **18**, 1617–1655 (2016)
7. Zhou, S., Zhao, M., Xibin, X., Wang, J., Yao, Y.: Distributed wireless communication system: a new architecture for future public wireless access. IEEE Commun. Mag. **41**(3), 108–113 (2003)
8. Sharma, A.: Joint power-domain and SCMA-based NOMA system for downlink in 5G and beyond. IEEE Commun. Lett. **23**(6), 971–974 (2019)
9. Wang, C.: Cellular architecture and key technologies for 5G wireless communication networks. IEEE Commun. Mag. **52**(2), 9122–9130 (2014)
10. Shafi, M.: 5G: A tutorial overview of standards, trials, challenges, deployment, and practice. IEEE J. Selected Areas in Commun. **35**(6), 1201–1221 (2017)
11. EMAZE: Introducción al estádar GSM. En: Línea. Available: https://app.emaze.com/@AQZCZZRQ#Introducci%C3%B3n%20al%20est%C3%A1ndar%20GSM. Último acceso: 04 06 2021
12. Milligan, T.A.: Modern Antenna Design. John Wiley & Sons, Hoboken, NJ, USA (2005)
13. Stutzman, W.L., Thiele, G.A.: Antenna Theory and Design. John Wiley & Sons (2012)
14. Charles, L.: Geometría Analítica. Limusa, Madrid (1980)
15. Rohner, M.R.D.C.: Antenna Basics. Rohde Schwarz (2015)
16. Giacomo, B.: Development of the fictitious sources method for stratified media and design of resonant cavities antennas. Université Paul Cézanne - Aix-Marseille III (2007)
17. Mohamed, E.B.: Diseño y Medición de una Wearable. Universitat Politècnica de Catalunya, Barcelona (2016)
18. A.H.S. Inc.: Practical Overview of Antenna Parameters. En línea. Available: https://www.ahsystems.com/articles/Practical-overview-of-antenna-parameters.php (2021)
19. Rahmat-Samii, Y.: «Reflector antennas» de In Antenna handbook, pp. 949–1072. Springer, Boston (1988)
20. Pdfslide: Reflectores (2021)
21. Valero, A.: Antenas. Universitat Politècnica de València, Valencia (2021)
22. Feichtner, T.: Evolutionary optimization of optical antennas. Phys. Rev. Lett. **109**, 127701 (2012)

23. I., Robinson, T.A.: Optical geometry. New Theories in Physics (1988)
24. Pathak, P., Burnside, W., Marhefka, R.: A uniform GTD analysis of the diffraction of electromagnetic waves by a smooth convex surface. IEEE Trans. Antennas Propag. **28**(5), 631–642 (1980)
25. Ufimtsev, P.Y.: Fundamentals of the physical theory of diffraction. John Wiley & Sons (2014)
26. Ansys: Antenna Design & Modeling Software. Ansys. En línea. Available: https://www.ansys.com/applications/antenna-design-and-placement (2021)
27. Yang, D., Zhai, H., Guo, C., Li, H.: A compact single-layer wideband microstrip antenna with filtering performance. IEEE Antennas Wirel. Propag. Lett. **19**(5), 801–805 (2020)
28. Weinstein, L.A.: Electromagnetic Waves. Radio i svyaz (1988)
29. Someda, C.G.: Electromagnetic Waves. Crc Press (2017)
30. Sarkar, T.K., Ji, Z., Kim, K., Medouri, A., Salazar-Palma, M.: A survey of various propagation models for mobile communication. IEEE Antennas and Propag. Mag. **45**(3), 51–82 (2003)

ICT in Education

Technological Resources in a Blended Learning Environment for the Teaching of Ornithology

Marcos Vera-Morales[1,3], Jaime Naranjo-Morán[1,3,4], and Andrea Pino-Acosta[2,3](✉)

[1] Escuela Superior Politécnica del Litoral, ESPOL, ESPOL Polytechnic University, Campus Gustavo Galindo Km. 30.5 vía Perimetral, P.O. Box 09-01-5863, Guayaquil, Ecuador

[2] Escuela Superior Politécnica del Litoral, ESPOL. Facultad de Arte, Diseño y Comunicación Audiovisual, FADCOM, Bosque Protector "La Prosperina", ESPOL Polytechnic University, Campus Gustavo Galindo Km. 30.5 vía Perimetral, P.O. Box 09-01-5863, Guayaquil, Ecuador

ypino@espol.edu.ec

[3] Red de Bosques Secos, REBS, Km. 30.5 vía Perimetral, P.O. Box 09-01-5863, Guayaquil, Ecuador

[4] Universidad Politécnica Salesiana, UPS, Facultad Ciencias de la Vida, Ingeniería en Biotecnología, Campus María Auxiliadora, Km 19.5 Vía a la Costa, 090901 Guayaquil, Ecuador

Abstract. Ornithology courses have traditionally been developed through face-to-face lectures complemented with field trips, but currently technologies offer new possibilities to the training process. The present study aims to provide useful guidance for the application of technological resources in teaching the fundamental principles of ornithology and bird watching in a blended learning environment. The course was aimed at students and professionals of the biological sciences through the combined modality of synchronous and asynchronous classes. In the synchronous classes, lectures were held and the asynchronous ones through the use of a virtual learning environment with practical and autonomous activities. The result of the anterior and posterior knowledge test had variations, although without significant differences. Regarding the development of the activities, the undergraduate students were those who carried out the content creation tasks using different technological resources, while the postgraduate professionals showed tendencies towards the development of jobs with greater cognitive demand. Among the most popular resources, EDpuzzle, Padlet and GoConqr were highlighted. Regarding the virtual environment, the participants attended mostly in the evening and at night, with the most frequent sessions being the introduction, the anatomical study and the systematics of mangrove birds. In conclusion, the students have used the different technological resources provided by the teacher to design their own learning experiences, but they were not enough for them to fully master the different topics that were addressed in the course.

Keywords: Technological resources · Blended learning · Teaching · Birds · Ornithology

1 Introduction

In the study of various branches of biological sciences, the incorporation of technologies is a challenge, because much of the knowledge must be complemented with field trips so

M. Zambrano Vizuete et al. (Eds.): CI3 2021, LNNS 511, pp. 65–77, 2022.
https://doi.org/10.1007/978-3-031-11438-0_6

that students can develop skills and work skills. Therefore, the curricular gap prevents in many cases from making the most of the available technological resources [1]. Therefore, there is a need to investigate resources and methods that allow learning the concepts of the branches of biology interactively [2].

Currently, the use of technologies has increased in undergraduate courses in the subject of biology [3]. Despite the challenges that virtuality presents, such as the infrastructure, the prior knowledge of the instructor and the diversity of the students [4]. Technologies offer a wide variety of accessible and relevant educational materials to complement the learning built in virtual environments [5]. For example, synchronous blended learning environments where students actively participate via video conferencing, web conferencing, or virtual worlds [6].

Due to the COVID-19 pandemic, 49% of university students consider themselves in favour of the use and continuous improvement of blended or blended learning [7]. However, faced with this situation, dealing with stress, fear, anxiety and depression can significantly affect the mental health of teachers and participants [8], for this reason the role of the teacher in the virtual environment should focus on teaching strategies in an integrated and pedagogical way [9], since they could incorporate conceptualization of the topics covered in a practical and autonomous way to mitigate risk factors for other types of stress.

Although different digital media have informal relevance in people's lives, they can be a potential element in the teaching process. For example, there are mobile applications that are a potential support to collect information and consolidate learning. One of the applications such as eBird, allows to register birds and define other field variables [10], motivating students towards understanding conceptual constructs [11]. Social networks have also been used as a means of becoming familiar with bird identification. [12].

It is important to consider that technologies are constantly changing, so new and better options must be identified to join the educational process, keeping up to date those digital media with which the student is regularly [13]. However, the success of incorporating technological resources in planning depends on the extent to which teachers are familiar with the functionality of technologies [14], considering the impact on student learning [15]. Therefore, it is imperative to promote the use of combined synchronous and asynchronous spaces as a complement to the study of the different sciences, particularly ornithology. Therefore, the objective of this research was the application of technological resources in an introductory course to ornithology and bird observation through a blended learning environment aimed at students and professionals of the biological sciences.

2 Materials y Methods

Participants. The participants in this study were students and professionals from the area of biological sciences and related. The course began with 29 people, three withdrew and 26 managed to complete the entire course. Of those enrolled, 11 corresponded to undergraduate students aged between 20–30 years, 11 third-level professionals and four fourth-level professionals aged between 35–60 years.

Call for the Curse. This study focused on a course offered by the Dry Forest Network (REBS) for students and professionals passionate about tackling the study of birds. The title of the course was "Introduction to ornithology and bird watching in the dry forest (BsT) and mangroves (MnG)". The total course was 40 h, divided into 10 synchronous hours and 30 asynchronous hours.

For this study, research based on design was selected, through which the theories and methodologies of learning currents were used to investigate the mechanisms by which learning is built [16]. This design involved the use of technological resources, didactic instruction and the theories on which learning is sustained [17, 18]. Teaching effectiveness was measured by a single field experiment in which the variables were controlled using a pre-experimental design with a single group with pre-test and post-test [19]. During the course, a considerable amount of work was sent to be carried out with technological resources, so a control group would be at a great disadvantage.

Learning Environment. The classes were developed in a blended learning modality, where the students participated in the activities in face-to-face classes through video-conferences using the Zoom platform. While the asynchronous classes were developed using a website, designed on the Wixsite platform (https://rebsecu.wixsite.com/aves), where the slides, the recorded lectures, the practical, freelance work and the teaching material were uploaded. bibliographic reference.

Given the conditions of the study, it was necessary to know the level of understanding about the science of ornithology that the participants had before and after the course, for this reason a pre and post-test was designed using the quizzes platform. The test was planned to take place over 30 min and the topics covered were chosen from the ten units corresponding to the course.

Units consisted of an introduction to the study of birds, anatomical features I and II, taxonomy, ecology, bird watching I and II, tropical dry forest birds, mangrove birds, and conservation (see Fig. 1). To pass each unit of the course, the participants developed different activities called practical and autonomous work, making use of the tools available on the web, such as EDpuzzle, Padlet, Cmad Cloud, GoConqr, Pixton Comics, PowToom, Quizizz. These tools were used for the participants to work individually and collaboratively (Table 1). The activities were designed in order to simulate field trips in real conditions, taking elements from the environment of each participant. During the development of the course, a WhatsApp group was used to clarify the doubts and questions of the participants.

The works were evaluated by the facilitators according to the fulfilment of the activities assigned in each class section.

Participants' Perceptions. Once the course ended, a questionnaire of 20 questions was sent to the students, divided into 18 closed and 2 open. The closed questions were directed to knowing the perceptions about the facilitators, the didactic process and planning, on a scale from 1 (minimum) to 5 (maximum). Instead, open questions were directed at knowing the relevance of the course and possible improvements.

Analysis of Data. Data from the pre and post-test were analysed using the Student's t test for related samples with p values < 0.05, using the SPSS version 20.0 statistical program.

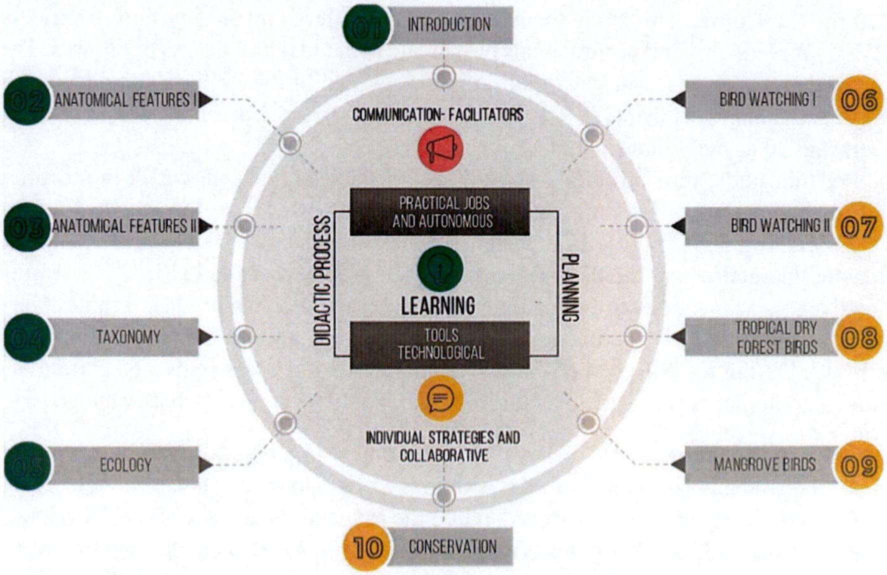

Fig. 1. Ornithology virtual teaching process through blended learning. Own elaboration of the authors

The data of the visits registered on the website were analysed with the InfoStat v. 1.2.0 (2017), by analysis of variance (ANOVA) and Tukey's test with p values < 0.05.

The perceptions data were descriptively analysed and then their statistical significance was calculated with the Chi-square test, using the statistical program SPSS version 20.0.

3 Results

Of the total participants registered at the beginning of the course (N = 29), 55% passed with an average greater than or equal to 70/100 points. To know the result of the teaching of the course in the learning of the students, a pre and post-test was used. Although the average scores of both tests varied by a 3% higher in the posterior knowledge test, no significant differences were found between the scores of both tests, as determined by the Student's t test for related samples ($p < 0.05$) (see Fig. 2). The results indicated that the teaching process of the bird's course, using different types of technological resources, was not enough for the students to fully master all the topics addressed. Which suggests that technological resources can facilitate the preparation of tasks, but not necessarily the understanding of the contents.

Table 1. Practical and autonomous work of the different work sessions in the ornithology course following the learning modality combined with technological resources.

Sessions	Practical jobs	Freelance jobs
Introduction	Reading about the history of ornithology (Summary of 300 words in Word) Reading of the orthomorphic figures (Synthesis of 50 words in Padlet)	Differences and similarities between vertebrates (Mind map in GoConqr) Story about a bird (Pixton Comics)
Anatomy I	The pose of a bird (Video in PowToon) Analysis of a video (Questions and answers in EDpuzzle)	Topography of a bird (Point out the parts of the birds and share it in Padlet) Reading on Bird Pigments (300 Word Summary in Word)
Anatomy II	Reading on Bird Diets (300 Word Summary in Word) Food Guilds (Group work on Cmad Cloud)	Field trip (Recording the song of birds) Xenocanto (Group work on Quizizz)
Taxonomy	Field trip (Photograph of the silhouettes of the birds) Taxonomic description (Using mind map in GoConqr to classify birds)	Drawing of a phenomenal bird (imaginary bird with its respective taxonomic classification uploaded in Padlet) Reading about Arquaeopterys (300 Word Summary in Word)
Ecology	Reading the ecosystem services of birds (Summary of 300 words in Word) Differences and similarities between tropical dry forest and mangrove birds (Use of mind map in GoConqr)	Differences and similarities between ecosystem, habitat and ecological niche (Use of mind map in GoConqr) Differences and similarities between bird migrations (Using mind map in GoConqr)
Bird watching, I	Field trip (Use of field diary)	Drawings of the birds observed in the field trip (Use of Padlet)
Bird watching II	Comments on behavior and ethics of people in bird watching (Use of Padlet)	Reading about birdwatching (Summary of 300 words in Word)
Tropical dry forest birds	Field trip (Use of mobile applications e-Bird, Naturalist, Merlin, OrnitO, Bird Data)	Recognition of the ecosystem, habitat and ecological niche of tropical dry forest birds (Use of mind map in GoConqr)
Mangrove birds	Field trip (Use of mobile applications e-Bird, Naturalist, Merlin, OrnitO, Bird Data)	Recognition of the ecosystem, habitat and ecological niche of mangrove birds (Use of mind map in GoConqr)
Conservation	Invasive bird reading (300 Word Summary in Word)	Bird conservation project proposal (Word)

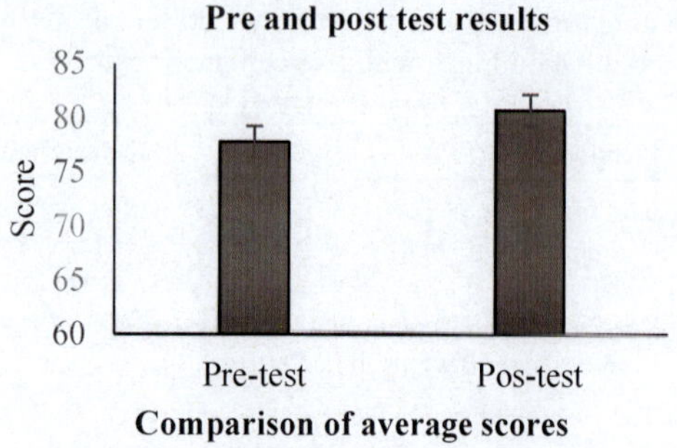

Fig. 2. Comparison of the mean scores (±SE) of the course participants (N = 26) in the pre and post-test of content knowledge. Own elaboration of the authors.

Visual inspection using a bispatial diagram revealed that undergraduate students tended towards the development of individual activities related to content creation through the use of technological platforms, while professionals tended towards group activities and postgraduate professionals towards the development of activities that demand greater cognitive understanding (see Fig. 3).

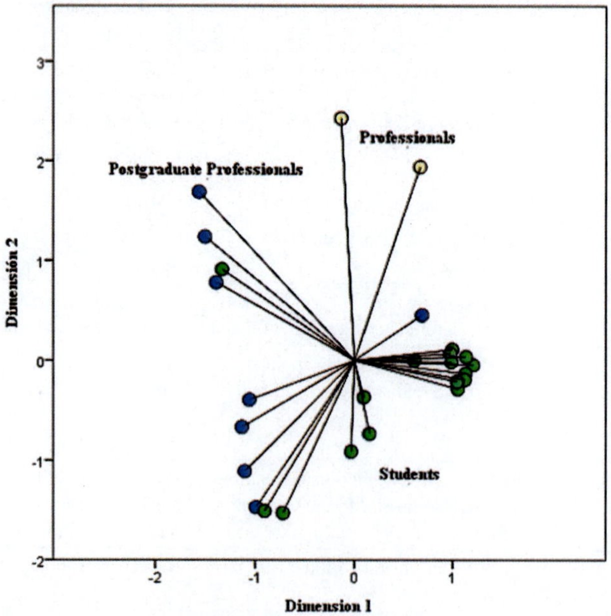

Fig. 3. Correspondence ordering diagram of the different activities developed during the introductory ornithology course, taking into consideration the academic level of the participants (students, professionals and postgraduate professionals). Green color-Creative individual activities developed on technology platforms. Yellow color-Group activities developed on technological platforms. Blue color-Individual activities of content analysis and synthesis. Own elaboration of the authors.

Six activities of the entire course were also evaluated, taking into account those of greater and lesser frequency. The results showed that the participants stood out in those that used some type of technology, such as video with a questionnaire on EDpuzzle, photographs on an interactive Padlet board and concept maps on GoConqr. While those of less frequency stood out the synthesis of 300 words in Word document (see Fig. 4). The findings suggest that students are inclined towards activities complemented with technological resources, while they scarcely develop those tasks that require a greater understanding of academic content.

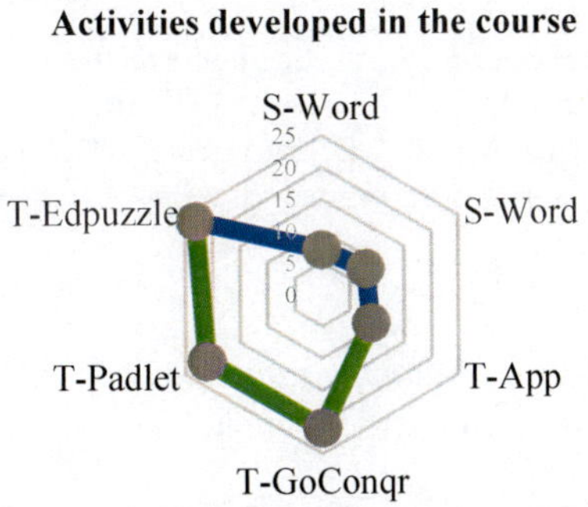

Fig. 4. Activities of greater and lesser frequency developed by the participants. Own elaboration of the authors.

The website designed as a dynamic repository, as a complement to the learning process, allowed observing the frequency of use of the participants. The visualization of the website had variations depending on the schedule and sessions of the course. The number of user visits registered significant differences, showing greater visits in the afternoon and evening (12–23 h) (see Fig. 5a). While the course sessions that most of the participants prioritized were the introduction, bird anatomy I and mangrove birds (see Fig. 5b). Therefore, the results show the interest of the participants in the recognition of birds at the taxonomic level.

Perceptions of the Facilitators, the Didactic Process and the Organization of the Course

Regarding the facilitators, there are no significant differences between the evaluated factors. However, 15% of the participants consider that time management should be improved to achieve the objectives set at the beginning of the course (see Fig. 6a).

Regarding the didactic process, women are the ones who prioritize the course content the most, in particular the contribution that each module made to the construction of their knowledge and professional activities ($\chi^2 = 4.846$; $p < 0.05$); and the design of the applied study methodology ($\chi^2 = 6.364$; $p < 0.05$) (see Fig. 6b). These results are similar to those found in the open questions, where several women mention the didactic process:

"The interaction of the facilitators and the duties in different applications was very interesting, since various methodologies were used" student.

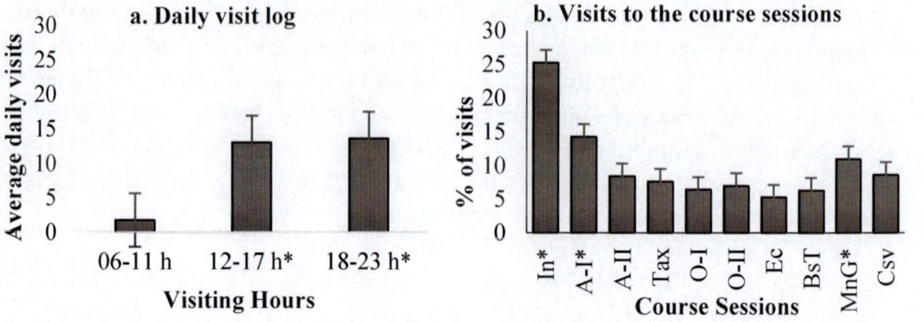

Fig. 5. Records of visits to the website. to. Visiting hours on the day. b. Percentage of visits to the course sessions In-Introduction, AI-Bird Anatomy I, AII-Bird Anatomy II, Tax-Taxidermy, OI-Bird Observation I, OII-Bird Observation II, Ec-Bird Ecology, BsT-Birds of the tropical dry forest, MnG-Birds of the mangrove, Csv-Conservation. *Significant statistical differences according to Tukey's test ($p < 0.05$).

> *"The interaction they had with examples and the tools currently available to help us with the birdwatching record"* professional.

Likewise, in the organization of the course, a higher percentage of female participants consider to be in strong agreement regarding the fulfilment of planning and programming, which shows significant differences ($\chi^2 = 4.870$; $p < 0.05$) (see Fig. 6c). These results are contrasted with the responses of some participants:

> *"The class schedules were accessible, as well as the general structure of the course with the use of applications and web pages"* female professional.

> *"(...) Some classes went over the scheduled time; it would be great to include that time in the total hours of the course"* male professional.

4 Discussion

The identification, biology and conservation of birds are subjects that must be taught with strong conceptual bases to students of biological areas. Through this research we propose that technological resources can help complement the learning processes in the area of ornithology. Therefore, a course with an innovative instructional design is approached to help teachers in the area of natural sciences to efficiently apply the tools available on the web in their teaching process. The intent of the design is not only for the technologies to be used appropriately, but also for a more detailed understanding of the fundamental constructs of ornithological science to be achieved. The proposed design is simple so that it can be expanded in other institutions of higher education and in other areas of biology.

Design-based research can be applied to the field of biology, especially to develop biological interpretation skills [20]. In the present study, the active participation of

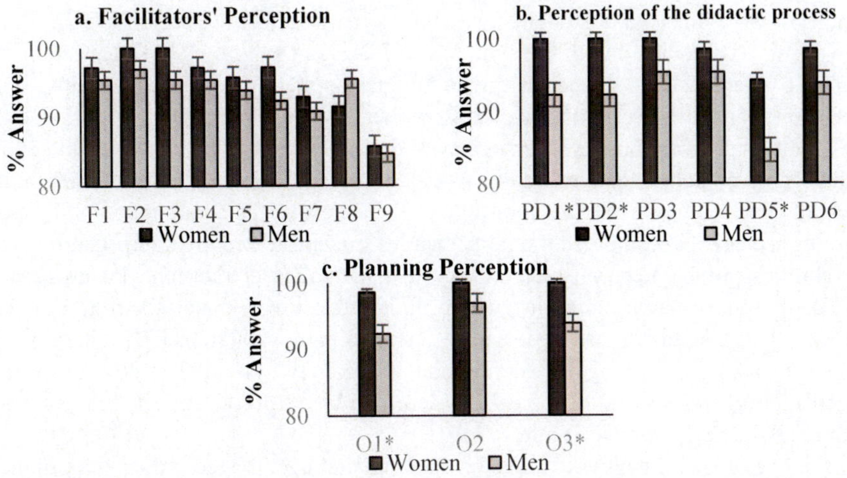

Fig. 6. Participants' perceptions of the facilitators, teaching process and planning. to. F1-Agenda of work and objectives, F2-Contents, F3-Knowledge update, F4-Answer to questions, F5-Interaction, F6-Quality, F7-Jobs, F8-Activities, F9-Time management. b. PD1-Contribution to knowledge, PD2-Methodology, PD3-Didactic and technological resources, PD4-Evaluation, PD5-Work functionality, PD6-Learning utility. c. O1-Management and organization, O2-Attention received, O3-Planning and programming. Statistical value $*p < 0.05$.

undergraduate students in individual activities that used technologies was evidenced. Although, due to their ages, this cohort is considered as "Millennials" generation that seeks to obtain immediate results without much cognitive demand [21], in our findings they were the ones that developed the most content creation activities, such as mind maps, comics, videos, audios, among others. In a previous study, it was stated that students should become the designers of their own learning experiences [22] with the teacher in charge of managing the content and evaluation [23]. Therefore, it is necessary to promote enriched learning environments that allow taking advantage of the opportunities offered by virtuality [24].

In addition, the results indicated that the participants tend to stand out in those activities where the different technological resources available on the web are used. For example, the applications available online help the learning process, in addition to being used in biological, environmental and bird conservation research [25], so they must be properly applied in the teaching processes and that students achieve develop skills in the domain of birds [26], as well as scientific process skills [27].

Another finding of this study was the high number of registered visits by students to the virtual learning environment both in the afternoon and at night. This shows that the flexible program allows students to review the content of the program several times, appropriating its curricular content. In a previous study, it has been shown that blended learning is effective in achieving good academic results and satisfaction [28], as well as the social and environmental attitude of students [29]. Furthermore, the blended learning modality allows both students and teachers to participate actively in the educational process [30, 31]. However, among the activities proposed, a low interest in writing synthesis was evidenced in students, which may be the effect of academic [32], digital or work stress associated with the symptoms of Burnout syndrome [33], caused by the current pandemic situation and the methodologies applied in combined learning, since recensions demand a great cognitive effort which can cause stress [34].

Traditionally, ornithology courses have been developed in person, with the application of varied and attractive strategies [35], however, achieving understanding in the study of birds requires a great deal of time [36]. Therefore, the combination of different learning scenarios in combined modality enables students to acquire interpretation skills and abilities, proving to be very satisfactory [37]. In this sense, traditional lectures can be replaced by constructivist activities, where students can create materials from the previous knowledge acquired, making use of different technological tools [38]. Therefore, to achieve meaningful learning in students about the biology, systematics and conservation of birds, it requires a series of logically ordered problems and solutions complemented with accessible educational resources [39].

The use of social networks played a fundamental role for the proper development of teaching-learning strategies, synchronous conferences, recorded videos, slides and complementary material allowed instant feedback of the knowledge acquired in each virtual section [40], simultaneously being in attunement with current society to ensure that students develop autonomy and collaboration among professional participants [41]. The learning environment, combining the activities, strengthened the emotional and cognitive abilities that make it possible to perceive, understand and regulate one's own emotions, being empathic with the opinions of other people [42].

5 Conclusions

The results demonstrated the applicability of the different technological resources in a blended learning environment. The resources that were most accepted by the students highlighted EDpuzzle, Padlet and GoConqr. Technologies as a complement to instructional design were used by students to design their own experiences and learning. In addition, the application of a blended learning environment made it possible to demonstrate that the activities aimed at personalized simulation are viable for the teaching of ornithology, which strengthens the skills for creativity, innovation and resilience, since the facilitator and The student has to adapt and use virtual tools effectively in order to avoid academic, digital or work stress caused by pedagogical instruction.

The results showed that the application of different technological resources can facilitate the development of tasks, making students more innovative and creative. However, they are not enough to achieve the desired learning in the subject of ornithology. Undergraduate students showed a greater tendency towards the development of digital content from the information provided, while professionals showed tendencies towards the development of activities of greater cognitive complexity.

Based on the students' perceptions, it is possible to conclude that the blended learning modality is effective in improving learning achievements in the ornithology course. Finally, social networks were tools of great importance for feedback and virtual coexistence among the participants to develop the slogans and strengthen teamwork. Therefore, new studies should be directed towards the use of didactic strategies that complement the academic effort in the construction of learning.

Special Thanks. The authors wish to extend they're thanks to Mg. Vanessa Sánchez and Ms. Katherine Morales for their help in logistics in the development of this research work. We also thank the institutional program of the Protected Forest "La Prosperina" for their help in disseminating the course on their social media platforms.

References

1. Wilson, M.A., et al.: Bioinformatics core competencies for undergraduate life sciences education. PLoS ONE **13**(6), e0196878 (2018). https://doi.org/10.1371/journal.pone.0196878
2. Wright, A.M., Schwartz, R.S., Oaks, J.R., Newman, C.E., Flanagan, S.P.: The why, when, and how of computing in biology classrooms. F1000Res. **8**, 1854 (2020). https://doi.org/10.12688/f1000research.20873.2
3. Chen, M.M., Scott, S.M., Stevens, J.D.: Technology as a tool in teaching quantitative biology at the secondary and undergraduate levels: a review. Lett. Biomath. **5**(1), 30–48 (2018). https://doi.org/10.1080/23737867.2017.1413432
4. Cummings, M.P., Temple, G.G.: Broader incorporation of bioinformatics in education: opportunities and challenges. Brief. Bioinformatics **11**(6), 537–543 (2010). https://doi.org/10.1093/bib/bbq058
5. Clausen, P.-H., Stelzer, S., Nijhof, A., Krücken, J., von Samson-Himmelstjerna, G.: Established and novel approaches for teaching and learning of veterinary parasitology in Berlin. Vet. Parasitol. **252**, 58–61 (2018). https://doi.org/10.1016/j.vetpar.2018.01.028
6. Bower, M., Dalgarno, B., Kennedy, G.E., Lee, M.J.W., Kenney, J.: Design and implementation factors in blended synchronous learning environments: outcomes from a cross-case analysis. Comput. Educ. **86**, 1–17 (2015). https://doi.org/10.1016/j.compedu.2015.03.006
7. Almuraqab, N.A.S.: Shall Universities at the UAE continue distance learning after the covid-19 pandemic? Revealing students perspective. Int. J. Adv. Res. Eng. Technol. (IJARET) **11**(5), 226–233 (2020). https://doi.org/10.34218/IJARET.11.5.2020.024
8. Valero, N., Vélez, M., Duran, Á., Torres, M.: Afrontamiento del COVID-19: estrés, miedo, ansiedad y depresión. Enfermería Investiga **5**(3), 63–70 (2020)
9. Peña, J., Rodríguez, M.C., Padilla, V.: Enfoques de concepciones de aprendizaje mixto, rol del profesor y estrategias de enseñanza de profesores de cursos mixtos. Revista Internacional de Ciencias Sociales y Humanidades, SOCIOTAM XXIV **2**, 179–198 (2014)
10. Surasinghe, T., Courter, J.: Using eBird to integrate citizen science into an undergraduate ecology field laboratory. Bioscene: J. Coll. Biol. Teach. **38**(2), 16–20 (2012)
11. Potenza, S.A.: Science is for the birds. Teaching Pre K-8 **37**(5), 50–51 (2007)
12. Hallman, T.A., Robinson, W.D.: Teaching bird identification & vocabulary with twitter. American Biology Teacher **77**(6), 458–461 (2015). https://doi.org/10.1525/abt.2015.77.6.458
13. Gleadow, R., Honeydew, M., Ford, A., Isaac, B., Abbott, K.: New tools for a new age: an evolution or revolution in higher education? F1000Res. **4**, 1502 (2015). https://doi.org/10.12688/f1000research.7533.1
14. Moore-Russo, D., et al.: A study of how angry birds has been used in mathematics education. Digital Experiences Math. Educ. **1**(2–3), 107–132 (2015). https://doi.org/10.1007/s40751-015-0008-y
15. Ketelhut, D.J., Schifter, C.C.: Teachers and game-based learning: improving understanding of how to increase efficacy of adoption. Comput. Educ. **56**(2), 539–546 (2011). https://doi.org/10.1016/j.compedu.2010.10.002
16. Scott, E.E., Wenderoth, M., Doherty, J.H.: Design-based research: A methodology to extend and enrich biology education research. CBE—Life Sci. Educ. **19**(3), es11 (2020). https://doi.org/10.1187/cbe.19-11-0245

17. Barab, S., Squire, K.: Design-based research: putting a stake in the ground. J. Learn. Sci. **13**(1), 1–14 (2004). https://doi.org/10.1207/s15327809jls1301_1
18. Bell, P.: On the theoretical breadth of design-based research in education. Educ. Psychol. **39**(4), 243–253 (2004). https://doi.org/10.1207/s15326985ep3904_6
19. Cohen, L., Manion, L., Morrison, K.: Research Methods in Education, 8th edn. Routledge, New York (2017). https://doi.org/10.4324/9781315456539
20. Zagallo, P., Meddleton, S., Bolger, M.S.: Teaching real data interpretation with models (TRIM): analysis of student dialogue in a large-enrollment cell and developmental biology course. CBE—Life Sci. Educ. **15**(2), ar17 (2016). https://doi.org/10.1187/cbe.15-11-0239
21. Moreira, M., Borrás, J., San Nicolás, M.B.: Educar a la generación de los Millennials como ciudadanos cultos del ciberespacio. Apuntes para la alfabetización digital. Revista de Estudios de Juventud **109**, 13–32 (2015)
22. Willis, P.: The bird box survey project: a field-based science inquiry investigation. Sci. Teach. **81**(2), 37–41 (2014)
23. Gleadow, R., Macfarlan, B., Honeydew, M.: Design for learning – a case study of blended learning in a science unit. F1000Res. **4**, 898 (2015). https://doi.org/10.12688/f1000research.7032.2
24. González-Sanmamed, M., Estévez, I.E., Souto-Seijo, A., Muñoz-Carril, P.: Digital learning ecologies and professional development of university professors. Comunicar. Media Educ. Res. J. **28**(1), 9–18 (2020). https://doi.org/10.3916/C62-2020-01
25. MaKinster, J., Trautmann, N., Burch, C., Watkins, M.: Where the birds live: using web-based maps to explore bird habitats and migration patterns. Sci. Teacher **82**(9), 42–49 (2015)
26. Sturner, K., Lucci, K.: The difference that data make: examining bird migration data to build scientific skills. Am. Biol. Teacher **77**(1), 56–61 (2015). https://doi.org/10.1525/abt.2015.77.1.8
27. Harahap, F., Nasution, N.E.A., Manurung, B.: The effect of blended learning on student's learning achievement and science process skills in plant tissue culture course. Int. J. Instr. **12**(1), 521–538 (2019). https://doi.org/10.29333/iji.2019.12134a
28. Ümit Yapici, İ: Effectiveness of blended cooperative learning environment in biology teaching: classroom community sense, academic achievement and satisfaction. J. Educ. Training Stud. **4**(4), 269–280 (2016). https://doi.org/10.11114/jets.v4i4.1372
29. Nair, T.S., Bindu, R.L.: Effect of blended learning strategy on achievement in biology and social and environmental attitude of students at secondary level. J. Sch. Educ. Technol. **11**(4), 39–52 (2016)
30. Alsalhi, N.R., Eltahir, M.E., Al-Qatawneh, S.S.: The effect of blended learning on the achievement of ninth grade students in science and their attitudes towards its use. Heliyon **5**(9), e02424 (2019). https://doi.org/10.1016/j.heliyon.2019.e02424
31. Naranjo-Morán, J., Vera-Morales, M., Sánchez-Rendón, V., Pino-Acosta, A.: Teaching botany for adults with the blended learning modality. In: Rodriguez Morales, G., Fonseca C., E.R., Salgado, J.P., Pérez-Gosende, P., Orellana Cordero, M., Berrezueta, S. (eds.) TICEC 2020. CCIS, vol. 1307, pp. 287–301. Springer, Cham (2020). https://doi.org/10.1007/978-3-030-62833-8_22
32. Berrío, N., Mazo, R.: Estrés académico. Revista de Psicología Universidad de Antioquia **3**(2), 65–82 (2011). https://revistas.udea.edu.co/index.php/psicologia/article/view/11369
33. Carlotto, M.S., Câmara, S.G., Borges, Á.M.B.: Predictores del Síndrome de Burnout en estudiantes de un curso técnico de enfermería. Diversitas **1**(2), 195 (2005). https://doi.org/10.15332/s1794-9998.2005.0002.07
34. Salinas, B.G.M., Chimal, A.M.: Afrontamiento del estrés en adolescentes estudiantes de nivel medio superior. Revista Digital de Investigación en Docencia Universitaria **9**(1), 9–20 (2015). https://doi.org/10.19083/ridu.9.380

35. Pecor, K.W., Lake, E.C., Wund, M.A.: Optimal foraging by birds: experiments for secondary & postsecondary students. Am. Biol. Teacher **77**(3), 192–197 (2015). https://doi.org/10.1525/abt.2015.77.3.7
36. Can, Ö.K., Lane, J.F., Ateskan, A.: Facilitating place-based environmental education through bird studies: an action research investigation. Environ. Educ. Res. **23**(5), 733–747 (2017). https://doi.org/10.1080/13504622.2016.1233389
37. Sancho, P., Corral, R., Rivas, T., González, M.J., Chordi, A., Tejedor, C.: A blended learning experience for teaching microbiology. Am. J. Pharm. Educ. **70**(5), 120 (2006). https://doi.org/10.5688/aj7005120
38. Compeau, P.: Establishing a computational biology flipped classroom. PLoS Comput. Biol. **15**(5), e1006764 (2019). https://doi.org/10.1371/journal.pcbi.1006764
39. Applegate, J.E.: An introductory exercise for courses in birding. Am. Biol. Teach. **44**(1), 14–20 (1982). https://doi.org/10.2307/4447385
40. Herrera, M.: Las Redes sociales como entornos académicos en la enseñanza universitaria. Revista Iberoamericana para la Investigación y el Desarrollo Educativo (12), (2014) ISSN: 2007-2619
41. Silva, J.: Un modelo pedagógico virtual centrado en las E-actividades. Revista de Educación a Distancia (RED) **17**(53), 1–20 (2017). https://doi.org/10.6018/red/53/10
42. Chiappe, A., Consuelo, J.: Fortalecimiento de las habilidades emocionales de los educadores: interacción en los ambientes virtuales. Educación y Educadores **16**(3), 503–524 (2013)

Analysis of Variables that Influence University Student's Perception About Virtual Education Adopted Due to COVID-19

Marcelo Calle(✉), Marlene Ullauri, Emilia Torres, and Nicole Peralta

Universidad del Azuay, Cuenca, Ecuador
marcalle@uazuay.edu.ec

Abstract. The changes in higher education due to the COVID-19 pandemic have forced the maintenance of a virtual education system, transforming face-to-face classes into online classes through different platforms available for this purpose. However, these classes, which can be synchronous or asynchronous, are new to many students, many universities, and several professors, so there is little information about their adaptation and which variables can affect students' better acceptance of this system. This study seeks to determine how the students' perception of the new educational system is altered by the access to technological tools to attend classes and, at the same time, how the adaptation of the university and professors to the new system influences said student perception. This study investigated the effects of these variables on students' perception by utilizing a questionnaire that obtained 617 complete responses, and the obtained data were analyzed using the linear regression model. The results found show that the student's perception of the new system improves as the capacity of the university and professors to adapt to the recent education model increases, and it also improves when students have access to technological tools. Furthermore, the variable of adaptation of the university and professors shows a greater incidence in improving student perception. These results lead to the search to implement strategies that will allow the capacity of these two variables to increase and achieve a better acceptance of the new system.

Keywords: Access to technology · Higher education · Student perception · University adaptation · Virtual education

1 Introduction

Nowadays, the world is suffering many devastating consequences in every aspect due to the COVID-19 pandemic. The mandatory restrictions in cities decreed by governments seem to have unleashed major economic problems, massive layoffs, physical and mental health problems, border closures, among others. The educational system was also affected by these decisions brought about by the COVID-19 virus and changed the teaching-learning process at different levels, leading in many countries to a non-face-to-face and rather virtual education system. Virtual education is considered 100% online

M. Zambrano Vizuete et al. (Eds.): CI3 2021, LNNS 511, pp. 78–91, 2022.
https://doi.org/10.1007/978-3-031-11438-0_7

teaching. It does not have any physical presence and can be carried out due to the rapid expansion of the Internet in learning and teaching in the education sector [1]. While in the United States, online courses are defined with at least 80% of the course content is taken through the online system, face-to-face instruction is understood to range from 0% to 29% of said content through the online system. The remaining alternatives, such as hybrid instruction (blended), have between 30% and 79% of the course content under the online system, and finally, it has 80% or more of its content in the online form, as shown in Table 1 [2].

Table 1. Classification of typical course instruction methods in the United States [2]

Amount of content offered online	Type of course	Typical description
0%	Traditional	Course in which online technology is not used: the content is delivered in writing or orally, Face-to-face
1 to 29%	Facilitated by Web	A course that uses web-based technology to facilitate what is essentially a face-to-face course. You can use a learning management system or web pages to publish the syllabus and assignments
30 to 79%	Hybrid/Blended	A course that combines online and face-to-face delivery. A substantial proportion of the content is delivered online, typically uses online discussions, and typically has few face-to-face meetings
80% to more	Online	A course in which most or all of the content is taught online. It usually doesn't have face-to-face meetings

In the United States, distance education is growing. According to Babson Survey Research Group data, between 2015 and 2016, the number of students in distance education rose by 5.6%, reaching the figure of 6,359,121 students who take at least one distance course in some institution of higher education within the United States, this figure represents 31.6% of all students enrolled in universities, of which, 14.9% represents those who exclusively take distance courses and 16.7% represents to students who take a combination of the distance and face-to-face modality [3]. Several aspects are involved in evaluating the quality of virtual education, such as the context where it is carried out, the groups involved, including teachers and students, the type of technological infrastructure through which the virtual course is offered, the pedagogy used to achieve its objectives, teaching strategies and the way to evaluate the student as well as the teaching materials and resources [4].

The objective of this study is to determine how the student perception towards the new virtual education system implemented during the Covid-19 pandemic is positively or negatively affected through the variables of the student's access to technological tools to receive virtual classes and the adaptation of the university and professors to the new virtual education system.

1.1 Adaptation of Universities and Professors to the New System

In March of 2020, the world took a new course. The COVID-19 pandemic forced people to take shelter in their homes to protect their lives, so Information and Communication Technologies ICT had to be included in educational processes [5]. Changes in online teaching require adapting universities and teachers in teaching practices and how modules are designed and assembled [6].

Through the International Institute for Higher Education (IESALC), UNESCO evaluated measures to avoid the stagnation of academic activities and institutional responses developed for the said purpose [7]. The Institute is aware of temporary stress implications caused by the cessation of in-person interactions that students have been left under a completely new situation and full of uncertainty as many people were economically harmed by the pandemic, which caused students from private universities to find themselves in a vulnerable position by not being able to assume costs and financial burdens, resulting in a delay in the continuity of their learning [7]. Furthermore, it is essential to mention that in this new education system, the synchronous/asynchronous dichotomy emphasizes the differences between the presence or absence of real-time interactions more than in the design of different types of instructional experiences to support learning pathways [8].

Consequently, several universities worldwide took specific measures to counter the health crisis and suspend face-to-face activities [9]. Despite the difficult circumstances, the professional training of the students had to continue; therefore, the universities have undergone a constant process of training, updating, and adaptation [10], as well as looking for ways to help students in situations of vulnerability, all of this to try to provide university communities with the best possible academic quality while ensuring their safety [11]. The students critique that the material was not adapted to the new virtual education system; it was the same as face-to-face-class [12]. For a faculty member, better infrastructure and training are essential factors in an e-learning system [13] and will focus on the student segment [14]. It also is necessary to promote policies to use resources with digital equity, invest in digital infrastructure to assure a quality system [15].

1.2 Student Perception of Virtual Education During the COVID-19

Quality university education is an idea that seems to go hand in hand with students' satisfaction as users of the services provided by an institution of academic instruction [16]. Therefore, strategies such as the following would allow increasing university student satisfaction: consolidate the educational offer at different levels, as well as the use of models and methodologies that promote quality and excellence; increase the presence of the university in low-reach cities with new distance education policies; teacher training along with monitoring and updating their improvement; analysis and updating of evaluation criteria and indicators that monitor teachers in addition to the care and maintenance of the current facilities [17].

In the virtual modality, the characteristics of a most valued teacher are responsibility, accessibility, availability, attention, good communicator, and friendly [18]. Based on perception, another study concludes that starting and completing an entire period of online teaching could change students' perception of the value of online education and its flexibility [19].

The assessment that students present in virtual distance education is based on an inverse relationship between dedication to study and perceived performance, and on the absence of concentration between teachers and students, it is specified in the lack of adaptation from the former to the personal circumstances of the students, making it a key to take into account the events in the pandemic, in which distance education was given [20]. The digital tools used are content adapted for the face-to-face mode to a virtual one without sufficient prior adjustment, a mostly asynchronous model was also chosen together with a teacher-student interaction based on communication via email, forums, or chats, this being a relevant aspect if the risks of abandonment associated with a disconnection between students and teachers are taken into account [20]. Finally, 48.4% of students preferred a mix of e-learning and face-to-face classes, 16.9% prefer e-learning exclusively, and 34.7% prefer face-to-face courses [21]. The student's perception is pessimistic regarding the online courses because they feel it is not adapted excellently to virtual education [22].

1.3 Student's Access to Technological Tools to Receive Class

By 2009, around 55 million students have taken online classes [23]. If we jump in time and travel to 2014, the Network Readiness Index (NRI) determined that Ecuador was in the 82nd position of 144 countries that made up a study of measures of the propensity countries have to take advantage of opportunities offered by information and communications technology [24]. Table 2 compares Ecuador with the European Union in 2015. At first glance, Ecuador has 27.5% and the EU 68% of households in each country with at least one computer; 28.3% for Ecuador and 79% for the EU of homes with internet access [24].

Table 2. Comparison: internet access indicators between Ecuador and the European Union [24]

	Ecuador	EU (28)
Households with at least one laptop/desktop PC	27.5%	68.0%
Households with internet access	28.3%	79.0%
Households with broadband internet access	24.9%	76.0%
People that use the Internet at least once a day	64.0%	62.0%
Young people from 16 to 24 who use the Internet	64.9%	84.0%
Digital illiteracy	20.0%	1.0%

In 2019, the National Institute of Statistics and Census (INEC) published exciting comparative data of 2018 and 2019 on ICTs. Table 3 shows the positive or negative variation in the percentage of people who had a desktop computer and a laptop: −1.2% and 4.3% respectively, which means that there was a decrease in possession of desktop computers, but there was an increase in notebooks. Regarding Ecuadorians who have internet access, one can appreciate that there was a positive variation for the three sectors, that is, national, urban, and rural [25].

Table 3. Access to information & communication technology. Ecuador 2018–2019 [25]

Indicators	2018 (%)	2019 (%)	Variation (%)
Desktop computer	24.5	23.3	−1.2
Portable computer (notebook)	24.2	28.5	4.3
Homes with internet access – National	37.2	45.5	8.3
Homes with internet access – Urban	46.6	56.1	9.5
Homes with internet access – Rural	16.1	21.6	5.5
People who use a computer – National	50.1	41.0	−9.1
People who use a computer – Urban	55.6	46.6	−9.0
People who use a computer – Rural	38.2	28.9	−9.3
People who use a computer – from 16 to 24 years old	75.7	67.7	−8.0
People who use a computer – from 25 to 34 years old	62.8	53.8	−9.0
People who use a computer – from 35 to 44 years old	49.8	39.3	−10.5
People who use a computer – from 45 to 54 years old	37.5	27.8	−9.7
People who use a computer – from 55 to 64 years old	21.9	19.9	−2.0
People who use the internet – National	55.9	59.2	3.3
People who use the internet – Urban	64.4	66.7	2.3
People who use the internet – Rural	37.9	42.9	5.0
People who have a smart cell phone – National	41.4	46.0	4.6
People who have a smart cell phone – Urban	50.0	54.0	4.0
People who have a smart cell phone – Rural	23.3	28.8	5.5
People with smart cell phone – from 16 to 24 years old	63.2	68.9	5.7
People with smart cell phone – from 25 to 34 years old	70.2	75.9	5.7
People with smart cell phone – from 35 to 44 years old	61.8	67.9	6.1
People with smart cell phone – from 45 to 54 years old	49.2	56.3	7.1
People with smart cell phone – from 55 to 64 years old	34.8	42.0	7.2
Digital illiteracy – National	10.7	11.4	0.7
Digital illiteracy – Urban	7.2	7.8	0.6
Digital illiteracy – Rural	19.0	20.0	1.0

Due to the COVID-19 pandemic, in 2020, the Ecuadorian government decreed that classes for schools, high schools, universities, and other educational centers should be taught in a non-presence modality, that is, online [26]. However, this caused several difficulties for the institutions, teachers, and students since, as seen previously, the vast majority of the Ecuadorian population does not have the necessary tools to follow the national plan, hence for this reason and many others, 81,982 children and adolescents decided to leave schools and high schools during the COVID-19 pandemic [27].

It should not be forgotten that online education was not born after the COVID-19 pandemic, but that several decades ago, many educational institutions were already venturing into this area [23]. Additionally, for teachers with low technological literacy, online classes become a challenge because, to face the health crisis and make sure the education of young people does not slow down, everything is done through the Internet, computers, tablets, and phones, others [28].

2 Methodology

A quantitative, cross-sectional, and descriptive investigation was carried out for this study. The study was developed among the students at the University of Azuay in Cuenca, southern Ecuador. The variables under study were a dependent variable, the student perception towards the new virtual education system (Perception) adopted by the universities due to the COVID-19 pandemic. The independent variables were the access to technological tools that students have to receive virtual classes (Access). The other independent variable was the adaptability of the university and professors to the new virtual education system (Adaptation). The goals of this research were: 1) Determine the impact the adaptability of the university and the teachers to the new virtual education system has on the students' perception of virtual education, 2) Determine how the student's access to technological tools alters the student's perception of virtual education, and 3) Identify which of the independent variables under study has the most significant impact on the dependent variable of student perception of the new virtual education system.

2.1 Hypothesis

In this study, we proposed two hypotheses to test: 1) A greater adaptability of the university and professors increases the positive student perception towards the new virtual education system caused by the COVID-19 pandemic. 2) Improving access of technological tools for students to receive classes improves student perception of the new virtual education system.

The applied research method collects information among the students of Universidad del Azuay from all the departments using a questionnaire as a research instrument developed to measure the variables under study. For a better understanding of the problem and development of the survey questions, telephone and zoom interviews were conducted with a student from each department of the Universidad del Azuay and a professor from the university, using open questions and a semi-structured questionnaire with questions such as "How has your experience with the virtual education system been," "How do you feel about the task of teaching through virtual media." The group of questions was constructed for each variable under study with said information. The Likert scale was used in each question under the variables in the analysis. A total of 36 questions were part of the questionnaire, with an estimated filling time range of 6 to 10 min. The questions were constructed and submitted through the Qualtrics® system.

The questionnaire was sent to all undergraduate students of Universidad del Azuay only once via email. The number of students who responded to our questionnaire was 849. In this survey, data was refined to ensure that the student had filled out all the questions corresponding to the three variables studied to be valid. This refinement led us to find that 617 helpful students answered the questionnaire fulfilling the previous condition. Student perception was measured as the dependent variable through 10 questions using the 5-point Likert scale where "totally disagree" equals 1 and "totally agree" equals 5. The variable seeks to measure the level of positive or negative perception the student has about the new teaching system being applied due to the changes caused by the COVID-19 pandemic. Among the questions asked through the questionnaire are "If this modality is maintained in the next cycle, will you enroll?".

The independent variables represented in this study were access to technological tools available to students to receive virtual classes, the university's adaptation skills, and its professors' adaptation skills during the first semester of compulsory courses in virtual form. The first independent variable was measured through 6 questions using the inverse 4-point Likert scale where 1 equals "always" and 4 equals "never" with questions such as: "Do you have an internet connection to access virtual classes?". The second independent variable was measured through 12 questions with a 5-point Likert scale, with 1 "totally disagree" and 5 "totally agree" with questions such as: "Do you consider the modality of online classes beneficial for your learning?". Once the responses were completed, the Stata® statistical software was used to process the data, perform the respective analyzes, create the descriptive tables, and perform the corresponding regressions.

3 Results

The students who responded to the questionnaire of this study are within the following characteristics. Table 4 shows the diversity of the origin of the responses concerning which department the students who responded to the survey belong to. It would even seem that the distribution of responses would have a proportional relationship to the number of students that each department has.

Table 5 shows the gender the students who responded to the questionnaire identify with. It should be noted that the difference in the total responses under study (617) is given by the students who did not answer this question or left it blank.

The questions used in each questionnaire variable adequately must represent the construct we are examining [29]. Since an exploratory factor analysis was performed for each variable studied, our questionnaire shows that the questions used measure what we want to measure. In the first place, a principal component analysis was performed with the questions of the dependent variable. The Kaiser-Meyer-Olkin (KMO) test was applied to measure whether the sample data's suitability to fulfill a factor analysis; this value is equal

Table 4. Department to which the responding students belong.

Department	Frequency	Percentage	Accumulated
Science of administration	149	24.15	24.15
Science and technology	137	22.2	46.35
Legal sciences	61	9.89	56.24
Medicine	60	9.72	65.96
Philosophy, letters, and education	103	16.69	82.66
Design, architecture, art	107	17.34	100
Total	617	100	

Table 5. Gender with which the student identifies as.

Gender	Frequency	Percentage	Accumulated
Feminine	331	54.35	54.35
Masculine	277	45.48	99.84
Prefer not to answer	1	0.16	100
Total	609	100	

to 0.7316; therefore, it is within the acceptable range to perform factor analysis. In this dependent variable of student perception, the factor analysis found three dimensions that group the questions used in the variable, and the variance explained equally to 57.08%. It is also shown that there is an internal consistency of the questionnaire since the result of performing Cronbach's Alpha analysis gave us a value of 0.61, which is acceptable, and we can say that the scale is reliable [30].

Table 6. Student's age

Age	Frequency	Percentage	Accumulated
Younger than or 18 years old	49	8.05	8.05
19–21 years old	323	53.04	61.08
22–24 years old	157	25.78	86.86
25 years old or more	80	13.14	100
Total	609	100	

In the same way, the same procedure was carried out for the independent variable "access to technological tools" to receive virtual classes. The KMO test gave a value of 0.7694, and two dimensions were found grouping the questions, with the variance explained equally to 61.16%. According to Cronbach's Alpha analysis, the scale's internal consistency gave a value of 0.7344, being acceptable and therefore reliable [30]. For the independent variable "adaptation that the university and its professors have shown," the KMO test gave a value of 0.9247. Again, it is represented in two dimensions or factors grouping the questions, with the variance explained equally to 59.91%. The value of Cronbach's Alpha analysis was 0.8970; therefore, as seen in the other variables under study, the analysis of the test used is reliable. Table 6 shows the ages of the students who responded and the diversity of these within the composition of the university.

It is essential to identify the variables under analysis within the results. Table 7 shows the correlations between the variables under study and the demographic questions. Said table shows a strong correlation between the variables of student perception and the adaptation of the university and professors to the new system. While students' access to technological tools to attend virtual classes has a lower correlation with the dependent variable and a similar correlation with the other independent variable, the other variables in Table 7 show a very low correlation and, in some cases, negative.

Table 7. Variable correlation matrix

	Perception	Adaptation	Access	Department	Age	Gender
Perception	1.0000					
Adaptation	0.5454	1.0000				
Access	0.2509	0.2799	1.0000			
Department	−0.0304	0.0331	−0.0406	1.0000		
Age	−0.0831	−0.0663	−0.0446	−0.1876	1.0000	
Gender	−0.0758	−0.1019	−0.0983	−0.1793	0.1609	1.0000

In Table 8, the descriptive statistics of the variables are shown, where the mean and standard deviation show us the behavior in the responses of the sample of students who responded in this study.

Table 8. Descriptive statistics of the variables

Variable	Obs	Mean	Std. Dev.	Min	Max
Perception	617	3.471799	0.564406	1	5
Adaptation	617	3.245462	0.859820	1	5
Access	617	3.333630	0.393316	1	4
Department	617	3.246353	1.859756	1	6
Age	609	2.440066	0.819241	1	4
Gender	609	1.458128	0.501941	1	3

When applying the regression between the dependent variable and the independent variables, according to Gujarati [31], the main objective of the regression analysis is to explain the mean or average behavior that the dependent variable Y has concerning the regressors or independent variables. Therefore, this operation explains how the dependent variable Y responds when the value of the independent variable X changes. That is why it should be emphasized that the causal relationship, if any, between Y, the dependent variable, and X, the independent variable, must be based on the relevant theory [31]. Thus, Table 9 shows the coefficient of the independent variable adaptation of the university and the professors about student perception. We can also see the constant and significant values given by the p-value and the 95% confidence intervals.

Table 9. Regression between student perception and adaptation of the university and professors

Source	SS	df	MS	Number of obs	=	617
				F (1, 615)	=	265.96
Model	59.2415	1	59.241510	Prob > F	=	0.0000
Residual	136.9878	615	0.222744	R-squared	=	0.3019
				Adj R-squared	=	0.3008
Total	196.2293	616	0.318554	Root MSE	=	0.47196
Perception	Coef	Std. Err	t	P > |t|	[95% Conf. Interval]	
Adaptation	0.360674	0.022116	16.31	0.000	0.317243	0.404106
Constant	2.301244	0.074249	30.99	0.000	2.155432	2.447056

Note. P < 0.01 Statistically Significant

Likewise, the regression analysis for the student perception variable and the student's access to technological tools to access virtual classes variable is reflected in Table 10.

Table 10. Regression between student perception and access to technological tools

Source	SS	df	MS	Number of obs	=	617
				F (1, 615)	=	43.69
Model	13.0143	1	13.01431	Prob > F	=	0.0000
Residual	183.2150	615	0.297911	R-squared	=	0.0663
				Adj R-squared	=	0.0648
Total	196.2293	616	0.318554	Root MSE	=	0.54581
Perception	Coef	Std. Err	t	P > |t|	[95% Conf. Interval]	
Access	0.369555	0.055913	6.61	0.000	0.259752	0.479358
Constant	2.239839	0.187683	11.93	0.000	1.871261	2.608418

Note. P < 0.01 Statistically Significant.

The results of the multivariate regression analysis with the two independent variables were also obtained. Table 11 shows the results of this operation.

Additional analyzes were carried out to verify the operation of the proposed models, and one of them was the analysis of the variance inflation factor (vif) to verify if there is multicollinearity in the multiple regression of Table 11, obtaining the value of vif = 1.09, so we confirm that there is no multicollinearity.

Table 11. Regression between student perception, access to tools, and adaptation of the university and professors

Source	SS	Df	MS	Number of obs.	=	617
				F (2, 614)	=	139.81
Model	61.4025	2	30.7012	Prob > F	=	0
Residual	134.8268	614	0.2196	R-squared	=	0.3129
				Adj R-squared	=	0.3107
Total	196.2293	616	0.3186	Root MSE	=	0.4686
Perception	Coef.	Std. Err.	t	P > |t|	[95% Conf. Interval]	
Access	0.157137	0.050091	3.14	0.002	0.058767	0.255508
Adaptation	0.340140	0.022914	14.84	0.000	0.295142	0.385139
Constant	1.844049	0.163325	11.29	0.000	1.523306	2.164792

Note. P < 0.01 Statistically Significant.

4 Discussion

In the search to determine the impact that accesses to technological tools to receive their virtual classes has on positive or negative student perception, we can affirm, ac-cording to Table 9, that the impact is positive. The coefficient has a positive sign, and the p-value is less than 0.05; therefore, it is statistically significant. It tells us that when the variable of adaptability of the university and teachers to the new virtual education system increases the student perception of the virtual class system increases positively. Therefore, the first proposed null hypothesis (Ho = 0) is rejected, observing that this variable influences student perception of the new virtual class system.

In the same way, the student's access to technological tools variable is analyzed. It determines that this positively impacts the student perception of the new education system variable due to the positive sign of the coefficient shown in Table 10 and the statistical significance of the p-value less than 0.05. Thus, null hypothesis 2 is rejected and determines that student perception of the new system increases when better access to technological tools increases. A positive effect of the independent variables can be observed in the multivariate regression. Through the coefficients that have a positive sign and the p-value, that is, in both cases, less than 0.05, their joint presence increases the positive student perception of the new system education adopted due to the COVID-19 pandemic. In Table 11, the university and professors' adaptation variable maintains a value of its coefficient like when it acts alone in the first model, and the access to technological tools variable decreases the value of its coefficient considerably, although maintaining the positive sign. It leads us to think that the adaptation of the university and professors variable has a more significant impact than the access to technological tools within the positive student perception of the new virtual educational system. A probable moderating effect on it is not analyzed in this study [32].

These results should be considered a first phase to develop more and better internal tools in the educational organization and its teachers to increase their adaptability, working on pedagogical training in information technologies that facilitate the teaching role in the new virtual modality. Distance learning models focus their efforts on the student by becoming more flexible and open and allowing them to be the leading actor in their self-regulated learning process [33]. Our found implications coincide with Day et al., [19], who conclude in their study that universities and colleges managed to change online and remote learning with surprising ease; however, the sudden change to distance and online education put many students and teachers at a disadvantage. Similarly, most students appreciated what the university professors did; however, they prefer to work in a face-to-face system [19]. Finally, our results are per Rapanta et al., [34], where the universities need to train their faculties in pedagogical and professional development using an online system.

5 Conclusions

The influence of the level of adaptation of the university and its professors in the perception that students have about the new virtual education system due to the pandemic indicates that this variable is significant in creating favorable perceptions about the new educational system. Due to the continuity of the pandemic and, therefore, of the system in many countries worldwide, the possibilities of improvement that various educational institutions can achieve on this variable in the short term will help increase this perception and probable better acceptance of the new system by the students. Furthermore, this variable is under the control of the educational system; therefore, its administration and efficiency can quickly alter the results towards a positive perception and hence a better acceptance of the new system among students.

Access to technological tools seems to be out of the universities' control. It is probably supported by a public policy of availability to provide these services. On the other hand, it could also respond to the household's economic capacity to acquire sufficient technological devices for this new educational system. Access to technological tools does determine and influence a positive perception by the student about the new system. Therefore, it should be considered to identify hidden student needs and create ways to facilitate this access through public entities or make funds of the educational entity to temporarily subsidize the access for those students who require this aid as an additional tool with the scholarships on tuition that are provided. This last idea could be, among others, to lend laptops to those who need to access their virtual classes and return the laptops to the educational institution at the end of the semester.

References

1. Bachelor, J.W.: El aula presencial, semipresencial, virtual e invertida: un estudio comparativo de métodos didácticos en la enseñanza de L2. Rev. Educ. **43**(2), 20 (2019)
2. Allen, I.E., Seaman, J., Straut, T.T., Poulin, R.: Online report card: tracking online education in The United States. Babson Surv. Res. Gr. **13**, 1–58 (2016)
3. Seaman, J.E., Allen, I.E., Seaman, J.: Grade increase: tracking distance education in the United States. Babson Surv. Res. Gr. **1**, 1–46 (2018)
4. Marciniak, R., Gairín, J.: Dimensiones de evaluación de calidad de educación virtual: revisión de modelos referentes. RIED. Rev. Iberoam. Educ. a Distancia **21**(1), 217–238 (2018). https://doi.org/10.5944/ried.21.1.16182
5. Vélez, R.M.: Retos de las Universidades Latinoamericanas en la Educación Virtual. Rev. Virtual Univ. Católica del Norte **59**, 1–3 (2020). https://www.researchgate.net/publication/339303761_Retos_de_las_universidades_latinoamericanas_en_la_educacion_virtual
6. Boling, E.C., Hough, M., Krinsky, H., Saleem, H., Stevens, M.: Cutting the distance in distance education: perspectives on what promotes positive, online learning experiences. Internet High. Educ. **15**(2), 118–126 (2012). https://doi.org/10.1016/j.iheduc.2011.11.006
7. Causa, M., Lastra, K.F.: Universidades públicas de la Región Metropolitana: algunas líneas estratégicas de acción para garantizar la inclusión en el contexto de la Pandemia Covid-19. Trayectorias Universitarias **6**(10), 029 (2020). https://doi.org/10.24215/24690090e029
8. Bryson, J.R., Andres, L.: Covid-19 and rapid adoption and improvisation of online teaching: curating resources for extensive versus intensive online learning experiences. J. Geogr. High. Educ. **44**(4), 608–623 (2020). https://doi.org/10.1080/03098265.2020.1807478
9. García-Peñalvo, F.J.: El reto de las Instituciones Educativas ante la pandemia COVID – 19. In: Congreso Iberoamericano de aprendizaje mediado por tecnología, p. 27 (2020)
10. Gutiérrez, J.R.: COVID-19 Filosofía e incertidumbre (Sobre mariposas, cisnes y rinocerontes). AnthropiQa 2.0 (2020)
11. García Gil, M.: Los efectos sociales del Covid-19 en la educación; Necesidad de nuevos planteamientos educativos. AnthropiQa 2.0 (2020)
12. Tejedor, S., Cervi, L., Pérez-Escoda, A., Tusa, F., Parola, A.: Higher education response in the time of coronavirus: perceptions of teachers and students, and open innovation. J. Open Innov. Technol. Mark. Complex. **7**(1), 1–15 (2021). https://doi.org/10.3390/joitmc7010043
13. Panda, S., Mishra, S.: E-Learning in a mega Open University: faculty attitude, barriers and motivators. EMI. Educ. Media Int. **44**(4), 323–338 (2007). https://doi.org/10.1080/09523980701680854
14. Boumadan, M., Soto-Varela, R., Poyatos-Dorado, C., Ortiz-Padilla, M.: What factors determine the value of an online teacher education experience from a teacher's perspective? Sustainability **12**(19), 1–16 (2020). https://doi.org/10.3390/su12198064
15. Vaillant, D., Rodríguez-Zidán, E., Zorrilla-Salgador, J.: Incidence of the age of access to information and communications technology and the use of the internet in science learning. Educ. e Soc. **40**, 1–25 (2019) https://doi.org/10.1590/es0101-73302019199206
16. Pérez, J.R., López, R.F., Peralta, M.D., Municio, P.: Hacia una educación de calidad: Gestión, instrumentos y evaluación, Primera Narcea Ediciones, Madrid (2000)
17. Cadena-Badilla, M., Acosta, A.M., Vega-Robles, A., Quiroga, J.V.: La satisfacción estudiantil universitaria: análisis estratégico a partir del análisis de factores. Industrial Data **18**(1), 9 (2016). https://doi.org/10.15381/idata.v18i1.12062
18. Castillo, M.A.S., García, J.J.V., López, R.I.G., Hernández, L.F.B.: Características Del Buen Profesor De Modalidad Presencial Y Virtual Desde La Perspectiva De Los Estudiantes. Eur. Sci. J., ESJ **13**(13), 78 (2017). https://doi.org/10.19044/esj.2017.v13n13p78

19. Day, T., Chang, I.C.C., Chung, C.K.L., Doolittle, W.E., Housel, J., McDaniel, P.N.: The Immediate impact of COVID-19 on postsecondary teaching and learning. Prof. Geogr. **73**(1), 1–13 (2021). https://doi.org/10.1080/00330124.2020.1823864
20. López, E.P., Atochero, A.V., Rivero, S.C.: Educación a distancia en tiempos de COVID-19: análisis desde la perspectiva de los estudiantes universitarios. RIED. Revista Iberoamericana de Educación a Distancia **24**(1), 331 (2020). https://doi.org/10.5944/ried.24.1.27855
21. Almulhim, A.Y., Almulhim, S.A., Almulhim, A.A., Khan, A.S.: The impact of e-learning modalities on medical students in KSA during the COVID-19 pandemic. J. Taibah Univ. Med. Sci. **15**(5), 437–438 (2020). https://doi.org/10.1016/j.jtumed.2020.08.001
22. Tejedor, S., Cervi, L., Pérez-Escoda, A., Jumbo, F.T.: Digital literacy and higher education during COVID-19 lockdown: Spain, Italy, and Ecuador. Publications **8**(4), 1–17 (2020). https://doi.org/10.3390/publications8040048
23. Picón, M.L.: Es Posible la Enseñanza Virtual? Foro Educ. **34**, 11–34 (2020). https://doi.org/10.29344/07180772.34.2357
24. Zambrano, D.M.M., Morueta, R.T., Gutierrez, I.M.: Niveles de acceso a Internet de los estudiantes del bachillerato en Ecuador. Chasqui. Revista Latinoamericana de Comunicación **2017**(134), 391 (2017). https://doi.org/10.16921/chasqui.v0i134.2722
25. INEC: Tecnologías de la Información y Comunicación. Encuesta multiporpósito - TIC 2019, pp. 1–41. Available: https://www.ecuadorencifras.gob.ec/tecnologias-de-la-informacion-y-comunicacion-tic/ (2019)
26. Ministerio de Educación Ecuador: ACUERDO Nro. MINEDUC-MINEDUC-2020–00020-A, pp. 1–4. Available: https://educacion.gob.ec/wp-content/uploads/downloads/2020/04/MINEDUC-MINEDUC-2020-00020-A.pdf (2020)
27. Ministerio de Educación Ecuador: Registros-Administrativos-2019–2020-fin-Ministerio-de-Educacion. Available: https://educacion.gob.ec/amie/ (2020)
28. De Vincenzi, A.: Del aula presencial al aula virtual universitaria en contexto de pandemia de COVID-19. Avances de una experiencia universitaria en carreras presenciales adaptadas a la modalidad virtual. Debate-Universitario **8**(16), 67–71 (2020)
29. Hinkin, R.T.: A brief tutorial on the development of measures for use in survey questionnaires. Organ. Res. Methods **1**(1), 104–121 (1998)
30. Cho, E., Kim, S.: Cronbach's coefficient alpha: well known but poorly understood. Organ. Res. Methods **18**(2), 207–230 (2015). https://doi.org/10.1177/1094428114555994
31. Gujarati, D.: Econometrics by Example, 1st edn. Palgrave Macmillan, London (2011)
32. Baron, R.M., Kenny, D.A.: The moderator-mediator variable distinction in social psychological research: conceptual, strategic, and statistical considerations. J. Pers. Soc. Psychol. **51**(6), 1173–1182 (1986). https://doi.org/10.1037/0022-3514.51.6.1173
33. Fardoun, H., González, C., Collazos, C.A., Yousef, M.: Estudio exploratorio en Iberoamérica sobre procesos de enseñanza-aprendizaje y propuesta de evaluación en tiempos de pandemia. Educ. Knowl. Soc. (EKS) **21**, 9 (2020). https://doi.org/10.14201/eks.23537
34. Rapanta, C., Botturi, L., Goodyear, P., Guàrdia, L., Koole, M.: Online university teaching during and after the Covid-19 Crisis: refocusing teacher presence and learning activity. Postdigital Sci. Educ. **2**(3), 923–945 (2020). https://doi.org/10.1007/s42438-020-00155-y

Evolution, Trends, and Challenges of University Innovation in Peru

Jessica Acevedo-Flores[1](✉), Marco Gutierrez-Aguilar[2], Victor Pulido[4], and John Morillo-Flores[3]

[1] Escuela Profesional de Medicina Humana – Filial Ica, Universidad Privada San Juan Bautista, Lima, Peru
jessica.acevedo@upsjb.edu.pe
[2] Universidad Nacional del Callao, Lima, Peru
[3] Instituto de Investigación en Humanidades, Universidad Privada San Juan Bautista, Lima, Peru
[4] Escuela Profesional de Medicina Humana, Universidad Privada San Juan Bautista, Lima, Peru

Abstract. This work analyzes innovation and its close relationship with productivity, poverty reduction and sustainable development and its sense of change in the search for solutions that contribute to improving reality. The article identifies the trends and weaknesses of the innovation indicators, as well as the development and evolution of university innovation in Peru. Innovation indicators from the competitiveness rankings presented by WIPO, Ricyt, Sunedu, Indecopi, Concytec, World Bank, Private Competitiveness Council, among other entities that evaluate national, regional, and global actions, were analyzed. As results indicate, some indicators of research, innovation, and quality in education had shown an irregular behavior. On the other hand, indicators such as the number of patents granted to Peruvian universities or investment in research have grown steadily in recent years. It ends with a proposal that recommends considering the evolution of Peru's innovation indicators, in order to develop policies that promote a true culture of innovation.

Keywords: Indicator · Innovation · University research · Patent

1 Introduction

Innovation is currently one of the most relevant elements when assessing the development of a country. Innovation processes comprise the state, the business sector, and the academy as main social players [1]. The state is responsible for promoting innovation activities through appropriate policies, investing in human capital formation in universities, providing appropriate incentives and recognitions to researchers, sponsoring research, developing mechanisms to define intellectual property rights, and financial incentives to finance research and development (R&D) and innovation in enterprises through tax reduction. In order to increase their productivity, businesses should invest in scientific production. And finally, the academy - represented by Peruvian universities and colleges - should increase their knowledge base to contribute new theories and products. That knowledge, theoretical and empirical, directly or indirectly increase productive capacity [2].

M. Zambrano Vizuete et al. (Eds.): CI3 2021, LNNS 511, pp. 92–105, 2022.
https://doi.org/10.1007/978-3-031-11438-0_8

Political reforms to improve the quality of higher education have been carried out in Peru, affecting research and innovation in the Peruvian university. The latest reform in Peruvian university system started in 2014 with the enactment University Law, that defines, among others, the importance of academic quality, and establishes research and innovation as values and objectives of every university [3]. However, there are no studies on the effect of these reforms on university innovation in Peru; although some authors have assessed the impact of the University Law on academic quality and growth in scientific research [4, 5]; thus, there is not literature neither with studies of the evolution and trends of Peruvian innovation, nor innovation in the Peruvian academy.

According to some Peruvian government entities such as the National Superintendency of Higher Education (Sunedu), the National Institute for the Defense of Competition and the Protection of Intellectual Property (Indecopi), the National Council of Science, Technology and Technological Innovation (Concytec), among others, Peruvian universities have shown improvements in scientific research and innovation [6]. However, in those studies it is scarcely mentioned that Peru's performance in innovation did not achieve the expected results; furthermore, they only point out some weaknesses and opportunities for improvement that represent challenges in public policies to be implemented in the future. The obtained results show that some innovation indicators did not have a regular performance and constant growth; consequently, an analysis based on indicators is necessary to measure and compare the development of innovation, particularly in Peruvian universities, extracting the most relevant data and trends to acquire an accurate assessment of Peru's performance in contrast to other countries in the region [7].

2 Methodology

The method carried out in this research consisted of a systematic search to collect all historical data concerning innovation indicators of Peru and other countries in Latin American region with similar characteristics. Furthermore, Peru's partners in the Pacific Alliance: Chile, Colombia and Mexico were considered for comparing evolution and trends results.

In this study, two important composite indexes were considered as a reference: the Global Competitiveness Index (GCI) to assess global competitiveness, measuring knowledge transfer through its indicators within the innovation pillar, and the Global Innovation Index (GII) to evaluate the development of innovation. The former is an annual report presented by the World Economic Forum (WEF) and it measures average productivity per worker. The GII is published annually by the group responsible for it: the World Intellectual Property Organization (WIPO), Cornell University, and the European Institute of Business Administration (INSEAD).

Additionally, the data collected by the Ibero-American Network of Science and Technology Indicators (Ricyt) and the Ibero-American Network of Higher Education Indicators (RedIndicES) are taken into account. Statistical information provided by Peru government organizations, such as Indecopi, the Private Competitiveness Council, Concytec, Sunedu were also considered. Such information - periodically updated – reports the advances in innovation, specifically those results concerning to universities regarding research products, such as innovations and inventions as well as other research-related values. Innovations products are registered as patents, which then are measured as a total number of patents formalized in the country.

The information collected is displayed in an orderly manner on charts in order to show trends based on time. Regarding the study period, it should be noted that in some cases no updated data was found in the last two or three years, and in some cases, for comparative purposes, data from the last twenty years or from an even earlier period were considered.

3 Innovation

The Economic Commission for Latin America and the Caribbean (CELAC) mentions three elements that greatly impact any country's level of productivity: science, technology, and innovation (STI) [8]. Moreover, according to the Organization for Economic Cooperation and Development (OECD), the relationship among these elements -STI- is knowledge; science generates knowledge, technology applies it, and innovation allows its commercialization.

The positive relationship between productivity and innovation is proved by analyzing the connection between the average productivity per worker and the development of innovation. Thus, the most productive countries are those with the greatest development in innovation [9].

The development of the STI-triad is a major factor for increasing productivity levels in a country's economy, which ultimately impacts its sustainable development. Consequently, STI plays indispensable roles in improving a country's productivity, development, and economic growth [10]. To provide a favorable environment and conditions for the growth of STI; that is to say, a proper promotion of the triad has a positive impact on economies, which in turn has a positive impact on social progress, reducing poverty and improving the citizens' quality of life [11, 12].

The series of statistical manuals known as the "Frascati family" or the "Oslo manual" defines innovation as "the implementation of a new, or significantly improved, product (good or service), or process, a new marketing method or a new organizational method in business practices, workplace organization or external relations" [13]. Innovation is based on learning, as soon as it is linked to the transformative action of the world; hence the intimate relationship between education and innovation [14].

The STI level of advancement is generally measured through R&D investment [15, 16]. However, when specifically talking about innovation, its development uses the number of patents as an indicator. A patent is the recognition of the intellectual property of an innovation or invention. There are two instances when assessing the quality of research, and the second one corresponds to the results or products, which is reflected in patents and publications [17]. Publications include all the scientific documents written and published in specialized journals, i.e., articles, books, books chapters, among others, detailing the conditions of the research, methods that were used, and findings obtained.

Once a patent is registered, the government bestows a document certifying the exclusive intellectual property right granted to the author or owner of an invention, marketing and production rights at a national or international scope and for a specified time, according to the laws and agreements governing the territory where the patent application is registered [2, 18]. Consequently, the patent corresponds to the legal order established as an official recognition to protect an innovation.

In addition, the time granted to patent protection depends on the type of patent; they are classified as inventions, utility models and industrial designs. The extent of an invention patent protection granted in Peru is twenty years; however, for utility model patent protection is ten years. The time starts from the filing date of the application. The Paris Convention establishes two types of protection: patents for inventions and for the utility model. The requirements for each type are different: inventions products or processes should meet three conditions of novelty (to be new), inventive level (not to be apparent), and industrial application (to be useful). For the utility model, products are required to be novel and to provide technical advantage (operation or utility other than the existing one) [19].

4 Innovation Indicators

Indicators are derived from statistical data to provide information that simplifies reality by extracting data and showing trends. Indicators arise in response to the need of measuring and comparing developed and developing countries. Indicators allow an accurate assessment of the performance of a country or a region. Furthermore, they provide warning signs to apply early actions and for decision-making, as well as regular, orderly classified comparisons [7].

Regarding the types of innovation indicators, as defined in the Oslo Manual, there is the innovation survey (product, process, organization, business model and sources of innovation). There are also those indicators built based on existing data and statistics specifically to measure the generation, flow, and distribution of knowledge (publications, citations, R&D, patents, human resources in STI, among others). Finally, composite indicators as GCI and GII, which are considered for this study, likewise the second type of indicators are considered.

The GCI comprises twelve pillars, innovation being the last one and it is divided into sub-pillars that measure R&D spending, the number of patents, scientific publications, among others. The GII consists of seven pillars that allow innovative activity, five pillars are related to the sub-index of resources for innovation, and the other two pillars are related to the sub-index of results.

The number of registered patents and the number of granted patents represent an innovation indicator to estimate the level of development and technological advancement achieved by a country. Peru has adhered to the Patent Cooperation Treaty (PCT), a patent system consisting in filing – not granting – patents, by the World Intellectual Property Organization (WIPO) which allows the rationalization of the processing of applications in 153 countries [20, 21].

Moreover, this study considers basic research and innovation indicators provided by Ricyt and Indice, concerning PCT patent applications with the participation of universities, as well as the participation of universities in publications in the Scopus database of indexed journals.

Finally, the number of national patent applications filed by Peruvian universities, as well as the evolution of percentage distribution according to the type of applicant, including universities, have been collected from Indecopi and Sunedu.

5 Results

5.1 Analysis of Results

In this section, the evolution of the compound indexes, GCI and GII is reviewed; results of both indexes have been taken from their annual reports. GCI is evaluated in the period 2007 to 2019, and GII has been analyzed from 2013 (its first evaluation) to 2020. For comparative purposes, the results of the four member countries of the Pacific Alliance have been taken, i.e., Colombia, Chile, Mexico, and Peru. The charts show the percentage of countries that, according to their position in the overall ranking, the four selected economies outperformed.

The GCI index includes twelve pillars, the ability to innovate is one of them, which in turn comprises a group of sub-pillars that measure different scopes that contribute to the development of innovation, for example, patents, R&D spending, scientific publications, among others. Figure 1 shows the evolution of the overall results of the innovation capacity pillar; Peru had the worst results compared to the other three fellow members; thus, being the weakest economy of the Pacific Alliance in innovation. The four members had irregular results during the period evaluated. Peruvian performance even had a continuous reduction from 2007–2017; however, its performance improves considerably from 2017–2018, when Peru lifted from 113th to 89th in the ranking, outperforming 100% more economies (18 to 36%) in one year. Chile and Mexico also improved their performances in 2018, both surpassing more than 60% of the economies evaluated.

It should be noted that Peru's results in macro-economic stability in recent years were outstanding; Peru reached the first place in this pillar in the 2019 GCI ranking. Nonetheless, these positive results have not been replicated in other pillars, specifically not in innovation.

The GII index "reflects a deeper and more elaborate understanding of the innovation process and the new trends of innovation in different countries of the world" [18]. The GII comprises two sub-indexes: resources and results; each one of them comprises pillars related to economy elements that help innovative activity. This index consists of seven pillars of innovation, two pillars related to results in innovation and five associated to innovation resources.

Figure 2 shows the evolution of the GII (global value) obtained by the four Pacific Alliance member countries for the period of 2013–2020, taking their position in the overall ranking as a reference, as well as the percentage of countries that are outperformed by each member. According to the results, the three South American members of the Pacific Alliance have shown a decline in their performances over the period considered for evaluation; in contrast, Mexico, despite its irregular evolution, showed sustained improvement over the past five years.

The overall performance of the three South American countries in the GCI and the GII is very similar. Regarding its partners, Peru is the weakest economy in the GII index, showing a particularly noticeable decline in its performance in the last report of 2020, in which Peruvian position descended seven positions in one year, being ranked 76th out of 131 countries. This position in the ranking means that Peru outperformed 42% of the economies participating in the evaluation. Chile ranked 54th in 2020, with a performance above 59% of the countries; while Colombia ranked 68th, surpassing 45%

of the evaluated economies. Chile's superior performance in technological innovation over the past eight years is noticeable, which places it in the upper third in the ranking; however, it should be noted that since its best assessment in 2015, its results have been declining steadily, no longer belonging to the upper third in innovation in the last five years. In contrast, Colombia and Peru have had almost constant behavior, without considerable variations in their positions; however, both countries achieved their best results in 2013 when they surpassed more than 50% of the economies under evaluation; after that year both member countries began to decline in their respective results. Mexico, for its part, had experienced a constant growth in its results over the previous four years.

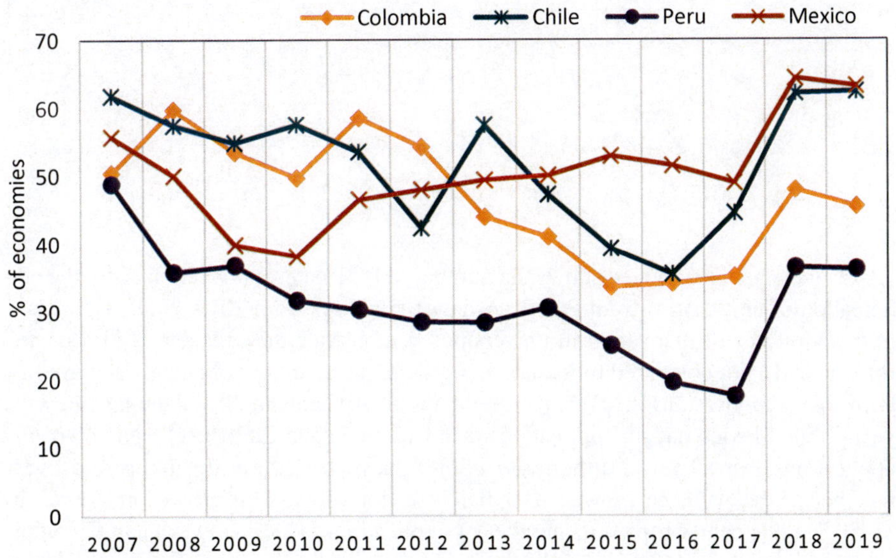

Fig. 1. GCI pillar of innovation: capacity to innovate of the pacific alliance countries 2007–2019. Source: GCI report.

Another important aspect in assessing a country's innovation is that related to research, as patents are the product of scientific activity. Figure 3 shows the publications of Peru indexed in Scopus in the period of 2010–2018, as well as those with the involvement of Peruvian universities. Figure 4 shows the evolution of the percentage of university contribution in the indexed publications in Scopus, compiling results from Peru and Latin American average in the period of 2010–2018. As it can be observed, Peruvian universities have presented sustained growth in their contribution to scientific research, reflected in a sustained increment of indexed publications in the Scopus database. The growth of publications of Peru and university contribution has grown proportionally, tripling both the total number of publications and the percentage of university contribution in the period evaluated. Additionally, in the last nine years, the reports show the evolution of the percentage contributed by Peruvian universities, which grew from 62% to 80%; nevertheless, it is still lower than the average contribution of universities in Latin America,

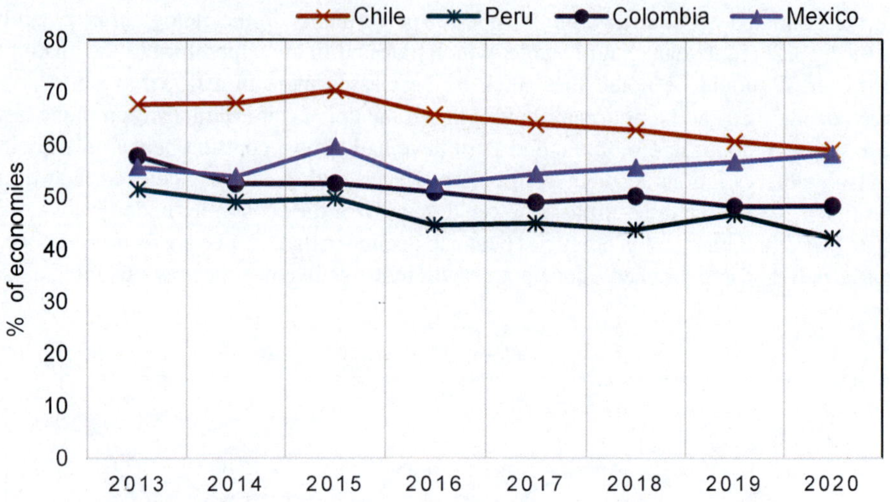

Fig. 2. GII of the pacific alliance countries 2013–2019. Source: GII report.

On the other hand, Fig. 4 shows the number of PCT patent applications with the participation of universities from Pacific Alliance countries from 2010–2018. The results in this indicator of innovation in universities demonstrate the considerably lower performance of Peru compared to its partners. The highest number of patents achieved by Peru was 5 applications in 2015, then reducing to 2 patents in 2018 - the last reported year- while Mexico had 46 applications, and Colombia and Chile had 56 and 60 patent applications, respectively. Furthermore, in the nine years of the evaluation period, every member of the Alliance showed irregular behavior, albeit their growth in the results of 2018. Chile nearly tripled its number of applications (173% growth), and Colombia quintuplicated them (409% growth), even Mexico upraised 18% in nine years; Peru is the only member that did not grow in this indicator.

The capacity to innovate of Peruvian universities is another important issue to assess and analyze; thus, official data regarding patent applications are considered for the period of 2005–2020. Figure 5 illustrates the evolution of patent applications filed by universities to the office responsible for regulating intellectual property in Peru, Indecopi. The total number of patent applications in Peru was less than 10 up until 2010, from then on there has been a sustained growth in the period of 2011–2016, increasing from 16 to 86, going through another irregular period (2017–2018) in which applications were reduced, then again increasing the number of applications in 2019 and 2020. In the last decade, the total number of patent applications filed by universities was 779, with an annual average of 78; i.e., 8.6% of total patent applications filed by residents and non-residents over the past decade, including invention patents and utility models, belonging to Peruvian universities [22].

It should be noted that Lima - the capital of Peru- and the Province of Callao concentrated most of the patent applications. Between 2011 and 2015, more than 90% of applications came from universities based in the Lima Metropolitan area and Callao [6].

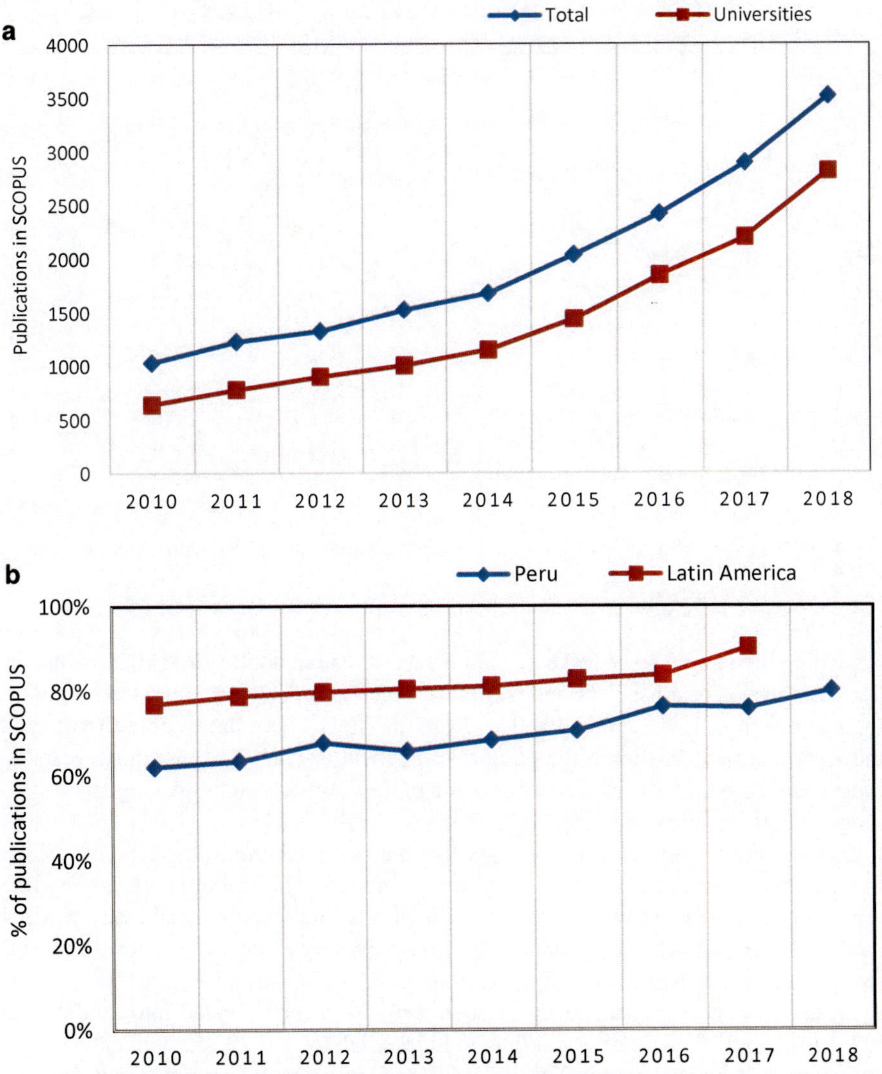

Fig. 3. **a** Total of Peruvian publications indexed in SCOPUS and publications with participation of universities 2010–2018. Source: Ricyt Portal and Indice Portal. **b** Percentage of Peruvian publications vs Latin American publications indexed in SCOPUS with participation of universities 2010–2018. *Data of Latin American is available until 2017. Source: Index Portal.

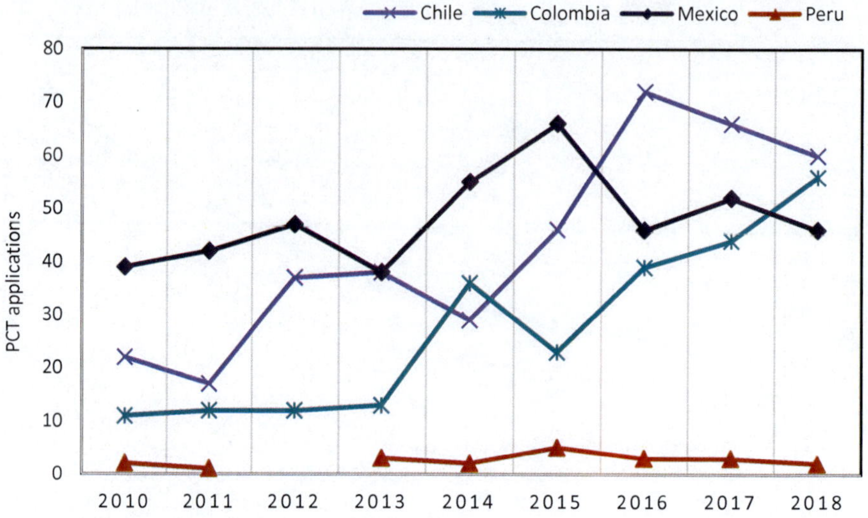

Fig. 4. PCT patent applications by universities in the pacific alliance countries. Source: Index portal.

According to Sunedu, since 2016 the percentage of patent applications filed by universities established in other Peruvian regions has increased; consequently, the number of applications from Lima and Callao has decreased. It is important to note that in 2016 and 2017 there was a drop in the number of applications; similarly, in those years the budget for Concytec was reduced by decision of the Government, observing an identical behavior to that presented in Fig. 5 [23, 24].

Finally, Fig. 6 shows the percentage distribution according to the type of resident applicant for invention patents and application models. The patent applicants by residents can be of different types, e.g., universities, a natural person, businesses, research centers. To simplify the analysis, in Fig. 6 only two percentages are considered, universities and other, which includes patent applications by businesses, natural persons, and research centers. For the analysis, the results have been divided into two periods: 1973–2014, and 2015–2020. According to Indecopi, a total of 29,852 applications were processed in Peru in the period of 1973–2014, filed by both foreigners and Peruvian citizens or residents; all of them were patent applications for inventions and utility models [19]. National patent applications reached 3,983 in that 42-year period representing 13.3%; approximately 58% (2,139) of this subtotal corresponded to applications for patents of invention and 42% (1,844) were applications for utility model patents. On the other hand, in the first period evaluated, 3% (119) corresponded to applications filed by universities and 97% (3,864) were filed by another type of applicant. In contrast, during the second period, a total of 9,099 patent applications were filed, 24.3% (2,207) of this number belonged to residents, and 35% (662) of this subtotal corresponded to applications filed by universities. It should be noted that, in the first period, natural persons applicants filed 72% of the national patent applications, and 24% corresponded to businesses applications, percentages that decreased for the second period [19, 22]. In addition, Peruvian businesses have shown no significant improvements in this indicator; economics behavior requires the business sector to make a stronger commitment to the development of innovation. If the businesses try to be sustainable and competitive, it is mandatory for them to be permanently innovative so that they are neither relegated nor expelled from the market [25].

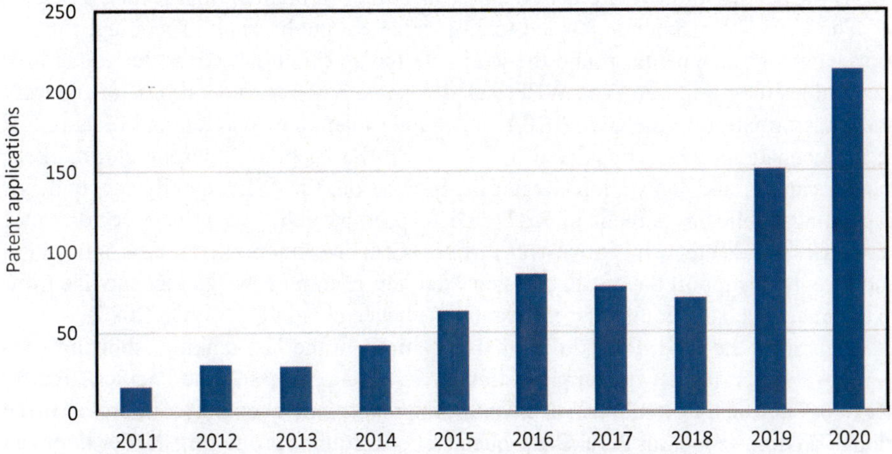

Fig. 5. Patent applications by Peruvian universities 2011–2020. Source: Sunedu.

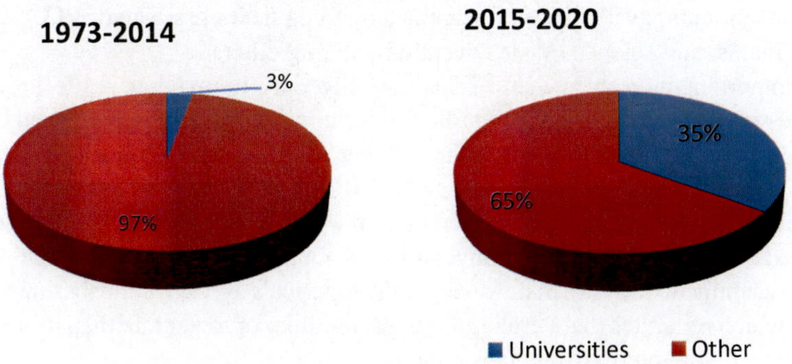

Fig. 6. Patent applications of Peruvian residents by type of applicants. Source: Indecopi.

6 Discussion and Conclusion

6.1 Discussion

According to the GCI competitiveness reports of the last thirteen years, Peru had a sustained decline in the innovation pillar referring to its ability to innovate in comparison with other economies under evaluation. This pillar comprises essential components for the development of innovation, which is very important in emerging economies; those components are the number of patents and patent applications granted, R&D spending, among others. It is important to note that the reduction in the innovation pillar experienced by Peru does not necessarily mean that there have not been improvements in the evaluated sub-pillar or components; actually, many of them show interesting growth. Peru's decline in innovative performance is represented in contrast to all the economies evaluated; that is, despite having improved, other countries also improved, albeit to a greater extent.

The Covid-19 pandemic has led to an increase in public health spending; furthermore, many governments around the world are facing economic crises due to the long quarantines they imposed. The WEF included some basic considerations for recovery and transformation in the GCI 2020 report, rather than an evaluation and ranking; the recommendations were considered in four areas: the enabling environment, markets, human capital, and innovation. Regarding innovation, the WEF mainly recommends expanding public investment in R&D and to promote it in the private sector, since "countries should create incentives that favour patient investments in research, innovation and invention, support the creation of new "markets of tomorrow" and incentivize firms to embrace diversity, equity and inclusion to enhance creativity". (WEF, 2020).

Regarding the GGI, Peru dropped six positions in the last ranking, showing gaps in innovation performance compared to the rest of its partners in the Pacific Alliance: Mexico, Chile, and Colombia; in fact, South American member countries improved in the global ranking, except for Peru. Consequently, Peru requires promoting the development of STI, reinforcing each sub-index that reveals weaknesses: expenditures in R&D as a percentage of GDP, number of scientific and technical articles per million inhabitants, university-industry research collaboration, number of patents and PCT by residents per million inhabitants, availability of scientists and engineers graduated, PISA scales in reading, maths, and science (basic education), among others.

An important indicator when evaluating university innovation is the products of scientific research; that is to say, scientific and technical publications in indexed journals, and patents. Regarding the former, the number of Peruvian publications in indexed journals in Scopus has grown steadily since 2010, the publications has tripled in eight years, while the contribution of universities in these publications has also increased, multiplied by a factor of 4.4. This university contribution in scientific research was 62% in 2010, reaching 80% in 2018; however, Latin America's average contribution was 90% in 2017, which reiterates that Peru is growing at a similar or lower rate than its neighbors, and that is why the pre-existing gaps remain.

With respect to PCT patents in the participation of universities, Peru has shown no improvement, compared to its Pacific Alliance partners which, although their irregular behavior, showed considerable growth. Hence, PCT applications with university contribution accumulated in the period of 2010-2018 were 387 by Chile, 246 by Colombia, 431 by Mexico, and 21 by Peru.

In contrast, one indicator that indeed reflected growth is the number of national patent applications by universities. In addition, according to Sunedu, there was also a steady growth in the percentage of participation of universities in other Peruvian regions, out of Lima and Callao; consequently, this shows a trend towards the decentralization of knowledge [6]. It is important to note that, after the latest University Reform began with the implementation of the new University Law 30220 of 2014, Peruvian universities have shown a greater commitment to scientific research and technological innovation. National policies have also been designed to promote research and innovation activities in universities through national contests to finance research projects, sponsored by public funds granted by Concytec and Fondecyt. Nevertheless, due to the coronavirus pandemic, Peru's budget for the 2021 has been reduced; previously the budget for research reached 0.1% of GDP, consequently, a slowdown would similarly be projected in this indicator.

According to Indecopi, the number of patent applications filed by residents in the category of Peruvian universities has increased considerably over the past six years; however, the percentage of applications from legal persons continues to prevail.

Finally, it is confirmed what the National Competitiveness Council stated in 2015, "the poor development of the driving factors of innovation is a critical condition that is affecting Peru's ability to progressively grow in the mid-term" [26].

6.2 Conclusions

Over the past decade, some improvements in the main indicators of innovation of Peru have been observed; however, this growth is not reflected in the narrowing of gaps with other Pacific Alliance partner countries. Therefore, it is important that Peru considers the recommendations of the WEF, OECD, National Competitiveness Council, and other national and international organizations that annually measure outcomes of innovation, as well as their impact on the economy of the assessed countries. These recommendations include driving innovation as a productivity tool through the creation of laboratories of innovation in public entities, incorporating incentives for all actors in the STI ecosystem (universities, researchers, businesses, government), expanding the budget for competitive grants and funds for STI programs, increasing the number of doctoral programs in areas prioritized by the OECD, developing a national R&D and innovation database and platform, designing viable and aspiring goals in a government-driven National Plan of Innovation with the participation of all actors linked to STI, and finally, strengthening university-industry collaboration, through the commitment of research centers and institutes, researchers, and highly skilled workers.

References

1. Consejo Privado de Competitividad: Informe de Competitividad https://www.compite.pe/publicacion/lanzamiento-informe-de-competitividad-2020/ (2020). Last accessed 13 May 2021
2. Superintendencia Nacional de Educación Superior Universitaria [SUNEDU]: Informe bienal sobre la realidad universitaria peruana. https://www.gob.pe/institucion/sunedu/informes-publicaciones/606251-informe-bienal-sobre-la-realidad-universitaria-2018 (2018). Last accessed 13 May 2021
3. Ley 30220: Ley Universitaria. Diario Oficial El Peruano, Lima (2014). https://www.sunedu.gob.pe/nueva-ley-universitaria-30220-2014. Last accessed 13 May 2021
4. Cervantes, L., Bermudez, L., Pulido, V.: Situación de la investigación y su desarrollo en el Perú: reflejo del estado actual de la universidad peruana. Rev. Pensamiento Gestión (46), 311–322 (2019) https://doi.org/10.14482/pege.46.761

5. Acevedo-Flores, J., Morillo-Flores, J., Tamashiro-Tamashiro, J., Mazzi-Huaycucho, V.: University reform and its impact on scientific production in Peru. Revista TURISMO: Estudos e Práticas (2021)
6. Superintendencia Nacional de Educación Superior Universitaria [SUNEDU] II Informe bienal sobre la realidad universitaria peruana. https://www.sunedu.gob.pe/informe-bienal-sobrerealidad-universitaria/ (2020). Last accessed 13 May 2021
7. Lizuka, M., Hollanders, H.: Indicadores de Innovación. DEIP Perú. Retrieved March 2, 2021, http://www.cies.org.pe/sites/default/files/files/actividades/presentacion_michiko_lizuka.pdf (2017). Last accessed 13 May 2021
8. Comisión Económica para América Latina y el Caribe (CEPAL): Innovar para crecer Desafíos y oportunidades para el desarrollo sostenible e inclusivo en Iberoamérica. Naciones Unidas, Santiago de Chile (2010)
9. Consejo Privado de Competitividad. Informe de Competitividad. https://www.compite.pe/publicacion/lanzamiento-informe-de-competitividad-2019/ (2019). Last accessed 13 May 2021
10. Consejo Privado de Competitividad. Informe de Competitividad. https://www.compite.pe/publicacion/lanzamiento-informe-de-competitividad-2020/ (2020). Last accessed 30 May 2021
11. Griliches, Z.: R&D and the Productivity Slowdown. National Bureau of Economic Research (1980)
12. Coe, D., Helpman, E.: International R&D Spillovers. Eur. Econ. Rev. **39**(5), 859–887 (1995)
13. Organización de Cooperación y Desarrollo Económicos (OCDE) Manual de Oslo. Guía para la recogida e interpretación de datos sobre innovación, 3ra ed. OECD Publishing, Tragasa (2005)
14. RICYT: El Estado de la Ciencia Principales Indicadores de Ciencia y Tecnología Iberoamericanos/Interamericanos. http://www.ricyt.org/wp-content/uploads/2021/02/ElEstadoDeLaCiencia_2020.pdf (2020). Last accessed 31 May 2021
15. Consejo Nacional de Ciencia, Tecnología e Innovación Tecnológica - Concytec. Política Nacional para el Desarrollo de la Ciencia, Tecnología e Innovación Tecnológica (2016). Lima, Last accessed 31 May 2021
16. Organización de Cooperación y Desarrollo Económicos (OCDE) Manual de Frascati. Guía para la recopilación y presentación de información sobre la investigación y el desarrollo experimental. OECD Publishing, Paris/FEYCT, Madrid (2015). https://doi.org/10.1787/9789264310681-es
17. Hackett, E., Amsterdamska, O., Lynch, M., Wajcman, J.: The Handbook of Science and Technology Studies, 3rd edn. Massachusetts Institute of Technology, Hong Kong (2008)
18. World Intellectual Property Organization – WIPO: Global Innovation Index (GII). Wipo.int: https://www.wipo.int/global_innovation_index/en/ (2021)
19. Villarán de la Puente, F.: Historia de las Patentes e Invenciones en el Perú. Indecopi, Lima (2015)
20. Edsberg, R., Truffer, I., Raimondo, E.: Los Indicadores De Patentes En Iberoamérica (1990–2000) (2001)
21. Castro Calderón, M.: El PCT en el Perú. Webinar sobre el Tratado de Cooperación en materia de Patentes (PCT) para empresas. Indecopi - WIPO (2020)
22. Instituto Nacional de Defensa de la Competencia y de la Protección de la Propiedad Intelectual - Indecopi. Informe anual sobre la situación de las universidades públicas y privadas en el Perú en materia de propiedad intelectual. Lima (2018)
23. Consejo Nacional de Ciencia, Tecnología e Innovación Tecnológica Doctorados: garantía para el desarrollo sostenible del Perú. CONCYTEC, Lima (2013)

24. Montoya, M.: El presupuesto del CONCYTEC pre-pandemia COVID 19 y cómo impacta en las instituciones de investigación. - Semillero científico Ceprecyt, STEM desde. https://modestomontoya.me/2020/09/12/el-presupuesto-del-concytec-pre-pandemia-covid-19-y-como-impacta-en-las-instituciones-de-investigacion/ (1992). Accessed 30 May 2021
25. Cabrales Muñoz, F.: Innovación Universitaria. Ingeniare. Revista Chilena de Ingeniería **16**(2), 280–281 (2008). https://doi.org/10.4067/s0718-33052008000200001
26. Chang-Chang, L., Franco, A., Soplín, P.: Veinte años del Perú en el Foro Económico Mundial. Consejo Nacional de la Competitividad (2015)

Neueducation in Social Networks Applied in a MOOC and Its Impact on Academic Performance

Alexandra Lorena Alajo-Anchatuña(✉), Rodolfo Matius Mendoza-Poma, Santiago Fernando Ramírez-Jiménez, and Angelita Azucena Falconi-Tapia

Technical University of Cotopaxi, Latacunga, Ecuador
alexandra.alajo@utc.edu.ec

Abstract. The present research aims to include Neuroeducation as an adaptive element of learning and social networks in a MOOC, as well as its impact on academic performance, incorporating in the teaching-learning process aspects such as motivation, emotion, curiosity, encouraging collaborative work, exchange of experiences, where the participant is the generator of his own learning. An experimental pilot was designed, with the participation of the Executive Management Secretarial Career students from Technical University of Cotopaxi, the control group and the experimental group were selected. For the collection of information, the descriptive transactional survey type was used, applying the questionnaire technique and observation. The results obtained from the application of the experimental pilot are based on the academic improvement of the experimental group in relation to the control group, which was evidenced by the interaction of social networks used as support for the teaching-learning process. Finally, in the experimental pilot meet the established objectives such as adaptive learning, Neuroeducation.

Keywords: Adaptive learning · Neuroeducation · Social networks · MOOC

1 Introduction

In the current context, education applies teaching methods that allow to achieve optimal results in students. Through Neuroeducation, it is possible to understand how the brain works and what are the brain mechanisms in which learning occurs in the student from the dimensions of affectivity, emotions, physical, ethics and culture. In this scenario, the teacher becomes a guide and advisor of adequate cognitive didactic processes. Technological progress has generated transformations in education, since it continues to be one of the great challenges of the society of the new century, achieving the union between education and science, where educational centers, teachers have to adapt quickly to the accelerated scientific changes offered by the Neuroscience, discipline applied to the teaching-learning process through the manipulation of social networks.

Currently, social networks are means of groups communication, communities linked to each other to provide information that was previously restricted. Therefore, it is

M. Zambrano Vizuete et al. (Eds.): CI3 2021, LNNS 511, pp. 106–119, 2022.
https://doi.org/10.1007/978-3-031-11438-0_9

considered important that educational institutions are aware of the use of social networks in an appropriate way for the teaching-learning process, where students can benefit from their training and personal development. One of the challenges of education today must be to reform the educational area to include a new society subject to continuous change. At this point, social networks in our society, their rootedness and the fascination they rouse for most of our students cannot be ignored, but rather studied for their pedagogical and didactic possibilities; that contributes to the interaction, responsiveness of a fast and efficient communication [1].

These contributions learning forces to rethink some aspects of the educational process in the classroom, as well as the role of the teacher and the student. Currently, educational systems are adopting a new vision of teaching through Neuroeducation in social networks, which is based on providing educational strategies and technologies focused on the functioning of the brain with the aim of improving the teaching and learning process. According to [2], the application of brain-compatible learning theory affects education in three aspects: 1) Educators for curriculum should design learning centered on students' interests and make learning interactive; 2) Teaching educators should enable students to learn in groups focused on real problems, with collaborative learning and students feeling motivated, stimulated to learn in environments outside the classroom; 3) Assessment should enable students to understand their own learning styles and preferences. In this way, students create their own learning style. Next, we mention that the Internet is an education model that allows people to prepare from any geographical location, accessing different web tools, one of them being MOOC courses (Massive Open Online Courses), which uses multimedia resources, presenting participants with dynamic, friendly environments, where students are no longer individual actors and become part of a wider learning community. The present research allows strengthening the teaching-learning process based on the theories of Neuroscience in social networks as adaptive learning in a MOOC, where students have access to technologies that permit them to interact with content adapted to the learning styles of students and, at the same time, establish communication with different media, promoting self-learning.

The objective of adaptive learning and Neuroeducation with the use of social networks in the teaching-learning process in a MOOC course, is that the student has a new perspective of self-learning in a simple, friendly way, that adapts to the needs, schedules, that serves as a tool for academic and work support. Currently, MOOCs have a considerable growth rate in different educational entities, where virtual education is seen as an opportunity that allows students to enrich their essential knowledge and self-learning with the support of social networks.

Motivation in education is considered as the fundamental pillar of social transformation, which awakens interest in learning curiosity, for what surrounds it, creating an environment of flexibility that allows fostering the creativity of human beings at all educational levels, strengthening the motivation of students with new knowledge, significant material based on the student and not on the teacher, awakening expectations of interest. Consider [3], motivation as the level at which students persevere to reach the academic goals they identify as achievements in meaningful learning. From the teacher's point of view, it means motivate the student to perform the activities, through the awareness, attitude and personal aptitude of each student; and [4] mentions motivate means to encourage the student so that he can have an active participation in the tasks assigned in the classroom.

The intention of motivation is to stimulate, generate interest and direct efforts towards self-learning. The teacher considers that the main objective of learning is motivation; its absence causes students to fail. The teacher-student relationship in the educational context is presented in a distant way so that learning is focused on the teacher and the content, that is where it is necessary to insert Neuroeducation and the use of social networks to create content, materials and appropriate resources so that through them the students are motivated and the knowledge can arrive in an adequate, simple way, promoting the taste for learning.

From a historical perspective, [5] refers to motivation as a phenomenon composed of several components, which appear and disappear according to the circumstances determined by social, cultural and economic phenomena, therefore, it must have a particular treatment for each of the subjects. Motivate is to aspire students to achieve the desired goal according to the proposal that the teacher imparts, not necessarily to achieve what they are expected to perform to what they have to develop.

Currently, Neuroscience is considered as a set of disciplines that studies the nervous system, the development of Multiple Intelligences, emotional intelligence, which allows people to process information, demonstrating that knowledge is achieved neither by memorizing nor by repeating, but by experiencing accompanied by emotions that are closely related. MOOC courses are those that allow access to thousands of students, professionals and ordinary people who wish to update their knowledge, improving the teaching-learning process, strengthening their job skills. However, according to research, the courses are not completed by many users and one of the reasons is that the courses become boring, monotonous and repetitive, losing interest in them and in many cases they finish the course more for the certificate than for the learning itself. The present research work aims to give a different approach to MOOCs, as it sometimes loses interest; for which adaptive learning, Neuro-education and use of social networks in a MOOC is raised, it is mentioned that motivation, emotion according to research conducted favor a deep, lasting, affordable learning that has a better benefit for the user. He points out that Neuroscience supported by Neuroeducation applies knowledge within the classroom, describing brain processes such as emotion, curiosity and attention, awakening knowledge mechanisms in the teaching-learning process [6].

The Web (World Wide Web) is one of the most used services on the Internet. However, it is important to point out that while the Internet is the technological medium that encompasses services such as e-mail, FTP (File Transfer Protocol), videoconferencing, chat, newsgroups, among others, the Web is the information hosted on the websites where the user accesses. That is why the challenges of education today is the use of technologies and one of them is the social networks of Web 2.0, which allows interaction and communication between the actors of education [7].

Web 2.0 technologies applied to education allow access to new means of communication, interaction scenarios incorporating a variety of resources in the teaching-learning process, so it is necessary that higher education institutions integrate spaces for knowledge exchange, taking advantage of the different virtues that Web 2.0 offers in the educational field [8]. The opportunities offered by Web 2.0 focus on the exchange of information through the network such as videos, documents, images, presentations, used in different applications such as YouTube, Slideshare, Google applications, Facebook, Scribd, Flickr, which allows not only communication between users but also the sharing of relevant information in collaborative work and autonomous learning [9]. The teacher focuses his efforts on expanding talents and competencies by applying new learning methods, becoming a guide, and the student on building his own learning [10].

The role of the student in the use of social networks is to develop the necessary skills to interpret, assimilate and reproduce information, since each person has different ways of perceiving, processing and capturing information according to learning styles, which can be visual, auditory or kinesthetic, theoretical, pragmatic, reflective and active. Alonso et al. (1994), Guild and Garger (1998), Kolb (1976), cited in García et al. (2009), "define learning styles as the set of aptitudes, preferences, tendencies and attitudes that a person has to do something that is manifested through a behavioral pattern and different skills that make him/her stand out from other people under a single label in the way he/she conducts, communicates, thinks, learns, knows and teaches". It is mentioned that networks allow learning to be dynamic and meaningful, where the actors can access other sources of information that help the construction of new knowledge and the development of students [11].

Social networks are the actors in the new educational spaces with the evolution of technology and society, allowing to change the traditional educational scenario to an environment that encourages self-learning with the inclusion of technologies in the classroom, where the teacher and student emits, receives information, shares knowledge and consolidates interpersonal relationships. "The growing popularity of social networks only highlights the need to incorporate their use as a platform for teaching and to investigate their potential in the academic and educational world. It is important to take advantage of the open attitude of students to relate to each other through the use of social networks, as well as to highlight their social character to generate synergies for knowledge exchange" [12].

Currently, MOOCs (Massive Open Online Courses) are being widely used nowadays and one of them is education, especially higher education, being considered as a new form of online study and training [13].

For [14], The important thing when designing a course is to have a friendly and flexible structure that responds to the learning styles of the target group. Therefore, the design should include the planning, the number of participants and ensure that the dropout rate is minimal. This study showed that the use of social networks as a learning strategy reduced the dropout rate and increased academic performance [15].

1.1 Neuroscience and Neuroeducation

For [16], "Education today receives fundamental contributions from neurosciences, and there are several entities that facilitate their knowledge, understanding the origin of conflicts in the classroom prepares us to obtain better results in the actions of innovation, development of talent and creativity, it is important to emphasize that neuroscience studies the human nervous system, the system that makes us what we are". Neuroscience applied to education is considered in the report of [17], who pointed out: "Imaging technologies allow the observation of the brain in operation which has meant an understanding of the perceptual, cognitive and emotional functions that have consequences for education where the actors involved in education include students, parents, teachers and policy makers to better understand the learning process and to structure environments that nurture it". "Neuroscience studies the nervous system, especially interesting in education is the division dedicated to the study of the structure and function of the brain. Neuroscience would not be in vogue today if we did not have the incredible experimental sources of neurological information and important techniques for the development of brain function" [18].

The elements that intervene in Neuroeducation are the following: a) Brain plasticity capable of generating neurons that connect with each other, promoting stimulation. b) Mirror neurons are activated when we perform an action, which are important for Neurodidactics. c). Emotions are affective processes that emerge in the interaction of the subject with the environment, and that lead to an expression from the physiological, motor, expressive behavioral and subjective experiential, on the basis of the assessment made by the subject of the stimulus and the visual content it possesses, like or dislike, in integration with their individual needs and motivations [19]. "These changes have implications for the biological, psychological and social functioning of the subject" [20].

Consequently, emotions interact in Neuroeducation as management of learning difficulties by creating an appropriate environment that attracts attention, avoiding sadness, anger and anxiety.

2 Methodology

2.1 Materials and Methods

The purpose of the research project is to achieve the specific objectives proposed and with them to reach the general objective that proposes a change in the traditional methodology, where the teacher continues to be a transmitter of knowledge with little student participation, to a motivational methodology that encourages student activity and interaction where they become the author of their own learning by developing their skills and abilities.

General Objective: Apply Neuroeducation and social networks in the teaching and learning process through a MOOC to improve academic performance.

Specific Objective.

- To identify the theoretical support of Neuroscience, social networks, MOOC in the teaching-learning process.
- To apply the experimental pilot in a MOOC virtual course, using social networks.
- To determine the impact on academic performance when using social networks.

Hypothesis: Does Neurosciences supported by social networks, improve the teaching learning process in students following a virtual course in a MOOC?

The methodology implemented in the research project is qualitative, experimental type, with the use of tools such as student surveys, data analysis for the proposal, the creation of the virtual course in a MOOC and the preparation of the necessary documentation for the course.

For this research, adaptive learning and Neuro-education with the use of social networks were proposed in a MOOC, which constituted a support tool for a Computer Science I subject for Higher Education students of the Technical University of Cotopaxi of the Executive Management Secretarial Career. The average time that the research project was executed was one year, in order to comply with the development of the activities proposed in the virtual MOOC course.

Adaptive learning and Neuroeducation with the of technologies in a MOOC for the teaching and learning process, the human being was considered by its emotions and then by the rational, encouraging students and teachers to have access to multiple technological tools, materials and information.

2.2 Adaptive Learning

Adaptability is adjusted to one or more learning characteristics, becoming formal educational environments supported by technology where the student's learning reaches the highest possible achievements, adapting the learning resources of each student considered as an individual person. Adaptive learning is supported by technologies, educational platforms that support the teaching-learning process, obtaining as a result information from students, which identifies strengths and weaknesses in a given subject in order to reinforce the missing content in the student's learning.

The student becomes autonomous with adaptive learning, in teaching and learning process, for which a platform was used as a means of communication between students and teacher, who must integrate learning experiences in the classroom, the platform suggests to each student their learning path and personalized content. The learning adaptation process had advantages for both the teacher and the students: Teacher a) With information and communication technologies, the teacher saves time in preparing didactic material, correcting and evaluating results, and resolving student doubts according to the student's requirements. b) The teacher receives data of interest in real time, such as activities carry out by the student on their performance, the time it takes to perform each activity, aspects that they master better and those in which they have more difficulties. c) Self-correcting activities save time in correction and preparation of materials. In this way, the teacher can make the most of the interaction and attention to each student. d) It allows the teacher to organize learning in different ways and individualize it, focusing on the key needs of each student or subject in real time. e) Freedom to the teacher opens a range of actions that can be performed by the teacher, who becomes, more than ever, a guide to the learning of their students. Students a) Learning is tailored to each student, facilitating the process of understanding and assimilation of concepts, helping them to feel more supported. Students can quickly move beyond the areas they already master to focus on those areas in which they need to improve. b) Students learn from their own mistakes, since they can review the activities and content as many times as necessary, as well as have access to the feedback that the teacher assigns to each task. c) Computer tools to support adaptive learning help the student to assimilate the proposed contents and to develop skills and abilities for their use. d) The student improves the digital competences, because they constantly use the educational platform, developing adaptive activities and assignments in a digital environment.

Adaptive learning systems in e-learning consist of the following modules: Expert model: it is the repository of information that stores the material being taught. It can be academic text, lessons, tutorials, problems with their solutions, interactive multimedia application, video fragment and even a set of clues to guide the learner in the learning process.

Student or user model: Stores user information, i.e. the system stores student data, such as profile, knowledge, previous qualifications and objectives. In addition, this information is completed with the data generated by the student in the learning process: activities performed, tests, results, among others.

Instructional model: It is in charge of providing information on appropriate activities to the user depending on the data stored in the user model, incorporating educational tools offered by technology such as multimedia presentations, with the expert advice of a teacher on presentation methods. The instructional model helps the learner to develop skills to accomplish certain tasks. Instructional interface or environment: It is the user interface through which students communicate with the system.

To achieve the research objectives, the ADDIE (Analysis, Design, Development, Implementation and Evaluation) model was used, which is presented below:

2.3 Analysis

The students participating in the research are located in Ecuador Cotopaxi province, Latacunga city, in the Technical University of Cotopaxi of the Executive Management Secretarial Career of the third cycles, a group that comes from distant provinces and low economic conditions. In the research we worked with an experimental group and a control group, allowing us to know how Neuroeducation and the use of social networks as adaptive learning in a MOOC. The participating in the research were randomly selected from the third cycles of the Executive Management Secretarial Career, they do not have adequate knowledge about the use of the program and are not adapted to it, so that by applying this methodology a friendly tool is provided, easy to use with adequate resources that allowed to motivate them, becoming the protagonist of their own learning.

We worked with the subject Computer Science I, which uses office automation tools for the performance of skills of a Managerial Executive Secretary. In order to know the learning styles, the Honey Alonso questionnaire was applied, obtaining results such as the behavior of the participants.

2.4 Design

Once the students, the subject, learning styles were identified, the design and creation of the MOOC course began, for which time was planned, adaptive contents were selected according to the learning styles of the students, resources, activities, motivations of the course, evaluations, all these elements were designed with the help of social networks and contents of the syllabus of the subject as support. For the design of the MOOC virtual course we used the Moodle Platform, version 3.2, taking into account essential elements of adaptive learning, Neuroeducation and the use of social networks, which help the teaching-learning process of the Computer Science I course, focusing on motivation, curiosity, communication and attention in the participants.

2.5 Development

Adaptive contents, resources and materials were developed with free web tools, which allow the student to interact in a friendly and easy to use way, entering from any geographical point, for the creation of these contents Neuroeducation and use of social networks as a support for the teaching-learning process in a MOOC course.

The following tools are used to incorporate adaptive learning, Neuroeducation and use of social networks in a MOOC:

Contents in the MOOC: Neuroeducation was used as a learning strategy, where the course activities aroused emotion, communication, curiosity, attention and games presented during the teaching-learning process.
Text content: The researcher created texts for each of the units in PDF, necessary resources for the development of activities that attract the attention of students, facilitating study and access to information.
Audiovisual contents: One of the strategies is that the teacher uses short videos of the topics of each unit, which are available on the web, strengthening the knowledge of the participating students.
Forums: It is a communication tool, where greater emphasis is given to debates and discussion of course topics. It is a synchronous activity since the participants do not have to access the system at the same time; the teacher must generate the discussion topic based on the programmed units.
Chat: It is a communication tool; the chat activity allows participants to have a synchronous discussion in real time within the Moodle course.
Evaluation test: The Moodle tool allows the creation of evaluation tests that provide a wide variety of question structures such as: multiple choice, false true, short answer and drag and drop text and images. These questions are stored in the question bank and can be re-used in different exams, their grade will depend on how the teacher configures the questionnaire, the course has an evaluation at the end of each unit.
Contenidos en el MOOC: Se utilizó la Neuroeducación como estrategia de aprendizaje, donde las actividades del curso despertaron la emoción, la comunicación, la curiosidad, la atención y los juegos que se presenta durante el proceso de enseñanza aprendizaje.

2.6 Implementation

Once the necessary contents were elaborated, the course was presented to the participants, who are responsible for assuming the execution of the proposed activities using elements of Neuroeducation, for its implementation relevant information was uploaded to the Informatics I course.

2.7 Evaluation

In this phase the Informatics I course, its design, creation, as well as the opinion of the students was evaluated, for which a satisfaction survey was applied at the end of the course.

3 Results

The results achieved in the experimental pilot meet the established objectives, such as adaptive learning Neuroeducation and use of social networks in the teaching-learning process, in a MOOC course, taking into account that the experimental pilot, can be applied to students from other careers to improve the teaching-learning process and thus strengthen the professional profile.

This phase allowed to know if the proposed model has as a result the strengthening of the teaching-learning process of the participants in the Computer Science I course, in relation to the group that is not in the MOOC course. The students use the MOOC course as an academic aid to their classes in a simple and fast way.

We approach the empirical part of the study with the bibliographic and documentary review about the elements that support our experiment at the theoretical level. The proposed methodology to be applied in the subject of Computer Science I of the third cycle of the Executive Management Secretarial Career. The MOOC of the Computer Science I course was proposed as a pilot test. With a number of 30 students, of which it is presented in Table 1, as follows:

Table 1. Population of participants experimental group in the pilot.

No. of Students	Men	Women
30	1	29

Source: student data career executive secretary of management

Table 2 Shows the student population in the control group.

Table 2. Population of participants control group.

No. of Students	Men	Women
30	3	27

Source: student data career executive secretary of management

For the experiment, the Honey Alonso questionnaire of learning styles was applied to the students of the experimental group of the Computer Science subject, which determine whether it is active, reflective, theoretical and pragmatic, the following data were obtained. These are shown in Table 3:

Table 3. Learning styles.

Learning styles	Frequency
Active	370
Reflexive	351
Theoretical	356
Pragmatic	358

Source: student data career executive secretary management

Table 3 Shows data obtained from Honey Alonso's questionnaire of learning styles applied to the experimental group:

Table 3 shows that students had learning style according to the questionnaire applied, where it was considered that students show interest in learning when ideas are original, when new knowledge is presented, new ideas are contributed, encourages teamwork, breaks the routine is, followed by pragmatic styles where theoretical contents are taught accompanied by practical examples of ordinary life, and shortcuts are promoted to reach the solution.

Following the survey applied to Honey Alonso's learning styles, a structured questionnaire was applied to the use of networks in learning from the point of view of the students. The questionnaire had a total of 13 items addressed to the students. The following is the analysis of the results obtained from the questionnaire applied to the use of social networks in the research focused on the students of the experimental group and the control group, information that is detailed below.

The students stated that they use the following devices to access social networks: cell phones, followed by laptops, although they use other media less frequently, such as desktop PCs and smartphones. The most used social network for their studies is YouTube followed by, WhatsApp, Facebook, google+. They use other social networks, less frequently, such as Instagram, twitter, blog, Yahoo! groups, Skype and others. The number of times per day they use social networks for study related matters is between 4 and 6 times per day. The number of people who are connected for study-related matters are 1 to 15 people. The number of times they exchange messages or information for study purposes per day is between 1 to 5 times. The use of social networks as a teaching tool for their studies they consider absolutely necessary. Students consider that it is necessary that social networks should be used to support learning. They considered that there are teachers who do not use social networks to support their learning.

In the research, the student has an active participation in their learning without waiting for the teacher to decide for them; spaces were created for the students to make their own decisions for their learning. In other words, activity-based learning where the student is the center of the learning process. The contents were adapted to the needs and interests of the participants, with current contents. With a collaborative work system centered on the student, rather than on the teacher, facilitating interaction, creativity, participation, communication, evaluation, and cooperation between teacher and students. For collaborative learning I use the following: blog, chat, wiki, google tools, office among others. With the support of ICT personalized learning, I facilitate participation and access on equal terms. The role of the teacher changes from transmitting knowledge to the students to being a facilitator in the construction of their own knowledge. In which the teacher acts first as a person and then as a content expert, facilitating learning and individual and group development.

For the evaluation of student participation in the virtual course is determined by qualitative and quantitative questions and guidelines, not only the designs of an evaluation with multiple answers reflect whether or not there has been a good learning, but also the details of assimilation and accommodation that the student has with the contents and the different activities proposed in the virtual course. Once the virtual course has been designed, documents related to the topic have been incorporated, as well as collaborative forums in each of the units of the development and evaluation process, in which the participation of the students is evidenced in a dynamic way, exposing their knowledge of the treated topic, feeding the forum with their objective and real opinions to the learning process to which they are exposed. Evaluation of learning, this activity reflects quantitative results of the knowledge acquired during the development of the activities in the virtual course. With the application of the satisfaction survey, students have the opportunity to express their opinions about the MOOC course in Computer Science.

The course was divided into units which have several sections such as Unit 1 deals with Microsoft Word with its sections such as PDF document, discussion forum on the topic, video on the topic, indications of the activity to perform and upload the activity to the platform, Word shortcuts game and finally a final evaluation of the Word Unit. Unit 2 deals with Microsoft PowerPoint with its sections such as PDF document, discussion chat on the topic, video, indications of the activity to be performed and uploading the activity to the platform, PowerPoint word search game and finally a final evaluation of Unit 2. Unit 3 deals with the Internet with its sections such as PDF document, discussion forum on the topic, video on the topic, indications of the activity to perform and upload the activity to the platform, motivation, alphabet sorting game and finally a final evaluation of Unit 3. For feedback we will generate a Blog with the theme of the Internet, build a Wiki on elements of Google applications, and finally we will make use of the WhatsApp web.

3.1 Experimental Group and Control Group

An evaluation was applied to the experimental group and the control group to obtain the average of the three units Microsoft Word, Microsoft PowerPoint and Internet, a physical template was elaborated applying in a traditional way, I should mention that the topics treated for the experimental group were also developed for the control group, considering that the treatment time for the experimental group and the control group were equal, materials, there was no favoritism between the groups participating in the computer science I course during the teaching-learning process. Figure 1 shows the performance of the control group, experimental group.

Figure 1 shows that the experimental group has a high academic average in comparison with the control group, it is evident that the experimental pilot by applying an appropriate methodology in the teaching-learning process, with significant materials and the use of technologies allowed the participants to raise their interest in learning

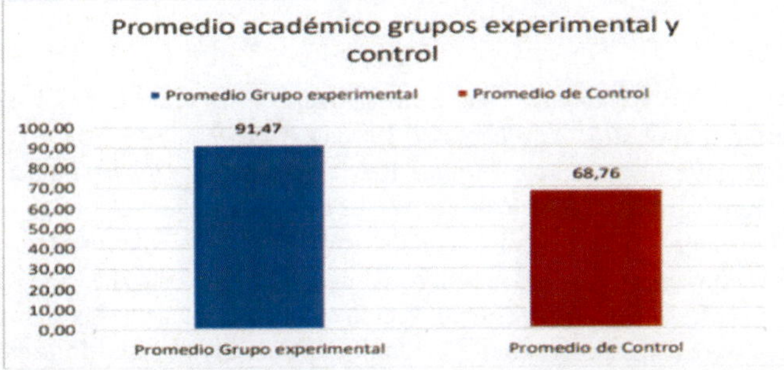

Fig. 1. Academic average experimental group/control group. Source: MOODLE platform data.

the subject of Computer Science I, feeling motivated, curious, provoking attention, concentration in the activities and becoming the protagonists of their own learning without geographical and time limits.

In the experimental pilot, adaptive learning was applied to Neuroeducation with the use of social networks in the teaching-learning process in a MOOC course, it is mentioned that students adapted quickly to the use and manipulation of the MOODLE platform, creating a habit of organization, responsibility at the time of fulfilling and developing their activities, acquiring new experiences and competencies that allow them to strengthen their professional and personal profile.

4 Discussion

In the experimental pilot it was possible to evidence relevant aspects on the part of the students such as economic, educational and geographic diversity, considering that many of them have not received computer science since in our country in the best of cases in educational institutions have reduced the number of hours of computer science and English, and others in other institutions do not have the subject of computer science or laboratories limiting the knowledge.

During the development of the experimental pilot we can state that all the facilities were given to the students during the process, supported by adaptive learning, Neuroeducation and the use of social networks in the MOOC course for the teaching-learning process, achieving an adequate result.

It is considered that the state of mind, the emotions of the individuals are very important for the teaching-learning process together with the contents, materials and resources, establishing an adequate communication and being able to share knowledge and information with the use of social networks where the role of the student is to generate his own learning.

The results achieved in the experimental pilot meet the established objectives such as adaptive learning, Neuroeducation and the use of social networks in the teaching-learning process, in a MOOC course, taking into account that the experimental pilot can be applied to students of other careers to improve the teaching-learning process and therefore their academic performance that strengthens their professional profile.

References

1. Artero, B.N.: La interación como eje de aprendizaje en las redes sociales. Retrieved from www.educaweb.com (2011). Last accessed 19 May 2011
2. Purpose Associates: Brain-based learning. Retrieved from http://www.funderstanding.com/brainbasedlearning.cfm (1998–2001). 19 April 2011
3. Polanco Hernández, A.: La Motivación en los estudiantes universitarios. Revista Electrónica "Actualidades Investigativas en Educación". Universidad de Costa Rica (2005)
4. Campanario, J.M,: ¿Cómo influye la motivación en el aprendizaje de las ciencias? Retrieved from http://www2.uah.es/imc/webens/127.html (2002). Last accessed 15 March 2005.
5. Abarca, S.: Psicología de la motivación. Editorial Universidad Estatal a Distancia, San José, C.R. (1995)
6. Mora, F.: Neuroeducación: sólo se puede aprender aquello que se ama. Alianza Editorial, Madrid (2013)
7. Claudia, I.T., María del Rocío, C.A.: Uso de las redes sociales como estrategias de aprendizaje. ¿Transformación educativa? Apertura **3**, 6–15 (2011). Retrieved from http://www.udgvirtual.udg.mx/apertura/index.php/apertura/article/view/198/213
8. García, S., Alonso, J.L.: Uso de las TIC de acuerdo a los estilos de aprendizaje de docentes y discentes. Revista Iberoaméricana de Educación **48**(2), 1–14 (2009)
9. Haro, J.: Educar para la comunicación y la cooperación social. In: Redes, vol. 5. Retrieved from http://www.webquestcreator2.com/majwq/public/files/files_user/31024/Embedding_Citizenship_Education_in_Engla.pdf#page=203 (2015). 15 Aug 2021
10. Meso Ayerdi, K., Pérez Dasilva, J.Á., Mendiguren Galdospin, T.: La implementación de las redes sociales en la enseñanza superior universitaria. Retrieved from http://dehesa.unex.es/bitstream/handle/10662/4555/1988-8430_12_137.pdf?sequence=1&isAllowed=y (2011). 20 Aug 2018
11. Imbernón, F., Silva, P., Guzmán, C.: Competencias en los procesos de enseñanza-aprendizaje virtual y semipresencial. Co - municar **36**, 107–114 (2011)
12. Ayerdi, K.M., Dasilva, J.P., Galdospin, T.M.: La implementación de las redes sociales en la enseñanza superior universitaria. Tejuelo: Didáctica de La Lengua y La Literatura. Educación **12**(12), 137–155 (2011)
13. Vázquez-Cano, E.: El futuro de los MOOC. Retos de la formación on-line, masiva y abierta. Educatio Siglo XXI **34**(1), 245–247 (2016)
14. Castrillo, Mª.: MOOC para el aprendizaje de lenguas extranjeras: claves para gestionar un curso online desde la masividad Retrieved from http://scopeo.usal.es/mooc-para-el-aprendizaje-de-lenguas-extranjeras-claves-para-gestionar-un-curso-online-desde-la-masividad/ (2013)
15. Fidalgo Blanco, A., Sein-Echaluce Lacleta, M., Borrás Gené, O., García Peñalvo, F.: Educación en abierto: integración de un MOOC con una asignatura académica. Educ. Knowl. Soc. (EKS) **15**(3), 233–255 (2014)
16. Victoria, M., Ayca, M.: La Neuroeducación En El Aula: Neuronas Espejo Y La Empatía Docente. Neuroeducation Classroom: Mirror Neurons Teach. Empathy **2014**(2), 9–18 (2014)
17. Della Chiesa, B.: La comprensión del cerebro. Ediciones Universidad Católica Silva Henriquez, El nacimiento de una ciencia. Informe de la OCDE CE/II. Santiago (2007)
18. Béjar, M.: Una mirada sobre la educaciíon. Neuroeducación. Padres y Maestros **355**, 49–52 (2014). https://doi.org/10.1017/CBO9781107415324.004

19. Guedes Gondim, S.M., Álvaro Estramiana, J.L.: Naturaleza y cultura en el estudio de las emociones. Revista Española De Sociología, (13). Recuperado a partir de https://recyt.fecyt.es/index.php/res/article/view/65164 (2010)
20. Fernández-Abascal, E.G. (Coord.): Emociones Positivas. Pirámide, Madrid (2011)

Methodology for the Improvement of the Learning Outcomes of the Public Universities of Ecuador Through the Support and Permanent Control Through Academic Tutoring

Karla Paola Negrete(✉), Jenyffer Yépez, Miguel Naranjo Toro, and Pamela Escobar

Universidad Técnica del Norte, Ibarra, Ecuador
kpnegrete@utn.edu.ec

Abstract. This paper details an investigation carried out at the Northern Technical University (Ibarra-Ecuador), for the improvement of the results of the teaching-learning processes, through the application of group academic tutorials for students who are repeating a subject (second or third registration). As a case, the methodology was applied in the Industrial Engineering Career of the Technical University of the North, during two academic periods (September 2019-February 2020 and June-October 2020), considering students who are in second or third enrollment, as well subjects, from second to tenth semester. Once applied as a results, an increase of 8.8% was obtained in the approval of subjects for students with second enrollment and increase of 8.3% for students with third enrollment, it should be noted that when planning the semester academic tutorials for students who are taking second or third enrollment subjects, an increase from 81.81% to 93,75% in the execution of academic tutorials.

Keywords: Process · Academic tutorials

1 Introduction

The UNESCO proposes that quality in higher education is the adaptation of the being and the work of higher education to its duty and recommends that: "The training provided by higher education institutions should respond to how to anticipate social needs" [1], while improving and promoting the growth of countries focused on social, economic and political factors.

Also, "Higher education has functioned as a social structure destined to the control of advanced technique and knowledge" [2], that is why, due to globalization and permanent change, institutions must adapt, modify teaching traditional, to carry out a follow-up that allows to improve the activities and performance indicators, for which processes such as the so-called academic tutorials have been implemented.

M. Zambrano Vizuete et al. (Eds.): CI3 2021, LNNS 511, pp. 120–132, 2022.
https://doi.org/10.1007/978-3-031-11438-0_10

In the field of Higher Education, academic tutorials are defined as a tool for the guided construction of learning, autonomous and comprehensive development of students, mitigating their weaknesses and promoting the development and continuous improvement of their capacities [20], allowing to reduce and mitigate the problem of student desertion that every time increases at a higher level.

On the other hand, it is described that academic tutorials should be understood as one of the teaching responsibilities in which a personalized interaction is established between the teacher and the student, with the aim of guiding their learning, adapting it to their individual conditions and style. to learn [16], which implies that by providing permanent accompaniment and monitoring to students, it allows knowing their criteria, points of view, weaknesses, fears and thus ensuring an increase in the percentage of approval of the subjects [21], this has made tutoring seen as a basic strategy to respond to the educational challenges facing university institutions.

To apply academic tutorials according to the needs of the students, it is necessary to know that academic tutorials are classified according to their function and can be individual and group tutorials.

"Group Tutorials: Since the educational process is eminently socializing, with this type of tutoring, the needs of the group are addressed in the educational process, and the recurrences and generalities of the students in academic and psychosocial aspects (either in class groups or due to specific needs or interests). Also, allow the optimization of both material, human and temporal resources.

Individual Tutorials: It refers to the personalized attention that is provided to the tutor to help them in any of the topics that influence their learning process. In this type of tutoring a planning of the sessions and, above all, a record of the results should be established as much as possible, which should be included in the group file" [6].

The incorporation of academic tutorials into teaching processes has become a necessity, given its close relationship with the quality of teaching-learning processes and recognizing the importance of targeted and personalized support for students, that promotes and guarantees an effective training process [18], this set of activities that are carried out are necessary to overcome the academic challenges faced by Higher Education Institutions, such as student desertion. According to a study presented by UNESCO and the Council of Rectors of Panama, at present, desertion and student repetition are among the main problems of Higher Education Institutions (IES), which, added to the lack of orientation and unified criteria qualification, directly affect the institutional integrity and the economic situation of students.

In Ecuador, according to the country's newspapers, desertion and repetition is between 26% in 2014, while in 2012 it was around 50% when the university admission system was implemented [19], which is not favorable for the country due to economic and productive losses since it is invested in each student, presenting an unfavorable scenario for the growth of society.

Also, according to Pozo and Hernández [17] "the lack of information about the university environment is one of the factors that hinders an adequate adaptation to said environment and, in many cases, brings with it little participation in academic life ultimately generating a loss of motivation and interest in the studies completed. Coupled with this, the expectations that students have about the degree they are going to pursue does not correspond to reality in many cases".

For the above this research focuses on the applicability and benefits of group tutorials within third level education, for which, a process called "Academic Tutorials" was developed and implemented, which aims to improve learning outcomes, as well as the quality and satisfaction of the students in the academic offer.

In order to effectively manage and concatenate all the procedures and sub-processes proposed for the "Academic Tutorials", this research was supported by the Bizagi Modeler software, which allowed the modeling and simulation of the process from its beginning to the end.

This document is divided into 4 parts: first, an introduction is made to the context of the research, the problem and the proposed solution; second, methodology, the design and development of the "Academic Tutoring" process and each of its activities are detailed; third, discussion and results, the results obtained after applying the process during two regular academic periods are described; and fourth, the conclusions and recommendations reached by the authors after this research are described.

2 Methodology

As a basis for the development of this research and the proposed process, the Deming Cycle or PDCA (Plan, Do, Check, Act) was taken as a reference, applying a criterion of permanent control and continuous improvement.

In the initial phase (phase I) planning, the exploratory methodology was applied where the bibliographic review, information gathering, surveys of teachers, students and authorities on the execution of academic tutorials was carried out.

Later, in the implementation phase II, the Bizagi Modeler software was used for the elaboration of the flow diagram, characterizations, formats and records.

In phase III verification, the results obtained in terms of the approval of subjects of students with second and third enrollment were compared, between the period June - October 2020, in which the process of "Academic Tutoring" was applied, and the previous period, September 2019 - February 2020, in which the aforementioned process did not yet exist.

In phase IV act once it has been verified that the implemented process is functional and beneficial for the University, proposals were made for its improvement and massification.

For the design and planning of the academic tutoring process, the legal regulations of Higher Education in force and applicable to Ecuador were considered, among which they stand out (Table 1).

Once the legal bases applicable to the academic tutoring process have been established and in accordance with the provisions of the scientific paper called "model for the implementation of the quality management system based on the requirements of the ISO 9001: 2015 standard" [15]. The Deming cycle was used (Plan, Do, Check and Act).

Table 1. Higher equation legal regulations.

N°	Regulation	Article
1	Organic law of higher education 2018	**Art 13.** "Functions of the higher education system"
2	Academic regime regulation CES 2019	**Art. 26.-** "Learning activities" **Art. 27.-** "Learning in contact with the teacher"

2.1 Phase 1 Planning

In the planning, the initial information gathering was carried out, through interviews with teachers and students that allow to have a base of the activities to be carried out in the individual and group tutorials.

Once the information on the activities and steps to follow was collected, the characterization and the flow diagram were designed where the steps to be followed, responsible, formats, records were established and this in turn was reflected in a document called procedure.

Characterization. With the information collected, the characterization corresponding to the academic tutoring process was elaborated which, according to the management by processes, must have the points detailed below (Fig. 1).

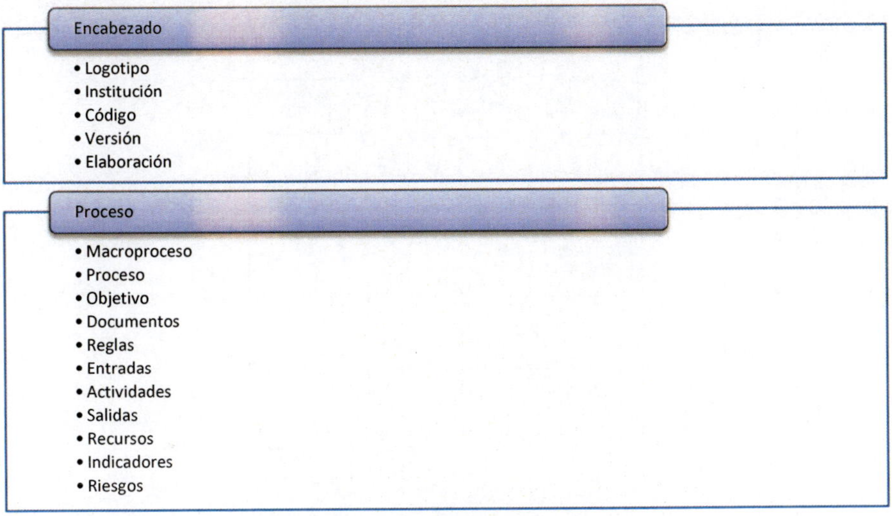

Fig. 1. Characterization format.

Process. Once the characterization of the process has been prepared, it is necessary to carry out the document called procedure which allows to know the process in detail and it describes (Fig. 2):

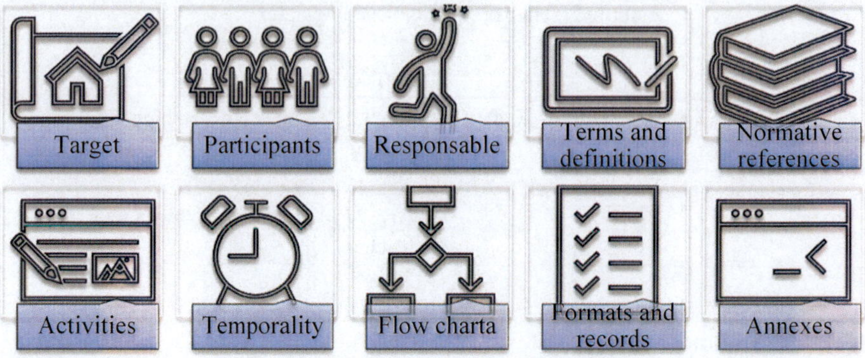

Fig. 2. Part of the academic tutoring procedure.

Process Diagram. In the diagramming of the academic tutoring process, the Bizagi Modeler software was applied, which allows diagramming the steps to be followed, establishing responsible persons, attaching formats and records necessary for the execution (Figs. 3, 4 and 5).

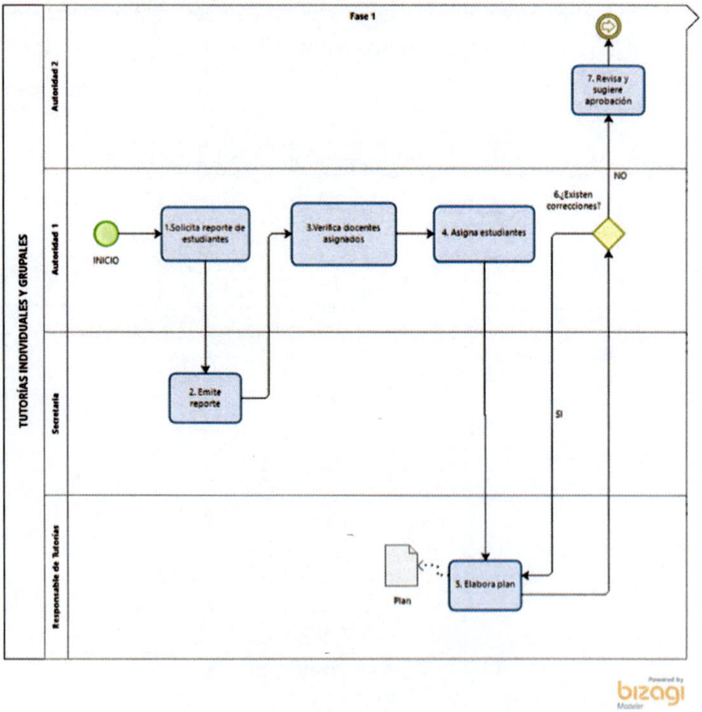

Fig. 3. Flow diagram of the academic tutoring process - phase I.

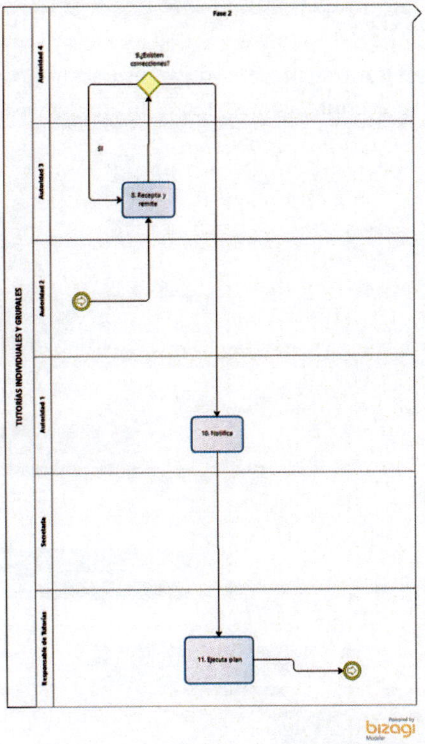

Fig. 4. Flow diagram of the academic tutoring process - phase II.

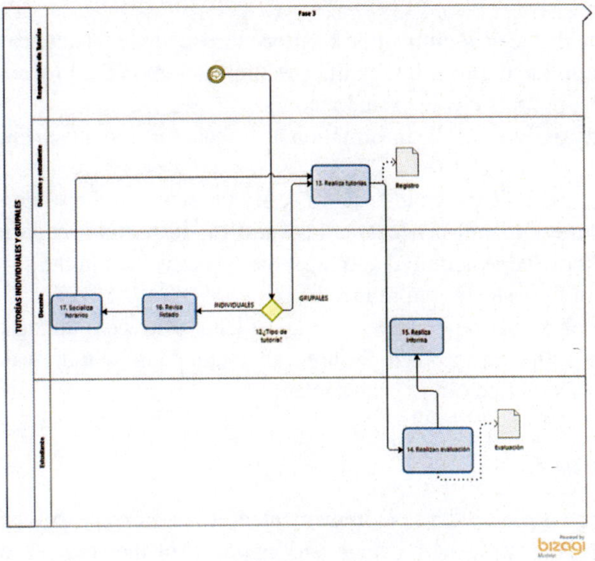

Fig. 5. Flow diagram of the academic tutoring process – phase III.

2.2 Phase 2. Execution (Do)

Fort the implementation of the academic tutoring process during the period June- October 2020, it was carried out in accordance with the following details of activities (Fig. 6).

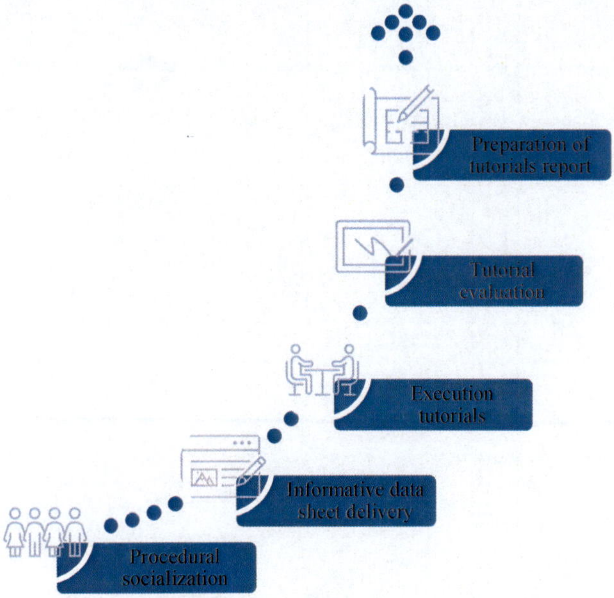

Fig. 6. Activities of the academic tutoring process.

Socialization of the academic tutorials procedure to students and teachers in charge of giving and individual tutorial, to publicize information on the procedure, applicable legal regulations, purpose, formats and records.

Subsequently delivery of the information data sheet for students to enter the required fields.

Execution of periodic group in order to verify the status of the student regarding the subjects he is studying and, if necessary, request the respective individual tutorials to strengthen the knowledge acquired and meet the objectives set in the subject.

Evaluation of tutorials by tutored students.

Preparation of a final report showing the students who were taking the second and third enrollment subjects, approval, failure, cancellation and withdrawal, as well as the individual tutorials carried out in each subject.

2.3 Phase 3. Check

As the academic tutoring process is implemented, it is necessary to verify if it is being applied properly and what the percentage of approval of the students was, for which, through the enrollment system, data has been collected from students who are taking subjects with second and third enrollment and at the same time review reports of academic tutorials in order to corroborate the percentage of students who passed the subjects (Tables 2, 3, 4 and 5).

Table 2. Students with second and third enrollment without having implemented the academic tutoring process.

September 2019-February 2020	Level										
	1st	2nd	3rd	4er	5th	6th	7th	8th	9th	10th	Total
II enrollment	17	4	7	1	0	5	6	1	1	4	46
III enrollment	2	0	1	0	0	1	0	0	1	1	6

Table 3. Report of students who passed or failed second enrollment courses.

Situation	II enrollment	III enrollment	Total
Approved	19	2	21
Disapproved	20	4	24
Retired	3	0	3
Canceled	4	0	4
Total	46	6	52

Table 4. Students with second and third enrollment when implementing the academic tutoring process.

June-October 2020	Level										
	1st	2nd	3rd	4er	5th	6th	7th	8th	9th	10th	Total
II enrollment	2	4	2	0	0	0	6	8	9	9	40
III enrollment	1	0	1	0	0	0	0	0	0	0	2

2.4 Phase 4. Act

With the data collected later when the academic tutoring process was implemented, in accordance with the activities established in the execution phase, a check list was applied where compliance is verified and at the same time the corrective actions to be implemented are established in order to improve the process (Table 6).

With the results obtained before and after the implementation of the academic tutoring process, it is established that there is an orderly execution, managing to obtain documented information that will serve as evidence for futures evaluations of the career, also, as an important point, there is evidence of an increase in the percentage of students who pass the subjects, as well as it is also necessary to consider the following corrective actions and improvements to the process (Tables 7 and 8).

Table 5. Report of students who passed or failed third enrollment subjects.

Situation	II enrollment	III enrollment	Total
Approved	22	1	23
Disapproved	13	1	14
Retired	2	0	2
Canceled	3	0	3
Total	40	2	42

Table 6. Check list.

N°	Activity	Compliance			
		September 2019-february 2020		Jun-October 2020	
		Yes	No	Yes	No
1	Socialization of tutorials		x	x	
2	Delivery of informative data sheet	x		x	
3	Execution of group and individual tutorials	x		x	
4	Follow up to students		x	x	
5	Evaluation of academics tutorials		x	x	
6	Academic tutorials report	x		x	

Table 7. Improvement plan

N°	Activity	Improvement
1	Socialization of tutorials	Make students aware of the legal regulations of academic tutorials, rights and obligations, process to follow, formats, records, planned activities
2	Delivery of informative data sheet	Send through official media the respective formats that are necessary documents for the fulfillment of the academic activities of the students, this will avoid confusion when filling out and sending the student files according to the indicated times
3	Execution of group and individual tutorials	Reach an agreement with the students regarding the weekly execution schedule, in such a way that it does not interfere with the other academic activities that the students must carry out, allowing permanent and continuous attendance at the planned academic tutoring meetings

Table 8. Improvement plan

N°	Activity	Improvement
4	Tracking students	Send communications through institutional mail in such a way that student participation is permanent and active
5	Evaluation of academic tutorials	Send through official media the respective formats that are necessary documents for the fulfillment of the academic activities of the students, this will avoid confusion when filling out and sending the evaluation sheets according to the indicated times

3 Discussions and Results

3.1 Results Obtained

As a result of the implementation of the process called academic tutorials, in the academic period June-October 2020 compared to the academic period September 2019-February 2020, it can be determined according to Tables 8 and 9 and Figs. 7 and 8, which has increased the percentage of students who complete the subjects in a positive way, achieving an 8.8% increase in the passing of subjects for students with second enrollment and an increase of 8.3% for students with third enrollment.

Table 9. Students II enrollment.

Condition	September 2019-February 2020	June-October 2020
Approved	19	22
Disapproved	20	13
Retired	3	2
Canceled	4	3
Total	40	40

3.2 Academic Tutoring Effectiveness Indicator

Applying the efficiency indicator, which is based on the planned and carried out tutorials, it is established that there was an increase in the percentage of execution of academic tutorials during the period June - October 2020, in relation to the academic period September 2019-February 2020, according to what is indicated in Table 10.

Where:

Pe = Percentage of the execution of tutorials (Table 11).

Tr = Number of tutorials carried out.

Tp = Number of tutorials planned.

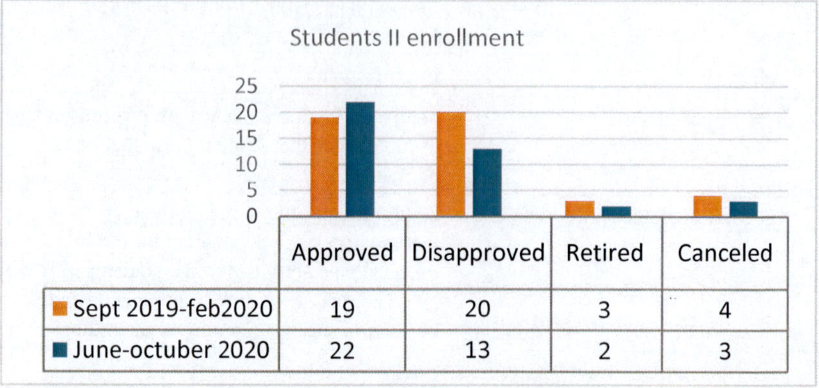

Fig. 7. Comparison of approved students for the semesters September 2019-February 2020 and June-October 2020 who were taking second – enrollment subjects.

Table 10. Students III enrollment.

Condition	September 2019-February 2020	June-October 2020
Approved	2	1
Disapproved	40	1
Retired	0	1
Canceled	0	1
Total	6	4

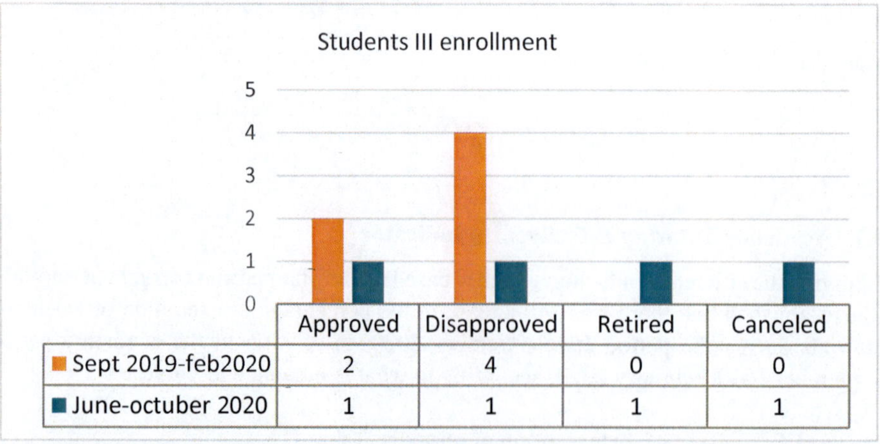

Fig. 8. Comparison of approved students of the semesters September 2019-February 2020 and June-October 2020 who were tasking third enrollment subjects.

September 2019-February 2020	June-October 2020
s $P_e = 100 * \left(\frac{T_r}{T_p}\right)$	$P_e = 100 * \left(\frac{T_r}{T_p}\right)$
$P_e = 100 * \left(\frac{18}{22}\right)$	$P_e = 100 * \left(\frac{15}{16}\right)$
$P_e = 81,81\%$	$P_e = 93,75\%$

Table 11. Tutorials carried out and planned academic period September 2019-February 2020 and June-October 2020.

Detail	September 2019-February 2020	June-October 2020
Number of tutorials carried out	18	15
Number of tutorials planned	22	16
Percentage of execution of tutorials	81,81%	93,75%

4 Conclusions and Recommendations

After having implemented the academic tutoring process, it can be seen that in the academic period September 2019-February 2020 there was an approval percentage of 32.5% and in the academic period June-October 2020 an approval percentage of 41.3%, which is why it is established that the increase in the percentage of approval of second enrollment subjects is 8.8%, on the other hand, in third enrollment subjects there was an increase in the percentage of approval of 8.3%, since in the academic period September 2019-February 2020 there was an approval rate of 25% and in the academic period June-October 2020 a rate of approval of 33.3%, which makes it possible to assert that the implementation of the process is positive, managing to improve student approval levels in the career.

The diagramming of the process through the Bizagi Modeler software allowed to simulate the execution of the academic tutoring process, to know the activities to be executed from start to finish, as well as the documents among which are formats and records to be presented throughout the process.

The proposed academic tutoring process allowed permanent monitoring of the activities to be carried out throughout the process, it is for this reason that the values presented in the efficiency indicator show an increase in tutorials carried out versus planned tutorials, in The academic period September 2019-February 2020 did not have said process, for which a percentage of 81.81% was obtained in the execution of planned practices, and in the academic period June-October 2020 a percentage of 93.75% denoting improvement in planned activities.

For the implementation of the processes within the career, it is necessary to have the support and contribution of students, teachers and authorities since they are all part of the process and generate the different activities, which allows good progress and progress.

Once the academic tutoring process has been implemented, an improvement plan has been drawn up which is recommended to be implemented in order to improve the activities carried out in it.

References

1. UNESCO. https://unesdoc.unesco.org/ark:/48223/pf0000113602_spa
2. Burton, R.: El Sistema de Educación Superior Una Visión Comparativa de la Organización Académica, México (1991)
3. Escobar, V.: Estudio sobre la Deserción y Repitencia en la Educación Superior en Panamá, Panamá (2005)
4. Zambrano, G.: Análisis de la deserción estudiantil en las universidades del ecuador y américa latina. Pertinencia Académica **8**, 1–28 (2018)
5. Peñalosa, E.: Reflexiones, Análisis y Experiencias sobre la Tutoría en la Educación Media Superior. Ediciones del Lirio, México (2017)
6. Gonzales, M., Mendoza, P., Arreola, A., Lozano, G., Flores, R., Rodríguez, P.: Tutorías en el Sistema de Educación Media Superior de la Universidad de Guadalajara. Editorial Universitaria, México (2017)
7. Secretaria de Educación Superior, Ciencia, Tecnología, Innovación y Saberes Ancestrales. https://siau.senescyt.gob.ec/estadisticas-de-educacion-superior-ciencia-tecnologia-e-innovacion/?doing_wp_cron=1612319212.5156009197235107421875
8. Instituto Nacional de Estadística y Censos. https://www.ecuadorencifras.gob.ec/educacion
9. Secretaria Central de ISO.: Norma Internacional ISO 9001:2015. Ginebra (2015)
10. Hitpass, B.: Business Process Management Fundamentos y Conceptos de Implementación. Editorial BHH, Chile (2017)
11. Vera, I.: Análisis y Diseño de la Fase de Planificación del Proceso de Titulación para la Carrera de Sistemas de la Información de la Facultad de Ingeniería Industrial de la Universidad de Guayaquil. Universidad de Guayaquil, Guayaquil (2018)
12. Presidencia de la República. https://www.ces.gob.ec/documentos/Normativa/LOES.pdf
13. Consejo de Educación Superior. https://www.ces.gob.ec/lotaip/2018/Enero/Anexos%20Procu/An-lit-a2-Reglamento%20de%20R%C3%A9gimen%20Acad%C3%A9mico.pdf
14. Consejo de Educación Superior. https://www.ces.gob.ec/lotaip/Anexos%20Generales/a3/Reformas_febrero_2020/REGLAMENTO%20DE%20CARRERA%20Y%20ESCALAFON%20DEL%20PROFESOR%20DE%20EDUCACION%20SUPERIOR.pdf
15. Negrete, K.P., Yépez, J., Maya-Olalla, E., Naranjo-Toro, M., Caraguay-Procel, J.: Quality at the university based on process management: design and implementation of the quality management system under ISO 9001:2015 standard applied to the industrial engineering degree of the northern technical university. In: Basantes-Andrade, A., Naranjo-Toro, M., Zambrano Vizuete, M., Botto-Tobar, M. (eds.) TSIE 2019. AISC, vol. 1110, pp. 183–192. Springer, Cham (2020). https://doi.org/10.1007/978-3-030-37221-7_16
16. Nieto, G.: La función tutorial de la universidad en el actual contexto de la educación superior. Interuniversitaria de Formación del Profesorado **22**, 21–48 (2008)
17. García, M.: Un proyecto de mejora en la orientación universitaria. Investig. Educ. **17**, 401–406 (1999)
18. Álvarez, P.: Los planes institucionales de tutoría y el desarrollo de competencias en el marco del espacio Europeo de Educación Superior. Perfiles Educativos **34**, 28–45 (2012)
19. El Telégrafo. https://www.expreso.ec/guayaquil/desercion-universitaria-paisalcanza-26-1456.htm
20. Tejeda, M.: La tutoría académica en el proceso de formación docente. Ciencias Humanas y Sociales **32**, 880–881 (2016)
21. Rúe, J.: Equipos Docentes y Nuevas Identidades Académicas, España (2010)

Multiclassification Models Applied to Medical Diagnoses Registered by the Ministry of Public Health in Ecuador

Oscar J. Alejo Machado[1(✉)], Maikel Yelandi Leyva Vázquez[2], Víctor G. Gómez Rodríguez[1], Tatiana Tapia Bastidas[1], and Luis Alberto Alzate Peralta[1]

[1] Instituto Superior Universitario Bolivariano de Tecnología, Guayaquil, Ecuador
oalejo@bolivariano.edu.ec
[2] Universidad Regional Autónoma de los Andes (UNIANDES), Babahoyo, Los Ríos, Ecuador

Abstract. The database of the Ministry of Public Health of Ecuador contains valuable information that can potentially be used to learn about the performance of the application of procedures concerning medical diagnoses. Considering that each element of the database corresponding to a patient is classified according to multiple labels, multilabel classification methods are applied in this research. The calculations are made using MEKA, a software that supports the implementation of this kind of method. The goal of this paper is comparing the different multilabel classification methods to select the most appropriate one(s) in the context of Ecuadorian public health, particularly in diagnosis, and to obtain rules to associate every diagnosis with a corresponding procedure.

Keywords: Multiclassification model · Multilabel · Medical diagnosis

1 Introduction

In automated learning, multilabel classification is one of the variants of classification problems. Each example or case is assigned multiple labels for this type of classification, see [1]. The classification task encompasses predicting the class or label to which a given instance corresponds, other than regression that seeks the prediction of a numeric value [2].

Today, multilabel classification has received bigger attention in different scenarios and application contexts within daily, business, educational, economic and political life. For example, in learning to rank, in the video indexing domain, in functional genomics [3].

Generally, the problem of classification as a concept is aimed at determining, given the instance of a certain data set, to which class or label it belongs. It can be binary or ordinal classification, where it can also involve probability prediction on two or more labels. It is an extension of single-label classification, where classes are not mutually exclusive. Each example can be allocated to several classes simultaneously [4] (Fig. 1).

M. Zambrano Vizuete et al. (Eds.): CI3 2021, LNNS 511, pp. 133–141, 2022.
https://doi.org/10.1007/978-3-031-11438-0_11

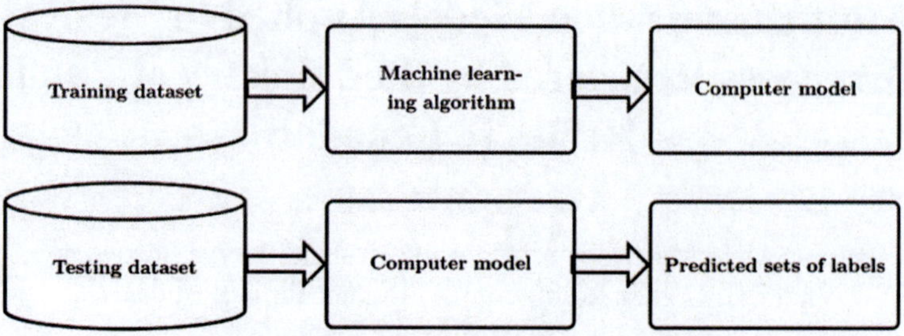

Fig. 1. Overview of machine learning system. Source: [4]

Multilabel classification is a generality of multi-class classification, which consists of categorizing an example with a label when there are more than two possible classes: in the multilabel problem, there are no restrictions on how many labels are assigned to the same case. The multilabel classification appeared with the problem of text categorization where each document could belong to different topics simultaneously and has been generalized to different areas, including the processing of structured, semi-structured, or unstructured data, see [5–8].

Multilabel classification analyzes within multi-label learning aims to develop models capable of labeling instances with sets of classes that are not mutually exclusive, therefore, there may be instances to which more than one label is associated at the same time. In this multilabel classification scenario, the labels associated with each instance are binary, the value 1 represents membership and the value 0 does not. Currently, there is great interest in this branch of the classification as it offers solutions to problems related to health, politics, economics and all types of research that can be modeled with such characteristics [9] (Fig. 2).

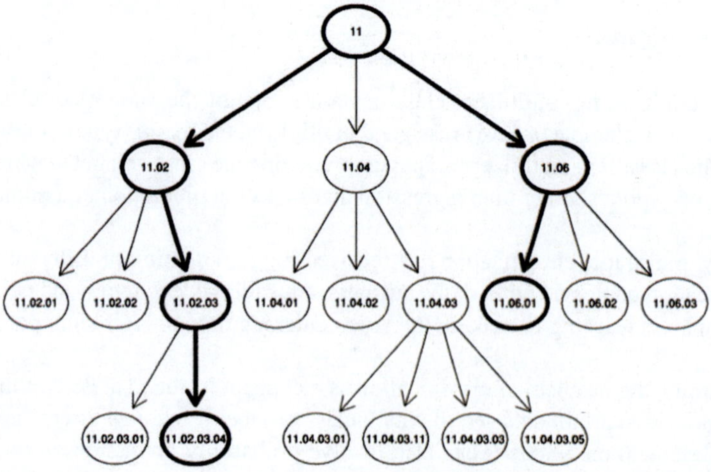

Fig. 2. An example of hierarchical multilabel classification networks. Source: [10]

Multilabel classification is the problem of finding a model that allows recognizing given a set of vectors or inputs represented by X, a binary vector Y (assigning a value 0 or 1 for each element or label of Y).

There are multiple methods to solve multi-tag assignment problems. In [11] they are classified into two fundamental categories: adaptation algorithms and problem transformation methods. The first of these categories generalizes adaptation methods to directly address the problem of multi-labeling, where well-known methods such as decision trees, classification rules, or artificial neural networks are adapted. While the problem transformation methods transform the multi-labeling problem into one or more unique labeling problems, then solve these by the methods used for this type of simpler problem.

Among the methods used to solve these problems can be listed: Binary Relevance (BR), Classifier Chains (CC), the one based on Back-Propagation Neural Network (BPNN), Multi-label k Neighborhood Nearest (MLkNN), the one based on Back-Propagation Multi-label (BPMLL) and the one based on the RAkEL method, which constitutes a substantial improvement to the method known as Label Powerset (LP), especially in domains with many labels, see [11–20].

The purpose of this paper is the Multivariable Study of the behavior of the patterns of prevalence and incidence of morbidity in Ecuador based on data mining techniques, especially the behavior study in the Province of Guayas, Ecuador.

The particular problem is the search for behavior patterns or rules of inference of procedures based on diagnoses and multiclassification models. For this, there is a database stored by the Ministry of Public Health of Ecuador during the years 2016 and 2017.

This study is of great importance to improve the quality of patient care in the public health system of Ecuador, especially to improve the application of procedures in medical diagnosis so that they improve in terms of medical efficacy and efficiency of material resources.

Additionally, the different methods of solving this problem are compared, and it is recommended which of them and how they should be applied in this context.

This article consists of a section of Preliminary Concepts where a summary of the main concepts and methods within the multilabel classification is made. The following section is devoted to studying the databases of the Ministry of Public Health of Ecuador, regarding the multilabel classification in the application of diagnostic procedures. Finally the last section contains the conclusions.

2 Preliminary Concepts

In the current section, we explain the problem of multilabel classification and some theoretical solutions to solve it.

Multilabel learning is related to learning with examples, where each one is associated with multiple labels, which can belong to a predefined set of labels. Multilabel classification's objective is to construct a predictive model that provides a list of relevant labels for classifying new examples given previously, see [13].

This problem will formally consist of the following elements:

A sample space χ consisting of tuples of data type values (boolean, discrete or continuous).

Formally, $\mathbf{x_i} \in \chi$, $\mathbf{x_i} = (x_{i1}, x_{i2},\ldots, x_{iD})$ where D is the size of the tuple or the number of attributes described.

A *space of labels* $\mathcal{L} = \{\lambda_1, \lambda_2,\ldots, \lambda_Q\}$ that consists is a set of Q discrete variables.

A set of examples E, where each example consists of two tuples, one for the example and one for the label. That is, $E = \{(\boldsymbol{x_i}, y_i)|\boldsymbol{x_i} \in \chi, y_i \in \mathcal{L}, 1 \leq i \leq N\}$ and N is the number of examples.

A *quality criterion* that rewards models with high predictive accuracy and low complexity.

The goal is finding a function h: $\chi \rightarrow 2^{\mathcal{L}}$ such that h maximizes q. Where $2^{\mathcal{L}}$ denotes the power set of the label set.

In this article, algorithms contained in MEKA will be applied to the given database. MEKA is an open-source framework programmed in Java with interfaces that facilitate practical applications in evaluation metrics and tools for conducting experiments and developing multilabel classifications (see [1, 21].

The binary relevance method (BR) is to apply a one-to-all strategy to convert the multilabel problem into several binary classification problems. This method is closely related to the *Classifier Chain* (CC) method, where Q binary classifiers are linked through a chain.

The characteristic space of each link in the chain extends to a 0/1 association of all previous links. Therefore, the prediction of the relevant tags in the CC method is the same as in the BR method.

The *Label power-set* (LP) method combines complete sets of labels into unique labels to form problems for a single label. That way the problem is simplified to a single label. This algorithm has the disadvantage that the space of subsets of possible labels can be very large, so it is required to apply a pruning method that considers only those labels that occur more times with respect to a threshold value.

The result of the *Multilabel k-nearest-neighbors* method is the same as in the traditional k-NN algorithm, the difference consists in determining the set of test labels.

The *RAndom k labEL* sets (RAkEL) method starts from small random sets of labels to create each base of the classifier. It learns from a unique label classifier for predicting each element in the power set. The computational complexity of the algorithm could be reduced by applying pruning methods. This method has proved efficient compared to others.

The use popular back-propagation algorithm used in Artificial Neural Networks, adapted to multi-labeling problems, is the basis of the methods of back-propagation neural networks (BPNN) and back-propagation neural networks Multi-Label Learning (BPMLL).

Basically, these algorithms replace the error functions with a new function to capture the characteristics of multilabel learning. Thus, labels that match the example are given a greater evaluation than those that do not match.

Others are based on the Multilabel C4.5 that adapt the C4.5 algorithm within the framework of multi-labeling. In this case entropy is used with the following equation:

$$entrop'ia(E) = -\sum_{i=1}^{N} (p(c_i)log(p(c_i)) + q(c_i)log(q(c_i))) \quad (1)$$

where E is a set of examples, $p(c_i)$ is the relative frequency of the class label c_i and $q(c_i) = 1 - p(c_i)$.

3 Results

The study carried out in this investigation is presented below.

For the processing of the files and obtaining representative models, the original database, attributes and instances per year were used, according to the disposition of the Table 1

Table 1. General characteristics of the data

Year	#Original attributes	# Selected attributes	# Original instances	# Selected instances
2016	92	48	3.428.706	3.421.433
2017	78	48	3.638.168	3.636.305

Source: the authors

In the transformation and structuring of the data, the attributes related to the diseases were grouped and coded according to ICD-10 [22], the acronym of the International Classification of Diseases, 10th edition. Especially in the investigation, the classification code AB to ND was used in in the coding for files of the Automated from Daily Registry of Attention and Outpatient Consultations or simply the 2013–2017 MSP RDACAA files [23].

Where CIE is the class attribute represented by the 22 labels that represent the groups of diseases collected in the case bases corresponding to the years 2016 and 2017.

The case bases are transformed to solve the multilabel problem posed. Each base is given in ARFF format, which is required by MEKA, see [1, 21].

Each case is made up of 22 binary attributes that represent feasible labels and that constitute the classes associated with groups of diseases presented by a patient and that identify the binary vector Y, plus 27 or 26 attributes of the nominal or symbolic type that represent the predictive features that characterize the case, to which the X input is associated.

The tool used for the evaluation is MEKA Release 1.9. Meka is very similar to the Weka framework, provided to support the development, execution and evaluation of multilabel and multi-instance classifiers, see [21, 24].

The evaluation method used is to partition the case base into two files, one training and one test, using 66% of the case base for training and 34% as evidence.

The measure suggested by several authors is *Hamming Loss* to measure the effectiveness of multiclassification [13, 25]:

$$HL(y_{ij}, z_{ij}) = \frac{1}{|\mathrm{N}||\mathrm{L}|} \sum_{\mathrm{i}=1}^{\mathrm{N}} \sum_{\mathrm{j}=1}^{\mathrm{L}} XOR(y_{ij}, z_{ij}) \tag{2}$$

where $\mathrm{y_{ij}}$ represents the actual values of the label, and $\mathrm{z_{ij}}$ the value that was obtained in the classification.

The following methods were used to compare the results of multiclassification: *Binary Relevance* (BR) and *Classifier Chains* (CC), BPNN and embedded implementations of the MULAN library, see [26, 27]. From the latter, the metaclassifiers based on *Multilabel and Nearest Neighbor* known as MLkNN (MULAN MLkNN), the one based on *Back-Propagation Multilabel* (MULAN BPMLL) and the one based on the *RAkEL method* (MULAN RAkEL1), which constitutes a substantial improvement to the *Label Powerset* (LP) method especially in domains with many labels. All these methods appear in MEKA and are suggested by [13].

The results obtained from applying the different methods for years are detailed below.

The case base comprises 3,421,433 cases described by 48 attributes, of which 22 correspond to the labels or classes that must be obtained.

2,258,145 cases were worked to train and test examples 1,163,288. The measure used was *Hamming Loss* [25].

To compare the results, the methods applied are: BR, CC, BPNN and MULAN RAkEL1, taking advantage of the implementations in MEKA, it was based on the variants of better results in precision and computational time according to all the experiments performed with the 2013 case base that has 1.8MM of examples.

Table 2. Comparison between the methods applied in terms of hamming loss function and computational time

Employed method	Hamming loss	Computational time (seg.)
BR_J48	**0,036**	**1.322,259**
BR_NB	0,048	**482,089**
CC_J48	**0,036**	29.455,427
CC_NB	0,059	27.881,879
BPNN_J48	0,062	5.159,619
MULAN_RAKEL_J48	**0,034**	**3.287,611**
MULAN_RAKEL_NB	0,045	5.113,792

Source: the authors

The best precision methods are: MULAN_RAKEL, CC and BR all using the J48 as a base classifier (See the ones highlighted in bold in Table 2). As regards to runtime, BR_NB, BR_J48 and MULAN_RAKEL_J48 are better.

They match both in precision and in time BR-J48 and MULAN_RAKEL_J48 (in the gray rows), but not CC_J48 which results one of the most computationally expensive but very suitable for interpreting results.

To compare the results, the BR, CC, BPNN and MULAN RAkEL1 methods are used, taking advantage of the implementations in MEKA, based on the variants of better results in precision and computational time according to all the experiments performed with the 2013 case base by Dr. María Matilde García who has 1.8MM of examples considering that this base doubles the previous one.

Table 3. Comparison between the methods applied in terms of the hamming loss function and computational time on a 2017 database.

Employed Method	Hamming loss	Computational time (seg.)
BR_J48	**0,039**	12.431,486
BR_NB	0,054	**484,018**
CC_J48	**0,039**	54.698,994
CC_NB	0,051	38.789,166
BPNN_J48	0,089	**6.229,901**
MULAN_RAKEL_J48	**0,037**	29.517,272
MULAN_RAKEL_NB	0,049	**5.676,212**

Source: the Authors

The methods with the highest levels of precision were: MULAN_RAKEL, BR and CC all using the J48 as a base classifier (See Table 3 for the values in bold).

Concerning runtime, BR_NB, MULAN_RAKEL_NB and BPNN_J48 are better, which are highlighted in red.

For this base, there are no methods that coincide in both aspects, they are the most complex computationally the Classifier Chains (CC), which have been maintained in the analysis given the results they provide, both in accuracy and interpretability to understand the results of the classification.

When the cause of the mismatch of the results in this 2017 base is analyzed with respect to the previous bases, it was evident that the feature Nnro_id -operative_ unit was numerical and as such, was treated. Given this situation, the feature was transformed to nominal, and under this new representation, the results obtained were:

By the results of the Table 4 it can be seen that once again the best results are achieved with the BR and MULAN_RAKEL methods using J48 as a base classifier, both in precision and in runtime.

Table 4. Comparison between the methods applied in terms of the Hamming Loss function and computational time of the 2017 database, with data type conversion

Employed method	Hamming loss	Computational time (seg.)
BR_J48	0,039	1.174,878
MULAN_RAKEL_J48	0,036	1.735,561

Source: the Authors

4 Conclusions

In this paper, authors studied the behavior of applying procedures based on medical diagnoses, according to the Ministry of Public Health databases in Ecuador i For this purpose, they were compared different multilabel classification methods programmed in MEKA.

The best results methods in both precision and runtime turn out to be Binary Relevance and MULAN RakEL using J48 as a base classifier in both cases. Although the results with the classifiers of chains of the CC type are expensive computationally, they are very suitable to interpret the results and throw a decision tree for the classification leaving in the first levels the features that most influence it.

Given the complexity of the case bases, higher preprocessing levels are required that improve the computational cost that the transformation methods entail, ranging from consistency analysis to processes of feature and instance reduction.

For future work with new historical bases, the effect of using SPARK to reduce the time to obtain results must be analyzed.

Grounded on the results obtained, it is recommended to create an interface associated with the RDACAA that allows loading and configuring such multi-classifiers with the MEKA library. Thus, offering doctors and specialists to perform behavioral analysis and prediction in real-time from of the data entered from a patient.

References

1. Modi, H., Panchal, M.: Experimental comparison of different problemtransformation methods for multi-label classificationusing MEKA. Int. J. Comput. Appl. **59**, 10–15 (2012)
2. Brownlee, J.:Multi-label classification with deep learning. Deep Learning (2020)
3. Yu, Y., Pedrycz, W., Miao, D.: Multilabel classification by exploiting label correlations. Expert Syst. Appl. **41**, 2989–3004 (2014)
4. Kanj, S.: "Learning Methods for Multi-Label Classification," Thèse présentée pour l'obtention du grade de Docteur de l'UTC, Technologie de l'Information et des Systèmes. Université de Technologie Compiégne, Francia (2017)
5. Zhang, M.L., Zhou, Z.H.: Multilabel neural networks with applications to functional genomics and text categorization. IEEE Trans. Knowl. Data Eng. **18**, 1338–1351 (2006)
6. Read, J., Pfahringer, B., Holmes, G.: Multilabel classification using ensembles of pruned sets. In: 2008 Eighth IEEE International Conference on Data Mining, pp. 995–1000 (2008)
7. Liu, S.M., Chen, J.-H.: A multilabel classification based approach for sentiment classification. Expert Syst. Appl. **42**, 1083–1093 (2015)

8. Koyejo, O., Natarajan, N., Ravikumar, P., Dhillon, I.S.: Consistent multilabel classification. In: NIPS, pp. 3321–3329 (2015).
9. Gopal, S., Yang, Y.: Multilabel classification with meta-level features. In: Proceedings of the 33rd International ACM SIGIR Conference on Research and Development in Information Retrieval, pp. 315–322 (2010)
10. Wehrmann, J., Cerri, R., Barros, R.: Hierarchical multilabel classification networks. In: International Conference on Machine Learning, pp. 5075–5084 (2018)
11. Tsoumakas, G., Katakis, I.: Multi label classification: an overview. Int. J. Data Warehous Min. **3**, 1–13 (2007)
12. Elissef, A., Weston, J.: A kernel method for multi-labelled classification. In: Dietterich, T.G., Becker, S., Ghahramani, Z. (eds.) Advances in Neural Information Processing Systems, vol. 14, pp. 681–687 (2002)
13. Madjarov, G., Kocev, D., Gjorgjevikj, D., Dzeroski, S.: An extensive experimental comparison of methods for multilabel learning. Pattern Recogn. **45**, 3084–3104 (2012)
14. Pestian, J.P., et al.: A shared task involving multilabel classification of clinical free text. In: Presented at the Workshop on BioNLP 2007: Biological, Translational, and Clinical Language Processing (2007)
15. Read, J., Pfahringer, B., Holmes, G., Frank, E.: Classifier chains for multi-label classification. Mach. Learn. J. **85**, 333–359 (2011)
16. Read, J., Pfahringer, B., Holmes, G.: Multilabel classification using ensembles of pruned sets. In: 8th IEEE International Conference on Data Mining, pp. 995–1000 (2008)
17. Read, J., Perez-Cruz, F.: Deep learning for multilabel classification. arXiv preprint arXiv: 1502.05988 (2014)
18. Tsoumakas, G., Katakis, I., Vlahavas, I.: Random k-label set for multilabel classification. IEEE Trans. Pattern Anal. Mach. Learn. **23**, 1079–108 (2007)
19. Zaragoza, J.C., Sucar, E., Morales, E., Bielza, C., Larranaga, P.: Bayesian chain classifiers for multidimensional classification. In: Presented at the Twenty-Second International Joint Conference on Artificial Intelligence (2011)
20. Zhang, M.L., Zhou, Z.H.: ML-KNN: a lazy learning approach to multilabel learning. Pattern Recogn. **40**, 2038–2048 (2007)
21. Read, J., Reutemann, P., Pfahringer, B., Holmes, G.: Meka: a multilabel/multi-target extension to Weka. J. Mach. Learn. Res. **17**, 667–671 (2016)
22. eCIEMapsv3.3.8.CIE-10 Disponible en: https://eciemaps.mscbs.gob.es/ecieMaps/browser/index_10_mc.html en español, Consultado: 20-agosto-2019
23. Páez-Medina, G.A.: Estrategias para el mejoramiento de la calidad del Registro Diario Automatizado de Atenciones y Consultas Ambulatorias (RDACAA) en la unidad operativa de salud augusto egas, provincia de santo domingo de los tsáchilas. Tesis de Grado Tesis de Grado, Facultad de Ciencias Médicas, Universidad Regional Autónoma de Los Andes, Ambato, Ecuador (2017)
24. Bouckaert, R.R., et al.: Weka manual for version 3-6-11 (2014)
25. Norouzi, M., Fleet, D.J., Salakhutdinov, R.R.: Hamming distance metric learning. In: Advances in Neural Information Processing Systems, pp. 1061–1069 (2012)
26. Tsoumakas, G., Spyromitros-Xioufis, E., Vilcek, J., Vlahavas, I.: Mulan: a java library for multilabel learning. J. Mach. Learn. Res. **12**, 2411–2414 (2011)
27. Mulan: Disponible en: htpps://Mulan.sourceforge.net, Consultado: 20-agosto-2019

Emotional Intelligence Within the Formation of Students of the Comprehensive Child Development Career

G. Enríquez(✉), V. Molina-Ipiales, G. Duque, A. Soto, and A. Puga

Instituto Superior Tecnológico José Chiriboga Grijalva, Ibarra 100110, Imbabura, Ecuador
rgenriquez@tecnologicoitca.edu.ec

Abstract. Emotional Intelligence at an educational level plays an important role in the academic formation of students in the Comprehensive Child Development career, since it allows us to respond to challenges on a personal, professional and social level. The responsibility acquired by the teacher in initial education creates the need to train professionals with theoretical and practical knowledge regarding the management and control of emotions. In this research a socio-demographic survey was used, an observation sheet to evaluate empathy and emotional control, the TMMS-24 test for the analysis and description of emotional intelligence, attention, clarity and emotional repair, as well as an interview with open questions addressed to the Career Coordination in order to know the feasibility of including emotional competencies within the curriculum of this career. The results obtained in the application of the instruments denote approximately fifty percent of development of Emotional Intelligence skills in the parameters analyzed, this result being a green light for the inclusion of content related to the subject, in the curriculum of the training of future teachers.

Keywords: Intelligence · Emotions · Emotional intelligence · Competencies · Empathy · Emotional attention · Emotional clarity · Emotional repair

1 Introduction

The influence of emotions in education, since the end of the 20th century, has been of interest to human beings. Many professionals who are immersed in the world of education have talked about the importance of emotions in the comprehensive development of teachers. When mentioning the influence of emotions, it is unavoidable to refer to the development of Emotional Intelligence (EI) considered as: the individual's ability to identify and understand their own emotions and their abilities to interact with others [5], the ability to perceive, assimilate, understand and regulate their own emotions and those of others, promoting emotional and intellectual growth; in addition, also stimulating the ability to reach and lead people and to act consciously in relationships with human beings [17].

The renowned psychologist and journalist Daniel Goleman is one of the main scholars of EI and alludes that, genetic inheritance endows a series of emotional traits that

M. Zambrano Vizuete et al. (Eds.): CI3 2021, LNNS 511, pp. 142–152, 2022.
https://doi.org/10.1007/978-3-031-11438-0_12

determine temperament, this statement should not be taken as an irreversible state that marks the life of the human being, but rather it should be considered that EI can be developed as a fundamental skill [6]. In this same sense, Howard Gardner in his theory of multiple intelligences, highlights the existence of interpersonal and intrapersonal intelligence (easily teachable) which refer to a social intelligence that gradually gave rise to the theory of EI [11].

In 1997, Mayer and Salovey constructed a model of EI in which they initially focused their studies of the construct on four dimensions: perception, facilitation, understanding and emotional regulation [7], later, these same authors emphasize that Self-perceived Emotional Intelligence (SPEI) consists of the beliefs that people have towards their emotional experience and the awareness of their own emotional abilities, deriving in the current model that for SPEI includes three dimensions that are: Emotional Attention (EA), Emotional Clarity (EC) and Emotional Repair (ER). EA refers to the degree to which people believe they pay attention to their emotions and feelings; that is, the ability to identify and recognize one's own feelings, as well as the physiological and cognitive states and sensations they entail; it involves paying attention to and accurately decoding emotional signals. EC refers to how people believe they perceive their emotions; that is, whether they experience their feelings clearly and understand how they feel, and involves breaking down the broad and complex repertoire of emotional signals, discriminating, labeling emotions, and recognizing into which categories feelings are grouped. ER alludes to people's belief in their ability to interrupt and regulate negative emotional states as well as prolong positive emotional states; this dimension includes the ability to be receptive to feelings and to reflect on them in order to benefit from or discard accompanying information based on its usefulness [7].

Having EI developed can increase the number of tasks that workers are expected to perform in the workplace, in addition to maintaining appropriate terms and relationships with the people around them; since feeling your feelings and those of others helps to improve interpersonal communication and therefore this leads to an increase in performance and transparency in the workplace [5].

The advantages of working on EI are multiple, such as avoiding conflicts having previously recognized moods, preventing negative thoughts from growing as a "snow ball", enjoying the achievement of small goals, being aware that violence is not a solution for irritation, using humor to face uncomfortable situations, seeing situations that provoke disturbing emotions differently, and seeking help when one's own resources are not enough [3]. In general terms, most studies on this subject show that emotionally intelligent people present an adequate psychological and social adjustment that is manifested by the presence of a series of indicators; among them, we can mention adaptive responses to negative life situations, low negative emotionality and alexithymia; low scores in psychological symptomatology (especially anxiety and depression); adequate levels of life satisfaction, empathy and optimism; appropriate interpersonal relationships, performance, and academic satisfaction, [4, 7]; in addition to adequate coping strategies to everyday stressful situations such as problem-focused coping, seeking social support, acceptance of emotions, and adequate management of rejection [14].

Teaching, as one of the most socially influential professions, represents in turn, the most evaluated and also observable by the constant interaction that maintains both with the students in their charge, as well as with their environment; this induces and requires a behavior that facilitates the development of empathy, optimism, satisfactory interpersonal relationships and adequate levels of life satisfaction that contribute to student progress.

Emotional education in every human being, and especially in teachers, allows us to respond to the new challenges posed by the society of the 21st century: to educate to enhance the critical capacity in a knowledge society, ability to tolerate frustration, postpone the reward and immediacy and, above all, to guide future adults towards competence understood as adaptation, improvement and intra and interpersonal realization and not to face everyday life as a mere search for rivalry and power [15].

The teacher is historically recognized as a reflective subject who makes decisions; it is a person who experiences teaching-learning situations, who gives personal meaning to what he/she teaches through reflection [1]. Teachers largely act in consequence of what they think; therefore, the idea of exploring teachers' performance and their conceptions is to recognize and understand a related set of beliefs, which are unfailingly related to their pedagogical praxis. These beliefs and approaches to teaching and learning are perhaps more important than their own theoretical knowledge [1]. It can even be affirmed that these beliefs and thoughts have a greater influence than the knowledge that teachers possess at the moment of planning their classes, the decisions they make and their performance in the classroom [1].

If the mission of education is to prepare for life, the school as a whole, considered not only as basic education, must fulfill the social task of forming the necessary qualities that allow each person to meet the demands of his or her time, and it is in this sense that conceptions about EI occupy a relevant place, since they are associated with the guarantees for good performance [3].

In a research conducted by Cejudo and López, in 2017 [4], with 196 early childhood and primary education teachers, the importance of EI dimensions was evaluated according to the educational stage in which they work and, according to their opinion, there are significant differences in the assessment of the degree of importance of the most necessary personal characteristics for the optimal performance of a teacher's work. In general, Early Childhood Education teachers recognize the importance of personal characteristics related to EI and it is possible that these results may be due, in the first place, to the importance in Early Childhood Education of the development of students' affective capacities, the acquisition of elementary guidelines for coexistence, the improvement of interpersonal relationship skills and the importance of fostering a positive and balanced self-image in students; secondly, these results are due to the educational needs that the students of Early Child-hood Education present according to their psycho-evolutionary characteristics. In addition to the above mentioned regarding the importance of EI development in early childhood education teachers, it can be highlighted that novice teachers with low levels of EI report a higher intention to leave teaching than their peers with higher levels of EI; being the most important dimensions in terms of occupational commitment, interpersonal perception and emotional assimilation [12].

Students, who are currently training to become teachers, have to their credit a great social responsibility in their future classrooms, since studies show that a teacher with developed EI encourages in his students: critical thinking, emotion management, motivation, commitment to learning achievements, interest, enjoyment of the learning process, comfort and positive attitudes towards the subject; besides increasing the students' perceptual skills [10]. While a teacher with under-developed EI carries problems in his teaching environment such as anxiety, frustration, discomfort, confusion, aggression and other emotional difficulties [10]. Therefore, it can be inferred that the social and personal skills that are part of EI are of vital importance when training highly competent teachers.

In this sense, Barbón and Cepeda, in 2017, after being impacted by a research conducted with medical students, report that, the balance and mastery of the empathic capacity are very important factors in the training of medicine professionals, and it should be considered that one of them is the teaching influence or the models that the student observes in his training stage, where the acquisition of cognitive-behavioral competencies that allow him to adequately channel all those complex situations in the professional occurrence is sought; this is not very different or less significant in the training of teaching staff; therefore, the empathic condition is one of the main factors to develop in higher education and that lies in the training of professionals not only with scientific skills, but with personal skills that position them as a reference capable of implementing their attitudes, aptitudes, skills and psychological resources to solve with excellence any situation that requires effective results. In this sense, competencies related to EI should be taken into account for the implementation of teaching-learning strategies that promote their formation from the first years of study [2].

A case study conducted at the University of Valencia-Spain highlights that work stress and emotional discomfort in teachers are considered a serious concern in the educational context and the same research shows the benefits of emotional skills on burnout and psychological problems of teachers included in the study sample [16].

In the Comprehensive Child Development career of the Instituto Superior TecnOlógico José Chiriboga Grijalva, the problematic situation lies in the fact that emotional competencies are not evidenced in the teaching-learning process of students, nor is EI content integrated in the curricula that could be considered fundamental in the training of future teachers; therefore, empathy skills, emotional control, motivation, self-regulation and social skills are not developed in a conscious and induced manner, which could have an unfavorable impact on their professional training.

This research intends to contribute to the development of EI skills in students of the Comprehensive Child Development Career, through the insertion in the curriculum of competencies related to the improvement of inter and intrapersonal relationships.

2 Methodology

This research is descriptive and field type. The population of participating students of the Comprehensive Child Development career belong to the Instituto Superior Tecnológico José Chiriboga Grijalva ISTJCHG and is represented by 301 students, 291 women and 10 men and the sample selected for the study was 168 students: 164 women and 4 men; the inclusion criterion was: students of the Comprehensive Child Development career in the last three semesters of the career and the exclusion criterion: students of other careers of the Institute and students of the Comprehensive Child Development career in the first three semesters.

The TMMS-24 test (Trait Meta-Mood Scale), which is a reduced version adapted to Spanish of a similar pre-existing test [17], was used for the analysis and description of EI. The measurement parameters that were analyzed were: EA, EC and ER; where the attention parameter evaluates the ability to attend to feelings adequately, clarity evaluates the adequate understanding of one's own emotional states and repair evaluates the ability to regulate emotional states adequately [17, 7].

An observation form designed as ad hoc was also applied, to evidence two important dimensions of EI, empathy and emotional control, of the students in their respective pre-professional internship placements and in direct contact with children.

The third research instrument used was an interview with open-ended questions, conducted with the career coordinator, regarding the opinion of the EI levels evidenced in the students of this career, the feasibility of training in emotional competencies within the curriculum and the possible addition of emotional intelligence contents to the subjects.

For the analysis of the obtained data, basic descriptive Excel tools were used.

3 Results and Discussion

The data obtained in this document were analyzed, discussed and contrasted.

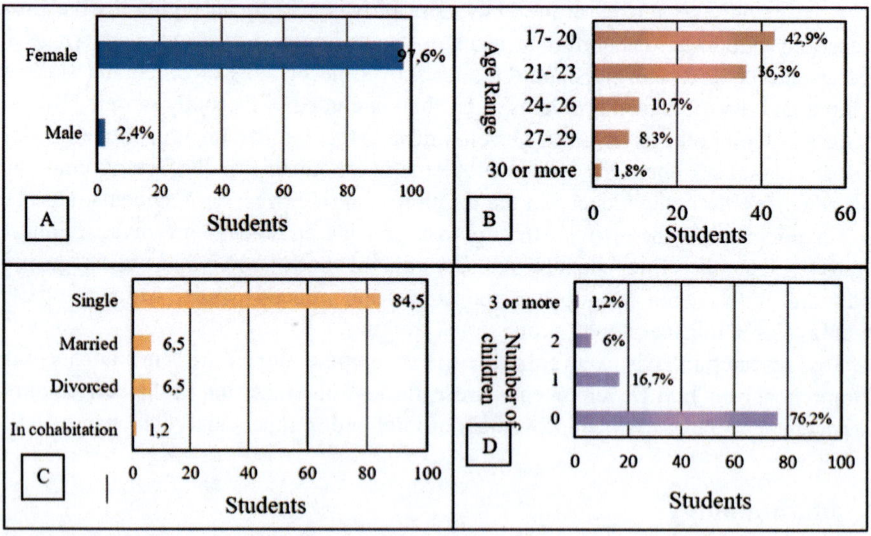

Fig. 1. Socio-demographic variables: (A) gender, (B) age range, (C) marital status, and (D) number of children.

Figure 1 shows the socio-demographic characteristics of the study sample, being mostly female (97.6%), with an age range between 17 and 20 years old (42.9%), single (84.5%) and childless (76.2%). It could be assumed that, since most of the participants in the study were single and childless, they had no previous experience in the recognition and management of children's emotions.

For the measurement of SPEI, the TMMS-24 (Trait Meta-Mood Scale) test was used considering the dimensions: EA, EC and ER (Fig. 2).

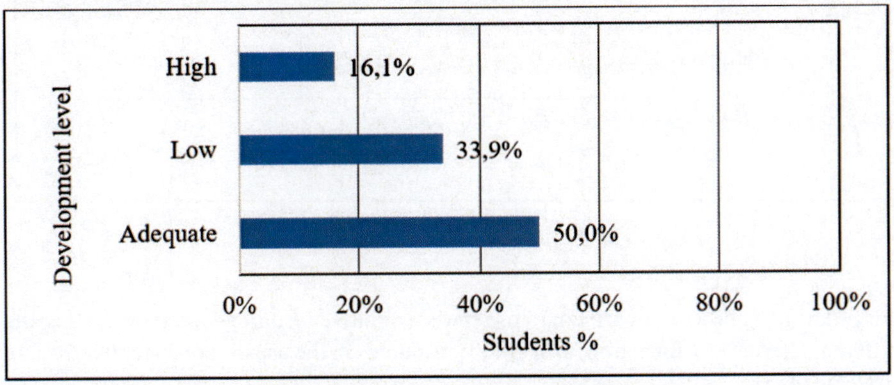

Fig. 2. Emotional attention development

Based on the total of 168 participants who were evaluated with the IEA TMMS-24 test, 84 students (50%) obtained adequate results in EA, 57 students (33.9%) reached an unfavorable result revealing a lack of attention skills to emotions, it should be emphasized that behavior, thinking and emotions are cognitive processes linked to attention, the lack of maturity and adequate development of these processes may be an intervening factor in the result of this item. Finally, 27 (16.1%) of the participating students pay too much attention to their emotions, which is also considered inadequate in this item specifically. 50% of students have correctly developed EC, while the other 50% need to work on EC. Several studies have shown that the attentional processes performed by human beings require a greater involvement of brain structures; as is the case when EC should be an important factor to be addressed within emotional competencies because of the possibility it provides to regulate moods when they are well developed in people [9]. It has been highlighted in publications that excessive attention explains part of the variance in Negative Orientation towards problems and that this fact occurs preferentially in women [13].

Therefore, it can be inferred that possibly the age of the students and the lack of formal EI stimulation could be linked to the results obtained in terms of the development of this item, in which 50% of those evaluated did not obtain adequate results; in addition, this could interfere negatively in the teaching performance in terms of emotional regulation (Fig. 2), the adequate management of conflicts (Fig. 3).

For the EC item, positive results are considered to be those that score with an adequate and high intensity of development, so it can be considered that 59.5% of the evaluated sample has a positive development in terms of EC, unlike 40.5% who need extra stimuli at the moment of clearly perceiving their own emotions. Studies have concluded that people with high EC have a greater Positive Orientation to Problems and a greater tendency to solve them rationally, so that they do not use the avoidance strategy to solve them;

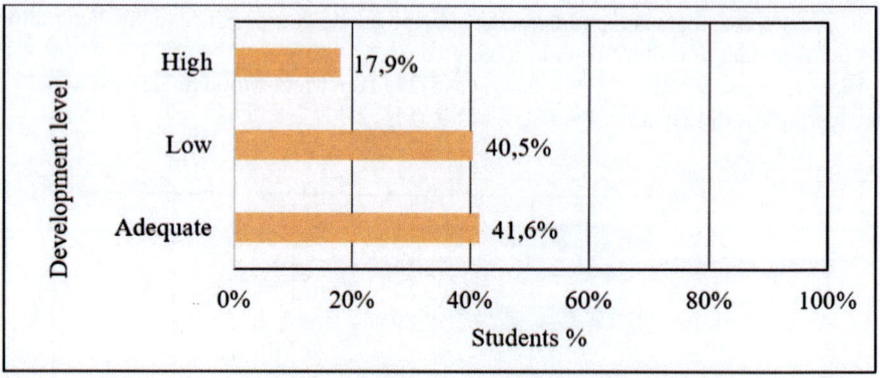

Fig. 3. Emotional clarity development

this emotional ability allows them to use fewer cognitive resources when evaluating the different alternatives for action, using those resources in the use of more adaptive coping strategies [13].

It can be inferred that possibly 40.5% of those evaluated will have difficulties related to the positive orientation of problems, conflict resolution using the avoidance strategy and the use of various cognitive resources when solving them, therefore, it is the task of higher formal education to overcome this type of problems so that they are not carried over to the teaching task in future classrooms.

It is evident that there is a need to establish strategies to increase the levels of ER, revealing that the students participating in the EI test do not have an adequate level of resilience (Fig. 4).

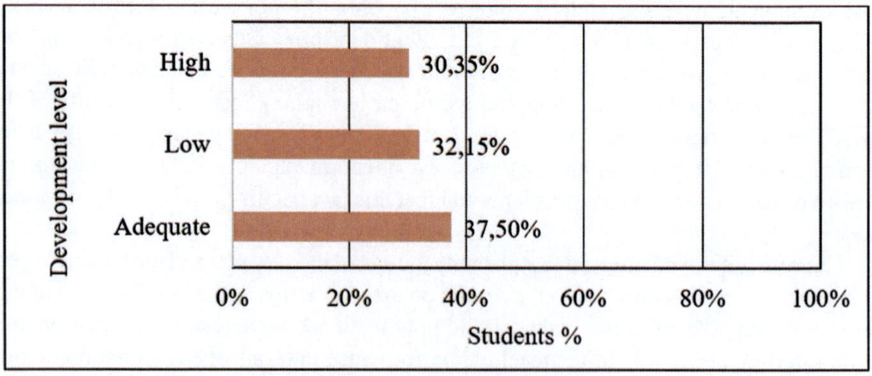

Fig. 4. Emotional repair development

According to the tabulation of the data obtained, 63 (37%) of the students have an adequate ER, 54 (32.15%) of the participants should improve their ER and 51 students (30.35%) have an excellent ER.

In a study conducted by Pena, Extremera and Rey in 2011 [13], it was concluded that in terms of repair the analyzed subjects who scored high levels in ER, had a better management of emotions, better problem-solving strategies being most of the regulation strategies, increasing positive emotions and minimizing negative ones.

Regarding the age of those evaluated, D'Zurilla, Maydeu-Olivares and Kant in 1998, cited by: Pena, Extremera and Rey, 2011 [13], have reported that young people between 17 and 20 years old present lower levels of problem solving than adults whose ages range between 40 and 55 years old and that the level of adaptive strategies in problem solving drops between 60 and 80 years old.

Emotionally intelligent people will find it easier to extrapolate the skills of attention, understanding and repair of emotions to those situations that require problem solving [13].

The low scores in each of the items show that there is a problem in at least 50% of the students regarding the development of EI. It is clear to determine that it is essential to propose alternatives that promote an improvement in the development of EI levels of students considering the remarkable need to establish in them high parameters of empathy, control, repair and clarity of emotions not only as a benefit at an educational level but also at a personal and social level.

Regarding the empathy parameter, with the application of the EI test, an observation form was applied to the students in their respective internship places. The presence or absence of the behaviors of the Child Development students in their interaction with the infants was observed (Fig. 5).

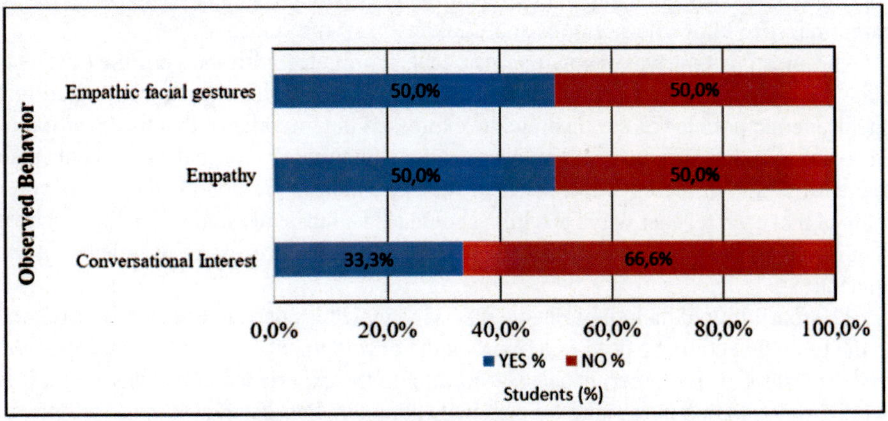

Fig. 5. Observed empathic behavior

33% of the observed students showed interest in the child's conversations, 50% were empathetic in their interactions and 50% showed empathic facial gestures. As can be seen in Fig. 6, regarding the emotional control parameter, it could be observed that 87.5% of the students were happy in their work, 50% of them reflected negative moods during the time of observation, and 66.6% are dynamic and active in relation to the group of children with whom they work.

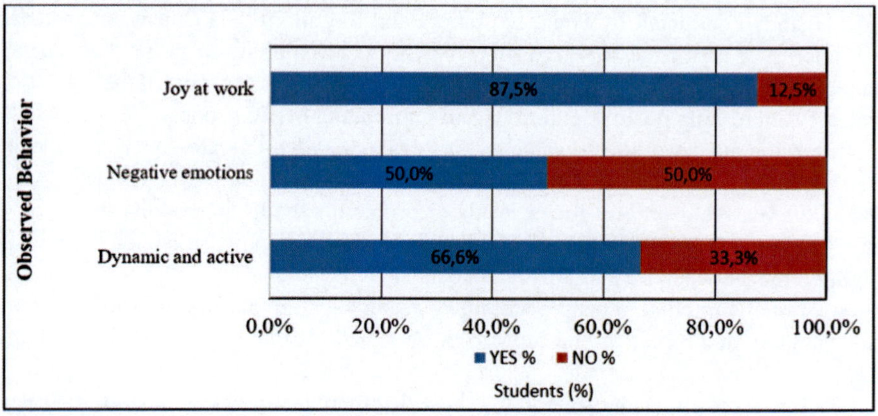

Fig. 6. Observed emotional control behavior

With the results obtained in the observation sheet, it is possible to corroborate the results of the EI test applied, since there is a clear need to apply, with the students of the Comprehensive Child Development career, formal EI stimulation strategies, which contribute to their future labor development, with greater emphasis on the control of emotions, this being one of the most important parameters for the teaching work.

An interview was also conducted with the coordinator of the career which ratifies the information obtained with the application of the test and the observation sheet, which contributed with a significant value to the research conducted, since as authority of the career refers to know the strengths and weaknesses of the students.

The meeting that was maintained for several minutes with the coordination, was quite gratifying since according to his teaching experience and knowing the institutional and academic policies of the Institute, he expressed the importance that the insertion of EI contents in the curriculum would be extremely favorable because it would contribute to the formation of these competencies in students, since at present it had been taken for granted that each teacher while teaching should also educate his students in the practical acquisition of EI skills, in order to guarantee a formative process of excellence in the students.

The feasibility of including EI contents is recognized, since there are subjects related to the topic that could be used as a basis for the development of EI skills, starting in the first semesters of the career; because according to the experience in the classroom, it is very common to find the absence of personal characteristics that denote the development of EI, in addition to the unawareness of the teaching role for the life of their students and the responsibility that falls on a trainer of initial education.

Future professionals who receive training in EI will be able to establish strategies for the control of emotions and interpersonal skills will be enhanced according to the demands required by society.

In the interview conducted with the coordination of the career, the information obtained with the application of the test and the observation sheet was ratified, which added significant value to the research carried out, since as authority of the career refers to knowing the student weaknesses and strengths.

4 Conclusions

The results obtained lead us to conclude that:

The inclusion of contents referring to EI in the curriculum of the Comprehensive Child Development Career is indispensable to achieve transcendental changes in the education of future teachers, in such a way that human beings with principles, values and emotional skills can be formed, which will allow them to develop in society in a harmonious and successful way.

Studies such as this research have shown that the level of development of EI is an indispensable aptitude for professionals in training in the Comprehensive Child Development Career.

It is necessary that the directors of the Comprehensive Child Development career know the importance of adding EI contents in the formation of students competencies.

The indicators of attention, clarity and ER are the fundamental basis of EI and it is necessary to work on the development of these parameters linked to EI to strengthen students skills.

The age of the students and the lack of formal EI stimulation are linked in terms of EI development, in the sample analyzed, which could interfere negatively in teaching performance.

Future professionals who receive training in EI will be able to establish strategies for the control of emotions and interpersonal skills will be enhanced according to the demands required by society.

By having a low development of EI skills, students will have difficulties regarding the positive orientation of problems, conflict resolution using the avoidance strategy and the use of various cognitive resources when solving them.

It is the task of formal higher education to highlight these types of problems so that they are not brought to the teaching task in future classrooms.

Studies conducted by several researchers on the subject of EI emphasize the need to generate a climate of security and positive emotions in the classroom with the unique purpose of facilitating the development and well-being of the individual.

Young people between the ages of 17 and 20 have lower levels of problem solving than adults between the ages of 40 and 55.

Emotionally intelligent people will have an easier time extrapolating the skills of attention, understanding and emotional repair.

References

1. Arancibia, M., Cabero, J., Marín, V.: Análisis factorial de una escala de creencias sobre la enseñanza y su relación con características personales y profesionales de docentes de Educación Superior. Revista Espacios **41**(02) (2020)
2. Barbón, O., Cepeda Astudillo, L., Garcés Viteri, L., Romero Rojas, H.: El pensamiento educativo de Fidel Castro Ruz como formador de valores internacionalistas en los profesionales de la salud. Educación Médica Superior **31**(1), 193–202 (2017)
3. Bello-Dávila, Z., Rionda-Sánchez, H.D., Rodríguez-Pérez, M.E.: La inteligencia emocional y su educación. VARONA **51**, 36–43 (2010)

4. Cejudo, J., López-Delgado, M.: Importancia de la inteligencia emocional en la práctica docente: un estudio con maestros. Psicología educativa **23**(1), 29–36 (2017)
5. Cui, Y.: The role of emotional intelligence in workplace transparency and open communication. Aggress. Violent. Beh. **58**, 101602 (2021)
6. Goleman, D.: Emotional intelligence. Why it can matter more than IQ. Learning, **24**(6), 49–50 (1996)
7. Gonzalez, R., Custodio, J., Abal, F.: Propiedades psicométricas del Trait Meta-Mood Scale-24 en estudiantes universitarios argentinos. Psicogente **23**(44), 1–26 (2020)
8. Hernández Vásquez, R.: La empatía: una necesidad en la educación superior. Educación Médica **20**(1), 190 (2019)
9. León-Domínguez, U., León-Carrión, J.: Modelo neuro funcional de la conciencia: bases neurofisiológicas y cognitivas. Revista de Neurología **69**(4), 159–166 (2019)
10. Li, C., Huang, J., Li, B.: Los efectos predictivos del entorno del aula y el rasgo de inteligencia emocional sobre el disfrute y la ansiedad de una lengua extranjera **96**, 102393 (2021)
11. Mercadé, A.: Los 8 tipos de inteligencia según howard gardner: la teoría de las inteligencias múltiples, transición a la vida adulta y activa. Monografía (2019)
12. Mérida-López, S., Extremera, N.: Cuando la falta de compromiso ocupacional del profesorado novel no es suficiente para explicar la intención de abandono: la inteligencia emocional importa. Revista de Psico Didáctica **25**(1), 52–58 (2020)
13. Pena Garrido, M., Rey, L., Extremera Pacheco, N.: Bienestar personal y laboral en el profesorado de infantil y primaria: diferencias en función de su inteligencia emocional y del género (2012)
14. Puigbó, J., Edo, S., Rovira, T., Limonero, J.T., Fernández-Castro, J.: Influencia de la inteligencia emocional percibida en el afrontamiento del estrés cotidiano. Ansiedad y Estrés **25**(1), 1–6 (2019)
15. Rueda, P.M., Filella Guiu, G.: Importancia de la educación emocional en la formación inicial del profesorado. Intercambio/Échange **1**, 212–219 (2016)
16. Schoeps, K., Tamarit, A., de la Barrera, U., Barrón, R.: Effects of emotional skills training to prevent burnout syndrome in schoolteachers. Ansiedad y Estrés **25**(1), 7–13 (2019)
17. Taramuel, J., Zapata, V.: Aplicación del test TMMS-24 para el análisis y descripción de la inteligencia emocional considerando la influencia del sexo. Revista Publicando **11**(1), 162–181 (2017)
18. Villacreces, J., Achi, V.: Aplicación del test TMMS-24 para el análisis y descripción de la inteligencia emocional considerando la influencia del sexo. Revista Publicando **4**(1), 162–181 (2017)

Evaluation of the Curriculum in Higher Education: Application Experience

L. Rodríguez-Cisneros(✉), N. Galárraga, E. Rodríguez, A. Pérez, and L. Chiliquinga

Instituto Superior Tecnológico José Chiriboga Grijalva, Ibarra 100110, Imbabura, Ecuador
lmrodriguez@tecnologicoitca.edu.ec

Abstract. This article presents the experience of evaluating the curriculum of an Ecuadorian Higher Technological Institute in the formulation and implementation of a system for monitoring the application of the curriculum. A descriptive and documentary research was carried out with a participatory action research methodology that allowed obtaining relevant information in each of the phases of the proposed model for curricular monitoring. The results of the application of the system showed that the career projects required curricular adjustments due to changes in national regulations, as well as adjustments in the micro-curricular description of some subjects of the approved projects to specify the scope of the contents and learning outcomes in relation to the graduation profile. Minimal differences were also found between the official curriculum and the operational curriculum in the teaching-learning process and the need to update the professional profile assessment instruments was determined to effectively verify satisfaction with academic training. It was concluded that monitoring the curriculum requires commitment from the educational community so that the systematized information in each of the phases of the proposed model can justify the decision-making necessary for the improvement of the curriculum.

Keywords: Assessment · Monitoring · Curriculum · Higher education

1 Introduction

Broadly speaking, the curriculum is synonymous with the educational process as a whole. More specifically, it is the study program of a career Bolívar Boitía (2000). From an organizational perspective, the curriculum integrates the plans, purposes, content, methodological guidelines and evaluation criteria that define a career (Fig. 1).

According to Posner (2005), translating the educational curriculum to and intentions into the classroom setting begins with the curricular design. This continues throughout the implementation and is completed with a curricular evaluation proving that educational objectives have been achieved.

Curriculum implementation requires follow-up processes to identify possible design errors due to institutional changes or updated national regulations Dirección de Armonización Curricular e innovación docente (2020). On the other hand, the implementation process uncovers unforeseen situations that can be improved Camilloni (2008);

M. Zambrano Vizuete et al. (Eds.): CI3 2021, LNNS 511, pp. 153–159, 2022.
https://doi.org/10.1007/978-3-031-11438-0_13

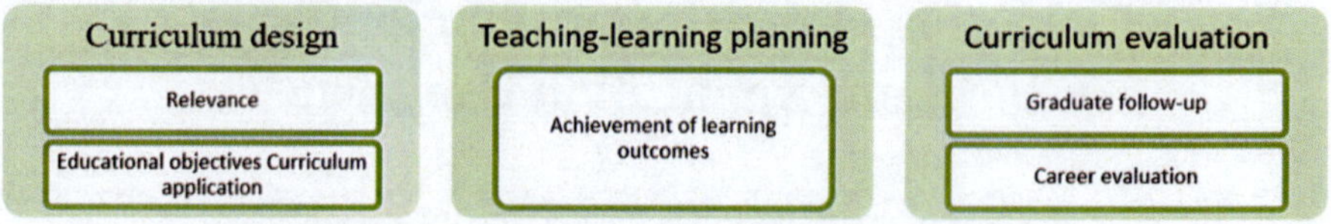

Fig. 1. Phases of the curriculum. Source: (Posner 2005)

Colmenares (2012). Curriculum application monitoring can be parallel to the implementation and part of the curricular evaluation. For Sánchez (2008), follow-up includes periodic evaluation of results, comparing them with the initial programmed indicators. Comparison allows necessary adjustments to achieve the proposed goal.

For Brovelli (2001), evaluation is considered an intrinsic part of the curricular process, which implies that coherence is required between the theoretical and methodological aspects of the curriculum with the conceptions that support it. However, it is common to observe in the curriculum the lack of consistency between pedagogical approaches and evaluation, according to Camilloni (2008). With these references, the need to systematize and verify the curricular evaluation processes is clear.

Given that monitoring is related to evaluation and evaluation can have different interpretations, the position of Stufflebeam (1985) will be assumed. Stufflebeam defines evaluation as the process of identifying, obtaining and providing useful and descriptive information about value. and the goals' merits Evaluation must provide relevant information for decision-making, solve liability problems and promote understanding of the phenomena involved.

To achieve this objective, in the evaluation of the curriculum, certain norms must be considered Stufflebeam (1985) points out that the evaluation must meet four main conditions (Fig. 2).

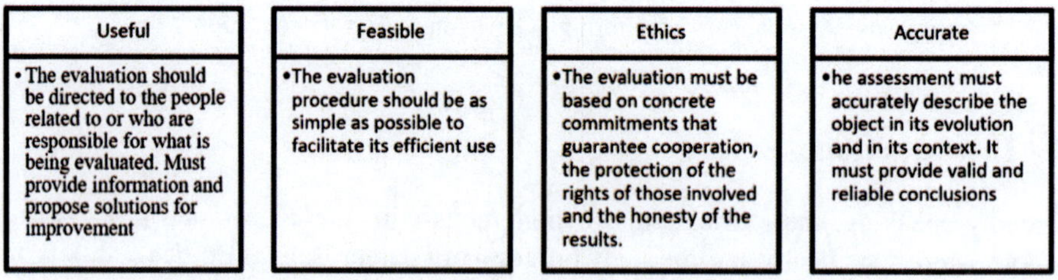

Fig. 2. Main evaluation conditions. Source: (Stufflebeam 1985).

On the other hand, curriculum evaluation has evolved from the pre-Tylerian period to the professionalism era where various evaluation models are proposed. For Stufflebeam (1985) cited by Mora Vargas (2004), evaluation models can be grouped into categories. The holistic category is relevant because its models consider evaluation as part of the overall evaluation. In this category, improvement-oriented evaluation would be limited and only yield results at the end of the process, negating timely problem resolution.

The CIPP (Context-Input-Process-Product) evaluation model is in the category of evaluation oriented towards improvement. In this model proposed by Stufflebeam (1985), evaluation is redefined as the process by which useful information is provided for decision-making. This model (Fig. 3) integrates four phases.

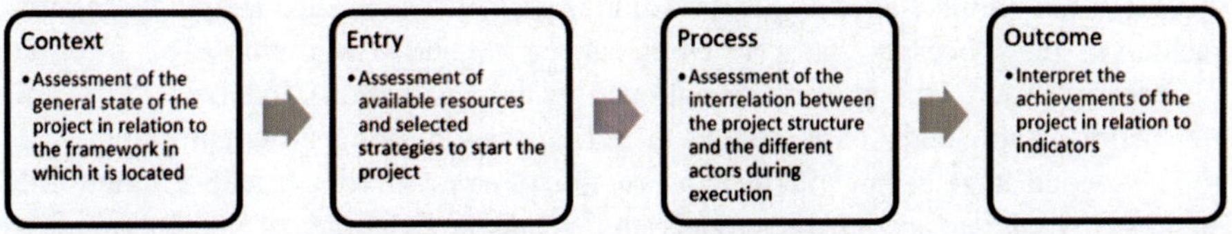

Fig. 3. Evaluation as a decision-making process. Source: (Stufflebeam 1985) cited by (Mora Vargas 2004).

With this frame of reference, curriculum implementation monitoring must be accompanied by a critical reflection on the articulation between the macro, meso and micro levels of the curriculum, this strategy allows inconsistencies to be detected and corrected. From this model, curriculum application is followed by improvements rather than testing.

In what corresponds to the Higher Technical and Technological Institutes (ISTT) that belong to the Ecuadorian higher education system, the monitoring of the implementation of the curriculum is beginning to be considered as an evaluation indicator in the Institutional Evaluation Model for ISTT in the process of accreditation Brovelli (2001).

With this background, the research objective is to present the application monitoring system of the proposed curriculum at the José Chiriboga Grijalva Higher Technological Institute. This curriculum This curriculum was developed based on participatory action research methodology.

2 Methodology

This research is descriptive in nature as it details the elements involved in the curriculum monitoring system and is documentary in nature as it is based on the review of curricular evaluation models proposed by researchers in academic publications. Participatory action research was used as a methodology, as proposed by Ander-Egg (2003) and Colmenares (2012) for the construction of the proposal of the monitoring system for the application of the curriculum.

This the research was developed throughout 2019 and 2020 at the José Chiriboga Grijalva Higher Technological Institute (ISTJCHG) in Ibarra, Ecuador. This is an accredited and self-financed institution within the Ecuadorian higher education system with academic offerings at a higher technical-technological level in face-to-face and semi-face-to-face modality that corresponds to the third level of academic training.

The first stage of the research began with the approach to the curricular monitoring as an object of study through an initial diagnosis, taking as a reference the fundamental elements of this indicator that appears in the Institutional Evaluation Model for ISTT in the accreditation process Brovelli (2001). According to what is established in the aforementioned model, the purpose of the monitoring system for the application of career curricula is to identify errors in the construction or failures in the process of applying the curriculum to establish corrective measures, if applicable. The established standard mentions that the basis of the monitoring process is the capture, processing and analysis of the necessary information about the student training process Brovelli (2001).

Within this initial stage, Academic Coordination was consulted about current documentation, the processes that were being carried out and at what time. This information contrasted with the evidence mentioned in the Evaluation Model for ISTT This information established a baseline for the curricular monitoring system proposal.

The second stage began with the construction of a system for curriculum monitoring This system took into account the models studied and the fundamental elements required by the Evaluation Model. Work meetings were organized between Internal Evaluation, Academic Vice-Rector, Institutional Planning, Academic Direction and Research, to analyze the information collected in the first phase.

Based on the CIPP model of Stufflebeam (1985), the research team formulated the methodological model for monitoring the application of the ITCA curriculum (Fig. 4).

Phases			
Context	**Entrance**	**Process**	**Outcome**
- Project Evaluation	- Evaluation of curricular planning	- Evaluation of the teaching-learning process	- Evaluation of the graduation profile
Feedback			

Moments			
Project Conditions	**Microcurricular Planning**	**Academic portfolio**	**Graduate awards**
prior to the opening of the cohort	beginning of the academic period	during the academic period	completion of academic training

Decisions			
Context	**Entry**	**Process**	**Outcome**
Decisions of the Project	Planning decisions	Implementation decisions	Continuity decisions

Fig. 4. ISTJCHG methodological model for monitoring the implementation of the curriculum.

In the first context phase, the model allows the evaluation of the approved project to establish that it corresponds to the educational model and verify that it complies with the updated regulations. In this phase, it is also assessed availabilities of the teachers and the minimum number of students enrolled for the opening of the cohort.

In the entry phase, subject planning is formulated in relation to the career project. the application of non-substantive adjustments is resolved and the micro-curricular planning is validated.

In the process phase, the teaching-learning activity specified in the teaching and student portfolio is evaluated. This allows critical reflection on executing the educational project in the classroom. In this phase, reports are assembled to analyze the curriculum at both the career and institutional levels.

In the final phase of the product, the Project achievements are validated based on the development of the learning results of the graduation profile that are specified in the satisfaction of graduates and employers with their academic training; in addition to the terminal efficiency that is estimated with the number of graduates with respect to the number of enrolled in that cohort. This phase provides additional information to make substantive adjustments to the project, with the authorization of the Higher Education Council (CES) and about the continuity of the academic offer.

This model proposed in the second stage of participatory action research is specified in a third stage that corresponds to the execution in the second academic period of 2020. For the application, a document was presented that describes the system for monitoring the application of the curriculum. starting with the normative reference, an overview of the curriculum, the theoretical foundation and the curricular monitoring model, the identification of the key monitoring actors, the components of the monitoring model and the procedure for implementing the monitoring of the curriculum of the careers.

3 Results and Discussion

Stage four, which corresponds to the final stage of the participatory action research method, involves reflection on the research development. in this section results from every phase of the monitoring system are presented and discussed.

In the first phase that corresponds to the project, the implementation results showed that career projects are highly related to the educational model. However, these projects are not aligned with current regulations, one example is, the total hours of the race. This happens because the projects in question were presented and approved with a prior regulation. This is due to the differences between the academic regime regulations (RAA) of September 2, 2015 and the RAA of July 15, 2020 (Table 1).

These results led the institution to analyze a possible curricular adjustment. This was considered according article 137 of the RRA, specifying that the adjustments can be substantive or non-substantive, and must be authorized or notified to the CES, as appropriate. Based on this analysis, the academic offerings were not changed.

In the second phase, where the curricular planning is evaluated, several micro-curricular plans were not aligned with the learning outcomes and contents described in the micro-curriculum of the career projects. The discrepancy is equivalent to approximately 10% of the career subjects This situation arose because the work teams that formulated the career projects were integrated with specialist teachers in specific areas. This explains the difference in terms of the depth or breadth of learning results and/or content of some subjects from other areas of knowledge.

Table 1 Provisions of the RRA for the preparation of the race.

Provisions for higher technological level careers	RAA – 2/9/2015	RRA – 15/7/2020
Ordinary academic period	18 weeks (art. 13)	16 weeks (art. 11)
Duration of academic periods	900 h (art. 17)	720 h (art. 11)
Duration of the careers	4500 h (art. 18)	Minimum 2880 h – maximum 3600 h (art. 18)
Degree unit (2015) Curricular integration unit (2020)	240 h (art. 22)	Minimum 96 h – maximum 240 h (art. 31)
Pre professional practices	Minimum 400 h (art. 89)	Minimum 240 h – maximum 480 h (art. 54)
Pre-professional community service practices	Minimum 160 h (art. 94)	Minimum 96 h – maximum 144 h (art. 54)

Source: regulation of the academic regime of 9/2/2015 and 7/15/2020.

Due to these results, some non-substantive adjustments of the career projects were made. ultimately, after reviewing reports, academic directors issued a resolution on their approval or refusal regarding career coordination.

For the third phase, the teaching-learning process was evaluated based on the advancement of planning, progress and achievement of objectives, achievement of learning results and pedagogical practice through the presentation of reports and the delivery of the teaching portfolio. The results showed differences in the official curriculum and the operational curriculum, especially in what corresponds to the application of didactic strategies and evaluation criteria. These results were useful for teachers to adjust them.

The fourth phase, which corresponds to the evaluation of the professional profile, is in progress with students who are finishing their academic training in the period October 2020 - March 2021. In preliminary results, it was established that there would be an acceptable satisfaction with the training However, it is necessary to ensure that the instruments used are updated based on the objectives, professional competencies and learning results that developed in career projects.

4 Conclusions

Based on the experience generated, the application of a follow-up process to the curriculum requires decisions and flexibility from of academic authorities. It also requires commitment and teamwork from career coordinators and teachers. The information collected in each of the phases of the proposed model can support decision making for curriculum improvement.

With a curricular monitoring system, it is possible to quickly identify differences between the contents, learning outcomes of the approved projects and the subject plans proposed by teachers. academic authorities can use this information to determine the cases in which it is appropriate to make non-significant adjustments to career projects. This research found that there are differences between the official curriculum and the classroom operational curriculum. This gives teachers flexibility to make adjustments. However, it is necessary to establish if these differences directly affect the educational objectives and proposed graduation profile to make the substantive or non-substantive adjustments that are required.

It is key that teachers who are involved in the design and redesign processes of the curriculum are involved in the formulation and application of the monitoring process to the curriculum. This Will give them the general evaluation process overview that will be implemented in new and updated projects.

The proposed curriculum monitoring model may be useful to other educational institutions that need to validate the execution of the official curriculum. In every phase, each institution can handle its own documents to show curriculum execution. They can also take appropriate corrective measures, when applicable.

References

Ander-Egg, E.: Repensando la Investigación-Acción-Participativa. Lumen Hvmanitas, País Vasco (2003)

Bolívar Boitía, A.: El currículum como un ámbito de estudio. En: Escudero Muñoz, J. (ed.), Diseño, desarrollo e innovación del currículum, pp. 27–34. Madrid (2000)

Brovelli, M.: Evaluación curricular. Fundamentos en humanidades **2**(4), 101–122 (2001)

CACES: Modelo de Evaluación Institucional para los Institutos Superiores Técnicos y Tecnológicos en proceso de acreditación 2020. Quito (2020)

Camilloni, A.: Estándares, evaluación y currículo. Archivos De Ciencias De La Educación **3**(3), 55–68 (2008)

Colmenares, A.: Investigación-acción participativa: una metodología integradora del conocimiento y la acción. Voces y Silencios: Revista Latinoamericana de Educación **3**(1), 102–115 (2012)

Dirección de Armonización Curricular e innovación docente. Modelo Metodológico de Seguimiento y Evaluación Curricular para la UNAP. Universidad Arturo Prat. Chile (2020)

Dirección de Armonización Curricular e Innovación Docente. Recursos Docentes. Universidad Arturo Prat. Chile (2020)

Ministerio de Educación. Página Web, https://educacion.gob.ec/curriculo/, acceso última vez 2021/05/31 (2021)

Mora Vargas, A.: La evaluación educativa: Concepto, períodos y modelos. Revista Electrónica "Actualidades Investigativas en Educación" **4**(2) (2004)

Posner, G.: Análisis del Currículo. McGraw-Hill, México (2005)

Sánchez, M.: Seguimiento y evaluación de planes y proyectos educativos: un reto y una oportunidad. Omnia **14**(3), 32–50 (2008)

Stufflebeam, D.: Evaluación Sistemática. Guía Teórica y Práctica. Ediciones Paidós Ibérica. Barcelona, España (1985)

Innovation Competencies in Higher Education Institutions (HEIs) Students: A Systematic Literature Review

Lizbeth Ximena Suárez Morales[1], Diana Rojas-Torres[2](✉), Wladimir Paredes-Parada[3], and Martha María Fernández Rodríguez[4]

[1] Instituto Tecnológico Superior Vida Nueva, Av. Maldonado y 4ta transversal, Quito, Ecuador
[2] Universidad de La Sabana, Autopista Norte Km 7 Puente del Común, Chía, Colombia
diana.rojas7@unisabana.edu.co
[3] Instituto Tecnológico Universitario Rumiñahui, Atahualpa 1701 y 8 de Febrero, Sangolquí, Ecuador
[4] Instituto Universitario Espíritu Santo, Av. Juan Tanca Marengo Km. 2.5 y Av. Las Aguas, Guayaquil, Ecuador

Abstract. The study has two objectives: 1) To identify the trends, main documents, and authors in innovation competencies and 2) To identify the innovation competencies that emerge from the literature review for Higher Education Institutions (HEIs). We performed a bibliometric analysis using Scopus databases, to do a systematic literature review in innovation competencies. The study found that innovation competencies is a nascent area with low degree of academic consensus. This article shows a general approach to improve conceptual specifications of innovation competencies. We identify trends in the literature review and connect them with innovation competencies. We proposed the following innovation competencies in this research: teamwork, sustainability, leadership, creativity, communication, collaboration, digital skills, flexible thinking, critical thinking, and goal oriented. The insights of this study can help universities to analyze this innovation competencies in the curriculum for their programs. We propose for future research the measure of the innovation competencies we defined for Latin American HEIs.

Keywords: Innovation competencies · Bibliometrics · Higher education institutions · Trends in education

1 Introduction

The twenty-first century society requires multidisciplinary and flexible professionals. The academy has new challenges and barriers to overcome regarding to the development of innovation competences in the classroom. For example, the challenge of solving relevant problems in an ethical manner in a local and global context is relevant for the real work. It is not enough to have knowledge and skills; is important to apply them creatively to solve real situations in a defined environment, preferably with regional impact. Today's professionals are invited to solve social problems, improve living standards and impact their environment.

M. Zambrano Vizuete et al. (Eds.): CI3 2021, LNNS 511, pp. 160–172, 2022.
https://doi.org/10.1007/978-3-031-11438-0_14

Because of this [1], formulates the third-generation university, where there is cooperation between public and private institutions with which traditionally did not relate to co-create solutions that impact the regions, as well as share capabilities with entities that complement them.

The open innovation initiative, where companies or institutions seek new opportunities to innovate both inside and outside their organization, ranging from the ideation process to accessing new markets [2]. This exercise at the level of Institutes of Higher Education (HEIs), aims to deepen the theories of the disciplines of each program, identify problems that can be addressed in the region, apply these theories to solve challenges in their communities and promote communication skills, both oral and written, of the solutions proposed.

Adjusting these concepts to educational processes requires a transformation in universities, starting by changing the traditional teaching model to models based on the development of competencies and particularly what we will refer to in this article as "innovation competencies".

We define innovation competencies as those competencies that develop transversal skills and knowledge for the solution of complex problems. The development of skills is not instantaneous, it takes time to develop competencies at different levels of mastery (incipient, intermediate, and advanced), as well as at different hierarchical levels in the organization where teachers and students are the key to the success of this model for the development of innovation competencies.

With respect to teachers, the competency-based model requires a transition from lecturing to facilitating and mediating. The professor is responsible for design learning modules that meet the proposed challenge and encourage students to explore, experiment and discover new scenarios to develop disciplinary and cross-disciplinary competencies at the same time [3].

On the other hand, according to Felix-Herran (2019) "the model requires students who are eager to learn, proactive and committed to the development of their own competencies. To develop professionals with the appropriate profile, the vision, mission, and values of universities must be aligned with the requirements of society" [4].

The gap between traditional university training and innovation competency-based training in Latin America is large and closing it requires the structure of a system that feeds back into the regional ecosystem. The regional innovation and entrepreneurship ecosystem must operate under the premise of trust, cooperation, diversity of talents, collaboration, and experimentation among the ecosystem's actors.

The key factors identified above are the ones that will allow the innovation and entrepreneurship ecosystem to fit as a promoter of the innovation culture in the region, allowing an exponential growth of entrepreneurship, start-ups, intra-entrepreneurship, impacting on the creation and formalization of companies, job creation, reputation, and wealth.

The article presents a review of the concept of competencies, with special emphasis on the educational context. Then, a bibliometric analysis of competencies in innovation is presented, and where the most cited competencies models in the literature are identified. Finally, on this basis, a set of innovation competencies to be developed in students in Higher Education Institutions is proposed, which can be used as a basis for future research.

The conclusions suggest that, although the concept of competencies is widely used by universities, there is still a lack of strategy to implement actions to train in innovation competencies for the development of entrepreneurship, intra-entrepreneurship and economic development of the region and the countries.

2 Theoretical Framework

2.1 Competence

Competence, by definition, is related to a position or function of the individual. Competence combines skills, knowledge, attitudes, and values [5]. Competencies are not seen as the task of the job, but rather as what enables people to perform the task.

Later [6] brought the term to the business context, where the American Management Association positioned it as the identification of characteristics that distinguish superior managerial performance from average. Competencies are not operational, such as the job task, but are defined as what enables workers to perform the task. These can be described in terms of essential element personal characteristics, skills, abilities, knowledge, and motives of the employee that lead toward superior managerial performance.

Boyatzis [6] developed his model of management by competencies through a study of more than two thousand managers, where he identified more than 100 potential managerial competencies. An important aspect of his work is the three levels of competencies: motivations, social role with self-knowledge and role transitions. This work of the American school focused on the concept of competence as an underlying characteristic of a person, which translates into action efficacy and/or superior performance in a job.

Mitchelmore [7], identified that entrepreneurs require both entrepreneurship and management competencies for businesses to have a broader vision and be able to find funding for their businesses. However, their study concluded that more research is needed to understand the impact of context on the development of competencies.

2.2 Innovation Competencies

To understand the development of innovation competencies, it is important to define the concept of innovation. In terms of this research, it will be understood as "a new or improved product or process (or combination of these) that differs from the unit's previous products or processes and that has been made available to users or put into use by the unit" [8].

The definition of innovation is part of the Oslo Manual, the first agreement of the global community of practitioners in the OECD national expert group, which was held in the city of Oslo in 1991, due to the need for countries to start measuring and conceptualizing business innovation as technological and innovation approaches [8].

Business innovation is defined as any business product or process that differs from any product or process that the company had which has been tested in the market or is in use by that company [8].

This innovation implementation can be divided into a business process which is defined as "a new or improved business process for one or more business functions that differs significantly from the firm's previous business processes and has been put into use in the firm" [8]. Likewise, in product innovations that refer to "a new or improved good or service that differs significantly from the firm's previous goods or services and that has been introduced to the market" [8].

In summary, innovation competencies are considered important for regional economic development as it promote growth and business success, and the development of these competencies can have important implications for practice. However, research in this line is still in its early stages [9]. Therefore, this article reviews the literature on innovation competencies research at the university level to:

- Review the state of the art and contributions related to innovation competencies defined by different authors.
- Propose the innovation competencies that emerge from this research for the identification of such competencies in programs under development.

3 Methodology

In this article we used bibliometric analysis to identify research fields, as well as to visualize some bibliometric indicators. The software used was Bibliometrix, a R tool for comprehensive analysis of scientific maps, quantitative research. The Bibliometrix package analyze data from SCOPUS, Clarivate Analytics' Web of Science, PubMed, and Cochrane databases, constructing data matrices for co-citations, scientific collaboration analysis and co-occurrence analysis (https://www.bibliometrix.org/).

3.1 Bibliometric Analysis

The bibliometric analysis was carried out using the scopus database. The search equation used was as follows:

- All fields: ((("innovation competencies" OR "entrepreneurship competencies") AND (universities OR "higher education" OR university OR college))).
- Time period: (until June 26, 2021).).
- Articles, books, book chapters, conference papers (Table 1).
- Journal: Excluded Journals in: Physics and Astronomy and Materials Science

We conducted two bibliometric indicators: citation analysis and co-occurrence analysis. Citation analysis refers to the evolution of papers in the research areas, categorized by country, journal and author. Co-occurrence analysis is understood as the study of co-occurrences, or joint appearances, of two terms in each text with the purpose of identifying the conceptual and thematic structure of a scientific domain. Once the terms to be analyzed have been selected, co-occurrence matrices are constructed, with which measures of similarity, or "similarity" are calculated [10].

Table 1. Descriptive analysis of the data downloaded in Scopus with the search equation.

Description	Results
Main data information	
Time period	1999:2021
Resources (journals, books, etc.)	599
Documents	992
Average number of publications per year	4,61
Percentage of citations per document	8,12
Percentage of citations per year per document	1,37
References	57530
Document type	
Article	650
Book	16
Book chapter	71
Conference article	205
Conference summary	3
Editorial	6
Content of the documents	
Keywords Plus (ID)	2720
Author's Keywords (DE)	2775
Authors	
Authors	2567
Author occurrence	2980
Single authors per document	133

4 Results and Discussion

This study is a bibliometric analysis. For example, the co-occurrence network of keywords, the citation and heat maps of the most used terms are the main indicators of this methodology.

4.1 Annual Citation and Most Cited Papers

Figure 1 shows the production dynamics in the field of innovation competencies from 1999 to 2021. The academic articles between 1999–2009 were 84 while between 2010–2021, we found a total of 908 papers. The increasing trend may be an indicator of the interest of the scientific community in this topic. The first article regarding innovation competencies was published in 1999, the paper called "the empowerment of employees in small and medium sized service firms" deals with the creation of corporate entrepreneurship as an induced empowerment process. It investigates two issues: how well firms succeed in developing entrepreneurship competencies; and the conditions for success [11].

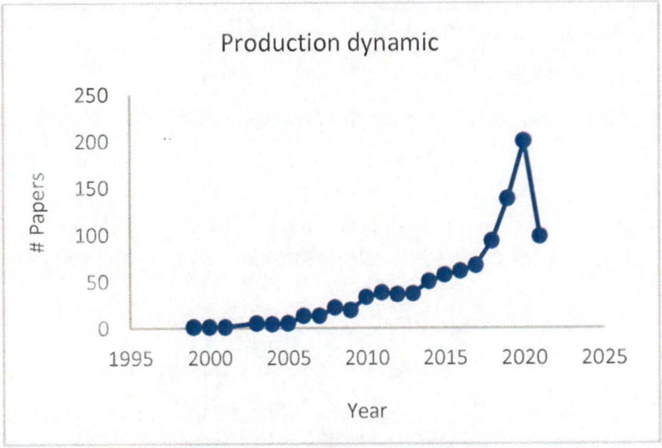

Fig. 1. Annual production dynamics in innovation competencies.

In 2017 a slight growth in publications show the relationship between innovation competencies at university students. The maximum number of papers (200) were published in 2020. In 2021, 98 papers had already been reported.

The Strategic Fit between Innovation Strategies and Business Environment in Delivering Business Performance. The study demonstrates the strategic fit between dynamism and product innovation strategy as well as between competitiveness and process innovation strategy [12].

In 2015, a paper with 80 citations called "Disentangling competencies: Interrelationships on creativity, innovation and entrepreneurship" develops the ongoing debate on the nature of entrepreneurial competencies critically analyzing the theoretical underpinnings to such interrelationships [13].

Regarding the journals in which it is published, it is found that the largest number of articles have been published in the journal Sustainability (35 Scopus Q1), Education and Training (20 Scopus Q2), Industrial Marketing Management (17 Scopus Q1), Technological Forecasting and Social Change (17 Scopus Q1) (Table 2).

4.2 Keyword and Cluster Analysis

Co-occurrence Analysis

Figure 2, present the conceptual structure about innovation competencies; it is drawn considering the most frequent keyword of the documents. For this case, the main terms are "Innovation", "Sustainable development" and "Students".

The cluster with the keyword "Students" is related to curricula, innovation competencies, education, and engineering education. For example, methodologies like Challenge

Table 2. Most important journals

# Articles	Journal
35	Sustainability (Switzerland)
20	Education and training
17	Industrial marketing management
17	Technological forecasting and social change
14	Proceedings of SPIE - the international society for optical engineering

Based Learning (CBL) (Table 3) or Project Base Learning (PBL) are considering the main experiential methodologies to develop innovation competencies in undergraduate students (Fig. 3).

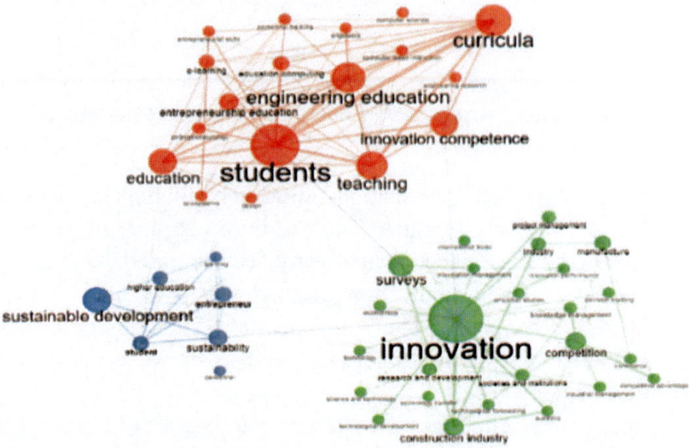

Fig. 2. Co-occurrence network for the topic of innovation competencies.

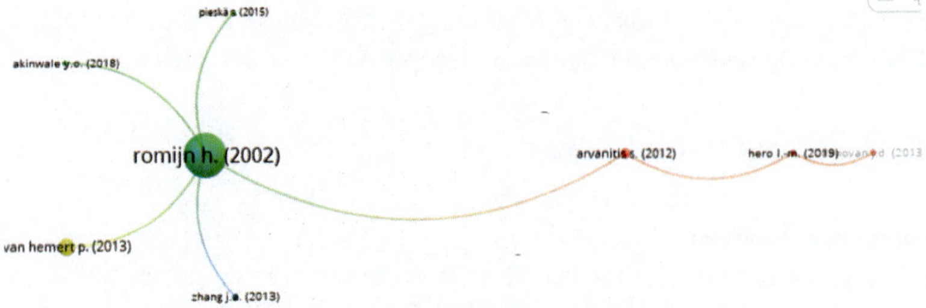

Fig. 3. Citation network using VosViewer software.

Table 3. Documents in the citation network.

Author	Title	DOI	Year	Journal
Pieskä, Sakaria; * | Kaarela, Jaria | Luimula, Mikab [14]	Enhancing innovation capability with cognitive info communications	https://doi.org/10.3233/IDT-140206	2015	Intelligent Decision Technologies, vol. 9, no. 1, pp. 67–78, 2015
Yusuf Opeyemi Akinwale; John-Felix Kayode Akinbami; Joshua Babatunde Akarakiri [15]	Factors influencing technology and innovation capability in the Nigerian indigenous oil firms	https://doi.org/10.1504/IJBIR.2018.089146	2018	International Journal of Business Innovation and Research (IJBIR), Vol. 15, No. 2, 2018
Patricia van Hemert, Peter Nijkamp & Enno Masurel [16]	From innovation to commercialization through networks and agglomerations: analysis of sources of innovation, innovation capabilities and performance of Dutch SMEs	https://doi.org/10.1007/s00168-012-0509-1	2013	Ann Reg Sci 50, 425–452 (2013)
Jing A. Zhang, Sam Garrett-Jones and Ricky Szeto [17]	Innovation capability and market performance: the moderating effect of industry dynamism	https://doi.org/10.1142/S1363919613500047	2013	International Journal of Innovation Management Vol. 17, No. 02, 1350004 (2013)
Henny Romijn Manuel Albaladejob [18]	Determinants of innovation capability in small electronics and software firms in southeast England	https://doi.org/10.1016/S0048-7333(01)00176-7	2002	Research Policy Volume 31, Issue 7, September 2002, Pages 1053–1067
Spyros Arvanitis, Tobias Stucki [19]	What determines the innovation capability of firm founders?	https://doi.org/10.1093/icc/dts003	2012	Industrial and Corporate Change, Volume 21, Issue 4, August 2012, Pages 1049–1084
Laura-Maija Hero, Eila Lindfors [20]	Students' learning experience in a multidisciplinary innovation project	https://doi.org/10.1108/ET-06-2018-0138	2019	Vol. 61 No. 4, 2019 pp. 500–522 Emerald Publishing Limited
Jerome Denis Donovan, Alex Maritz &Andrew McLellan [21]	Innovation training within the Australian advanced manufacturing industry	https://doi.org/10.1080/13636820.2013.783614	2013	Journal of Vocational Education & Training Volume 65, 2013 - Issue 2. Pages 256–276

Co-citation Analysis

Our research is built on the existing literature about innovation competencies, focusing on HEIs students. During the time of analysis, we discovered some topic relations. For example, sustainable development is a keyword that emerge from the analysis, and we will include it in the innovation competencies.

We conduct a search document about the description of the innovation competencies and how academics measure it using quantitative research. We show in Table 4, some papers that use instruments to measure innovation competencies.

4.3 Papers Regarding to Innovation Competencies

A review of the literature revealed the following articles oriented exclusively to innovation competencies.

Table 4. Articles that use instruments to measure innovation competencies.

Author	Name of the paper/document	Description of the innovation competencies it measures
Keinänen, et al. (2017) [22]	Students' perception of learning innovation competences in activity-based learning environment	Creativity (9 items). Critical thinking (6 items). Initiative (6 items). Teamwork (7 items). Networking (6 items)
José Sánchez. (2013) [23]	The impact of an entrepreneurship education program on entrepreneurial competencies and intention	Self-efficacy (20 items). Proactivity (10 items). Risk-taking (…). Intention to be self-employed (3 items)
Watts, Aznar-Mas, et al. (2013) [24]	Innovation competency development and assessment in higher education	**A 3-level model is identified**: individual, interpersonal, relational Competencies at the personal level: independent thinking, decision making, goal-oriented, executing actions, creativity, working methods **Interpersonal competencies**: ability to cooperate in diverse groups, proactivity, responsibility, ability to work in research and development projects, ethics, social responsibility, and effective communication **Relationship level competencies**: Creating and maintaining connections with people, ability to network, ability to cooperate in multicultural and multidisciplinary environments, ability to co-communicate and interact in an inter-national environment

(*continued*)

Table 4. (*continued*)

Author	Name of the paper/document	Description of the innovation competencies it measures
Keinänen, et al. (2018) [25]	Researching learning environments and students' innovation competences	Creativity (9 items). Critical thinking (6 items). Initiative (6 items). Teamwork (7 items). Networking (6 items). Active learning and teaching methods (4 items). Life orientation for work (6 items). Multidisciplinary learning environments (3 items). Flexible curriculum (1 item). Internationalization (1 item). Entrepreneurship (1 item)
Keinänen and Butter (2018) [26] Bacigalupo M., Kampylis P., et al. (2016) [27] Proyecto de la Unión Europea	Applying a self-assessment tool to enhance personalized development of students' innovation competences in the context of university-company cooperation EntreComp: The Entrepreneurship Competence Framework	Creativity (9 items). Critical thinking (6 items). Initiative (6 items). Teamwork (7 items). Networking (6 items) **Includes a 3-area model**: Ideas and opportunities, resources, Action **Ideas and opportunities**: Opportunity hunting. Creativity, Vision, evaluating ideas, Ethical and sustainability thinking **Resources**: Self-care and self-efficacy. Motivation and perseverance. Resource mobilization. Financial and eco-economic illiteracy. Influencer **Action**: Initiative. Planning and management. Coping with uncertainty, ambition, and risk. Teamwork. Learning through experience At the center of this project is the development of FINCODA's innovation barometer evaluation tool. This is a psychometric tool that measures the innovation capacity of individuals. It divides innovation into 5 core areas and assesses the individual's capacity in each of these areas separately

In this research, we propose an innovation competence based on the literature review and search documents about how to measure innovation competencies in undergraduate students. Table 5 define the competencies.

Table 5. Competencies proposed from the research developed

Innovation competencies	Definition
1. Teamwork	1. To identify entrepreneurial characteristics at individual level to empower themselves and put them at the service of building a solution to a problem
2. Sustainability	2. Ability to identify sustainable strategies, establish priorities to access and assign resources (natural, human, technological, financial, etc.) to achieve an objective
3. Leadership	3. Recognize their own capabilities to be managed itself. Develop skills to inspire and motivate a teamwork
4. Creativity	4. Ability to generate new ideas from associations between known ideas to offer solutions
5. Communication	5. Convey information effectively and clearly
6. Collaboration	6. Find common work areas and different points of view that promote diversity
7. Digital skills	7. Critical and safe use of Information Society Technologies for work, leisure, and communication
8. Critical thinking	8. Development of a sense of urgency and sensitivity to the context
9. Flexible thinking	9. Analyze and evaluate information to structure your own thinking and build your own conclusions, in addition to assuming responsibility for them
10. Outcome/goal oriented	10. Increase and maintain high levels of work execution, successfully achieving goals with a view to sustainability

5 Conclusion

This article offers a broad overview of the state of the art and the research that has been carried out on innovation competencies, showing how this area of research is still incipient globally. In Latin America we have a great opportunity to develop research on innovation competencies, considering that universities and technological institutes in developing countries have the responsibility to train professionals with integrity who contribute to the solution of local, regional, and global problems, positively impacting the quality of life of their communities and the economic development of the country.

In quantitative terms, the bibliometric indicators analyzed show how incipient this research topic is, although with the current challenges facing education, the commitment to develop innovation competencies can be a differential that allows HEIs to be more competitive.

The co-occurrence analysis demonstrates the trends and relationships that are emerging. For example, we discover the keyword "sustainable development" from the analysis.

The innovation competencies might be based on environmental, economic and social sustainability. It is important for the development of future professionals to make the innovation and sustainability agenda compatible to harmonize regional development from any perspective.

The proposal of this research is given in the definition of 10 innovation competencies, Teamwork, sustainability, leadership, communication, creativity, collaboration, digital skills, critical thinking, flexible thinking, and outcome/goal oriented. In fact, the innovation competencies we are proposing are key competencies in post-pandemic.

Acknowledgment. The authors would like to thank to Corporación Ecuatoriana para el Desarrollo de la Investigación y Academia - CEDIA for the financial support given to the present research, development, and innovation work through its CEPRA program, especially for the Project "Modelo para el desarrollo de competencias de innovación que promueva el desempeño profesional en la Educación Tecnológica del Ecuador". The researchers want to thank Nathaly Puentes Ramírez. Head of Entrepreneurship office at UNIMINUTO, Kenny Mauricio Gómez – Coordinator of Entrepreneurship office at Universidad Autónoma de Manizales for their valuable contributions in the construction and feedback of innovation competencies and Camila Padilla, researcher at ELANET network at University of La Sabana for her feedback in the bibliometric analysis.

References

1. Wissema, J.G.: Towards the Third-Generation University, Managing the University in Transition. Edward Elgar Publishing, Cheltenham (2009)
2. Chesbrough, H.: Open innovation: a new paradigm for understanding industrial innovation. Open innovation: Researching a new paradigm **400**, 0–19 (2006)
3. Argudín, Y.: Educación basada en competencias: Nociones y antecedentes. Trillas (2005)
4. Félix-Herrán, L.C., Rendon-Nava, A.E., Nieto Jalil, J.M.: Challenge-based learning: an I-semester for experiential learning in mechatronics engineering. Int. J. on Interact. Des. Manuf. (IJIDeM) **13**(4), 1367–1383 (2019). https://doi.org/10.1007/s12008-019-00602-6
5. Gale, L.E., Pol, G.: Competence: a definition and conceptual scheme. Educ. Technol. **15**(6), 19–25 (1975)
6. Boyatzis, R.E.: The Competent Manager: A Model for Effective Performance. John Wiley & Sons (1982)
7. Mitchelmore, S., Rowley, J.: Entrepreneurial competencies: a literature review and development agenda. Int. J. Entrepreneurial Behav. Res. **16**, 92–111 (2010)
8. de Oslo, M.: Manual de Oslo. Recuperado de http://gestiona.com.br/wpcontent/uploads/2013/06/Manual-de-OSLO-2005.pdf (1997)
9. Brinckmann, J.: Competence of Top Management Teams and Success of New Technology-based Firms. Gabler, Wiesbaden (2008)
10. Galvez, C.: Análisis de co-palabras aplicado a los artículos muy citados en Biblioteconomía y Ciencias de la Información (2007–2017). Transinformação **30**(3), 277–286 (2018). https://doi.org/10.1590/2318-08892018000300001
11. Sundbo, J.: Empowerment of employees in small and medium-sized service firms. Empl. Relat. **21**, 105–127 (1999)
12. Prajogo, D.I.: The strategic fit between innovation strategies and business environment in delivering business performance. Int. J. Prod. Econ. **171**, 241–249 (2016)

13. Edwards-Schachter, M., García-Granero, A., Sánchez-Barrioluengo, M., Quesada-Pineda, H., Amara, N.: Disentangling competences: interrelationships on creativity, innovation, and entrepreneurship. Thinking Skills Creativity **16**, 27–39 (2015)
14. Pieskä, S., Kaarela, J., Luimula, M.: Enhancing innovation capability with cognitive infocommunications. Intell. Decis. Technol. **9**(1), 67–78 (2015)
15. Akinwale, Y.O., Akinbami, J.F.K., Akarakiri, J.B.: Factors influencing technology and innovation capability in the Nigerian indigenous oil firms. Int. J. Bus. Innov. Res. **15**(2), 247–268 (2018)
16. Van Hemert, P., Nijkamp, P., Masurel, E.: From innovation to commercialization through networks and agglomerations: analysis of sources of innovation, innovation capabilities and performance of Dutch SMEs. Ann. Reg. Sci. **50**(2), 425–452 (2013)
17. Zhang, J.A., Garrett-Jones, S., Szeto, R.: Innovation capability and market performance: the moderating effect of industry dynamism. Int. J. Innov. Manag. **17**(02), 1350004 (2013)
18. Romijn, H., Albaladejo, M.: Determinants of innovation capability in small electronics and software firms in southeast England. Res. Policy **31**(7), 1053–1067 (2002)
19. Arvanitis, S., Stucki, T.: What determines the innovation capability of firm founders? Ind. Corp. Chang. **21**(4), 1049–1084 (2012)
20. Hero, L.M., Lindfors, E.: Students' learning experience in a multidisciplinary innovation project. Education + Training **61**, 500–522 (2019)
21. Donovan, J.D., Maritz, A., McLellan, A.: Innovation training within the Australian advanced manufacturing industry. J. Vocat. Educ. Training **65**(2), 256–276 (2013)
22. Keinänen, M., Oksanen, A.: Students' perception of learning innovation competences in activity-based learning environment. Ammattikasvatuksen Aikakauskirja **19**(4), 48–61 (2017)
23. Sánchez, J.C.: The impact of an entrepreneurship education program on entrepreneurial competencies and intention. J. Small Bus. Manage. **51**(3), 447–465 (2013)
24. Watts, F., Aznar-Mas, L.E., Penttilä, T., Kairisto-Mertanen, L., Stange, C., Helker, H.: Innovation competency development and assessment in higher education. In: Proceedings of the 7th International Technology, Education and Development Conference, pp. 6033–6041 (2013)
25. Keinänen, M.M., Kairisto-Mertanen, L.: Researching learning environments and students' innovation competences. Education + Training **61**, 17–30 (2019)
26. Keinänen, M., Butter, R.: Applying a self-assessment tool to enhance personalized development of students' innovation competences in the context of university-company cooperation. J. Univ. Pedagogy **25**(2), 18–28 (2018)
27. Bacigalupo, M., Kampylis, P., Punie, Y., Van den Brande, G.: EntreComp: The entrepreneurship competence framework. Luxembourg: Publ. Off. Eur. Union **10**, 593884 (2016)

Approaches and Meanings: The Place of Student Well-Being and Formative Accompaniment in Higher Education in Ecuador

Catalina Vélez Verdugo(✉) and Lorena Araujo Silva

Pontificia Universidad Católica del Ecuador, 12 de Octubre y Roca, Quito 1076, Ecuador
catalinavelezv@yahoo.es

Abstract. This document analyzes the approaches and meanings through which the Universities and Polytechnic Schools in Ecuador have conceived and implemented the concepts of student well-being and formative accompaniment in higher education in Ecuador. Analyzed and understood as a set of tools, the relevance of the welfare approach helps, on one hand, to equate structural inequalities and, on the other hand, contribute to the scope of the purposes of university academic training. This approach tends to face student dropout while strengthening retention strategies during training courses, thus promoting greater terminal efficiency. Student well-being and training support as axes of higher education propose the challenge of incorporating support structures within educational models. This makes it possible to build the necessary conditions for the maximum development of students' capacities. Well, it is not only a question of guaranteeing access to higher education, but, above all, has to do with enabling all students to continue studying and successfully complete their training processes.

Keywords: Student well-being · Formative accompaniment · Higher education · Inclusion

1 Introduction

Analysis developed on the quality of academic management in higher education, as well as student performance, including the development of strategies that prevent dropout and their relationship with approaches to student well-being and formative accompaniment to students, have grown exponentially in recent years in the region.

The body of research studied [1–3] emphasizes the need to identify a series of factors that range from vocational guidance, individualized tutorials, support programs, analysis of socio-economic conditions, family contexts and psycho-emotional conditions, as elements that influence the successful transition of students into the higher education system. In this way, it is clearly identified the fact that desertion and retention are multi-causal processes in which personal, family or institutional variables intervene. All of these approaches have contributed to the development of a systematic debate and to deepen the role of higher education understood not only as a space for the generation and reception of knowledge but also as a social activator. Therefore, higher education

M. Zambrano Vizuete et al. (Eds.): CI3 2021, LNNS 511, pp. 173–186, 2022.
https://doi.org/10.1007/978-3-031-11438-0_15

institutions can apply tools that would allow inclusion processes in societies with wide structural inequalities and social asymmetries, in which student dropout has not been sufficiently analyzed and, above all, weakly managed.

Along these lines, this document analyzes the approaches and meanings through which the Universities and Polytechnic Schools in Ecuador have conceived and implemented the concepts of student well-being and formative accompaniment. To start, according to a systemic approach, student well-being and training support implemented through various tools, plans and programs by higher education institutions allow to work against structural inequalities. In addition, they contribute to the scope of the purposes of university academic training, insofar as they tend to face student dropout by strengthening retention strategies during training courses, thus promoting greater terminal efficiency.

The approach to well-being and training accompaniment, through which a praxis was set from the management of programs and projects by the Institutions of Higher Education (IES), will allow students to transition efficiently, or not, the entire educational circuit. As such, it is not only a question of guaranteeing access to higher education, but, above all, making it possible for all students to remain in it and successfully complete their training processes.

2 Methodology

This study follows two traditions of qualitative research: documentary review and critical discourse analysis. Their articulation allows to explore the approaches and meanings through which the Universities and Polytechnic Schools in Ecuador have conceived and implemented the concepts of student well-being and formative accompaniment. To start, according to a systemic approach, student well-being and training support implemented through various tools, plans and programs by higher education institutions allow to work against structural inequalities. In addition, they contribute to the scope of the purposes of university academic training, insofar as they tend to face student dropout by strengthening retention strategies during training courses, thus promoting greater terminal efficiency.

Broadly stated, documentary review is an academic research tool that furthers the construction of knowledge, amplifies the hypothetical constructs from which an investigation starts, and provides the historical, spatial and temporal elements to interpret a reality from any discipline.

Documentary research plays an essential role in academic research. It allows us to understand events through a systematic review of the information to analyze the meanings manifested in the documents. In this way, documentary review enables a deep understanding of what "is happening" and "how it is happening" in specific social settings. In this way, it is possible to overcome the exclusive focus on mere perceptions and to reveal social practices that are normally "hidden" from the everyday view.

On the other hand, the interest of discourse analysts lies in elucidating how the semantic content of concepts and meaning structures establish their direct relation to the social, political and cultural structures of societies. In other words, we define a discursive dimension as a unit of signs (homogeneous or heterogeneous) syntactically and paradigmatically ordered, which have the purpose of containing discourses and producing actions (praxis). Therefore, discourse analysis seeks to elucidate the complex network of signs in language systems conditioned by social dynamics and power structures [4]. Discourse, then, fulfills a series of functions beyond the mere transmission of information and it is in this way that subjects and objects exist, are built, represent, connect and are given a meaning and relative status within social dynamics [5].

The combination of both traditions of qualitative methodologies within academic research, documentary review and critical discourse analysis, enables us to study, through their approaches and meanings, the identified praxis of Universities and Polytechnic Schools in Ecuador and how they have conceived and implemented the concepts of student well-being and formative accompaniment. To this end, our documentary approach provided the context through lengthy fieldwork. Critical discourse analysis provided a rigorous analytical framework focused on the study of meanings through which communication politics work and student welfare are applied within the institutions of Higher Education in the country.

3 Theoretical-Conceptual Approach to Well-Being and Formative Accompaniment in Higher Education

The challenge for education in general and higher education in particular is to develop the capacities of students to the maximum. In unequal societies, this is not just the fundamental objective but also a real mechanism for the construction of equal opportunities, as the seed for democracy and the construction of citizenship and social mobility.

The capability to recognize higher education as a generator of opportunities and equity also leads to monitor the socio-structural factors that affect it, from a point of view that highlights complexity, since educational institutions, as indicated [6], can be spaces for cultural reproduction.

In this regard, a welfare and training support approach can further an understanding of the socio-structural factors that condition crucial aspects of higher education. Some authors, such as Amartya Sen [7], complicate this discussion. In this standpoint, well-being cannot be reduced or equated with pleasure, happiness or satisfaction. Other values lead people to transform their resources into freedom to achieve their own ends. In this way, it is important to recognize community management and participation.

Along the same lines of reflection, Martha Nussbaum [8] proposes a broader analysis on education and its mission of creating capacities as a key methodological and conceptual tool. The goal is to improve educational systems and the corresponding academic training paths.

This approach to well-being and comprehensive accompaniment to the student body that takes up what was proposed by A. Sen and M. Nussbaum [9], is based on the concept of human development as an expansion of capacity. That is, as the possibility that people have (in this case students, teachers, actors that make up a university community) to fulfill themselves and achieve the life that they value thanks to their human capacities. In this sense, well-being is understood in terms of what a person can "do" or achieve, insofar as they can make decisions and act freely.

Over the last decades and, from the viewpoint of various disciplines, the interest of the study of well-being and training accompaniment have been strengthened as strategies to guide and solidify the learning experience of students in higher education.

The understanding of the concept of well-being is directly linked to recognizing the transition that it has had as a category of analysis throughout history. From its conception as a State policy through the so-called Welfare State in the 19th century, its adaptation to the Latin American context in the 20th century, the crisis (economic, social policy) of the welfare model and the corresponding questions about the modes of distribution and redistribution for the common welfare.

In turn, the application of the notion of well-being to the field of higher education does not exempt itself from this socio-historical heritage, and it allows us to reflect on the ways in which educational institutions arrange the issues that affect experiences witnessed in the academic and professional training courses.

In this sense, the starting point is to understand well-being in the sense of fullness, of feeling good about oneself. It regards a state of personal satisfaction that considers as positive and/or adequate aspects such as health, success (social, economic, professional), pleasure and harmony with oneself and with the environment.

Undoubtedly, the concept of well-being has a primary relevance within the educational field (at all levels), since it is important to note that educational institutions should be concerned not only with the academic training of their students, but also with their integral development.

Therefore, in higher education institutions, it is necessary to start from understanding students as subjects/agents rather than as mere recipients of knowledge. This requires contemplating, analyzing and understanding the subjective and behavioral framework of youth in the third-level educational context. It is important to take into account their personal situation, their particularities and the socio-structural conditions of their biographies.

The components of their psychological structure, identity, temperament, character, personality, feelings, attitudes, values and principles, as well as the interpersonal modalities of socialization and social relations that they build and sustain, greatly influence their psychic/psychological development as well as in the way they develop and perceive the world.

From a psychological and socio-anthropological perspective, the context that encompasses the particular situation of a student includes the relationship and personal experiences that the student has with his family, his peers, teachers, authorities and other members of the educational community and society. Furthermore, the effect that these, in turn, exert on their subjectivity and socio-emotional development are noted.

Every subject considers, apprehends and internalizes each experience or relationship in a particular way of understanding and facing the world. Students are in constant construction, in terms of interaction with their environment. This process of self-recognition and of recognizing the environment support the development of the human being and its well-being.

As individuals signify their experiences, they will develop capacities to generate their own autonomy and integration with their environment. Culture is an important part of the structuring of subjects. It is a dynamic and moving relationship, which becomes more complex as the subject grows and culture transforms. This is evident during the passage through university classrooms.

During academic training, the institutional environment contributes to the formation of its own individual processes and enables the construction of competencies for individual, family and citizen development. These competences are usually identified in decision-making, actions for conflict resolution, peaceful coexistence and recognition of risk situations, among others.

Likewise, from the perspectives of Psychology and Pedagogy, in order to broaden the outlook regarding the conception of the formative process in higher education, well-being has been incorporated from a broader perspective, less reductionist, in which it must be understood linked to its psychosocial components. Social well-being arises from a notion of subject that highlights interpersonal and intergroup relationships within the social context [10] and refers to the assessment that a person makes of circumstances and functioning within the social environment of which it is a part.

In this sense, the concept of scaffolding, developed by Brunner [11] proposes the construction of tools that start from recognizing the educational needs of the subject and identify resources and tools so that students can complete a task, count on methodological guidance among other circumstances that underpin their personal and social cognitive development.

Some authors propose various statements to serve groups in higher education applied to processes of inclusion of vulnerable groups in society. Other analysis show that there are a series of variables that go from vocational orientation, study habits, socioeconomic conditions, family contexts and psycho-emotional conditions, as well as students who enter the higher education system, "with disadvantageous cultural capital, difficulties in their academic biographies and relatively more uncertain future expectations "[12]. All of these aspects influence the successful transition of students facing the higher education system.

In this line of analysis, student well-being and training support as axes of higher education propose the challenge of incorporating support structures within the educational models of higher education institutions. This allows institutions to build the necessary conditions for the maximum development of the students' capabilities.

Hence, what is proposed is a comprehensive perspective of training support linked to well-being. Institutional academic managements should incorporate these aspects to the center of the educational processes themselves capable of detecting individual, socio-structural, psychosocial and other particularities. Attended in a timely manner to make an impact on educational processes, the idea is to minimize the risk of student dropout. Here are some elements to consider as part of this comprehensive perspective:

i. Constant evaluation and review of curricular designs, achievements, learning objectives and evaluation systems;
ii. Evaluation and review of learning environments, implementation of active methodologies, including infrastructure, scholarship programs, accompaniment;
iii. Application of support mechanisms that allow identifying the achievements, challenges and individual difficulties of the students during their educational trajectories, to incorporate opportune mechanisms of attention and monitoring of student performance;
iv. Implementation of routes and protocols for effective pedagogical interventions for the population with potential risk to enable, as mentioned before, the maximum

development of students' capacities and the successful completion of their academic and professional training processes.

The training accompaniment, and its link with well-being, mainly seeks to enhance the academic performance of students in each area of knowledge during the training process. The objective is to promote the meaningful learning of the knowledge that will guarantee their human and professional development. It is a relevant tool of academic management, as various investigations indicate when proposing that distributed leadership in the educational community aims to improve student performance [13].

In assisting higher education management, this approach considers the difference between "teaching and learning". It seeks to build knowledge through dialogic and quality relationships, which, going beyond the transmission of content, provoke interest in learning and promote the personal and professional development of the student body within shared projects of coexistence, respect for diversity and commitment to education as a mechanism for transformation and social mobility.

This is directly related to the role of teachers in their capacity as guides and trainers. For this reason, the relationship established between the teaching staff and the student body is of great importance. Through timely processes of pedagogical mediation and effective didactic communication strategies, the goal is to provide feedback on the educational trajectories of the students and promote greater and better learning.As stated by García-Pérez and Mendía [14] "Accompaniment is considered a fundamental dimension in the development processes of individuals and groups (Hsieh, 2014); relevant to promote the itinerary of personal and social growth of each individual. In addition, if the accompaniment is carried out in the process of building a shared project, it is informed through the quality of the relationship and the understanding of a social meaning in the experience is made possible (Vigotsky, 1997)".

Accompaniment must be present throughout the student's training journey. That is, in the admission processes to higher education. During leveling as a strategy to match basic knowledge and skills. Along the training trajectories to promote the permanence/retention of students and avoid dropping out. As well as in the stage of completion of studies to promote timely graduation, in addition to monitoring graduates and their inclusion within the field of work.

Formative accompaniment, then, requires efficient follow-up strategies, which are articulated from the academic-institutional management, in order to know how the students are learning, what are their main achievements, challenges and difficulties. The idea is also to strengthen the integration and development of the student body within the institution and its link from it to society.

At the academic level, among the most frequent strategies, we can mention leveling activities, monitoring, tutorials, reinforcement workshops, counseling, etc. These activities are to be complemented with mechanisms that promote student participation in higher education institutions, such as cultural, social, sports, exchange and recreational activities.

As a sustained, systematic and pertinent process, it promotes better learning and, in this sense, involves reflection and analysis on the didactic circuits. It helps to identify if the objectives are achieved in the different training itineraries that make up higher education. In sum, accompaniment is a key element that contributes not only to the personal, academic and professional development of the student body. It also furthers the achievement of the missionary goals of higher education institutions: to build knowledge and train professionals with quality, efficiency, relevance and social responsibility. On the other hand, accompaniment, as an integral dimension of academic training, can have a significant impact on reducing student dropout, a situation that significantly affects Ecuadorian as well as regional universities.

4 Discussion and Results

The Ecuadorian Higher Education System is a mixed system made up of public and private institutions. It is made up of 60 universities and polytechnic schools, 33 public, 8 private individuals who receive income and assignments from the State, and 19 private individuals who are self-finance. In addition, it has 241 higher-level technical and technological institutes (112 public and 129 private) (SENESCYT, 2018). As a whole, they offer around 3 770 degrees nationwide (1270 technical and technological degrees, 1979 third-level university degrees, and 521 fourth-level programs) [15]. (Table 1).

Table 1. Distribution of 2018 enrollment of universities and polytechnic schools, according to financing mode.

IES type	Number	Percentage
Public	357 358	60,70%
Co-financed particulars	152 076	25,80%
Self-financed particulars	79 489	13,50%
Total	588 923	100%

Source: ministry of education, higher science, technology and innovation (SENESCYT) - higher education information system (SIIES) - (2018)

According to figures from SENESCYT, in 2018, 588,923 students enrolled in undergraduate and graduate programs in the 60 universities and polytechnic schools. 60.7% of enrollment is concentrated in public universities and polytechnic schools, 25.8% in co-financed private institutions and 13.5% in self-financed universities.

During the last years, the Higher Education System had important advances in relation to access, coverage, quality and research processes. However, other problems persist and their deep treatment with formative accompaniment and training are pending academic praxis such as dropout, retention, terminal efficiency.

As reference, in a 2017 World Bank study, it is noted that approximately 50% of students in Latin America who entered higher education at some point did not complete their studies [16]. In the same study, for Ecuador, the data ranges between 35% and 40%.

If we consider that higher education is a social activator, equity becomes the way to build equal opportunities. Therefore, it becomes relevant to incorporate cross-cutting mechanisms to battle against inequalities in issues such as access, retention and successful titling. For this reason, it is necessary to promote educational inclusion policies that are not exhausted in access to higher education, but are accompanied throughout the student's journey to effective graduation. This process tends to avoid educational dropout, especially in the early years and between those populations with a historical backwardness that become more vulnerable to access, continue and finish higher education. Considering processes of "inclusion under conditions of inequity has the consequences of increasing repetition, dropping out of studies and low mastery of knowledge and skills" that will immediately have an impact in job placement and, as the ultimate goal, improve life quality for people who access higher education. [17].

The problems that students most frequently face throughout their educational journey are associated to difficulty in relating to the environment, problematic use of narcotic and psychotropic substances, eating or behavioral disorders, situations of violence and/or sexual violence, unplanned pregnancies, vocational disorientation, lack of study skills, feelings of frustration, difficulty in handling pressure, frustration, abandonment, financial difficulties.

In this sense, as Table 2 shows, there is a robust regulation that positions training accompaniment and well-being as fundamental axes to achieve the principle of equal opportunities. Nonetheless, higher education institutions have not been able to implement programs with a systemic approach, as well as policies and tools that allow monitoring and accompanying students who face these psychosocial and psycho-educational issues.

Table 2. Regulations on student well-being, comprehensive support and curricular adaptation

Norm	Theme	Content
Higher Education Organic Law. Art: 86	Attributions of the student welfare units	It establishes the obligation of the IES to maintain an administrative unit of student welfare. Among the most important attributions it must have are: promoting respect for the rights and physical, psychological and sexual integrity of the university community; implementing policies and programs to prevent and care for victims of sexual violence; generate integration projects for historically excluded groups. Developing vocational guidance processes

(*continued*)

Table 2. (*continued*)

Norm	Theme	Content
Regulation to Guarantee Compliance with Free Public Higher Education. Arts: 8, 10	Gratuity and prohibition of charge in student welfare services	Institutional student welfare services are within a long list that includes library, computer services, scientific infrastructure, etc. These services should not be charged to students. Likewise, in public higher education institutions the principle of gratuity is applied to student welfare services, among others
Regulation for the Granting of National Scholarships for Postgraduate Studies 2017. Arts: 15, 18, 20	Attributions of the administrative units of student welfare	It focuses on public IES and establishes the important role that student welfare units play. These units are in charge of monitoring the scholarship recipients at the fourth level
Regulation to Guarantee Equality of All Actors in the Higher Education System. Art: 17	Designation of units	IES have the obligation to establish in their administrative structure one department or area exclusively focused on student welfare to safeguard the right to education
Academic Regime Regulations. Arts: 85, 85A	Curricular Accommodations Division	In order to meet special educational needs that may or may not be related to disability, IES can propose: affirmative action programs, policies and plans. The changes made are divided into significant adaptations, when they modify the program or the object of study, and not significant when they only modify the duration, methodology, extracurricular activities or assessment instruments

(*continued*)

Table 2. (*continued*)

Norm	Theme	Content
Resolution 079–2021 of the CES. Art: 1	Curricular adaptations	Authorizes IES to make adaptations such as increase in the number of courses, increase in the number of students per course and per cohort. Between 40% and 200% according to various criteria
Prevention and care protocol in cases of conflict, violence, harassment and discrimination based on gender and sexual orientation in public higher institutes	Coordination, creation and operation of welfare units	It is a binding action protocol for Higher Education Institutions. It regulates the creation, competition and purposes of the student welfare units. This protocol establishes the routes of action in cases of gender violence, from detection, intervention and attention to victims. Likewise, it also establishes criteria for the prevention of violence and comprehensive support for the victims

When analyzing the regulatory frameworks and public policies in higher education issued in Ecuador in recent years, student well-being and training support appear as rather recent concepts, particularly from the transformations carried out in the higher education system in recent years. Both, the Ecuadorian Constitution and the Organic Law of Higher Education (LOES) and its Regulations, have established clear principles and guidelines from which higher education institutions should tend towards ensuring the quality of education. In this context, well-being and support are key axes.

For the purposes of this study, after the documentary review regarding the information on formative accompaniment and student welfare of the Universities and Political Schools of the country, a systematization and categorization is carried out. The purpose is to understand the meanings from where well-being and training support are conceived and the standpoint from where the implementation of the different programs and services are executed.

4.1 Places, Meanings and Gender Roles of Well-Being

All the Universities and Polytechnic Schools have a department of student welfare. However, the place that these departments hold within the organizational structure are mostly third-order, that is, subordinate to other administrative areas and with little capacity for decision-making. Their link is given mainly from the administrative component and with little incidence and relationship with the academic areas.

Likewise, 80% of the welfare departments are headed by women, due to the "natural" association of the scope of the functions of these departments as a task assignable to the feminization of care. That is, in this case, gender roles as a sociocultural pattern for the configuration of identities, representations, discourses and social positions. Therefore, well-being and the tasks that derive from it fall into the reproduction of patterns and stereotypes. This fact leads to certain segregationist conceptions regarding the choice made by women and men in their professional performance. In this way, certain professions chosen by women and socially assigned to them as natural, legitimate and reproduce sexual or arbitrary cultural habits established as universal, within the framework of the trans-historical constancy of the relationship of male domination that perpetuates existing divisions. This framework established around gender identities which in turn nourish the spheres of symbolic reproduction [18].

4.2 Types and Approaches of Well-Being and Formative Accompaniment

In the majority of the cases reviewed, about 60% of the higher education institutions implement traditional programs of well-being and student accompaniment that are aimed primarily at offering a set of assistance services, focused mainly on subsidizing needs, student materials (food, transport, housing or study grants, connectivity grants, etc.). This view is anchored in conceptions that propose independent, unsustainable and non-articulated logics of construction of well-being and training support surrounding substantive functions of higher education.

However, several of these problems are preferably faced from an additive approach in principles and in their practice, restricted to access and/or granting of goods and services - in many cases, from a purely welfare perspective.

In other cases, 30% of higher education institutions carry out actions strongly linked to an administrative vision. These actions rely on bureaucratic logic. In other cases, there is a tendency to implement assistance for access to material goods and services. In a large percentage of cases, actions are carried out closely linked to the implementation of "extracurricular" activities as their main form of operation.

Of all the Universities and Polytechnic Schools analyzed, only 10% have taken important steps towards the incorporation of training accompaniment programs from a wellness approach with a significant impact on the teaching-learning processes through accompaniment programs, tutorials or mentoring.

Due to the aforementioned, it is essential to analyze and debate the relevance of strengthening the praxis of well-being and support in higher education. They work as valid strategies within the framework of the implementation of programs that aim to strengthen the quality of higher education through strategies that favor greater inclusion and access to higher education. In addition, they articulate mechanisms that effectively confront dropouts and promote student retention/permanence as well as the successful completion of academic and professional training processes (graduation - terminal efficiency).

This is an innovative approach to student well-being and formative accompaniment. It is based upon human development and the realization of capacities. That is to say, a proposal that goes beyond a strategy focused on the sole provision of goods and services to the student body, as is frequently found in the regulations of some higher education institutions.

In this context, both student well-being and training support must be addressed while taking into account the recognition of its multiple meanings (or dimensions). These are associated with the plurality of actors involved, regarding quality as a guiding criterion of academic training activities, diversity of ways of relating to the environment and the intercultural richness of those who make up higher education.

Our proposal, then, is to reconsider well-being and formative accompaniment as mechanisms (rather than ends in themselves) that contribute to excellence in the functions and substantive processes of higher education. In order for an institution to be considered of excellence, it is essential that its actors feel good, are well and can access, use, enjoy and participate in the academic, social, human and material satisfaction associated to environments conducive of human fulfillment (personal and collective).

In the same sense, the possibility of generating better conditions and levels of well-being would be associated with the achievement of quality and academic excellence. For a higher education institution to guarantee and provide levels of well-being to its community, it is necessary that it recognizes quality as a criterion of its institutional-academic management processes, of its training and curricular activities, as well as in the access and quality of its services. The configuration of the institutional habitat is very important too.

From the human development perspective, well-being and formative accompaniment require, institutionally, the deployment of activities and processes that promote the realization of human capacities and that guarantee the material, social, affective, spiritual, academic conditions for each person or university group.

In this context, wellness and support programs should be designed and implemented within forms of organization and institutional articulation that can be translated into sustainable, healthy and harmonious practices and lifestyles. It is crucial to take into account both the diversity of beliefs, cultural and social representations of university actors, such as the "mosaic" of educational, socioeconomic and intercultural territories that are expressed and intertwined in every single institution of higher education.

In this sense, well-being and support are closely related to the perspective of quality with responsible autonomy and inclusion. Substantive functions of higher education cannot be measured only by acquisition/accumulation of knowledge and linked to the cognitive process or acquisition/implementation of specific infrastructures. These are usually measured through arbitrary parameters that remain distant from the diverse realities that coexist in higher education institutions. On the contrary, they will be addressed by social responsibility, linked to the creation of opportunities towards those groups with historical disadvantages, as the basis of social justice.

In this way, through a set of principles, management modes, policies and services oriented towards quality, it will not only contribute to the excellence of higher education, but also to the generation of institutional environments and academics that promote, foster and enhance institutional values. The life quality of its actors and groups, as the construction of academic communities are means to ensure that all members develop the appropriate quality of higher education as a joint commitment and task.

5 Conclusions

Despite the existence of a broad regulatory framework that makes the importance of well-being and support visible, Ecuadorian higher education has some problems related to the educational path of students that still need to be solved, like: dropout, low academic performance and repetition, low graduation rates and limitations when implementing curricular adaptations.

Several studies show that the price of not considering the processes of access to education linked to conditions of inequity and socioeconomic inequality, is the increasing repetition of dropping out of studies. The consequence is a low mastery of knowledge and skills, which will immediately have an impact on the insertion of employment. The ultimate goal is the improvement of life quality for the people who access higher education [17].

Therefore, the construction of programs from a welfare approach must guarantee equal opportunities. It is of great importance and must be a crosscutting process regarding issues such as access, retention and successful qualification, coverage and diversification of the offer and linkage with productive and labor sectors.

There is a need for educational inclusion public policies that are not solely centered in access to higher education. Further, they should be accompanied throughout the student's journey to effective graduation, avoiding educational dropout, especially in the early years and among those populations with historical lag that become more vulnerable when it comes to accessing, continuing and successfully completing the higher education training experience.

In this case, we return to the approach of well-being and training support as two axes of academic management. Such an approach seeks to favor the development of students' capacities and, thus, contribute to maximizing their performance during the training, human, academic and professional processes. In other words, if the multi-causality of retention is considered, the tools and programs that Higher Education Institutions can implement to identify, derive and manage personal or environmental difficulties of a student, will have a direct consequence an impact in retention.

References

1. Tinto, V.: Dropout from higher education: a theoretical synthesis of recent research. Rev. Educ. Res. **43**(1), 89–125 (1975)
2. Donoso, S., Donoso, G., Arias, Ó.: Initiatives of retention of higher education students. Qual. Educ. (33), 15–61 (2010). https://doi.org/10.31619/caledu.n33.138. Accessed 12 Sep 2021
3. Donoso, S., Schiefelbein, E.: Analysis of the explanatory models of student retention in the university: a vision from social inequality. Pedagogical Studies **XXXIII**, (1): 7–27 (2007)
4. Foucault, M.: Archeology of Knowledge. Siglo XXI, Mexico (1970)
5. Gee, J.P.: An Introduction to Discourse Analysis: Theory and Method. Routledge (2014)
6. Bourdieu, P., Passeron, J.C.: Reproduction: elements for a theory of the educational system. Les Éditions de Minuit, Paris, Paris (1971)
7. Sen, A.: Development and Freedom. Oxford University Press, United Kingdom (1999)
8. Nussbaum, M.: Building capacities: proposals for human development. Paidos, Barcelona (2012)

9. Nusbaum, M., Sen, A.: Comp.: Quality of Life. Fondo de Cultura Económica, México D.F. (1998)
10. Blanco, A., Valera, S.: The basics of psychosocial intervention. In: Blanco, A., Rodríguez Marín, J. (eds.), Psychosocial Intervention. Madrid, Prentice Hall (2007)
11. Brunner, J.: Higher Education in Latin America. Changes and Challenges. Fondo de Cultura Económica, Santiago de Chile (1990)
12. Landinelli, J., et al.: Scenarios of diversification, differentiation and segmentation of higher education in Latin America and the Caribbean. In: https://www.researchgate.net/publication/242514102_ESCENARIOS_DE_DIVERSIFICACION_DIFERENCIACION_Y_SEGMENTACION_DE_LA_EDUCACION_SUPERIOR_EN_AMERICA_LATINA_Y_EL_CARIBE (2008). Last accessed 12 Sep 2021
13. Leithwood, K., Louis, K.S.: Linking Leadership to Student Learning. Jossey-Bass, San Francisco (2011)
14. García Pérez, A., Mendía, R.: Educational accompaniment: the role of the educator in learning and solidarity service. Curriculum Magazine and Teacher Training **9**(1) (2015). http://www.ugr.es/~recfpro/rev191ART3.pdf. Last accessed 10 Sep 2021
15. National Secretariat of Higher Education, Science and Technology (2018)
16. Ferreyra, M., Avitabile, C., Botero, J., Haimvocih, F., Urzúa, S.: Turning Point: Higher Education in Latin America and the Caribbean. World Bank Group, Washington (2017)
17. Aponte-Hernández, E.: Inequality, inclusion and equity in higher education in Latin America and the Caribbean: trends and alternative scenario in the 2021 horizon. UNESCO-IESALC, Caracas (2008)
18. Bourdieu, P.: Male domination. Anagrama, Barcelona (2000)

Electronics and Automation

PID-Dahlin Polynomial Speed Controller Optimized by Ant Colony Algorithm on an ARM Platform

Sofía Torres, Miguel Melo, and William Montalvo(✉)

Universidad Politécnica Salesiana, UPS, 170146 Quito, Ecuador
wmontalvo@ups.edu.ec

Abstract. A polynomial PID controller does not have the same transient response as a conventional PID, since it tends to be relatively slower, but at the same time much more stable. To demonstrate the above, this paper proceeds to develop two speed controllers for a permanent magnet DC motor. One is a PID controller with standard structure tuned by Euler using the PID Tuner tool of Matlab, and the other is a polynomial PID controller based on the Dahlin algorithm, which is tuned using Ant Colony Optimization (ACO). The controllers are implemented on a card with Advanced RISC Machine (ARM) technology of the STM32F4-Discovery family, programmed in graphical language using Simulink from Matlab. The performance of the controls is compared using the Wilcoxon statistical method of IBM's Statistical Package for the Social Sciences (SPSS), which determines the best control based on the Integral of Absolute Error (IAE).

Keywords: Proportional Integral Derivative (PID) · Dahlin · Wilcolxon · STM32F4-discovery · Integral Absolute Error (IAE)

1 Introduction

In the industrial area, as well as in the automotive and robotics fields, there are several variables that are controlled for different processes, such as speed, temperature, flow and pressure as the most common ones. For this particular case, the variable to be controlled is the speed, since it is one of the most typically controlled and stabilized variables in the fields of automation and control. For this variable, the Proportional Integral Derivative (PID) controller is one of the most commonly used in process control.

Due to its simplicity, in addition to its high performance, the PID controller is the most popular closed-loop controller in the field of automatic control. Used in various processes applied in the industrial field, the traditional PID controller currently has an approximate usage of 90% or even more within control loops [1]. In addition, the performance of a PID controller is very satisfactory in most industrial control applications [2], however, in specific areas, or in special performance demands, the traditional PID controller is not enough because it does not meet the required characteristics, requiring a more optimal control, such as the Dahlin.

M. Zambrano Vizuete et al. (Eds.): CI3 2021, LNNS 511, pp. 189–202, 2022.
https://doi.org/10.1007/978-3-031-11438-0_16

The Dahlin controller is a modification of the Deadbeat controller that produces a smoother exponential response. Its main objective is to bring the controlled variable to a steady state in a minimum sampling time [3]. For this case, the variable to be controlled is the speed of a 12 Volt permanent magnet DC motor, which is part of an EPC control trainer module, which has a linear voltage amplifier to convert the logic voltage from 0 to 5 V control voltage into a power voltage from 0 to 12 V to feed the DC motor, as well as a pulsed signal generator for recording the speed of rotation of the same, being necessary for the implementation of the controller [4]; as the motor speed is a fast variable, it is essential to reduce the sampling time, being necessary an optimization of the controller, once the Dahlin algorithm is performed.

One of the controller optimization methods is the Ant Colony Optimization (ACO), which has been implemented by M. Dorigo in 1990, as a resolution method to optimize complex algorithms [5]. This controller bases its principle of operation on the behavior of these insects when searching for food around them, and when they find it, they evaluate it and leave a pheromone trail as a guide for the other ants. Previous research shows that ants communicate by pheromone trail, which helps them to choose a better path from the colony to their food [6]. The ACO needs a cost function for its operation. It can be delay time, settling time, steady state error or Integral Absolute Error (IAE).

This paper worked with the IAE index, which is one of the most used parameters in the measurement of performance in control loops, for its easy implementation and ability to measure the deviation of the process variable (PV), in relation to the Set-Point (SP), which allows to characterize complete records of data in control loops [7]. Considering the reduction in sampling time, response time of the controller, and the increase of data processed within it, a controller with good processing performance must be used, while it must have sufficient capacity to handle the control data, being in this case the STM32F4-Discovery card one of the most suitable instruments due to its technology.

The STM32F4-Discovery board consists mainly of an Advanced RISC Machine (ARM) processor, which is based on the Reduced Instruction Set Computer (RISC) architecture, which is designed to reduce the execution time, load and storage of various logical sequences, obtaining a better execution of data [8], where each of its pins are programmable and can be initialized as general-purpose digital input or output, in addition to pins with additional functions, that can be programmed to function as analog inputs, PWM analog outputs, serial communication interfaces. Each of its pins can work according to the need of the user or developer. In turn, the STM32F4 card is superior in data processing compared to an Arduino card, as well as in sending and receiving data through UART interface by having independent modules within the same encapsulation, and processing latency is reduced, while it is more convenient compared to a Raspberry Pi card by not having an operating system to load before being operational, having a boot time of less than 1 s [9].

2 Method

The implementation methodology for the PID-Dahlin polynomial speed controller tuned using Matlab PID Turner and ACO software is described below:

2.1 Methodology for Data Collection Using the ARM STM32F4-Discovery Card

In order to obtain the speed data of the permanent magnet DC motor from the STM32F4-Discovery card, the card's internal PWM Capture module was used, while the Matlab software was used to collect and process the data obtained, which receives the data from the card through the card's serial port.

For communication and subsequent collection, programming and control of the STM32F4-Discovery card, the WAIJUNG library for Matlab was installed, which allows, through the Simulink graphic interface, to create, control and modify the parameters of the PID controller. Different modules integrated into the board and in the Simulink library were used to obtain data such as: ADC, Serial Host, PWM Capture, etc. and tools that allowed the control such as: Basic PWM, Gain, etc.

Figure 1 shows the physical elements used in the development of the controller.

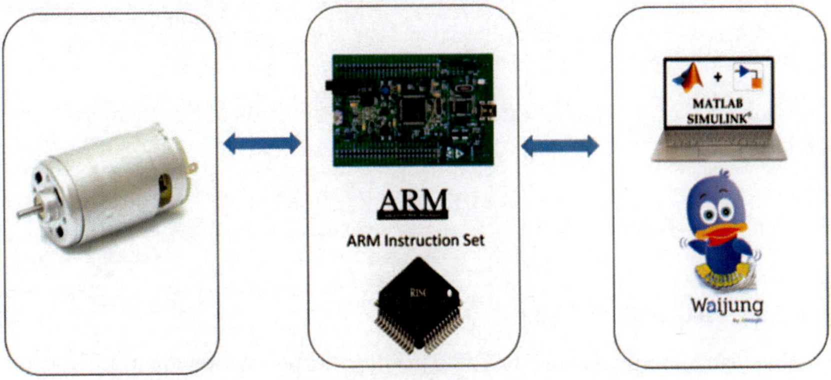

Fig. 1. Block elements for control implementation

2.2 Methodology for System Identification

The black box methodology was used within the Matlab Software version 2015b to take the samples from the plant, and within the software, 672 samples were taken with a time of 10 ms between samples.

In the identification process, the Set-Point input value, given by a potentiometer, is represented in percentage of desired speed (0-100%), where its maximum value is 3 V, while at the output is the DC motor speed with a maximum value of 2220 rpm. Figure 2 shows the data collected.

The "System Identification" tool was used to obtain the transfer function of the system.

The transfer function identified by the software with the motor data in Fig. 2 is Eq. (1):

$$G(s) = \frac{46.1}{s + 46.96} \tag{1}$$

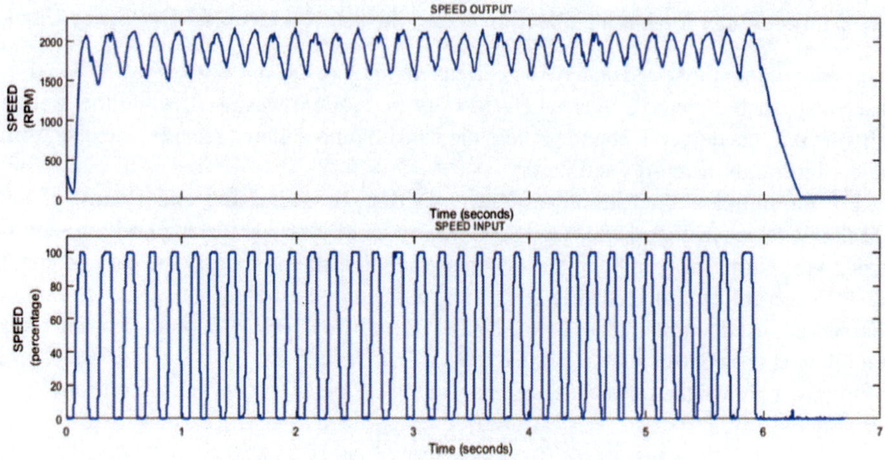

Fig. 2. Experiment input and output signals.

Subsequently, using Matlab software, the transfer function of Eq. (1) was discretized to obtain Eq. (2)

$$G(z) = \frac{0.3679}{z - 0.6253} \tag{2}$$

2.3 Tuning Methodology Using PID Tuner

The tuning was performed using the PID Tuner application within the Matlab software, as can be seen in [10], the transfer function obtained in Eq. (2) was entered. Then the tuning of the controller was performed, resulting in the graph shown in Fig. 3.

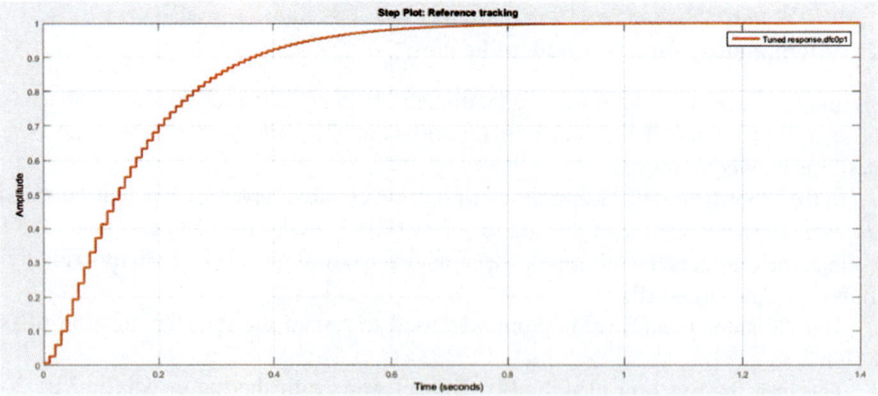

Fig. 3. PID controller step signal response.

With this application, the gain values shown in the Table 1 were obtained.

Table 1. PID gain values with PID Tuner

Gains	Values
Kp	0.05909
Ki	5.909
Kd	0.0001477

2.4 Tuning Methodology Using ACO

Ant Colony Optimization (ACO) represents the way in which ants communicate by leaving a pheromone trail. In [11], the algorithm represented in Eq. (3) is developed.

$$p\left(c_{ij}|s^{p}\right)=\frac{T_{ij}^{\alpha}.\eta(c_{ij})^{\beta}}{\sum_{c_{il}\,\epsilon N(s^{p})}T_{ij}^{\alpha}.\eta(c_{ij})^{\beta}},\ \forall c_{ij}\epsilon\ N(s^{p}) \tag{3}$$

where:

- **T_{ij}:** Is the pheromone value associated with each component c_{ij}.
- **η:** Is a function that assigns to each step of the construction a heuristic value for the feasible solution component $c_{il}\,\epsilon N(s^{p})$.
- **α y β:** Are the values that determine the importance of the pheromone.

This algorithm was implemented in a Matlab script, where experimental parameters were set, denoted in Table 2.

Table 2. Valores asignados para el algoritmo ACO

Parameters	Values
Number of iterations	2
Number of ants	30
Alpha value	0.2
Beta value	0.8
Alpha value	0.2
Pheromone evaporation rate	0.7
Number of parameters for pheromone	3

The PID values were obtained from Table 3.

2.5 PID-DAHLIN Controller Methodology

Its goal is to bring the output to a steady state in the shortest possible time [3]. The development methodology was as follows.

Table 3. PID gain values with ACO

Gains	Values
Kp	3.1272
Ki	8.6612
Kd	1.2539

With the sample data in Fig. 2, a transfer function with a time delay was obtained, as shown in Eq. (4)

$$G(s) = \exp(-0.01s) * \frac{1.514e06}{s^2 + 8560s + 4.137e04} \tag{4}$$

From Eq. (4), the transfer function was discretized by Zero-Order Hold (ZOH) obtaining Eq. (5)

$$G_{zoh} = G(z) = \frac{1.708Z + 0.01972}{Z^3 - 0.9528Z^2 - 1.058e - 16} \tag{5}$$

For the Dahlin controller, a desired transfer function is established from Eq. (6):

$$T(s) = \frac{e^{-Ls}}{Ts + 1} \tag{6}$$

where:

- L is the plant stabilization time
- T is the approximate constant time in sampling

For the development of the Dahlin controller, it starts from a transfer function within the z-domain given by Eq. (7):

$$T(z) = \mathcal{L}\{\mathrm{G_{zoh}(s)T(s)}\} = \mathcal{L}\left\{\frac{1 - \mathrm{e}^{-\mathrm{Ts}}}{\mathrm{s}} \frac{\mathrm{e}^{-\mathrm{Ls}}}{\mathrm{Ts} + 1}\right\} \tag{7}$$

For the development of the controller, Eq. (6) is modeled by assigning a value to T, obtaining Eq. (8):

$$T(s) = \frac{e^{-Ls}}{3s + 1} \tag{8}$$

Replacing Eq. (8) in Eq. (7), Eq. (9) is obtained.

$$\begin{aligned} T(z) &= \mathcal{L}\left\{\frac{1 - \mathrm{e}^{-\mathrm{Ts}}}{\mathrm{s}} \frac{\mathrm{e}^{-\mathrm{Ls}}}{3\mathrm{s} + 1}\right\} \\ &= \left(1 - z^{-1}\right) z^{-\frac{L}{T}} \mathcal{L}\left\{\frac{1}{\mathrm{s(3S + 1)}}\right\} \end{aligned}$$

$$= \left(1 - \frac{1}{z}\right) z^{-\frac{L}{T}} \frac{z\left(1 - e^{-1/3T}\right)}{(z-1)\left(z - e^{-1/3T}\right)}$$

$$= z^{-k} \frac{\left(1 - e^{-1/3T}\right)}{\left(z - e^{-1/3T}\right)} \tag{9}$$

where:

- T is the time period, in this case 0.01 s
- $\frac{L}{T} = k$

The Dahlin Transfer function is given by Eq. (10)

$$D(z) = \frac{1}{G(z)} \frac{T(z)}{1 - T(z)} \tag{10}$$

Equation (5) and Eq. (9) are replaced in Eq. (10), obtaining Eq. (11)

$$D(z) = \frac{1}{\frac{1.708Z + 0.01972}{Z^3 - 0.9528Z^2 - 1.058e - 16}} \frac{z^{-k} \frac{\left(1 - e^{-1/3T}\right)}{\left(z - e^{-1/3T}\right)}}{1 - z^{-k} \frac{\left(1 - e^{-1/3T}\right)}{\left(z - e^{-1/3T}\right)}}$$

$$D(z) = \frac{Z^3 - 0.9528Z^2 - 1.058e - 16}{1.708Z + 0.01972} \frac{z^{-k} \frac{\left(1 - e^{-1/3T}\right)}{\left(z - e^{-1/3T}\right)}}{1 - z^{-k} \frac{\left(1 - e^{-1/3T}\right)}{\left(z - e^{-1/3T}\right)}} \tag{11}$$

For the controller to be feasible: the degree in the denominator equation must be greater than the degree in the numerator equation, while k must have a value of:

$$3 - k \leq 1 \Rightarrow k \geq 2$$

The Dahlin function was obtained in a Matlab Guide and is shown in Fig. 4.

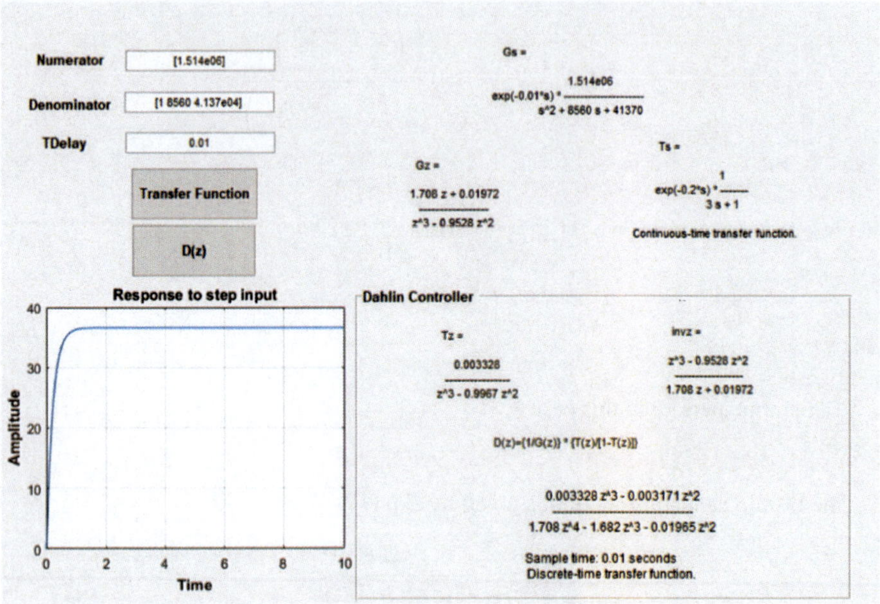

Fig. 4. Dahlin control function in Matlab Guide

3 Results

The ARM board can be programmed using Matlab Simulink Software to implement the equations and parameters calculated above. Figure 5 shows the base scheme of the Dahlin algorithm to be implemented on the ARM board to control the permanent magnet DC motor.

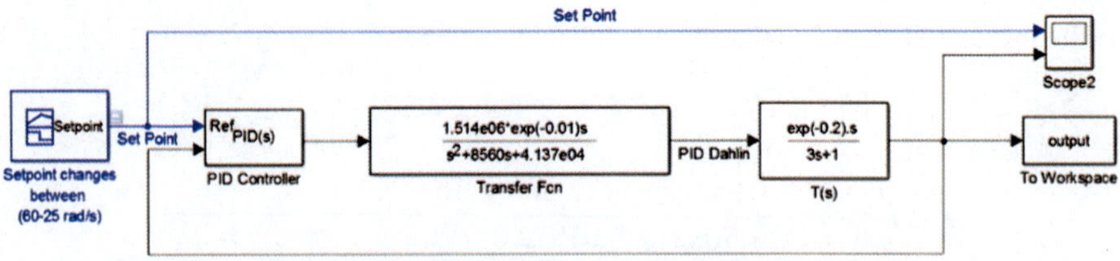

Fig. 5. Basic scheme of Dahlin PID control algorithm in Simulink.

3.1 Discrete-Time System Response of a Conventional PID Controller for Speed Control of DC Motor Tuned with Matlab

Within the Simulink software, the PID control algorithm was implemented using both the modules integrated in the Waijung library, as well as the modules of the Simulink program, as shown in Fig. 6, implementing the PID controllers as the transfer function module, once received the Set-Point data, the controller sends the output value to the DC motor so that it reaches the desired speed. The PWM Capture module returns the current speed value of the DC motor to the ARM card, so the algorithm can close the control loop, and make corrections to the output according to the PID control algorithm.

In this case, Eq. (2) was used with the PID parameters of Table 1.

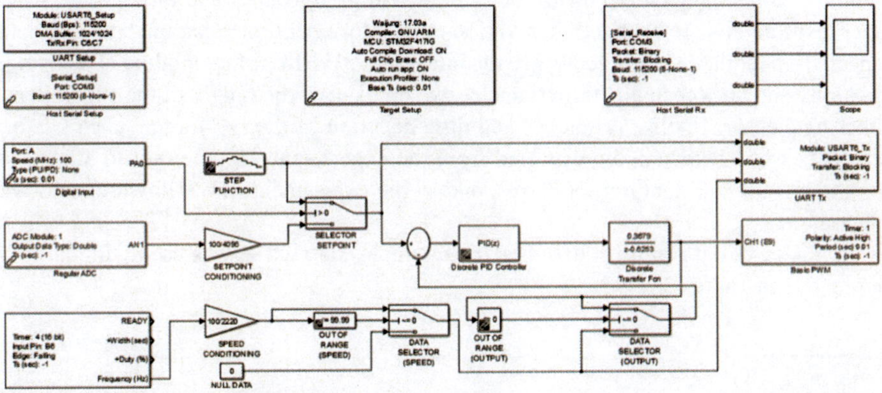

Fig. 6. Diagram for practical implementation of conventional PID parameter control tuned with Matlab.

Once the controller was implemented in the ARM card, the input data and the motor speed response were collected, as shown in Fig. 7. The red color shows the Set-Point value and the blue color shows the real time speed response. The ARM board while controlling the DC motor by means of its loaded PID algorithm, also delivers the Set-Point and speed input and response values, respectively, for later visualization, collection, storage and analysis.

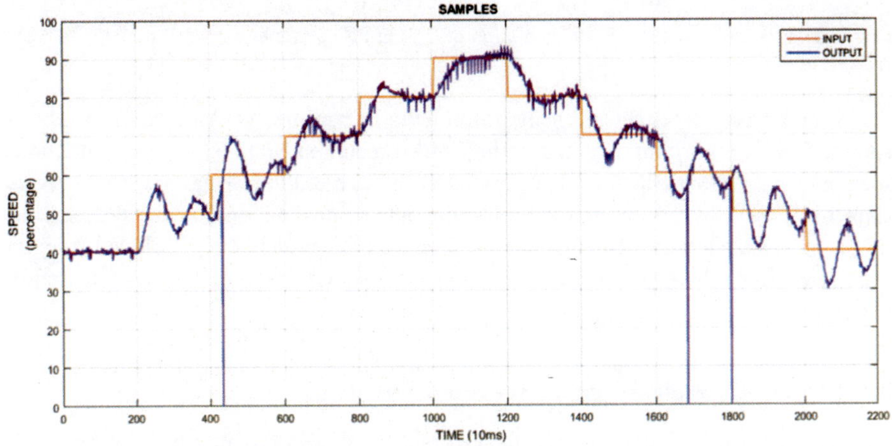

Fig. 7. Set-Point input signal and speed response of traditional PID controller.

3.2 Discrete-Time System Response of a Dahlin PID Controller to Regulate the Speed of the DC Motor Tuned with ACO

In order to test the performance of the Dahlin PID controller, the controller shown in Fig. 8 was implemented in the Simulink platform where the same blocks of the traditional PID controller are used, however, the parameters of the transfer function and the PID parameters, obtained through the Dahlin algorithm and through the ACO tuner respectively, are changed. In addition to the aforementioned parameters, two data cancellation modules are added, since the integrated PWM Capture module of Waijung, randomly and unexpectedly, throws one or two data out of the ranges of the experiment, either by electrical noise, or unexpected disconnection of the DC motor speed sensor, such data when detected, are canceled from the experiment. In order not to affect the controller with data that do not correspond to the experiment, the following steps are taken.

In this case, the Dahlin equation given by the Matlab guide was used with the PID parameters in Table 3.

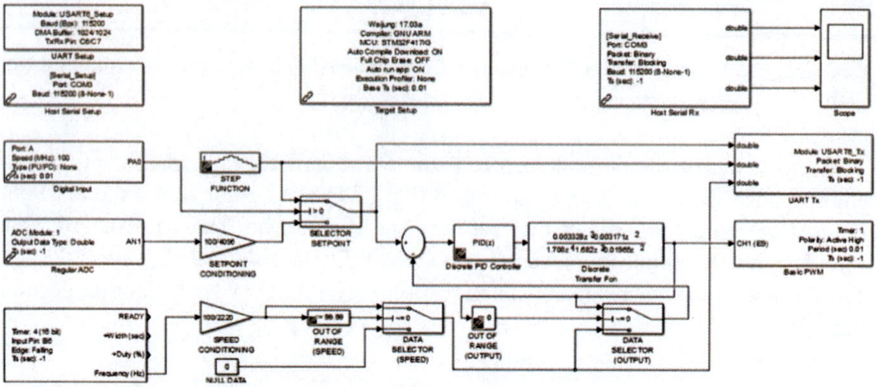

Fig. 8. Diagram for practical implementation of Dahlin PID parameter control tuned with ACO.

When loading the PID Dahlin algorithm with the parameters obtained from Table 3, and the Dahlin algorithm obtained in the Guide created in Matlab, the input and motor speed response data collection is performed again, marked in red and blue respectively, shown in Fig. 9, observing in addition to the new response signal of the Dahlin algorithm, two cancelled data because of out of range data, which were excluded from the experiment immediately detected, continuing with the real response values of the Dahlin algorithm.

3.3 Statistical Results Using the Wilcoxon Test

To compare the traditional PID with the Dahlin PID, the Wilcoxon test was used, which has a confidence level of 95% as explained in [12], as well as its development in the IBM SPSS program. The Integral Absolute Error (IAE) performance index, which determines

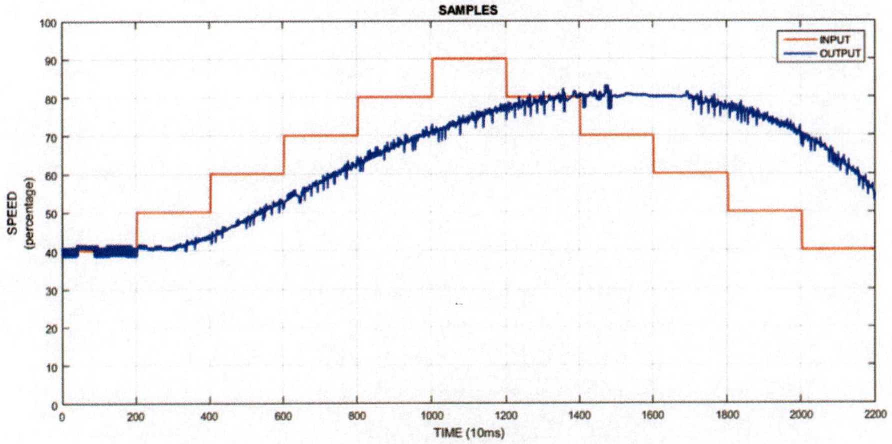

Fig. 9. Set-Point input signal and speed response of Dahlin PID controller.

the amount of wasted energy of the traditional PID controller with respect to the Dahlin PID, was used for this purpose. An estimation of the optimal IAE performance of PID controllers and servomechanisms is shown in [13].

A total of 30 experiments of the IAE of the two controllers were performed and analyzed using the Wilcoxon test. The tools used for the collection and analysis process were IBM SPSS and Microsoft Excel.

The following hypotheses were proposed.

- **Null hypothesis** (H_o): "The IAE performance indices of the two PIDs are equal in the process of speed control of a DC motor".
- **Alternate Hypothesis 1** (H_a)**:** "The IAE performance indexes of the Dahlin PID control tuned by ACO is lower than the conventional PID tuned by Matlab in the process of speed control of a DC motor".

Within the SPSS program, when performing the Wilcoxon test, a comparison is made between the IAE results of the Dahlin PID controller with respect to the traditional PID controller, resulting in the level of reliability (p), which is compared to the minimum level used by researchers in the area of education that corresponds to $p < 0.05$. This value is delivered in each and every one of the test experiments as shown in Fig. 10. This value determines whether the hypothesis of the experiment is accepted, or on the contrary is rejected, accepting the null hypothesis.

The statistical model obtained in the performance tests of the traditional PID and Dahlin PID controller, show that in all tests, the P value is less than 0.05 which indicates that the null hypothesis is rejected, accepting the alternative hypothesis.

4 Discussion

According to previous research in [14] and [15], it is concluded that the Dahlin synthesis added to the traditional PID control algorithm, ideally, and simulated have a better

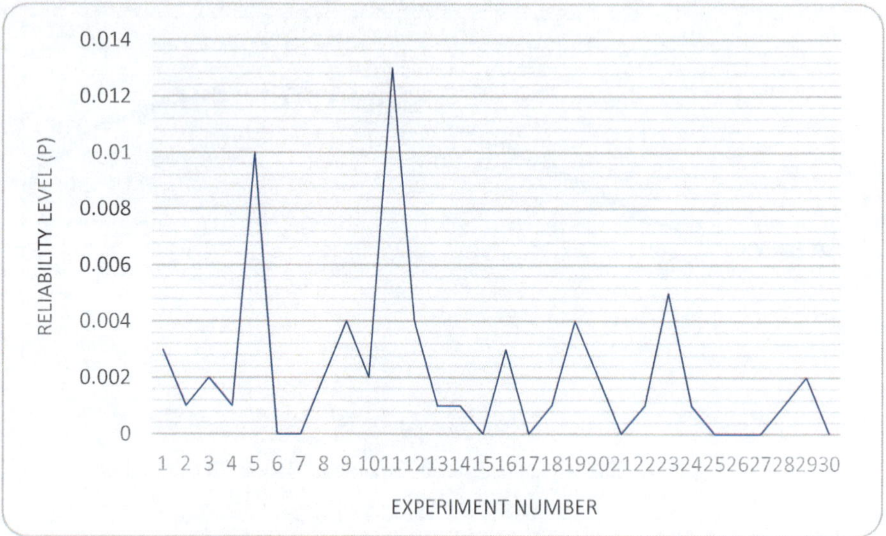

Fig. 10. P result with Wilcoxon, PID vs Dahlin PID test

performance, in addition to the response time to disturbances improves, as detailed in [16]. Demonstrating its practical application on a permanent magnet DC motor of the EPC trainer plant.

The research of [17] shows that an adaptive PID controller, based on Dahlin synthesis, has a longer settling time than a traditional PID controller, resulting unfavorable in the control of voltage output of a fuel cell, however, the Dahlin PID controller has a better response to disturbances than a traditional PID, while decreasing possible overshoot observed in Figs. 7 and 8. The Dahlin PID is ideal for controlling fixed variables, where its settling time is negligible, for example in the paper production industry, as well as conveyor systems in continuous manufacturing lines, can couple the new Dahlin PID parameters to the PID controller already established in the production plant.

With the performance results in [18], favorable results were obtained in the overshoot attenuation of the traditional PID controller when tuned by means of ACO, taking this control as a basis for the tuning of the Dahlin controller, given its favorable performance. On the other hand, ACO tuning requires a considerable amount of processing, being unfavorable to apply the ACO algorithm directly on a controller with limited resources, being the Dahlin controller more feasible to apply.

In the comparison made in [18], if a polynomial order Dahlin PID controller had been used instead of a Discrete PID, a considerable similarity in performance would be found despite handling different cost functions in the optimization algorithm.

With background from the STM32F4, Arduino Uno and Raspberry Pi Comparison Chart from [8], using an STM32F4-Discovery board that has an ARM architecture will decrease the latency in processing speed control, versus the latency that the Arduino board would have, as well as a reduction in boot time versus the boot time that the Raspberry Pi board normally has.

5 Conclusions

The Dahlin algorithm presented a satisfactory performance compared to a traditional PID speed control, since it is much more stable against physical disturbances, such as sudden braking or an unexpected increase in the mechanical load of the DC motor.

By analyzing the IAE performance index tests and the Wilcoxon test, the alternative hypothesis is accepted, since it manages to decrease the amount of energy wasted throughout the setup time.

The Dahlin PID controller within the industrial field, being stable against mechanical disturbances could be applied in production lines within medium to small companies, since having limited budgets they are not able to renew their production lines on a large scale. Instead, they can upgrade their existing production and manufacturing lines with a minimum investment cost.

References

1. Åström, Hägglund: PID Controllers: Theory. NC, USA (2013)
2. Alfaro, V.M., Vilanova, R., Arrieta, O.: A Single-Parameter Robust Tuning Approach for Two-Degree-of-Freedom PID Controllers (2009)
3. Ahmed, M.N.A.: Discrete Controller Design (Deadbeat & Dahlin Controllers), https://mnourgwad.github.io/CSE421/lectures/CSE421DigitalControlL10.pdf. Ultimo acceso 04 Jan 2021
4. DataLights National Instruments: Entrenador de Planta de Control EPC. 3rd edn. DataLights Cia. Ltda, Cuenca – Ecuador (2016)
5. Dorigo, Gambardella.: Sistema de colonia de hormigas: Un enfoque de aprendizaje cooperativo para el problema del vendedor ambulante. IEEE Tr. Evol. Comp. **1**, 53–66 (1997)
6. Alonso, S.: La metaheurística de optimización basada en colonias de hormigas: modelos y nuevos enfoques. Granada, Departamento de Ciencias de la Computación e Inteligencia Artificial, e.t.s (2004)
7. Cárdenas, D.: Desarrollo de software para inspección técnica de una aplicación CPM, pp. 25–26. Universidad de Concepción (2017)
8. Puello. C.: Computación de Alto Desempeño: Entorno de ejecución pasarela basado en procesadores de arquitectura ARM, pp. 14–16. Universidad Tecnológica de Bolívar (2015)
9. Lopez. E.: Hapkit: Interfase Háptica Con Control de Fuerza, p. 49. Universidad Carlos III de Madrid (2015)
10. Kolaj, W., Mozarin, J., Sifere, M.: PLCPIDTuner: application for PID tuning with SIMATIC S7 PLC controllers. In: 21st International Conference on Methods and Models in Automation and Robotics (MMAR), pp. 306–311 (2016)
11. Arito, F.: Algoritmos de Optimización basados en Colonia de Hormigas aplicadas al Problema de Asignación Cuadrática y otros problemas relacionados. Universidad Nacional de San Luis, San Luis (2010)
12. Gamarra, Pujay, Ventura.: Aplicación de las pruebas estadísticas de Wilcoxon y Mann-Whitney con SPSS, Revista De Investigación Multidisciplinaria CTSCAFE, https://ctscafe.pe/index.php/ctscafe/article/view/51. Ultimo acceso 05 Jan 2021
13. Alfaro, V.: Estimacion del Desempeño IAE óptimo de Reguladores y Servomecanismos PID. Universidad de Costa Rica (2005)
14. Wei, S., Rui, Huihe, S.: Flow Control Based on Dahlin Control Algorithm in ATM Networks, p. 6. Shanghai (2002)

15. Clitan, I., Abrudean, M., Muresan, V.: The control of an industrial monipulator's positioning system using Dahlin algorithm. In: 5th International Conference on Information Technology and Electrical Engineering, pp. 1–6. Rumania (2015)
16. Osorio, L., Mendes, J., Araujo, R., Matias, T.: A comparison of adaptive PID methodologies controlling a DC motor with a varying load. IEEE Journal of Systems Engineering and Electronics, 1–6 (2014)
17. Darjat, Sulistyo, Triwiyatno, A., Julian, E.: In: 5th International Conference on Information Technology and Electrical Engineering. Design of Adaptive PID Controller For Fuel Utilization in Solid Oxide Fuel Cell, pp. 1–6. Indonesia (2018)
18. Palacios, S., Chiriboga, S., Montalvo, W.: Controlador de velocidad PID-2DOF-ACO para motor DC sobre plataforma ARM, pp. 217–228. Revista Ibérica de Sistemas e Tecnologias de Informação (2020)
19. Aviles, E., Ortega, M., Montalvo, W.: Performance comparison of two electronic controllers on an ARM platform, pp. 1–12 (2020)
20. Mathworks: System Identification Toolbox. https://www.mathworks.com/products/sysid.html. Ultimo acceso 07 Feb 2019
21. Aimagin: Waijung Library. https://www.aimagin.com/waijung-1-stm32-target.html. Ultimo ac-ceso 02 March 2019
22. Mathworks: GUI de Matlab. https://es.mathworks.com/discovery/matlab-gui.html. Ultimo acce-so: 08 Aug 2019
23. Mathworks: Simulink. https://es.mathworks.com/products/simulink.html. Ultimo acceso: 01 Dec 2019
24. IBM SPSS software: https://www.ibm.com/analytics/spss-statistics-software. Ultimo acceso 01 Sep 2019

Assisted Pedagogy: Robot Alpha for Gross Motor Learning in Sublevel I

Byron Machay(✉), Carlos Ruiz, Natalia Contero, and Diana Nogales

Instituto Superior Tecnológico Vida Nueva, Quito, Ecuador
byron.machay@istvidanueva.edu.ec

Abstract. The present research project examines the integration of the Alpha 1 Robot as an assisted pedagogy for gross motor learning, where the main objective is the use of educational robotics in early education. The problem is the non-existence of robotics for teaching based on modern education and this is still using traditional education strategies and not based on the cutting edge of technology. In order to introduce the Robot Alpha, I as a technological tool, quasi-experimental research with qualitative and quantitative approaches was used, in which a pretest and posttest are used to children, in order to obtain a control group and an experimental group, in which the different behaviours will be studied; the items evaluated are in three stages: initiating, in process and acquired. The study sample was thirty students of initial education, divided into two groups, in which their study planning, attendance and even worse their grades will not be affected; all this execution activity is based on the guidelines of class planning requested by the Ministry of Education. During the research process, the children show great motivation, interest, and enthusiasm when they see and interact with the robot, which is considered another friend in their environment. The results are totally satisfactory, so it can be mentioned that the Alpha 1 robot does strengthen the teaching-learning of gross motor skills.

Keywords: Initial education · Assisted pedagogy · Educational robotics · Learning · Gross motor skills

1 Introduction

The research topic is related to the field of education, not only at the national level but also at the global level; this allows the improvement of the quality of education linked to science, innovation, sustainability and inclusion to prevail. [1] support that "Innovation policies, articulated with science and technology policies, are necessary to efficiently link the efforts of companies, governments and academic sectors."

The importance of the research topic is based on the innovation and relationship with robotics in the area of Sublevel 1 Education. Therefore, it will allow the change of methodological strategies in the teaching-learning process, which aims to adapt new techniques in the development of children's motor skills. It will be possible to obtain a modern education, motivated teachers in the techno-logical area, an adequate exploitation of gross motor skills and related to the cognitive area. Within the curriculum planning

M. Zambrano Vizuete et al. (Eds.): CI3 2021, LNNS 511, pp. 203–215, 2022.
https://doi.org/10.1007/978-3-031-11438-0_17

of sub-level 1 education is the exploitation of the children's skills and abilities, as well as the [2] one of its objectives is "To explore the movements of the child's body in order to develop gross and fine motor skills that allow for coordinated movements and actions to improve the body scheme".

The presence of robots in the classroom does not replace the teacher, if not an aid in the teaching process, also robotics does not try to train students in this discipline, but to take advantage of its functionality to generate learning environments according to modern education and give an excellent use of global trend technologies. For some authors, educational robotics is An experience that contributes to the development of new skills, new concepts, strengthens the student's systemic, logical, structured and formal thinking, while developing their ability to solve concrete problems [3].

Worldwide, there is an educational institution dedicated to integrating new technologies and robotics in the classroom so that students can develop their imagination and creativity. One of these institutions is GALILEI [4] emphasizes that the application of robots in the formal school program is an interdisciplinary teaching method that facilitates the development of these essential skills from a playful-formative perspective. Robotics and Technology in the classroom encourage innovative and collaborative learning while fostering project-based learning, allowing students to work interdisciplinary [5].

The Alpha 1 robot is applied in the initial sublevel 1 with the objective of strengthening the teaching-learning process in gross motor skills, considering that the technical and technological aspects of this humanoid are related to education. This will help it to be a mirror element, which the child can imitate its programmed movements within its capacity of functioning and internal memory. UBTECH [6, 7] emphasizes that "Alpha can be instructed to move forward, backward, right and left in the living room, on obstacle courses in the home or elsewhere." (p.01). [8] This humanoid can also be programmed by computer with advanced 3D software to play soccer, sing, dance and perform yoga exercises.

2 Content Development

2.1 ICT

The abbreviation ICT is based on the meaning of Information and Communication Technologies in the areas of information technology and telecommunications, which aim at the production, access, processing, and communication expressed in codes. The Internet and the computer are the main elements of the new technologies that influence the change in the way of knowing or relating ICT to mankind [9].

ICT is also known as all the resources and technical support tools that help to process, manage, and share information, they can be through cell phones, computers, televisions, laptops, etc. They are currently used in the field of education, both asynchronous (collaborative communication elements such as chat, e-mail, forum) and synchronous (collaborative communication tools, but with immediate access and in real time, where information can be shared).

2.2 Educational Robotics

Educational robotics is focused on technology, science and society that aims to include robotics in the educational curriculum, this is not an easy task for educational institutions, for the reason that there is a lack of knowledge about new technological strategies in the teaching-learning process. This is where we start betting on humanoid robots in the classroom, which will introduce the world of technology and a new digital era.

Educational robotics also known as pedagogical robotics is facilitating in the educational field to improve the teaching-learning process, in which allows the acquisition of new knowledge and projects thinking in both children of early childhood education and high school students. The breakdown of the traditional models of education and the beginning of modern education is due to robotics, which has become a resource and educational tool in the teaching-learning of science and technology [10].

Educational robotics, designated as an integrative discipline in the area of knowledge, consists of two independent processes. The first one is focused from the point of view of automation and engineering, which is aimed at the study and process of conceiving, designing, and building robotic mechanisms. The second is based on didactics, which is focused purely for educational purposes involving the disciplines of knowledge of education, teaching, and learning [11].

2.3 Alpha Robot 1

The company UBTECH conceptualizes that the Alpha 1 robot is a programmable humanoid that is employed in the area of education and entertainment, the 3D visual programming software and the PRP function (pose, record, and play) allows the robot to perform its high precision movements based on its servos and its controls are performed through an application from a cell phone or Tablet.

The humanoid robot is designed in resemblance to the human being, which allows to emulate the movements and communication methods through sounds and gestures. It is designed so that its movements are more curved than linear, this facilitates its imitations to a person, in addition its construction elements are of high quality from plastic to metals. The humanoid's movements are executed by 16 high quality servos, thus providing impressive stability when walking and maintaining balance in its programmed functions. The movements performed by the Alpha 1 robot depend on the user's programming.

The control of the robot is through an application of the humanoid downloaded freely from the official UBTECH website and is operated through a mobile device compatible with Android and iOS. The connectivity of the Alpha 1 for the control of the programmed movements is via Bluetooth 4 of the cell phone or tablet. The programming screen and the robot's software are fully interactive to create the Alpha robot's operating movements (Fig. 1).

2.4 Initial Sublevel I Education

Early childhood education is the education that children receive in their first years of life, being one of the most important stages of the human being, which begins to strengthen the way of reasoning, learning, and thinking. At this stage, children develop physical and

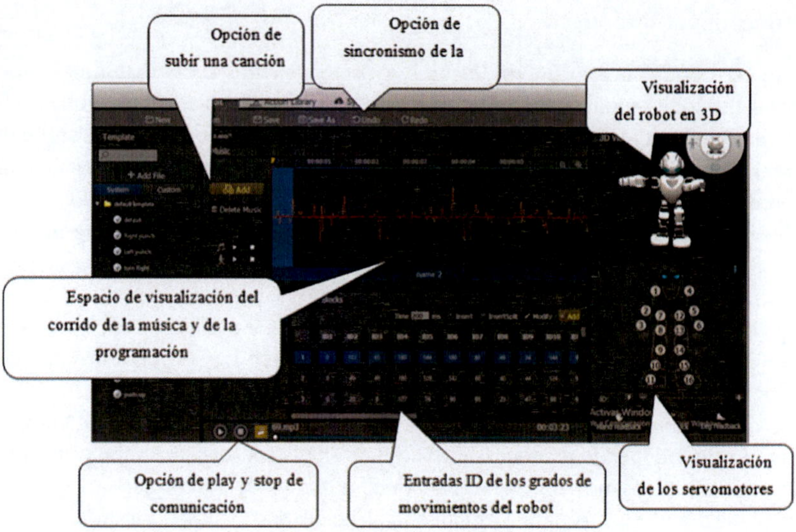

Fig. 1. Programming screen.

psychological skills that help them to foster creativity, be autonomous, recognize their bodies, among others. In Ecuador, the Ministry of Education is responsible for providing and offering quality education for the Initial I sub-level and creating a curriculum for the permanent guidance of the teaching and learning processes [2].

2.5 Robotics as a Didactic Strategy in Early Education

Currently, teacher-guided learning is a traditional practice in the teaching of early education, focused from the same educational curriculum, considering in the child a vertical learning, being favourable in the acquisition of knowledge, so that in education every day new didactic strategies are used in the teaching-learning process.

Looking for new improvements that are a fundamental pillar that aims to motivate students using in this way the Robot Alpha 1 as a new teaching strategy that assumes the role of the teacher and assimilates learning from a horizontal perspective and generates in children the aspects of learning gross motor skills such as: laterality, movements, being a playful learning that significantly favours learning.

2.6 Pedagogy and Robotics

It is also necessary to understand that robotics turns out to be a science that from technology each time increases its use and when talking about educational robotics it is not a new topic, at present it is indispensable the insertion of ICT in education, that is why in primary education the role of the teacher is directed to the use and insertion of current technologies, robotics over time has been a useful tool in the educational process as follows [10] pedagogical robotics, as a means of learning in which students who are motivated by the design, construction and creation of robotic prototypes for pedagogical purposes participate, considering of course robotics as another way of acquiring skills.

If pedagogy is considered a science, then it must be sustained by means of procedures and methods that study the problems of education in the teaching and learning process [12]. Therefore, it is feasible that the teacher considers integrating ICT in the teaching of the child even from early education as an essential element of belonging to the new virtualized and technological era that makes use of various strategies and uses them as a tool to consolidate learning.

2.7 Assisted Pedagogy with the Alpha 1 Robot

ICTs are increasingly friendly, accessible, adaptable tools that schools assume and act on personal and organizational performance. These schools that incorporate the computer with the purpose of making pedagogical changes in the traditional teaching towards a more constructive learning [13], therefore its use is important in early childhood education.

To apply ICT in education, it is vital that both teachers and students become familiar with ICT in order to improve the teaching-learning process of gross motor skills, which is a very important subject for children, considering of course that the term pedagogy is linked to the study of education and teaching. [14] indicates that pedagogy is a set of knowledge used around education with specifically social and human traits. Therefore, he deduces pedagogy as an applied science and conformed by sociology, economics, psychology, history, philosophy, or medicine.

2.8 Didactics and Robotics as a Teaching Tool

When teaching gross motor skills, it is vital that the teacher works directly with the children, especially in early childhood education. In this regard [15] didactics is the art of teaching, and to teach is to instruct, and to instruct is to systematically communicate knowledge. This implies that the teacher must generate methodological strategies, so that the student acquires skills, abilities, and the capacity to solve problems, thus generating self-determination, independence, decision making, among other factors.

It is therefore vital that ICTs in education are increasingly necessary, [16] indicates that the correct integration of ICT in the teaching-learning process is that the teacher has the ability to apply didactic strategies when specifying the contents and potentials of the technologies. Hence, the teacher's way of teaching is useful in the teaching-learning process [11]. Robotics can be considered one of the most booming technological areas today, based on the study of robots, which are systems composed of mechanisms that allow them to make movements and perform specific, programmable, and eventually intelligent tasks.

2.9 Gross Motor Learning

In teaching, the activities that are developed to explore the skills of children within the gross motor skills have behavioural areas, so it is defined as a motor education. The motor education is a type of learning, which are based on bodily activities and generate fundamental aspects of motor skills in relation to the evolution. This type of education helps the child to discover and create new movements that can be performed with proper body coordination [17].

3 Methods

3.1 Design

This research contains both qualitative and quantitative approaches. Qualitative because of the attempt to obtain and understand the information expressed by people, through interpretations and meanings from different points of view. In addition, it is selected because the subject of study is little explored and is related to a social group, in which it is directed to the teaching-learning process of gross motor skills. The qualitative part characteristic has a broad sense, in that it produces descriptive data such as: people's own words, whether written or spoken, and observable behaviour [18].

The quantitative part is due to the collection of statistical data and the analysis of the evaluations of the aspects of motor skills influenced by the ludic strategies used by kindergarten teachers, which will make it possible to know the level of acceptance of the robot in the area of education; in which it contains validated instruments previously designed for the development of the research. The quantitative part approaches the data in a static way and assigns a numerical value through statistics, in which data collection is used to test a hypothesis [19].

The research is quasi-experimental, so the Alpha 1 robot will be put into practice in the classrooms of the children of Initial Education Sub Level 1, to check if there is an improvement in the teaching of gross motor skills through a pre-test and post-test. There will be two groups of students, the first considered as a control group and the second considered as an experimental group. This quasi-experimental design identifies a comparison group, i.e., the closest thing to a pre-treatment group prior to a baseline study [20]. For the verification and reliability of the data entry, it will be done through Cronbach's alpha. In addition, the results obtained will be presented in an organized and systematized way to know the reality of the environment where the research is carried out, so that the objective is pursued in order to have a scientific and human knowledge.

The data will be analysed using the non-parametric test, this is given by the comparison of data between the control and experimental groups. In addition, it will be verified if the results are feasible using the Mann Whitney U test where the median of the pretest and posttest is tested, therefore, this will facilitate to deduce if the hypothesis is null or alternative.

3.2 Participants

A total of 30 children participated, divided into two groups: one group of 15 children considered as a control group and the other group of the same number considered as an experimental group. The study universe is small, so the research is carried out with a hundred percent and no sample is taken for statistical evaluation. Two kindergarten teachers are also involved in this research for the collection of qualitative information (Fig. 2).

Fig. 2. Alpha Robot Socialization

3.3 Ethical Considerations

All parents and guardians received information about the objectives and characteristics of the study and gave their verbal and written informed consent on behalf of the participant. In this socialization, it was necessary to request the respective self-registration and the consent form, in which the children could be photographed. The consent form contains explicit information about the project, objective, personal information, scope and supporting signatures.

4 Results

The application of the Alpha 1 robot in children for the strengthening of motor skills occurs in three stages: beginning, development and closing; in which the most relevant actions visualized in the control group can be interpreted, such as:

1. The reception of the robot as a technological tool for the classroom is very satisfactory, so it awakens many curiosities in the children, for example: How does the robot work? Where do the robots live? What colour are the robots? all these questions are answered with certainty to each of the children and at the end of this activity the children mention that they want to see the robot.

2. When the robot is presented for the first time in a physical and tangible way, and to see how the robot moves and dances, most of the children show excitement and a face of admiration for what Alpha 1 does. But, on the other hand, there are children who show shyness and fear when they hear the sounds of the servomotors. In which it is interpreted that the presentation of these humanoids with the children of initial education, it is necessary to have a lot of patience to be able to introduce a new element and even more if it is a humanoid. For the analysis of the results, quantitative evaluations of the children have been used, where the posttest and pretest evaluations used in the data collection can be used, as shown in the following tables (Tables 1 and 2).

3. At the end of the data collection with the experimental group, it is observed that the children have already made it part of their classroom and that some of them say

Table 1. Pretest experimental group

		Recognizes the right side			Recognizes the left side			Recognizes above			Recognizes below			Balance on one foot			Raise the right foot			Raise the left foot		
Nº	Name	Started	In process	Acquired	Started	In process	Acquired	Started	In process	Acquired	Started	In process	Acquired	Started	In process	Acquired	Started	In process	Acquired	Started	In process	Acquired
1	A	x			x				x			x			x				x			x
2	B		x			x				x		x			x			x				x
3	C			x			x			x		x			x			x			x	
4	D		x		x					x		x			x				x			x
5	E	x			x				x			x		x					x			x
6	F			x			x		x			x			x			x			x	
7	G	x			x				x				x		x				x			x
8	H			x			x		x			x			x				x			x
9	I			x			x		x				x		x		x			x		
10	J			x			x			x			x		x				x			x
11	K		x			x				x			x		x				x			x
12	L			x			x		x			x				x			x			x
13	M			x		x			x			x			x				x			x
14	N			x	x					x			x		x		x			x		
15	O		x			x			x				x			x		x				x

Table 2. Posttest experimental group

		Recognizes the right side			Recognizes the left side			Recognizes above			Recognizes be-low			Balance on one foot			Raise the right foot			Raise the left foot		
Nº	Name	Started	In process	Acquired	Started	In process	Acquired	Started	In process	Acquired	Started	In process	Acquired	Started	In process	Acquired	Started	In process	Acquired	Started	In process	Acquired
1	A			x			x			x			x			x			x			x
2	B			x			x			x			x			x			x			x
3	C			x			x			x			x			x			x			x
4	D			x			x			x			x			x			x			x
5	E			x			x		x				x		x			x				x
6	F			x			x		x				x			x			x			x
7	G		x			x				x			x		x				x			x
8	H			x			x			x			x			x			x			x
9	I			x			x			x			x			x			x			x
10	J			x			x			x		x				x			x			x
11	K			x			x			x			x			x		x				x
12	L			x			x			x			x			x			x			x
13	M			x			x			x			x			x		x			x	
14	N			x		x				x			x		x				x		x	
15	O			x			x			x			x			x		x			x	

goodbye as if he were a very close relative of them. It can be interpreted that it did have a great impact on each of the children and that it is a very good tool to be applied in the strengthening of gross motor skills. For validation, Cronbach's alpha was used, using IBM SPSS software, and the results shown below were obtained (Table 3).

Table 3. Cronbach's alpha

Cronbach's alpha	Cronbach's alpha based on standardized items	N of items
,784	,793	2

Cronbach's alpha contains a value of 0.784 in the feasibility validation for the execution of the pretest and posttest, i.e., there is a favourable feasibility statistic, because it is a value greater than 0.700.

Once this reliability has been verified, the comparison of the results between the control groups and the experimental group is developed, considering that they are non-parametric tests and with two independent samples. The pretest and posttest are analysed in a general way to verify the total results on the assisted pedagogy of the Alpha 1 robot in the teaching of gross motor skills in early childhood education I. Where is presented the statistical data that is centred on the Mann Whitney U. In addition, the application of the Alpha robot is tested as a tool in the pretest and posttest test statistics in relation to the Mann Whitney U and Wilcoxon's W test (Tables 4 and 5).

Table 4. Ranges

	STUDY GROUP	N	Average range	Sum de ranks
Pretest	Control group	15	17,33	260,00
	Experimental group	15	13,67	205,00
	Total	30		
Posttest	Control group	15	12,63	189,50
	Experimental group	15	18,37	275,50
	Total	30		

The analysis shows that the average range of the post-test with the experimental group is 18.37 and the average of the pre-test in the experimental group has a range of 13.67.

According to the data found in a general way the Mann Whitney U mentioned that the average range of the post-test of the experimental group is greater than the average range of the pre-test, so it is deduced that the pedagogy assisted by the Alpha 1 robot for the teaching of gross motor skills does have a positive effect in the initial education of sub-level 1. The data are also supported by Barrera [21]. The sensitization activities prior to the playful activities with educational robots proved to be fundamental

Table 5. Mann-Whitney U and Wilcoxon W statistic

Test statistics		
	PRETEST	POSTEST
Mann-Whitney U	85,000	69,500
Wilcoxon W	205,000	189,500
Z	−1,149	−1,827

to cause an emotional tension in the students, achieving a significant learning attitude, since the students were motivated when explicitly and implicitly indicated the use of the knowledge addressed and these uses were of interest to them.

In addition, to corroborate this information, the statistical table is interpreted, where the Wilcoxon W has a value of 189.500, which exceeds the Mann-Whitney U of 69.500, giving a totally favourable result (Fig. 3).

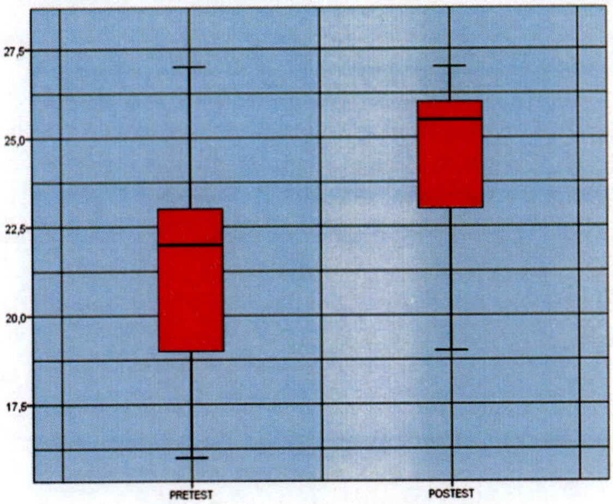

Fig. 3. Median of pretest and posttest.

By means of this general box diagram, it can be more clearly evidenced that there is a positive effect in the application of the robot. Therefore, it can be safely deduced that the Alpha 1 robot-assisted pedagogy does strengthen the teaching and learning of gross motor skills in early childhood education level I.

5 Discussion

By observing the data and contrasting the theories of the cited authors, it can be identified that the contribution of pedagogy assisted with the Alpha 1 robot does improve the teaching process of gross motor skills, considering that educational robotics is currently breaking paradigms in traditional education. [11] It proposes educational robotics as a didactic alternative, which, in parallel to the already established methods, favours new approaches that promote in students' interests that contribute to the creation of learning environments in which students find favourable circumstances for the construction of concepts and their personal interpretation of reality. [13] In order for ICTs to have a real impact on the configuration of new ways of teaching and learning, an integrating vision of educational policies, the organization of the institution, material resources and the actors involved is required, all of which are part of the development of a clearly defined and shared educational project.

Robotics in early education goes hand in hand with the integration of technological tools, so it is also determined that the incorporation of ICT in early education lies in the use that is made in the formation of the child, therefore, [22] teachers as well as parents are involved in the development of educational processes, and it is estimated that the new generations are primarily using ICTs; there is even talk of a digitalized education. Therefore, assisted pedagogy in the teaching of gross motor skills in early childhood education responds to the current reality, digital tools become essential resources in the teaching-learning process.

6 Conclusions

Assisted pedagogy is considered as a teaching strategy using components, tools, and technological applications for the teaching-learning process in early education, therefore, educational robotics is being integrated as a tool in the training process of children and youth, the integration of robots in early education requires teachers to update their knowledge in the use and management of technology.

The use of educational robotics such as the Alpha 1 robot strengthens the teaching and learning process of gross motor skills in children of the initial sublevel 1, which was analysed by means of a pretest and posttest between the control and experimental groups.

The technological tools, especially the Alpha 1 robot does not replace the teaching of the teacher, but rather becomes a technological tool for the teaching of knowledge, so it will help to maintain motivation and participation in the classroom, considering that children must be changeable in the mood.

References

1. Económicas, F.D.C.: Revista Electrónica de Investigación en Ciencias Económicas Abriendo Camino al Conocimiento Mercado Laboral y Género Labor Market and Gender **6**(12) (2018)
2. Ministerio de Educación, E.: Currículo Educación Inicial 2014. Currículo Educ. Inicial 2014, 21–36 (2014). [Online]. Available: www.educacion.gob.ec
3. Bravo Sánchez, A., Ángela, F.: Forero Guzmán. La Robótica como recurso, **13** (2012)
4. Marte, V.A.: PROJECT
5. Es, Q.U.É., Robótica, L.A.: Fundamentación Plan Provincial de Robótica Educativa en Escuelas Primarias ÍNDICE

6. Guzhñay Lucero, A.D.: Diseño y construcción de un asistente robótico para el soporte en la enseñanza del lenguaje Braille en niños de 6 a 8 años (2018). [Online]. Available: https://dspace.ups.edu.ec/handle/123456789/16324
7. Torres, M., Enrique, E.: Desarrollo de secuencias psicomotrices mediante el uso de un robot humanoide, en los niños del centro de estimulación temprana 'Baby Place' (2017)
8. UBTECH Robotics: https://www.ubtrobot.com/?ls=en. Accessed 06 Jun 2021
9. Comunicación, T.I.C., De Valencia, U.: Tecnologias De La Informacion Y Comunicacion, pp. 1–7 (1998). [Online]. Available: https://es.wikipedia.org/wiki/Tecnologías_de_la_información_y_la_comunicación
10. L.P.Q.S: LA ROBÓTICA : OTRA FORMA DE APRENDER Robotics : another way of learning why can we bring robotics closer to early childhood education ? Liliana Patricia Quiroga S, pp. 51–64 (2018)
11. Pinto Salamanca, M., Barrera Lombana, N., Pérez Holguín, W.: Uso de la robótica educativa como herramienta en los procesos de enseñanza. Ing. Investig. y Desarro. I2+D **10**(1), 15–23 (2010)
12. Valenzuela, M.: CONCEPTOS BÁSICOS EN PEDAGOGÍA BASICS CONCEPTS IN PEDAGOGY Jairo Enrique Rojano Mercado Universidad Nacional Experimental Rafael Maria Baralt – Venezuela. Guía con las bases Metod. e Investig. para una Mejor. la Educ., 75 (2010). [Online]. Available: www.eduinnova.es
13. Gavilanes Sagñay, M.A., Yanza Chavez, W.G., Inca Falconi, A.F., Torres Guananga, G.P., Sánchez Chávez, R.F., Las TICs en los procesos de enseñanza y aprendizaje. Cienc. Digit. **3**(2.6), 422–439 (2019). doi: https://doi.org/10.33262/cienciadigital.v3i2.6.575
14. Barea, G.R.: LA PEDAGOGIA EN LA EDUCACIÓN, pp. 1–9 (2007). [Online]. Available: https://archivos.csif.es/archivos/andalucia/ensenanza/revistas/csicsif/revista/pdf/Numero_15/GUSTAVOADOLFO_ROMERO_2.pdf
15. Zarzar, C.: La didactica grupal (2001)
16. Camilli, C.: Didáctica general y formación del profesorado (2016)
17. Esteves, Z.I., Toala Santana, V.N., Poveda Gurumendi, E.E., Quiñonez, M.: La Importancia de la Educación Motriz en el proceso de enseñanza de la lecto – escritura en niños y niñas del nivel preprimaria y de primero. INNOVA Res. J. **3**(7), 155–167 (2018). doi: https://doi.org/10.33890/innova.v3.n7.2018.896
18. Lecanda, R., Garrido, C.: Introducción a la metodología de investigación cualitativa. Rev. Psicodidáctica **14**, 5–39 (2002). [Online]. Available: http://www.redalyc.org/html/175/17501402/%0A http://www.redalyc.org/resumen.oa?id=17501402
19. Del Canto, E., Silva Silva, A.: Metodologia cuantitativa: abordaje desde la complementariedad en ciencias sociales. Rev. Ciencias Soc. **0**(141) (2013). doi: https://doi.org/10.15517/rcs.v0i141.12479
20. White, H., Sabarwal, S.: Diseño y métodos cuasiexperimentales. Sinop. Metod. Sinop. la evaluación impacto **8**, 1–16 (2014). [Online]. Available: https://www.unicef-irc.org/publications/pdf/MB8ES.pdf
21. Barrera Lombana, N.: Uso de la robótica educativa como estrategia didáctica en el aula. Prax. Saber **6**(11), 215 (2015). doi: https://doi.org/10.19053/22160159.3582
22. Fombona, J., Roza Martín, P.: Uso de los dispositivos móviles en educación infantil Mobile devices in early childhood education Uso de los dispositivos móviles en educación infantil Mobile devices in early childhood education. Iber. Rev. d'études ibériques ibéro-américaines **20**(1), 313–334 (2016). [Online]. Available: b-learning; childhood education; e-learning, m-learning.paris-sorbonne.fr/?p=1177%5Cn http://iberical.paris-sorbonne.fr/wp-content/uploads/2014/04/05-05.pdf%0A http://artnodes.uoc.edu/articles/abstract/10.7238/a.v0i18.3049/

Digital Image and Video Processing to Motivate Physical Exercise Using a Kinect-Based Module

Margarita Ruiz[1], Cristian Tasiguano[2(✉)], and Ana Rodas[1]

[1] Departamento de Automatización y Control Industrial, Escuela Politécnica Nacional, Quito, Ecuador
danny.zea@ister.edu.ec
[2] Instituto Tecnológico Universitario Rumiñahui, Sangolquí, Ecuador
cristian.tasiguano@ister.edu.ec

Abstract. This work presents the development and results of a prototype to motivate physical exercise in adults through image and video digital processing. The system has a sensor responsible for image and video acquisition (Kinect), a user interface, a dispenser mechanism, and a microprocessor system. By using the Skeletal Tracking of Microsoft Kinect sensor, this prototype identifies the person located in front of the module, recognizes and validates the appropriate execution of predefined physical exercises (squats or lunges), and counts the number of times that the user performs the exercise in a base of time. Once a specific number of exercise repetitions has been properly executed (at least 20 in 120 s), the interface shows the counter result, and the system delivers a water bottle as an incentive. With the use of this module, it is intended to promote physical exercise in an unconventionally and strikingly way.

Keywords: Image-processing · Interface · Kinect · Matlab · Microprocessor system · Physical exercise

1 Introduction

According to the World Health Organization (WHO) [1], obesity has tripled worldwide since 1975, reaching more than 1.9 billion overweight adults over 18 years of age in 2016, of whom of which, more than 650 million are obese, which implies a prevalence in adults over 18 years of age of 39% overweight and 13% obese. Concerning the child population, more than 41 million children under 5 years of age are overweight or obese worldwide. One of the factors for being overweight and obese is a sedentary lifestyle, which characteristic is the lack of physical activity.

Globally, 1 in 4 adults does not reach the recommended levels of physical activity and more than 80% of the world's adolescents have an insufficient level of physical activity. People with an insufficient level of physical activity have a 20% to 30% higher risk of death compared to people who achieve a sufficient level of physical activity [2].

Some of the main aggravating factors of these rates include little time to exercise, high gym costs, indiscriminate use of electronic devices (cell phones, laptops, computers, video games, etc.), and the excessive consumption of sugars and saturated fats [3].

M. Zambrano Vizuete et al. (Eds.): CI3 2021, LNNS 511, pp. 216–228, 2022.
https://doi.org/10.1007/978-3-031-11438-0_18

Exercising regularly helps people remain healthy and fit. According to the World Health Organization (WHO), it is recommended at least 150 min of moderate-intensity aerobic physical activity in a week, or at least 75 min of vigorous-intensity physical activity practiced in bouts of a minimum of 10 min. The WHO also indicates the advantages of exercise, for instance, fewer cases of strokes, reduced rates of coronary heart diseases, reduced mortality rates, and an overall health enhancement [4].

Among the resources used to face the lack of physical activity in the population, *exergames* have been developed to encourage and guide physical activity, in support of rehabilitation and improved fitness [5]. These games have attracted the attention of the research community since full-body motion interaction devices such as the Wii and the Kinect™ have become widely available.

Literature shows that virtual reality and image processing systems have especially focused on senior exercising and physical rehabilitation. For instance, J. W. Burke et al. [6] developed a visual tracking system suitable for upper limb stroke rehabilitation at home by using color segmentation, motion detection, and game design principles to ensure patient motivation and engagement. Furthermore, T. Lin et al. [7] presented a Kinect-based rehabilitation system to perform the "seated Tai Chi" exercises and help patients with movement disorders. By using a baseline/intervention strategy to evaluate the effectiveness of the system, they demonstrated the significant improvement of two patients in performing the Tai Chi exercise. In 2014, W. Zhao et al. [8] presented a feasibility study for using a single Kinect sensor to assess the quality of rehabilitation exercises. The singularity of this work is the use of a rule-based approach with automated patient feedback. The definition and application of correctness rules allowed determining whether or not the Kinect motion-sensing data is accurate enough. In 2018, V. Fernandez-Cervantes et al. [5] presented VirtualGym which is a system that supports exercise experts to specify exercise routines for older adults. By employing an exergame, a coaching avatar guides seniors through a personalized version of different routines. The data obtained from the Kinect™ sensor give information to evaluate the performance and provide feedback.

In the field of exercise for maintaining good health at any age, there are commercial products such as the one developed by New York-based startup Mirror [9]. This machine not only functions as a mirror but also includes an LCD screen to show live and on-demand fitness classes that are monthly paid. The machine by itself could cost around $1500 in the USA market and even more in Ecuador. Another example is FightCamp [10]. This is a connected home boxing gym that offers hundreds of on-demand classes (for an additional monthly payment). It uses punch-tracking technology to keep the user motivated.

There have also been different campaigns to motivate the exercise in exchange for economic revenue since it has been shown that financial incentive-based approaches help to lose weight [11]. The most famous campaign was developed in Moscow as part of a one-time promotional campaign for the 2014 winter Olympics in Russia. This campaign started three months before the Olympics inauguration and allowed passengers to pay fare tickets by doing squats in a Moscow subway station [12]. Similarly in 2019, the insurance company Prudential in partnership with SMRT placed a booth located in

Tampines Station (Singapore) that gives passengers a single-trip ticket for the train if they can accomplish 20 squats in proper form under the given time [13].

Although literature describes interesting technological advances in the physical exercise field, to our knowledge few or almost none of them have been applied in Ecuador. Considering the aforementioned high rates of overweight and obesity in the Ecuadorian population and the low income that prevents people from accessing sophisticated exercise machines, this project aims to develop an electronic prototype to promote physical exercise by delivering an incentive (bottle of water) when predefined exercises have been correctly performed a certain number of times. This system uses a Kinect sensor for image acquisition and Matlab for its processing. For the prototype design, squats and lunges were chosen because they exercise more muscles of the lower extremities, increase the resistance of the user, work with flexibility and increase the strength of the muscles [14].

The remainder of this paper is organized as follows: in Section II, we provide details of the system architecture; in Section III, we describe the design and implementation of the prototype; in Section IV, we give information on the experimental methods; in Section V, we show and discuss the obtained results. Finally, in Section VI, the main achievements of this work are summarized.

2 System Architecture

2.1 System Overview

A system flowchart of the Kinect-based exercise system is illustrated in Fig. 1. The computer allows the operation of Matlab and sends all orders to the application to work according to the required specifications. Besides, it shows the graphical interface so that the user can interact with the application. The Kinect sensor is responsible for image and video acquisition. The microcontroller receives the signals from Matlab and activates a motor that starts the dispensing system. It also receives the signal from the infrared sensor that allows the detection of water bottles in the reservoir. A schematic connection diagram of the microcontroller is shown in Fig. 2.

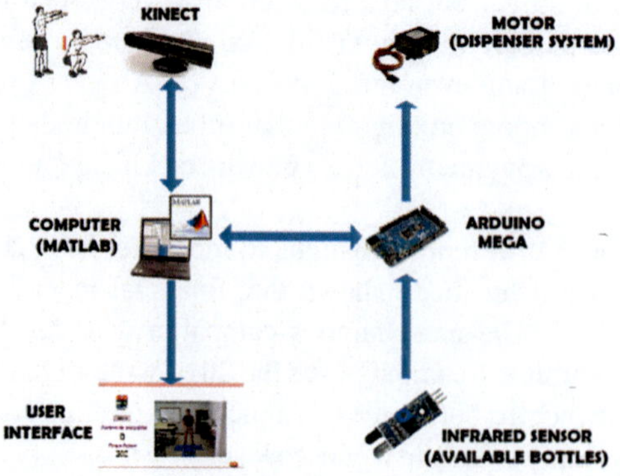

Fig. 1. System architecture

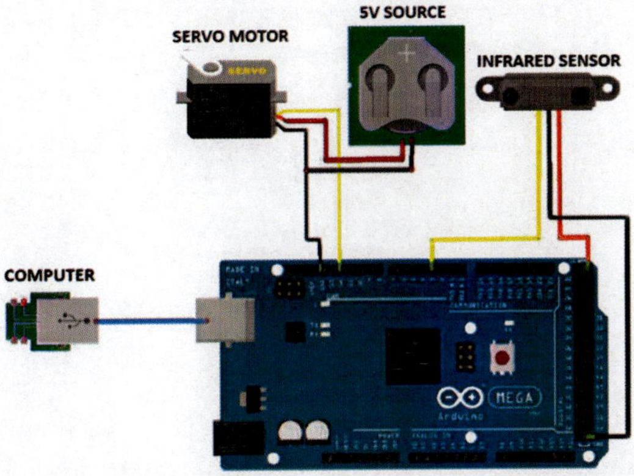

Fig. 2. Schematic diagram of the microprocessor system

2.2 Kinect Sensor

The Kinect sensor is a device developed by Microsoft as an accessory for the Xbox video game console. Later, Microsoft launched a Kinect for Windows to support the development of academic and professional applications. Microsoft also created the Software Development Kit (SDK) which allows direct controlling of the Kinect sensor to ease the application programming [15]. The Kinect sensor captures depth and color images simultaneously at a frame rate of approximately 30 [fps]. By recording the consecutive depth images, a higher density of points can be obtained, possibly in real-time, which allows monitoring the human body and its movements [16]. This device incorporates rigorous advanced detection hardware, a depth sensor, a color camera (RGB), and a four-microphone array that provides full-body 3D motion capture capabilities, facial recognition, and voice recognition. It also has a motorized tilt at its base as shown in Fig. 3 [15].

The RGB color camera is responsible for capturing the spatial resolution, has an 8-bit resolution and VGA of 640 × 480 pixels. The depth sensor consists of an infrared projector and a monochrome CMOS sensor. The sensor can capture, under any light condition, 3D video data. The monochrome video channel is 16-bit deep, VGA resolution of 640 × 480 pixels with up to 2048 sensitivity levels.

The limit of the visual range of the Kinect sensor is between 1.2 and 3.5 m away, with a viewing angle of 57° horizontally and an angle of 43° vertically, while the pivot can be oriented up or down extending up to 27°.

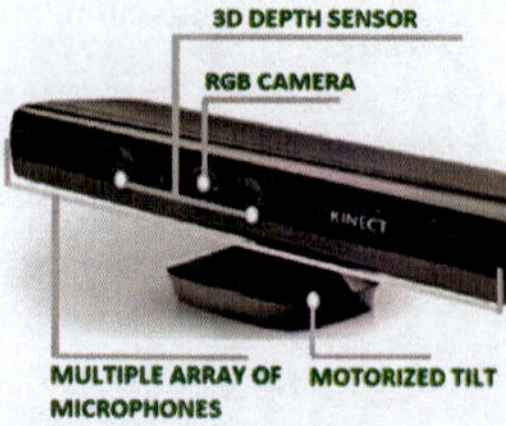

Fig. 3. Kinect sensor with its main components

2.3 Skeletal Tracking

The Kinect sensor has a resource called *Skeletal Tracking*. It is capable of recognizing the joints of the human body using an internal algorithm, which is responsible for recognizing the human body by segmenting its parts when working with the depth sensor and the pixels of the acquired image [15]. Any pixel is taken from the image and compared with two other nearby pixels. Depending on the result of their depth, it is known whether they are part of the same object or not, after that, another two pixels are chosen to continue with the comparisons and so on several times. Depending on the route of comparisons that the first pixel has traveled, a value can be assigned to determine which segment of the human body it belongs to. These routes constitute a decision tree and several of them form a decision forest. In Skeletal Tracking, a human body is represented by a series of segments that capture body parts such as the head, neck, shoulders, arms, etc., and the joints are represented by their respective coordinates as shown in Fig. 4. Using the Skeletal Tracking resource, a data matrix called JointImageIndices is obtained. It projects the coordinates of 20 joints of the human body in real-time. Figure 5 shows the order of the joints registered by the Kinect sensor and an image that places these joints graphically in a human skeleton.

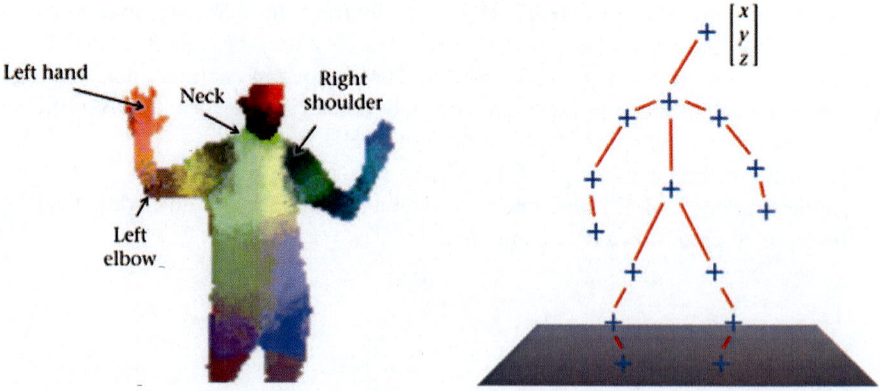

Fig. 4. Representation of several parts of the human body with the Skeletal Tracking. Figure modified from [15].

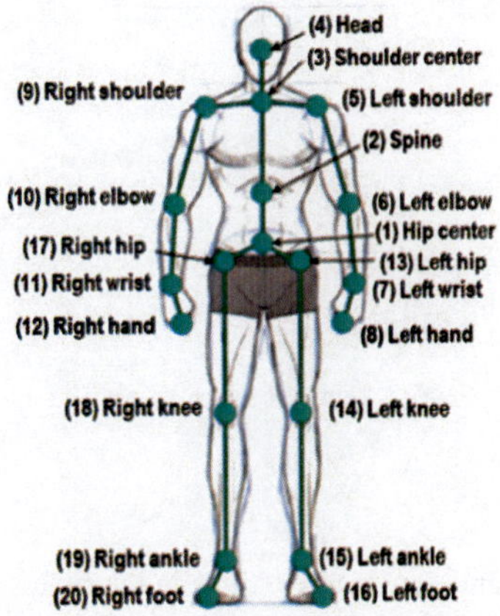

Fig. 5. Human joints and its order by Skeletal Tracking

2.4 Squats Exercises

With all the data provided by the Skeletal Tracking resource, the different coordinates of the joints in the "X", "Y", and "Z" space are identified. These coordinates allow implementation of the algorithm used in this application which consists in find distances between each other and the necessary angles to make the logic control. To determine if the exercise is appropriately executed is necessary to calculate: 1) the distance between the right hip and knee, 2) the distance between the left hip and knee, 3) the inter angle of the two knees, 4) the distance between the ankles, 5) the distance between the shoulders, and 6) the distance between the module and the user.

The first two points are used to find the internal angle of each knee by applying the law of cosines. This value must be less than 140° to be counted as a squat. Moreover, the distance between ankles must be less than 80 units, because the ideal posture of this exercise requires it in this way. The distance between shoulders must be greater than 60 units (it was experimentally determined) when the person is appropriately facing the module. Finally, the distance between the module and the user is used to verify that the person is within the detection area, which is located at a distance between 2.4 m to 2.6 m from the module. If the user is not correctly located (the distance between the ankles is also verified) the counter is not allowed to increase and the squat is invalid.

Figure 6 shows a person placed in the squat position with three of their joints represented, *17* refers to the right hip, *18* to the right knee, and *19* to the right ankle.

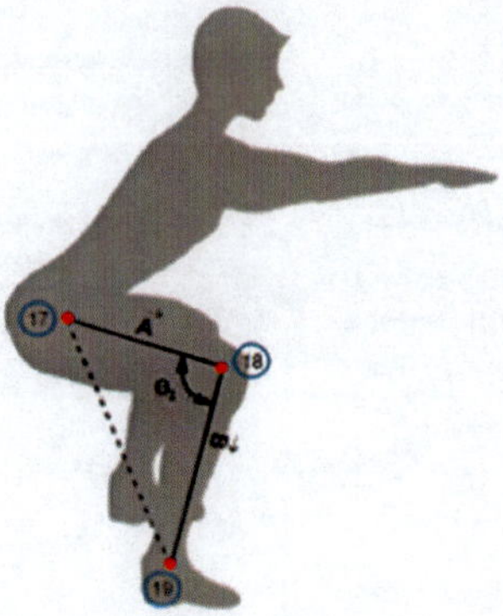

Fig. 6. Representation of three joints in the right side of a person

To determine the right knee angle, first, the modules of vectors A and B are calculated by using the following equations:

$$|A| = \sqrt{ax^2 + ay^2} \tag{1}$$

$$|B| = \sqrt{bx^2 + by^2} \tag{2}$$

Second, the dot product of the two vectors is calculated with Eq. 3 and finally, the right knee angle (1) is obtained through Eq. 4. The left knee angle and the other distances used in the required restrictions of this exercise are similarly extracted.

$$\vec{A}.\vec{B} = (ax.bx) + (ay.by) \tag{3}$$

$$\theta_1 = \cos^{-1}\left(\frac{\vec{A}.\vec{B}}{|A|.|B|}\right), \quad 0 \leq \theta_1 \leq \pi \tag{4}$$

2.5 Lunges Exercises

The same measured parameters in the squats (from 1 to 6) are also useful to validate a correct lunge. To increase the counter of this exercise, the value of the angle of each knee must also be less than 140, the distance between ankles must be greater than 120 units (it was experimentally determined) and the distance between shoulders must be less than 60 units since the user is located in its side view. It is important to highlight that also in this exercise the user must be located inside the detection area otherwise the lounge will be not taken into account.

2.6 User Interface

The user interface was designed in the GUIDE of Matlab and has a cover screen, a home screen, two instruction screens (one for each exercise), and two counting screens. The *cover screen* indicates the title of the project in which this prototype was developed and the main data of the authors. This screen is displayed for 10 s and automatically moves to the *home screen* where the user has the option to choose the type of exercise wanted to perform (squats or lunges). If the incentive is over, an alert message appears and the user can continue doing the exercises without incentive or leaving the system. In the case of selecting the squat option, an *instruction screen* with the indications to correctly perform the exercise shows up. This screen has a duration of 15 s. After that, the scenario where the user is going to be located, the number of squats, initially at zero, and a timer are shown in the *squat screen*. The interface similarly works if the user selects the lunges exercises.

3 Prototype Design

The prototype was designed according to the application requirements. As shown in Fig. 7, it has the space for the Kinect sensor, a screen for the user interface, an area where the incentive comes out; and a cabinet that contains the water bottles, the computer, the microprocessor system with the infrared sensor, and the dispensing mechanism. In the prototype design, a bottle of 16 cm × 6 cm and a capacity of 300 ml was considered as an incentive.

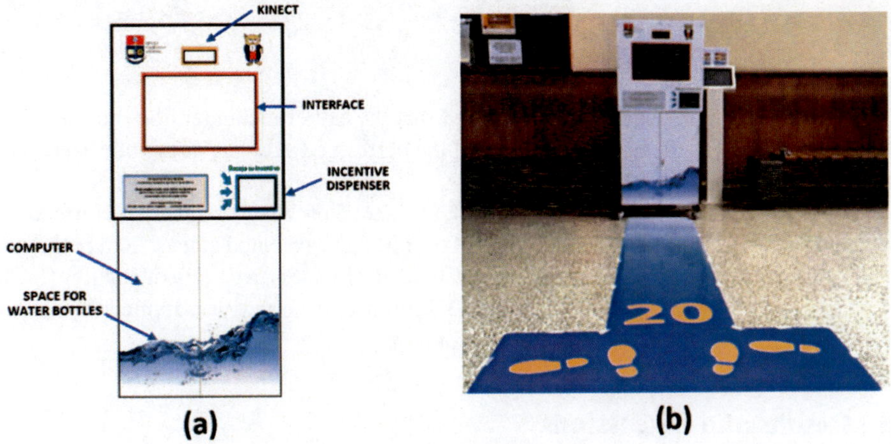

Fig. 7. Prototype design (a) and implementation of the complete module (b).

3.1 Dispensing Mechanism

The design of the structure has two ramps with a suitable inclination to allow the descent of the bottles by gravity until reaching the dispensing mechanism as illustrated in Fig. 8. The mechanism used to deliver the incentive is called a revolver. It has six compartments that are designed according to the diameter of the water bottle and are located in a rotating cylinder, which is mounted on a support. To successively aligned the compartments with the structure and rotate according to the position established within the control, this mechanism includes a coupled motor (see Fig. 8 (b)).

(a) (b)

Fig. 8. (a) Bottle sliding system and (b) dispenser mechanism coupled to the servo motor.

4 Experimental Methods

During the testing procedure, it was verified that the exercise counter does not increase if the user is outside the detection area or if the performed exercise does not correspond to the selected one.

To validate the total operation of the module, a total of thirty tests were performed on different adult users. They chose the type of exercise and executed at least 20 repetitions. Participants exhibit gender diversity (9 females and 21 males), a variety of statures (from 1.55 m to 1.88 m), and ages (from 20 to 31). Moreover, a survey was applied to users to have feedback on the functioning of the prototype.

5 Results and Discussions

Figure 9 demonstrated that the counter increases just when the user is inside the detection area and executes the selected exercise (squats). The prototype similarly works with the lunge exercise, but the results are not shown for the sake of brevity.

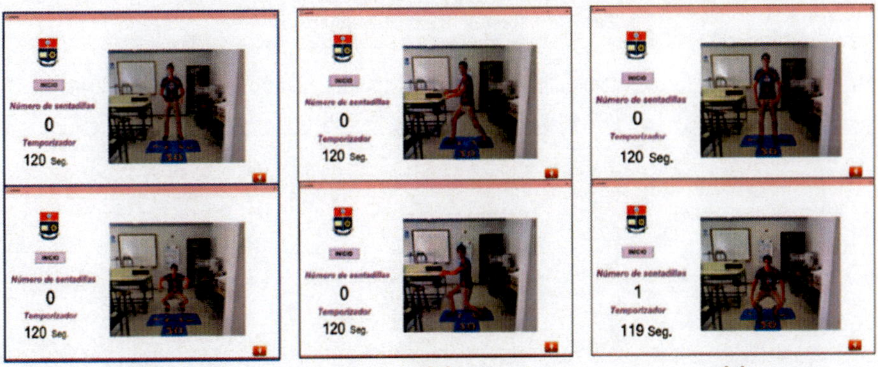

(a) (b) (c)

Fig. 9. (a) An squat performed outside the detection area or (b) the execution of the incorrect exercise do not modify the counter. (c) A proper squat increases the counter.

Moreover, Table 1 shows the results of the thirty tests including the number of repetitions of the manual count and the repetitions detected by the module for comparison and calculation of the success percentage. These data allowed obtaining the percentage of system reliability, which is 92.5%.

When users executed the exercises, it was found that the Kinect sensor does not properly detect the joints if users wear black/dark clothing, in the lower part of the body (pants). Microsoft attributes this to the infrared light absorption of some black garments, which makes it difficult for the Kinect to track the user correctly. This problem appeared in the users of tests 4 and 5 (color fail) who did not have a correct detection of the lower joints because they wore black clothes, so they could not execute the exercises.

The last ten tests were carried out in a place with a lot of sunlight. This caused a reflection of the floor and the Kinect sensor did not properly track the joints of the human body (light fail). Microsoft recommends that the sensor or user should not be in a place where they are exposed to direct sunlight. This problem occurred in the users of tests 26 and 27. Therefore, although they were detected, the counter did no correctly increase.

Furthermore, some users performed the exercise very quickly and the system failed to capture the human structure effectively. This can be classified as an error of each individual and was presented in users 6 and 7.

Finally, the survey results show that the prototype is easy to use (100%), its design is motivating to exercise (100%), the incentive is satisfactory (86.7%), the number of exercise repetitions is adequate (93.3%), and participants recommend the use of the module in public spaces (86.7%).

Table 1. Results from different users

User	Type of exercise	Manual repetition counting	Module repetition counting	Individual success rate %
1	squats	25	20	80
2	squats	23	20	87
3	squats	25	20	80
4	squats	0	0	Color fail
5	squats	0	0	Color fail
6	squats	30	20	67
7	squats	29	20	69
8	squats	20	20	100
9	squats	21	20	95
10	lunges	22	20	91
11	squats	26	20	77
12	squats	20	20	100
13	squats	20	20	100
14	squats	20	20	100
15	squats	22	20	91
16	squats	20	20	100
17	squats	21	20	95
18	squats	20	20	100
19	squats	21	20	95
20	squats	20	20	100
21	squats	20	20	100
22	squats	20	20	100
23	squats	24	20	83
24	lunges	20	20	100
25	lunges	20	20	100
26	squats	0	0	Light fail
27	squats	0	0	Light fail
28	squats	20	20	100
29	lunges	20	20	100
30	lunges	21	20	95
Total success rate				**92.5**

6 Conclusions

This paper proposed a Kinect-based prototype to motivate physical exercise practice in adults. We use the incorporated Skeletal Tracking of the Microsoft Kinect sensor and the software Matlab to acquire, process, and validate the appropriate execution of two predefined exercises (squats and lunges). The experimental results obtained from 30 users indicates that the distance between the Kinect sensor and the user is an important parameter for exercise detection. Therefore, the user position must be inside the detection area to ensure the counter increase. Moreover, two inconveniences with the recognition of the joints of the human body were identified. The first one is that the Kinect sensor does not properly detect joints in users who wear black clothing, in the lower part of the body (pants). The second one is that sunlight affects the tracking and detection of user movements. Both problems are related to the intrinsic limitations of the Kinect sensor. Additionally, although the results may vary since the users execute the exercises in their way; the developed prototype exhibits a promising success rate of 92.5%. Finally, the users found the module adequately design, easy to use, and appropriate to incentive physical exercises in public spaces.

References

1. World Health Organization (WHO): Fact Sheets Obesity and Overweight (June 2021). Available at: https://www.who.int/news-room/fact-sheets/detail/obesity-and-overweight
2. World Health Organization (WHO): Fact Sheets Physical Activity (November 2020). Available at: https://www.who.int/news-room/fact-sheets/detail/physical-activity
3. Cigarroa, I., Sarqui, C., Zapata Lamana, R.: Efectos del sedentarismo y obesidad en el desarrollo psicomotor en niños y niñas: Una revisión de la actualidad latinoamericana. Univ. y Salud. **18**, 156 (2016). https://doi.org/10.22267/rus.161801.27
4. Ganesan, S., Anthony, L.: Using the kinect to encourage older adults to exercise. In: Proceedings of the 2012 ACM annual conference extended abstracts on Human Factors in Computing Systems Extended Abstracts - CHI EA '12, p. 2297. ACM Press, New York, New York, USA (2012)
5. Fernandez-Cervantes, V., Neubauer, N., Hunter, B., Stroulia, E., Liu, L.: VirtualGym: a kinect-based system for seniors exercising at home. Entertain. Comput. **27**, 60–72 (2018). https://doi.org/10.1016/j.entcom.2018.04.001
6. Burke, J.W., Morrow, P.J., McNeill, M.D.J., McDonough, S.M., Charles, D.K.: Vision based games for upper-limb stroke rehabilitation. In: 2008 International Machine Vision and Image Processing Conference, pp. 159–164. IEEE (2008)
7. Lin, T.-Y., Hsieh, C.-H., Lee, J.-D.: A kinect-based system for physical rehabilitation: utilizing tai chi exercises to improve movement disorders in patients with balance ability. In: 2013 7th Asia Modelling Symposium, pp. 149–153. IEEE (2013)
8. Zhao, W., Feng, H., Lun, R., Espy, D.D., Reinthal, M.A.: A kinect-based rehabilitation exercise monitoring and guidance system. In: 2014 IEEE 5th International Conference on Software Engineering and Service Science, pp. 762–765. IEEE (2014)
9. Mirror, C.P.I.: (2020). https://www.mirror.co/workouts
10. Hykso, I.: FightCamp 2017–2020, FightCamp. https://joinfightcamp.com/lifestyle/
11. Volpp, K.G., John, L.K., Troxel, A.B., Norton, L., Fassbender, J.: Financial incentive – based approaches for weight loss. JAMA **300**, 2631–2637 (2008)

12. Forbes: Moscow Subway Station Lets Passengers Pay Fare In Squats. https://www.forbes.com/sites/andrewbender/2013/11/14/moscow-subway-station-lets-passengers-pay-fare-in-squats/#6d8f2597b0d8
13. Independent, T.: Do 20 squats in 40 seconds and you get a free MRT ride. http://theindependent.sg/do-20-squats-in-40-seconds-and-you-get-a-free-mrt-ride/
14. Keogh, J.: Lower-body resistance training: increasing functional performance with lunges. Strength Cond. J. **21**, 67 (1999). https://doi.org/10.1519/1533-4295(1999)021%3c0067:LBRTIF%3e2.0.CO;2
15. Zhang, Z.: Microsoft kinect sensor and its effect. IEEE Multimed. **19**, 4 (2012). https://doi.org/10.1109/MMUL.2012.24
16. Khoshelham, K.: Accuracy analysis of kinect depth data. ISPRS - Int. Arch. Photogramm. Remote Sens. Spat. Inf. Sci. XXXVIII-5/, 133–138 (2012). https://doi.org/10.5194/isprsarchives-XXXVIII-5-W12-133-2011

Prototype Development for Automation of Irrigation System Based on IoT for Small and Medium Rural Producers

Jorge Jimenez[1], Marcelo Zambrano[1,2], and Juan Minango[1](✉)

[1] Instituto Superior Universitário Rumiñahui, Sangolqui, Ecuador
{jorge.jimenez,juancarlos.minango}@ister.edu.ec
[2] Universidad Técnica del Norte, Ibarra, Ecuador

Abstract. Over the last year, the Internet of Things (IoT) has been applied in different areas one of them is the agricultural area. The use of Internet of Things in agricultural area is principally for automating the water pumps, fans, light bulbs, among others resources. Thus, in this paper, we develop and build a prototype for automation of low-cost irrigation systems based on Internet of Things. In this way, our prototype, which is based on the use of Arduino Mega 2560, proposes to turn up small and medium gardens into smart gardens in order to improve the production process as well as optimize natural resources like water and energy. The prototype works in conjunction with the aid of a mesh of different sensors such as humidity sensor, soil moisture sensor, and temperature sensor. The measures provided by these sensors are management in real-time through a computer or mobile device over an internet connection.

Keywords: IoT · Arduino · Sensor network

1 Introduction

The technological revolution 4.0 has completely changed society's ways of working. The automation and robotization of production process and systems enabled greater productivity and profitability with less manual effort. In this scenario, Internet of Things (IoT) has become a fundamental constituent in autonomous management and monitoring for a multitude of applications, such as agriculture, industry and medicine [1].

In this context, agriculture has been occupying an important space in the country's economy. Thus, technological innovations must be focused on this field. With the aid of IoT, irrigation systems use in agriculture can be automated. These systems can be benefited from using water efficiently, that is, it ensures that the plant receives the ideal amount water for it s growning, avoiding waste. Automation also optimizes energy consumption and reduces labor costs [2]. There are automation systems for irrigation on the market, but they only focused into large companies. Small producers continue to use ineffective methods to irrigate their plantations, using water incorrectly and employing more labor [3,4].

M. Zambrano Vizuete et al. (Eds.): CI3 2021, LNNS 511, pp. 229–239, 2022.
https://doi.org/10.1007/978-3-031-11438-0_19

On the other hand, IoT can be explained as a network that connects computing and communication devices to provide intelligent service [1,5]. In general, the architecture of IoT consists of edge devices, gateways, server and applications. The way in which the elements communicate with each other is defined by communication protocols [6].

The Message Queue Telemetry Transport (MQTT) communication protocol is already quite popular. This protocol employs the publish/subscribe standard, where the edge devices, sensors and actuators, can send data and receive performance requests through the MQTT server, also known as the MQTT broker. In [7], it is proposed to collect data from sensors, soil moisture and pump status, and send these data to users using ESP 8266 as a gateway server to connect to the MQTT broker.

In this work, a prototype to automate the irrigation system of a plantation, using the Arduino, mesh of sensors and the MQTT protocol was developed.

The remainder of this paper is organized as follows: Sect. 2 describes the irrigation system. Section 3 shows an overview of the IoT and a brief explanation about the MQTT protocol. Section 4 describes the design and implementation phases of the developed automate prototype. The results and discussion are shown in Sect. 5. Finally, the conclusions are presented in Sect. 6.

2 Irrigation System

Irrigation is the artificial way of watering plants. When it is done in a conventional way, it consumes more water than necessary, more energy, and requires more manpower. In recent years, more efficient ways of carrying out this activity have been found. Through the automatic control of the irrigation system, it is possible to provide the ideal amount of water at the best times for its absorption. This reduces work in the field and optimizes the irrigation process [8].

Surface irrigation and localized irrigation have been the common practice when it comes to technology for irrigation systems. In surface irrigation, water is distributed over the surface through the effect of gravity. Surface irrigation techniques are best suited for basic crops. Localized irrigation is a type of precision agriculture and requires high water efficiency, so techniques are needed to ensure the ideal amount of water for irrigation. The methods most used in localized irrigation are drip irrigation, sprinkler irrigation, and channel irrigation [9].

Crops need specific amounts of water to develop. Thus, adequate irrigation can significantly improve crop yield. Drip irrigation is one of the best ways to save water during irrigation. In this system, water lows through the main pipe and it is divided into several adjacent pipes and drips slowly at the root of the plants through valves and pipes. The high cost is a disadvantage of this irrigation system in relation to the others [8]. Small and medium producers face difficulties to introduce technologies in the cultivation of their crops and to increase, considerably, the production. Considering the importance of this class for the domestic market in Ecuador, it is necessary to develop effective low-cost technologies for the agriculture field. On the other hand, the use of information and communication technologies is spreading in agriculture and one of these technologies is IoT where there are a variety of devices that communicate using different protocols.

3 Overview of Internet of Things (IoT)

The Internet of Things (IoT) refers to the connection between devices within a given system to provide service intelligently. This means adding computing resources and connectivity to the cloud to systems that until then were not connected, allowing the analysis and control of any application. IoT systems are composed of four basic elements common to practically all IoT solutions: edge devices, gateways, cloud platforms, and applications. The edge devices are hardware made up of sensors and actuators, that is, they are able to collect data and perform some action. Gateways collect, pre-process and transfer sensor data to the cloud or provide actuation requests from the cloud to devices. The cloud or server platform provides services such as data acquisition and analysis, device management, and performance. Finally, applications can range from data visualization panels on the WEB to specific mobile applications [10].

Currently, applications rely on the Hipertext Transfer Protocol (HTTP) for transfer information. This protocol is quite complex for small IoT devices [11]. The vast majority of IoT applications are simple elements, such as lamps, sockets, temperature sensors, or valves, so they do not require complex protocols and do not need operating systems. Thus, protocols such as Message Queue Telemetry Transport (MQTT) can well meet the needs of the system. Communication protocols configure the way information is exchanged between gateway and internet, update users with the most recent data, and transmit application commands to edge devices [12].

3.1 MQTT Protocol

MQTT is a network protocol widely used in IoT due to it is simple and easy to implement, ideal for remote communication between devices that will transfer little data and on networks with limited bandwidth. It provides asynchronous communication between the parties. MQTT is based on the Transmission Control Protocol/Internet Protocol (TCP/IP) stack and it has a publish/subscribe architecture. Furthermore, MQTT is an open standard Organization for the Advancement of Structured Information Standards (OASIS) [12].

The MQTT comprises two entities: the server, also known as a broker and the clients. The MQTT broker is responsible for storing customer data and providing that information to interested customers. The client is any element capable of interacting with the broker, it can be an IoT sensor or an application in a data center. MQTT messages are organized into topics. Therefore, communication occurs as follows: the client connects to the server via TCP/IP and subscribes to the topic of interest, the client publishes the messages in a topic and sends it to the broker, who is responsible for forwarding these messages to clients who subscribed to this topic [13].

3.2 IoT in the Agriculture

The use of devices in the Agriculture Industry is grown, some solutions and equipment like Geographic Information System (GIS), Global Position Service (GPS), Remote Detection, Video Intelligence, Sensors, among others are been employed to transform the traditional agriculture in smart agriculture. These systems aid in the management and the monitoring of soil and climate conditions with the principal purpose that the farmer knows in real-time the status of the products, thus being able to improve the use of resources, fail corrections, minimize costs and reduce losses, with the goal of obtaining better production and high rentability [14].

IoT has a necessary presence in the agriculture sector due to its integrates with the aid of humidity sensors, temperature, rain, and drone mapping, all devices are connected to the Internet to inform the production and situation of agricultural fields. According [15], more than 75 million IoT devices will be implemented in the agriculture sector by the final of this decade, with these devices' growth there is an estimation that the farmers produce and manage around 4 million of data in 2050.

4 Implementation of the Prototype

The development of our prototype is divided into the two parts. The first one corresponds to the embedded part which is built by employing Arduino board, while the second part corresponds to a Web application to manage and control an irrigation system. The main characteristics of our prototype are:

- Soil moisture reading,
- Automation of the irrigation system,
- History of readings performed,
- Real-time readings,
- Panel for manual activation of the irrigation system.

The above characteristics are supported by the description of each electronic device used in the implementation of the proptotype which is presented in this section.

4.1 Soil Moisture Sensor

This sensor was made to detect variations in moisture in the soil. In the first case when the soil is humid, the sensor output is in a low state and dry soil put the sensor output in a low state. The potentiometer included in the module permits adjusted the limit between dry and wet, regulating the digital output D0. However, to have a better resolution, it is possible to use the analog output A0 and connect to an AD converter, like the one present on the Arduino.

4.2 Temperature Sensor LM35DZ

The LM35DZ Temperature Sensor is accurate, in addition to have easy communication with microcontrollers like Arduino, PIC, ARM, and Raspberry Pi. This sensor is widely used for home or even industrial automation. The communication between the LM35DZ uses only one pin. The voltage output is proportional to temperature and it has a reading range from 0 °C to 100 °C with an accuracy of 0.5 °C.

4.3 Temperature and Humidity Sensor DHT22

DHT22 is a sensor that was composed of two parts. The first one is a humidity capacitive sensor, and the other is a thermistor. The Arduino module of this sensor incorporates an integrated circuit that performs analog to digital conversion. The range operating of DHT22 is to temperature between −40 °C to 80 °C and for humidity is from 0% to 100% of relative humidity (RH). This sensor has a resolution of 0.1 °C and 0.1% RH. DHT22 offers low consumption of energy and its maximum transmission capacity is 20 m distance.

4.4 Relay Module

The relay module works like an electrically operated switch that can be turned on or off, leaving or not flow current. The principal characteristic is that it can control electronic devices with high voltage (i.e. light bulbs, small-medium motors, water pump) with the signals of low voltages provides by Arduino.

4.5 Organic Light-Emitting Diode (OLED)

The OLED of 0.96-in. display with 128×64 pixels is used to shows the measurements of the sensors. The OLED is composed of organic material like carbon which emits light when electricity is applied, this type of screen offers better efficiency in energy terms. The module SSD1306 of this screen permits connect with the Arduino board by SPI or I2C communication.

4.6 Arduino Mega 2560 R3 Built-in ESP8266

The Fig. 1 shows the Arduino board used, this board integrates two chips, one is the classic microcontroller Atmega2560 known as Arduino Mega, and the other one is the microcontroller esp8266 where the two are connected by a dip switch. The esp8266 has 4 Mb of flash memory and the ATmega2560 has 256 kb. The board can works independently with each chip or together. The configuration will depend of the user necessities.

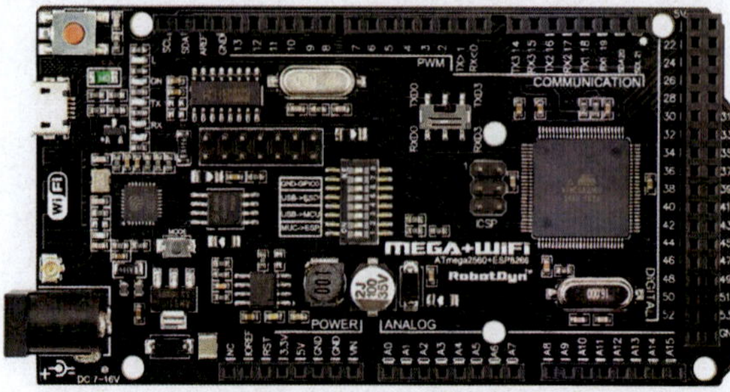

Fig. 1. Arduino Mega 2560 R3 built-in ESP8266

4.7 Thinger IO

The Thinger IO is an open-source IoT platform that permits the connection and management of IoT devices. Thinger IO offers a scalable cloud infrastructure to connect several devices. The administrator console allows control and visualization of all devices connected in real-time by computer or Android and iOS applications for mobiles.

Figure 2 shows the implementation flow of our prototype. Firstly, the sensors stage is responsible for all soil moisture measurements, external temperature, humidity, and temperature close to the soil. These measurements are stored in a unique variable which is sent by serial communication from Atmega2560 to esp8266. The esp8266 was previously configured with an SSID and password to provide an internet connection. The sensor variable is read by the microcontroller and send by MQTT protocol to the Thinger IO platform which stored the data in a database and published it into a dashboard that permits a real-time data monitoring. Furthermore, Thinger IO allows control by computer or mobile the on/off of water pumps.

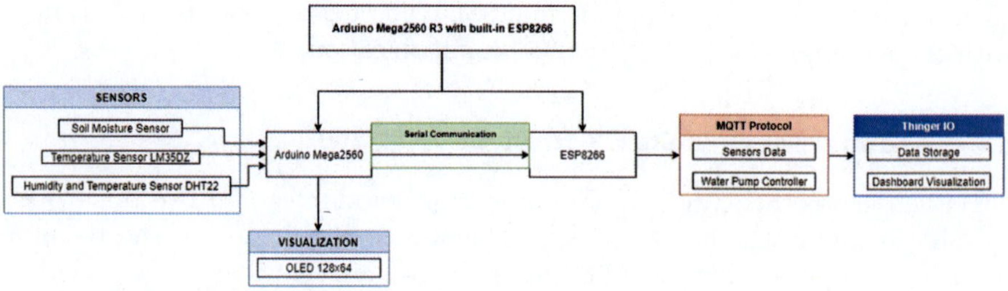

Fig. 2. Steps of the workflow for automation of irrigation system based on IoT.

5 Results

In this section, the process of configuration and implementation of each component in the smart garden prototype is shown. Besides, the connection between the Arduino and the Thinger IO for visualization and management on real-time is presented.

5.1 Hardware

The hardware stage is composed of all electronics devices explained in Sect. 4. Figure 3 shows the implementation and connections of the sensors to the Arduino board. The power supply provides $12V_{DC}$ to the operation of the board. The sensors stage is operated with the internal power supply offered by the Arduino board which is $5V_{DC}$. The soil moisture sensors are connected through an extended wire to take the measurement values of the garden at different distances.

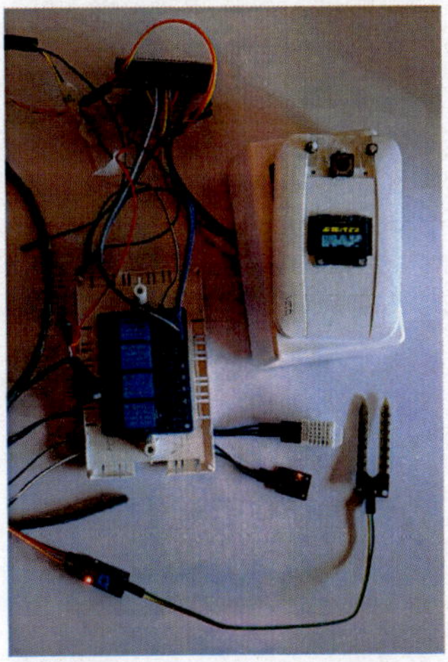

Fig. 3. Implementation of irrigation system based on IoT.

The LM35DZ (external temperature) and DHT22 (humidity and temperature) sensors are located close to the principal board. Furthermore, two relay modules are implemented to control the water pumps. Further, a power bank is connected to the Arduino board to supply energy in case of fall of the electric supply energy.

5.2 Software

Two scripts are upload to the Arduino board. The first one is upload to the microcontroller Atmega2560, which contains the configuration and the variables that store the measurements from sensors. Four soil moisture sensors are configured to take measurements each two meters inside the garden. Once all measurements are obtained, their values are visualized in a LCD. Later, these values measures are saved in a unique variable which is sent by serial communication to the esp8266.

The second script is upload to the esp8266, which is configured to read the values send the microcontroller Atmega2560. In this script, we configure the SSID and password in order to have an internet connection. Furthermore, the relay modules are controlled by the following three conditions:

1. The first condition follows that if the soil moisture sensors send a high value that indicates the soil is dry, then, the water pumps are activated.
2. The second condition allows to control the activation of the water pumps by a push button.
3. The third condition is based on the control of water pumps through the Thinger IO platform.

Furthermore, in the second script is configurated the MQTT protocol which sends the information obtained by esp8266 to the Thinger IO dashboard.

The Thinger IO platform allows the connection to esp8266 device. Figure 4 shows the connection of esp8266 to Thinger IO, which provides information about if the device is connected, the quantity of data transmitted and received, the geolocation of the device, and other settings related to the connection.

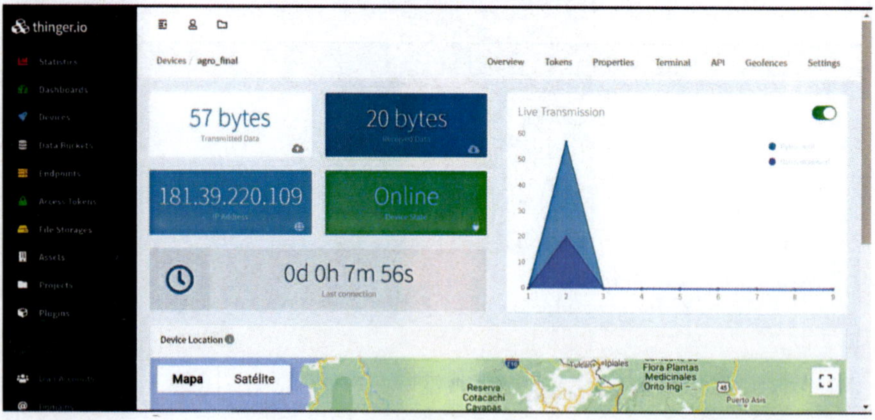

Fig. 4. Connection between esp8266 and Thinger IO platform.

The database configuration is shown in Fig. 5, where the measurements values from the sensors together with the system date are stored in the Thinger IO cloud. In this window, it is possible to determine with which frequency the values are stored in the database. In our prototype, it is configured to store the measurements every 5 min.

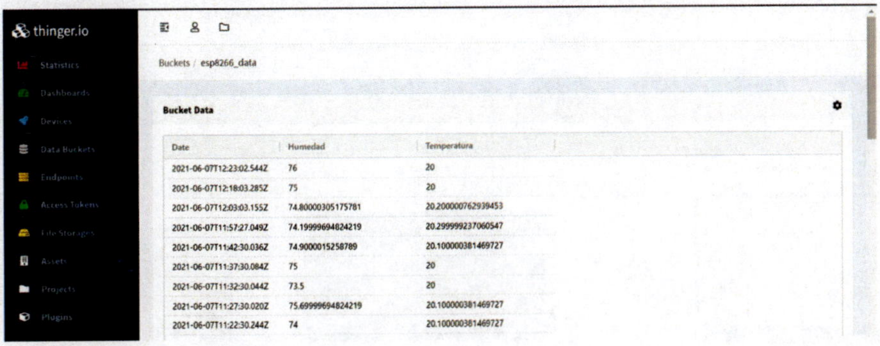

Fig. 5. Database configuration in Thinger IO.

Figure 6 shows the dashboard build into the Thinger IO platform, where the temperature and the humidity are displayed in a tachometer and in a time series chart. The external temperature is displayed in a numeric value, the soil moisture is shown in a gauge display. Note that all these measurements are configured in the dashboard to refresh the displayed values every ten minutes. Two push-buttons are configured to control the water pumps.

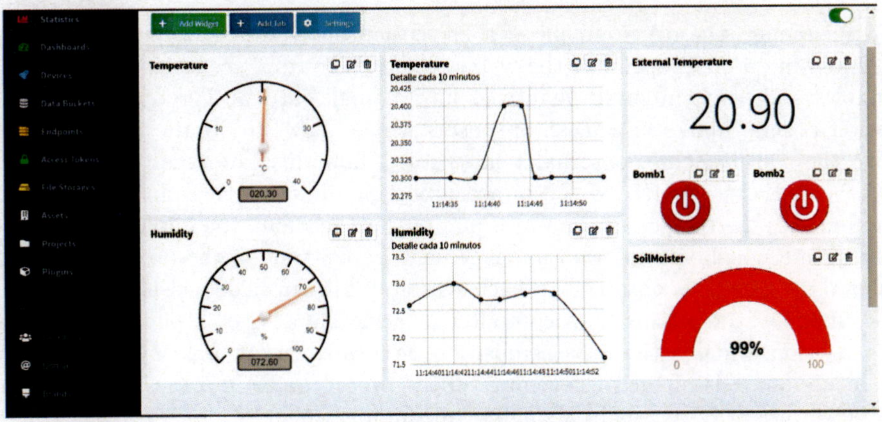

Fig. 6. Dashboard into Thinger IO.

Finally, Fig. 7 shows the design implementation into the garden. The garden is divided into two sections in each one has two soil moisture sensors. Besides, there are two water pumps which each one supply each section of the garden. An plastic box is used to protects the principal board connections. The software and hardware are proved into a real scenario in a garden.

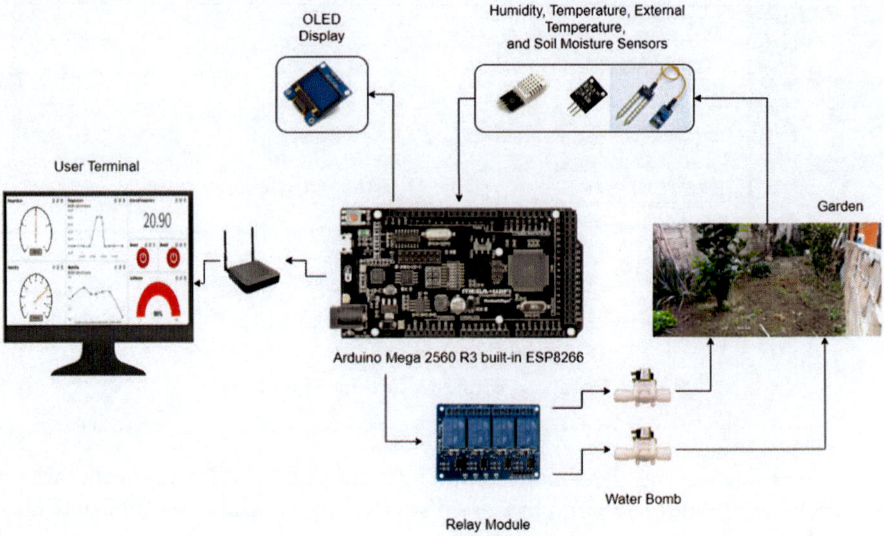

Fig. 7. Design garden implementation.

6 Conclusion

In this paper, the development of a low-cost prototype is proposed and implemented in order to optimize the water resources in rural scenarios to yield vegetables and fruits in small quantities into a small garden. The optimization of water occurs thanks to a mesh of sensors which allows to obtain measurements like the soil moisture, external temperature, humidity, and temperature close to the soil. Once obtained these measurements, it is designed an algorithm to automatize the turn on of the water pumps. Furthermore, the implementation of Thinger IO dashboar allows a management in real-time of the prototype which shows the measures obtained by each sensor and it can take decisions to turn on or off the water pumps by a computer or a mobile device through an internet connection. Finally, the data acquisition performed in our prototype will allow a future analysis of the performance of the prototype to improve it and having a better use of resources in a small or medium garden.

References

1. Navulur, S., Sastry, A.S.C.S., Prasad, M.N.G.: Agricultural management through wireless sensors and internet of things. Int. J. Electr. Comput. Eng. (IJECE) **7**, 3492–3499 (2017)
2. Abaya, S., De Vega, L., Garcia, J., Maniaul, M., Redondo, C.A.: A self-activating irrigation technology designed for a smart and futuristic farming. In: 2017 International Conference on Circuits, Devices and Systems (ICCDS), pp. 189–194 (2017)

3. Caetano, F., Pitarma, R., Reis, P.: Advanced system for garden irrigation management. In: Rocha, A., Correia, A.M., Costanzo, S., Reis, L.P. (eds.) New Contributions in Information Systems and Technologies. AISC, vol. 353, pp. 565–574. Springer, Cham (2015). https://doi.org/10.1007/978-3-319-16486-1_55
4. Tripathi, A.D., Mishra, R., Maurya, K.K., Singh, R.B., Wilson, D.W.: Chapter 1 - estimates for world population and global food availability for global health. In: Singh, R.B., Watson, R.R., Takahashi, T. (eds.) The Role of Functional Food Security in Global Health, pp. 3–24. Academic Press (2019)
5. Sisinni, E., Saifullah, A., Han, S., Jennehag, U., Gidlund, M.: Industrial internet of things: challenges, opportunities, and directions. IEEE Trans. Industr. Inf. **14**(11), 4724–4734 (2018)
6. Lin, J., Yu, W., Zhang, N., Yang, X., Zhang, H., Zhao, W.: A survey on internet of things: architecture, enabling technologies, security and privacy, and applications. IEEE Internet Things J. **4**(5), 1125–1142 (2017)
7. Santhi, P.V., Kapileswar, N., Chenchela, V.K.R., Prasad, C.H.V.S.: Sensor and vision based autonomous AGRIBOT for sowing seeds. In: 2017 International Conference on Energy, Communication, Data Analytics and Soft Computing (ICECDS), pp. 242–245 (2017)
8. Villarrubia, G., Paz, J.F.D., Iglesia, D.H.D.L., Bajo, J.: Combining multi-agent systems and wireless sensor networks for monitoring crop irrigation. Sensors **17**(8), 1775 (2017)
9. Hong, G.-Z., Hsieh, C.-L.: Application of integrated control strategy and bluetooth for irrigating romaine lettuce in greenhouse. IFAC-PapersOnLine **49**(16), 381–386 (2016). 5th IFAC Conference on Sensing, Control and Automation Technologies for Agriculture AGRICONTROL 2016
10. Khanna, A., Kaur, S.: Evolution of internet of things (IoT) and its significant impact in the field of precision agriculture. Comput. Electron. Agric. **157**, 218–231 (2019)
11. Yassein, M.B., Shatnawi, M.Q., Aljwarneh, S., Al-Hatmi, R.: Internet of things: survey and open issues of MQTT protocol. In: 2017 International Conference on Engineering MIS (ICEMIS), pp. 1–6 (2017)
12. Raut, R., Varma, H., Mulla, C., Pawar, V.R.: Soil monitoring, fertigation, and irrigation system using IoT for agricultural application. In: Hu, Y.-C., Tiwari, S., Mishra, K.K., Trivedi, M.C. (eds.) Intelligent Communication and Computational Technologies. LNNS, vol. 19, pp. 67–73. Springer, Singapore (2018). https://doi.org/10.1007/978-981-10-5523-2_7
13. Al-Fuqaha, A., Guizani, M., Mohammadi, M., Aledhari, M., Ayyash, M.: Internet of things: a survey on enabling technologies, protocols, and applications. IEEE Commun. Surv. Tutor. **17**(4), 2347–2376 (2015)
14. Muxito, E., Silva, A.M., Duarte, C., Aragão, A., Sambongo, E.: IoT na agricultura - automação de pivôs e canais de irrigação com arduíno e webservice, September 2018
15. Smart agriculture: 13 trends to watch out for, June 2017

Design and Construction of a Variable Autotransformer for Laboratory Applications

Juan Ramírez[1], Abraham Loja[1(✉)], Miguel Argoti[2], and Joselito Murillo[2]

[1] Escuela Politécnica Nacional, Escuela de Formación de Tecnólogos, Ladrón de Guevara E11-253, Quito, Ecuador
{juan.ramirezd,abraham.loja}@epn.edu.ec

[2] Instituto Superior Universitario Central Técnico, Av. Isaac Albéniz E4-15 y El Morlán, Quito, Ecuador
rargoti@istct.edu.ec, joselito.murillo@halliburton.com

Abstract. A three-phase autotransformer with a variable output voltage was built for the Electrical Machines Laboratory of the National Polytechnic School. The study for the selection of each of the components of the autotransformer is carried out based on theoretical and practical knowledge of electromechanical technology, bibliographic references and with the experimental support of a transformer factory.

The autotransformer is built on a ferromagnetic core, with an approximate power of 3 (kVA) between the windings of the autotransformer in which the connections will be made, in addition to using it for open and short circuit tests in transformers, the primary winding it supplies 208 (V) between phases and has a variable output voltage between 0 (V) to 311 (V).

Through the calculations based on the dimensions of the core and the voltages that need to be obtained, the three coils are built and subsequently mounted on the columns of the ferromagnetic core of the three-phase autotransformer.

Furthermore, a mobile mechanism is implemented to obtain the variable output voltage, by means of three brushes that slide on the windings of the columns of the three-phase autotransformer. Said manual mechanism is based on the translational movement of a linear rheostat, which by varying its position slides the brushes on the turns of the windings and a variable voltage is obtained at the output of the autotransformer terminals.

Keywords: Three-phase autotransformer · Variable voltage · Moving mechanism

1 Introduction

The National Polytechnic School (EPN) has a Technologists Training School (ESFOT) whose goal is to prepare professionals with a solid academic, humanistic, theoretical and practical training. Such that these professionals interact in the society to get a contribution to the technological development of the country [1]. The university technical education has been the key that opened the migration door that allowed developed countries to change a manufacturing-based workforce system to a knowledge-based one [2].

M. Zambrano Vizuete et al. (Eds.): CI3 2021, LNNS 511, pp. 240–254, 2022.
https://doi.org/10.1007/978-3-031-11438-0_20

Moreover, industries need more trained employees that can use their creativity to develop new products that meet the needs of society [3].

Technical education institutes need their students to develop their practical skills. For this, the laboratories are the connection bridge between theory and practice, and allow the application of theory to solve problems through constructions and designs while allowing students to develop interactive learning [4]. The laboratory is important in preparing students for practice beyond university to develop in a sustainable society that requires professionals who are familiar with experimentation and laboratory work [4].

With this background, this article presents an example of practical application of learning in a technical education institute. The work begins with an exploratory study that is used to observe and understand the needs within the current Laboratory of Electrical Machines of the ESFOT-EPN, where the problem to be solved is the absence of a three-phase autotransformer with variable output voltage. The autotransformer is needed as a three-phase power source for the execution of the short-circuit and open-circuit tests to a 50 (kVA) distribution type didactic three-phase transformer available in the laboratory. These tests are commonly applied in the industrial setting locally and globally.

Subsequently, a design is established and the calculations and selection of the necessary components are made for the construction of the three-phase autotransformer and the implementation of the mobile mechanism for voltage variation. Of the elements necessary for construction, such as the ferromagnetic core, they can be used from recycled equipment and the other elements that are available in the national market.

Finally, the construction of the proposed equipment is carried out and the functional tests are carried out to verify that the equipment built is functional. In this way, this article presents how theoretical-practical knowledge is applied for a solution to a specific need.

2 Materials and Methods

2.1 Techniques and Instruments Used in the Autotransformer Building

The initial phase of the project consisted of determining the necessary requirements for the building of the three-phase autotransformer by analyzing the needs of the students in the laboratory. With the information obtained by the students and teachers in charge of the subject, a series of important data was obtained about the installed capacity in terms of test autotransformers and transformers. Thus, the voltages and current intensities available in the laboratory were obtained, to achieve the proposed objectives by applying the short-circuit and open-circuit tests.

In a second phase, all the necessary calculations were carried out to determine the dimensions of the core, considering that the length of its column is high enough to host the primary and secondary windings in a single layer. In addition, the type and gauge of wire to be used, and the number of turns for each winding, are calculated.

In the third phase that corresponds to the process of building the mechanical and electrical part of the autotransformer, the use of the tools and molds necessary for the elaboration of the primary and secondary windings is detailed; each of these parameters depends on the dimensions of the selected core.

In the fourth phase, the entire process is detailed step by step regarding the assembly of the windings in the core, location of the fastening elements, insulations, necessary connections, assembly and adaptation of the mobile mechanism that holds the brushes in charge of the voltage variation in the three-phase autotransformer.

Finally, in the fifth phase, the three-phase autotransformer performance tests are carried out, such as: insulation test, winding resistance test, open circuit test; each of these from the experimental point of view, in order to verify the pursued objectives.

2.2 Component Requirements for the Autotransformer

After the gathering of information about the needs of the equipment to be designed and built, it was determined that the electrical characteristics for the construction of the three-phase autotransformer with variable output voltage are:

- Configuration: Three phase
- Power: 3 (kVA)
- Input Voltage: 208 (V)
- Variable Output Voltage: 0–311 (V)
- Voltage Variation Mechanism: Manual
- Connection: The primary and secondary windings are connected as an autotransformer in YNyn.

The core of the autotransformer is made of magnetic sheets of grain-oriented silicon steel. This steel offers great resistance to aging and can be worked at high inductions due to its high permeability and very low losses in the rolling direction. In Fig. 1, the dimensions of the core for the building of the transformation equipment are illustrated; where AN is the height of the core, HVN is the height of the core window, AVN is the width of the core window, DEP is the distance between columns, LN is the length of the core base and E is the core thickness.

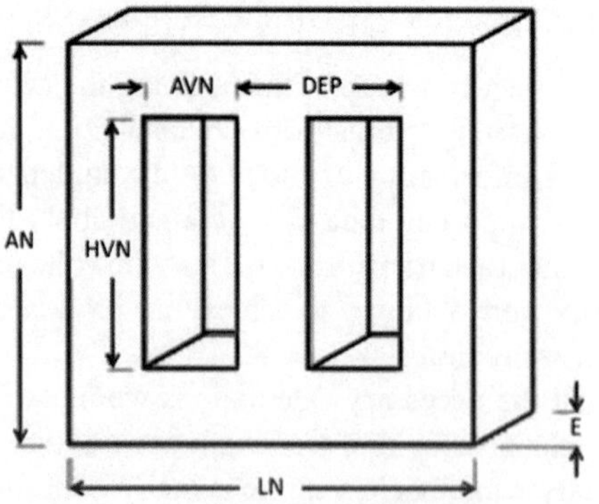

Fig. 1. Stacked core main dimensions

Table 1 presents a summary with the dimensions of the available core for the design of the three-phase autotransformer.

Table 1. Core dimensions

Parameter	Value
Sheet width (cm)	5
Sheet thickness (mm)	0,26
Core base length (cm)	25
Core thickness (E) (cm)	7
Core height (cm)	53
Core window width (AL) (cm)	5
Core window height (cm)	43
Distance between columns (cm)	10
Core weight (Kg)	46

2.3 Governing Equations

The crossed-section area of the core column depends on the width of the AL sheet (in [cm]) and the thickness of the reinforced core E (in [cm]), according to Eq. 1.

$$A_{Core} = AL * E \tag{1}$$

The magnitudes of voltages and currents to be considered for the calculations in the design of the three-phase autotransformer are presented in Fig. 1:

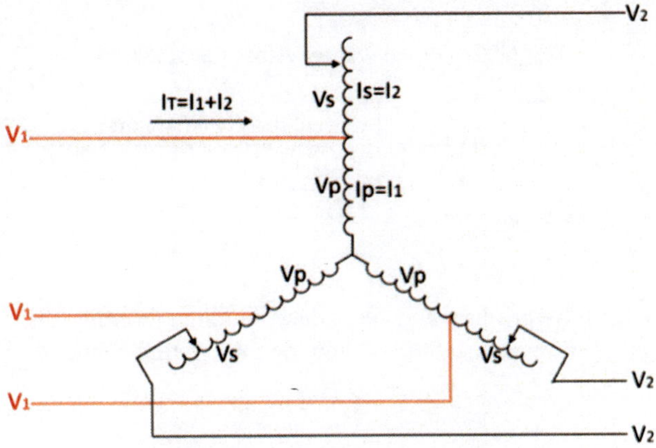

Fig. 2. Windings voltages and currents

where V1 is the supply phase-phase voltage expressed in [V], V2 is the maximum phase-phase voltage at the output of the autotransformer expressed in [V], Vp is the voltage per phase in the primary winding expressed in [V], Vs is the voltage per phase in the secondary winding, Ip is the current in the primary winding expressed in [A], is the current in the secondary winding expressed in [A], IT is the total current absorbed by the windings primary and secondary expressed in [A].

The primary winding of the autotransformer is in the "Y" configuration; therefore, its phase voltage is calculated with Eq. 2. The phase voltage of the secondary winding of the autotransformer is calculated with Eq. 3.

$$V_p = \frac{V_1}{\sqrt{3}} \tag{2}$$

$$V_s = \frac{V_2 - V_1}{\sqrt{3}} \tag{3}$$

The maximum power P_{max} expressed in [W] that can be obtained from the core is given by Eq. 4 [5].

$$P_{max} = \left(\frac{A_{Core}}{1,152}\right)^2 \tag{4}$$

The calculation of the maximum currents available per each phase I_{max} expressed in [A] of the windings is calculated with Eq. 5, considering the maximum output voltage V2 per phase

$$I_{\max} = \frac{P_{max}}{V_2} \tag{5}$$

To calculate the conductor gauge, it is necessary to take the current density, which depends on the type of transformer to be designed as presented in Table 2 [5, 6].

Table 2. Current density values for windings

Transformer type	Current density δ (A/mm^2)
Oil	2,5 to 3,5
Dry	1,5 to 2,5

The S section of the primary and secondary windings in [mm^2] is calculated with the Eq. 6 for a current I expressed in [A] flowing through the conductor [5, 6].

$$S = \frac{I}{\delta} \tag{6}$$

The Eq. 7 is used to calculate the number of turns per volt. for a frequency f expressed in [Hz] and a core magnetic flux density B expressed in [Webber/m^2] and the core area [5, 6].

$$\#_{\frac{turns}{volt}} = \frac{1}{4,44 * f * B * 0,0001 * A_{Core}} \tag{7}$$

Equation 8 is used to calculate the number of turns N carried by each winding for the voltage level of the winding V expressed in [V] [5, 6].

$$N = V * \#_{\frac{turns}{volt}} \tag{8}$$

The height of the autotransformer coils considers to have a voltage that varies from 0 (V) to 311 (V) and that the coil will have a single layer to be able to couple the voltage variation mechanism with brushes. The dimensions of coil lengths are presented in Fig. 3.

Equation 9 is used to calculate the electrical height that the winding will have in each column of the autotransformer considering NT total turns per layer. A tolerance T of 5%, the thickness L of the conductor in [mm] and the amount C of conductors necessary to complete the required section.

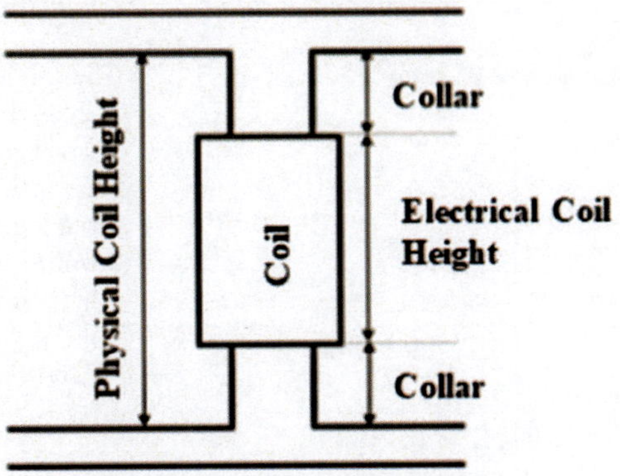

Fig. 3. Height dimensions of the Coil

$$HEB = [N_T + 1] * T * L * C \tag{9}$$

Equation 10 is used to calculate the power per phase of the autotransformer.

$$S = \frac{V_2}{\sqrt{3}} * I_{\max} \tag{10}$$

3 Execution and Results

3.1 Design Results

The requirements and equations described in Sect. 2.2 determine the construction parameters that are summarized in Table 3.

Table 3. Autotransformer calculation results for its building.

Parameter	Value	Comment
Core Section A_{Core}	35 [cm^2]	
Input Three Phase Voltage V_1	208 [V]	
Maximum Output Three Phase Voltage V_2	311 [V]	
Primary Coil Voltage V_p	120 [V]	
Secondary Coil Voltage V_s	60 [V]	
Core Maximum Available Power P_{max}	923.06 [W]	
Windings Maximum Current I_{max}	5.12 [A]	
Wire gauge S from Primary and Secondary Coils	2.08 [mm^2]	The result is 2.048 [mm^2], so it is chosen a #14 AWG enamelled wire of Sect. 2.08 [mm^2]. A current density for a dried transformer of 2.5 [A/mm^2] (Table 2) is considered
Turns per Volt	1.34 [turns/volt]	For a frequency $f = 60$ [Hz] and $B = 0.8$ [$Webber/m^2$]
Turns number N	$N1 = 161$ [turns] for the primary coil $N2 = 81$ [vueltas] for the secondary coil $N = N1 + N2 = 242$ [turns]	
Winding Height HEB	42 [cm]	
Power S per phase	0.9 [kVA]	Three Phase Power of 2.7 [kVA]

3.2 Autotransformer Structure

In Fig. 4 it is possible to see the proposal of the metallic structure, which has a rectangular base that will be useful as a support to hold the autotransformer. Two vertical and parallel rails are added, which will attach the brush holder base. The movable mechanism with the brushes for the voltage variation in the three-phase autotransformer slides on the rails.

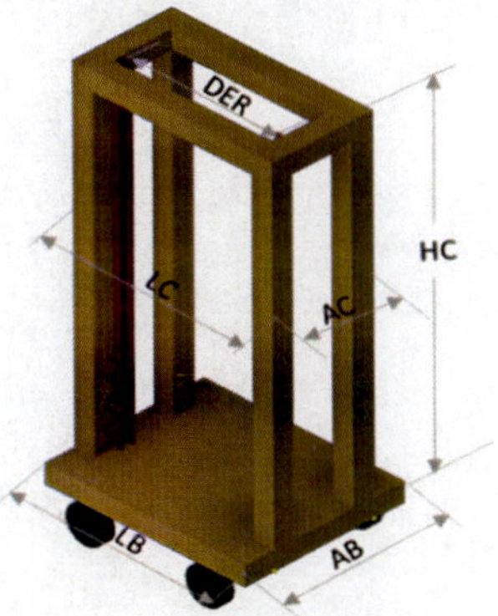

Fig. 4. Metal structure

Finally, two axes are attached to the base of the metal structure that will be used to mount four wheels for transport.

The Table 4 shows the metallic structure dimensions of the Fig. 4.

Table 4. Metallic structure dimensions

Parameter	Value
Base length (LB) (cm)	40
Base width (AB) (cm)	25
Body length (LC) (cm)	40
Body width (AC) (cm)	18,50
Distance between rails (DER) (cm)	13,50
Body height (HC) (cm)	65

3.3 Variable Connection System

The system is made of a brush holder plate that contains the carbon brushes, a threaded rod that passes through the body of the plate that when rotating clockwise or counter-clockwise transmits a translation movement through its axis; thus the holder plate brushes ascends or descends through the rails placed in the metallic structure; in turn, the carbon brushes are responsible for bringing the voltage at the outputs of the three-phase autotransformer.

In Fig. 5, the most important elements that constitute the movable brush system or mechanism for the voltage variation in the three-phase autotransformer are presented.

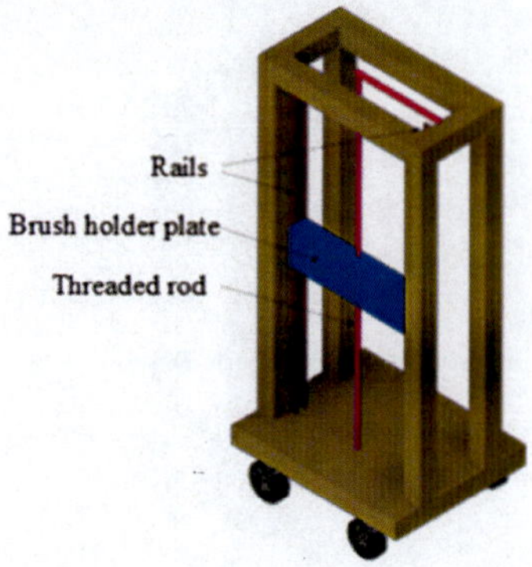

Fig. 5. Moville mechanism for the voltage variation

3.4 Electric Connection Diagrams

To vary the output voltage of the autotransformer, the carbon brushes slide linearly on the turns of the windings arranged on the columns of the three-phase autotransformer. In Fig. 6, the one-line connection diagram of the three-phase autotransformer with variable output voltage is shown. This is connected to the power supply of the three-phase network and the fixed voltage outputs as well as the variable output towards the terminal blocks on the terminals.

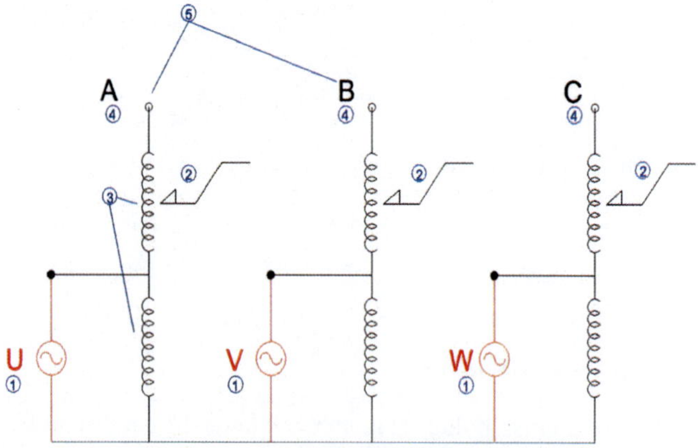

Fig. 6. On-line connection diagram

where:

1: Three-phase input voltage supplied by the network. Phases U, V y W (120 (V) / 208 (V)).

2: Manual voltage variator mechanism, built on the principle of movement of the linear rheostat.

3: Autotransformer windings, according to the design of the present study.

4: Maximum single-phase voltage at the output of the autotransformer A, B y C (180 (V)).

5: Maximum three-phase voltage at the output of the autotransformer AB, AC, BC (311 (V)).

3.5 Autotransformer Components for Building

The autotransformer consists of many parts which are divided into two groups, a main and an auxiliary. They are detailed below:

Main components:

- Silicon magnetic core (35 (cm^2) of total section, height of 55 (cm)).
- Enameled copper conductor # 14 AWG for primary and secondary winding.

Auxilary components:

- Wooden formwork.
- Terminal blocks.
- Isolating material NOMEX 220 (ºC)
- Dielectric varnish type H 220 (ºC)
- Fiberglass tape 220 (ºC)
- Cylinder heads.
- Galvanized iron square tube of 1″ x 2 (mm) for the mobile mechanism.
- Metal-wood brush holder base.
- Carbon brushes Hitachi M17 - 7 x 13 x 17 (mm) with spring, cable y connector for 10 (A).
- Threaded rod of ½" x 80 (cm), diametrical pitch
- Four wheels M60R of 3"
- Hex Head Self Drilling Bolts of 5/16″ x ¾"
- Clear acrylic plate 4 (mm) of 65 (cm) x 42 (cm).
- Three Pole Thermal Circuit Breaker of 10A for Din rail.
- Din rail of 2".
- Flexible cable # 14 AWG for connections between terminal blocks and brushes.

3.6 Components Installation

In Fig. 7, it is possible to see the fully assembled magnetic core.

The Fig. 8 shows the mechanism and the necessary elements for the building of the autotransformer windings. Once each winding has been made, it is removed from the winding machine. Then, the wooden formwork is extracted and subsequently the coil is mounted on the three-phase core.

To mount the windings to the core, the core clamps and clamps are removed to allow entry of the coils into each leg of the core. The coils are then carefully inserted into the core legs. In Fig. 9, it is seen the assembly of the windings in the two columns of the three-phase autotransformer.

Fig. 7. Assembled core

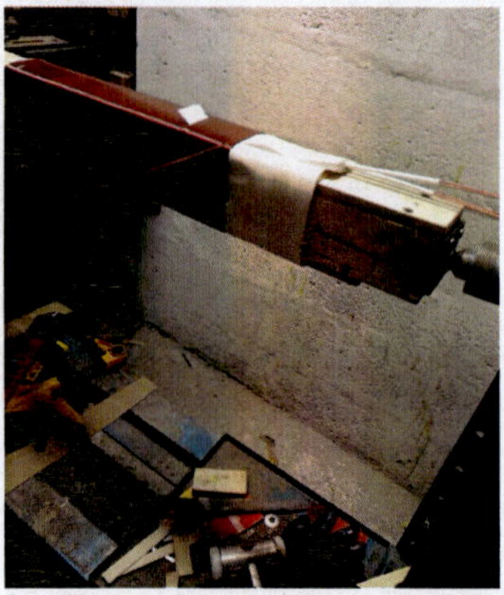

Fig. 8. Winding building

Fig. 9. Mounting the windings on the core

Once the three windings have been assembled on the columns, the core closures and the presses (cylinder heads) are placed to be able to firmly hold and transport the core with the coils that make up the transformation equipment.

The Fig. 10 (a) shows the fully assembled three-phase autotransformer.

(a) (b)

Fig. 10. (a) Three-phase autotransformer built. (b) Wiring and mounting of terminal blocks

Then the identification of all the connections that allow the correct operation of the autotransformer is carried out. In Fig. 10 (b) the connections of the windings and the mounting of the terminals in the three-phase autotransformer are illustrated.

The voltage variation mechanism is based on the operation of a linear rheostat, whose translational movement of the brushes is carried out by means of a shaft composed of a threaded rod. This, when rotating in one direction or another, moves the brushes on the surface of the previously sanded windings. In addition, for the coupling of the mobile mechanism, a metallic structure must be built, which works as a support for the autotransformer and for the voltage variation system.

Then the windings are sanded on the back of the three-phase autotransformer, for which a # 1200 fine sandpaper is used. The objective is to remove a fringe of enamel from the surface layer of the primary and secondary windings. This is done in order to create a correct contact between the enamel-free turns of the windings and the brushes at the moment of coupling the mobile mechanism.

In Fig. 11 (a) the stripes on the winding columns to be sanded are illustrated. For this process adhesive tape is placed around the stripes to be sanded and with total care, it is gently sanded to remove the enamel along the column of each winding.

Figure 11 (b) shows the main constituent components of the mobile mechanism in the three phase autotransformer.

(a) (b)

Fig. 11. (a) Sanding the windings. (b) Assembly of the mobile voltage variation mechanism

3.7 Functionality Test

With the finished prototype, the performance tests based on [7] are carried out and they are summarized in the Table 5.

Table 5. Autotransformer performance test results

Test	Result
Transformation relation	1.5 at the nominal position of the coils
180 [V] Coil insulation resistance to ground	IRH1 = 24.00 [MΩ] IRH2 = 24.20 [MΩ] IRH3 = 24.40 [MΩ]
120 [V] Coil insulation resistance to ground	IRX1 = 23.20 [MΩ] IRX2 = 23.50 [MΩ] IRX3 = 23.90 [MΩ]
No-load losses	5.85 [W]
180 [V] Winding resistance	RH1 = 1.13 [Ω] RH2 = 1.19 [Ω] RH3 = 1.49 [Ω]
120 [V] Winding resistance	RX1 = 0.45 [MΩ] RX2 = 0.44 [MΩ] RX3 = 0.56 [MΩ]
Short circuit losses	41.85 [W]
Winding performance	$\eta_1 = 96, 4\%$ $\eta_2 = 96, 3\%$ $\eta_3 = 95, 5\%$

Once the autotransformer has been tested both mechanically and electrically and the functional tests have been carried out, the identification and connection data are labeled, which are detailed on the plate on the front of the transformation equipment. Finally, the three-phase autotransformer with variable output voltage is located in the ESFOT Industrial Technology Laboratory for its use and disposal.

Figure 12 illustrates the finished module with all the elements both for its operation and for its protection.

Fig. 12. Final view of the autotransformer

4 Conclusions

A dry type three-phase autotransformer was built, which is complemented with a metallic structure for mounting the movable brush mechanism. This mobile mechanism is based on the operation of a linear rheostat for the vertical translation of the carbon brushes, which allows the operator to manually rotate the threaded rod and thus obtain different voltage values at the variable output of the autotransformer.

For the construction of the three-phase autotransformer with variable output voltage, a magnetic iron core (not common in the market) with a height of 52 (cm) was used. On this core, the primary and secondary windings are mounted forming a single layer that is the main characteristic that was considered in the design of the device. This in order to better couple the movable mechanism of brushes that slide on the turns and thus obtain values of variable voltage at the terminal outputs, from 0 (V) in the minimum position, up to 311 (V) in the maximum position of the transformer equipment.

One of the solutions to the lack of a variable output voltage source is the construction of a three-phase autotransformer with variable output voltage. This allows the EPN students and, in particular the students from the Electromechanics Superior Technology career have at disposal a transformation device for the pertinent practices in the curriculum of the Electrical Machines Laboratory.

References

1. ESFOT-EPN: ¿Quiénes Somos? - Escuela de Formación de Tecnólogos. (2020). https://esfot.epn.edu.ec/index.php/quienes-somos Accessed 04 Jun. 2021
2. Venkatraman, S., de Souza-Daw, T., Kaspi, S.: Improving employment outcomes of career and technical education students. High. Educ. Ski. Work. Learn. **8**(4), 469–483 (2018). https://doi.org/10.1108/HESWBL-01-2018-0003. Jan.
3. Lucianelli, G., Citro, F.: Financial conditions and financial sustainability in higher education: A literature review. In: Financial Sustainability in Public Administration: Exploring the Concept of Financial Health. Springer International Publishing, pp. 23–53 (2017)
4. Winberg, C., Winberg, S.L.: The role of undergraduate laboratories in the formation of engineering identities: a critical review of the literature. J. Transdiscipl. Res. South. Africa **17**(1), 12 (2021). https://doi.org/10.4102/td.v17i1.962. Apr.
5. Kulkarni, S., Khaparde, S.: Transformer engineering: design, technology, and diagnostics (2017). Accessed 04 Jun. 2021. [Online]. Available: https://books.google.es/books?hl=es&lr=&id=aynOBQAAQBAJ&oi=fnd&pg=PP1&dq=Transformer+Engineering:+Design,+Technology,+and+Diagnostics,+Second+Edition&ots=HI64X_ffhX&sig=ydt38mGjZbHigVEWVUq6i5CqNfM
6. Sudhoff, S.D.: Power Magnetic Devices: A Multi-Objective Design Approach vol. 9781118489994. Wiley-IEEE Press (2014)
7. "IEEE C57.12.00–2015 - IEEE Standard for General Requirements for Liquid-Immersed Distribution, Power, and Regulating Transformers." https://standards.ieee.org/standard/C57_12_00-2015.html. Accessed 04 Jun. 2021

Quantitative Analysis of the Dehumidification Parameters of Compressed Air for CNC Plasma

Beltrán Leonardo[1], Jumbo Anderson[1], Cusicagua Lenin[1](✉), Santamaría Tania[1], and Gómez Víctor[2]

[1] Instituto Superior Universitario Central Técnico, Quito 170138, Ecuador
lbeltran@istct.edu.ec, leninxx_8@outlook.com
[2] Instituto Superior Universitario Bolivariano, Guayaquil, Ecuador

Abstract. In the following investigation, a quantitative analysis of the dehumidification process that compressed air submits in a CNC plasma cutting machine was carried out. The research focused on the plasma cutting process and how much it affected the use or not of dry compressed air, the dehumidification type of the machine and its quality is important for a successfully operation. Through the experimentation method, it was possible to describe the physical effects such as penetration, cutting and heat deformation, caused by compressed air in a CNC plasma cutter in different conditions, whether they are with dry compressed air or with wet compressed air, by means of the descriptive method through the use of different types of materials and tests in specimens, the effects of using dry air and its advantage over humid compressed air were better determined.

Keywords: Dehumidification · Filtration · Quality · Plasma · Transformation

1 Introduction

Currently, the knowledge about the air quality management of a compressor in an industry manufacturing line is mainly focused on the transforming process that it undergoes in different stages of demand in its performance, one of them is the dehumidification of the compressed air. This research has a previous analysis of the air behavior in the different transforming stages it faces inside an air compressor before the itself dehumidification and output for supplying different machines. The quantitative analysis of the air dehumidification process has been focused on the compressor air output conditions, its advantages and disadvantages at the time of operating together [1] with the machine, due to the fact that there are defects in its physical structure such as the humidity excess in the air compressed which does not allow the proper performance of the CNC plasma machine.

The Air compressed is an air body with a higher pressure than the atmospheric one, when being compressed it has an energy increasing when it is expanded, it is being used as a source of a mechanical work development [1].

The Air compressed is an essential source and of wide use in industrial processes and applied in all sorts of industries, its main purpose is to supply air at a suitable pressure, flow rate and quality [2].

M. Zambrano Vizuete et al. (Eds.): CI3 2021, LNNS 511, pp. 255–269, 2022.
https://doi.org/10.1007/978-3-031-11438-0_21

The influence of humidity in the facilities causes a series of inconveniences such as the rusting appearance and pre-mature abrasion due to the water dragging the lubrication among other disadvantages [3].

The risks produced by compressed air were taken into account, such as: compressor overload and leaks, noise nuisance, particle emission [4].

The compressed air has gained in recent years great utility value and has been a reason for observation and study for this reason special attention has been paid to the contamination immersed in the air, since it contains contaminants in the atmosphere such as dust, moisture and hydrocarbons. In addition to those contaminants generated in the process of air compression [5].

In his Hypertherm user manual (2018) mentions that the air quality must be clean, dry, free of oil.

A closely related research that helped us is the comparative analysis of the effects produced by the application of OFC, PAC, WEDM cutting processes in low-carbon steel specimens, where it refers to the gas supply [6].

In the research of [7] makes reference that the plasma cutting process the cutting speed increases by 25% with respect to traditional cutting, but has disadvantages such as increased rusting and electrode abrasion, using humid dry air. As Perez (2020) researched, the gas supply for a Hypertherm plasma must be clean, dry and oil-free.

Considering the common unfiltered air supply increases the consumption of the flow required by a plasma cutter by 30% [6].

Oxygen increases cutting speeds in the air by 25% over traditional dry plasma cutting, however, it also leads to a highly rusty cutting surface and a rapid electrode erosion inside the nozzle [8]. The success of the application of plasma cutting lies in the proper gas choice or gas mixtures, including air [9].

Nowadays there are several methods of air dehumidification, however in a screw compressor dehumidification by cooling is the most indicated for this process, as described below: "Dehumidification can be caused by cooling the air to a temperature below the dew point, by absorption, by absorption or by compression followed by cooling" [10].

> "Air can be dehumidified with systems of: conventional vapor compression air conditioning. These cool the air at a constant pressure to a temperature below the dew point temperature, thereby condensing part of the water vapor present in the air [...] it must work at a lower temperature than that which is required to extract the sensible cooling heat load from the conditioned space, this causes the system to have low levels of condensation in the dry air" [11].

> "Dryers are equipment designed to treat compressed air or gases to reduce their water vapor content, lowering the dew point of the compressed air to a satisfactory level so that the moisture and oil vapor are reduced before entering the distribution grids, avoiding water condensation n at the usage points" [12] (Fig. 1).

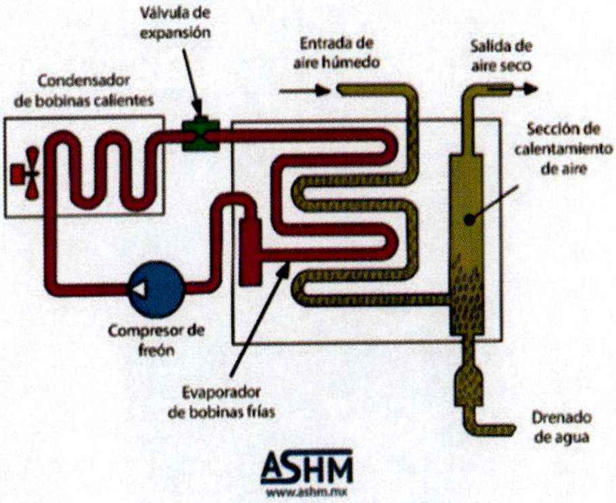

Fig. 1. Cooling drying diagram. Source: ASHM

For a better understanding of the importance of dehumidifying the air compressed in the CNC plasma machine to generate a clean cut on the material, the following information is taken into account, among those cited:

> "Plasma is known as one of the four states of the matter aggregation, along with liquid, solid and gaseous. When it is spoken of a substance in plasma state it is related to an ionized gas, or in other words, to a gas whose atoms have been removed from part of their original electrons, and being electrically charged" [13].

Realizing that plasma production is a mixture of ionized air and electricity, this is where air compressed complies a prime function as cited below:

> "Air compressed is atmospheric air under pressure, usually composed of 78% Nitrogen, 21% Oxygen and 1% other gases; physically it is odorless, colorless and tasteless" [9].
>
> "Compressing air causes a faster movement of molecules, which increases the temperature. This phenomenon is called "heat of compression." Compressing air force, it to enter into a smaller space and, as a result, molecules are attracted to each other." [14] (Fig. 2).

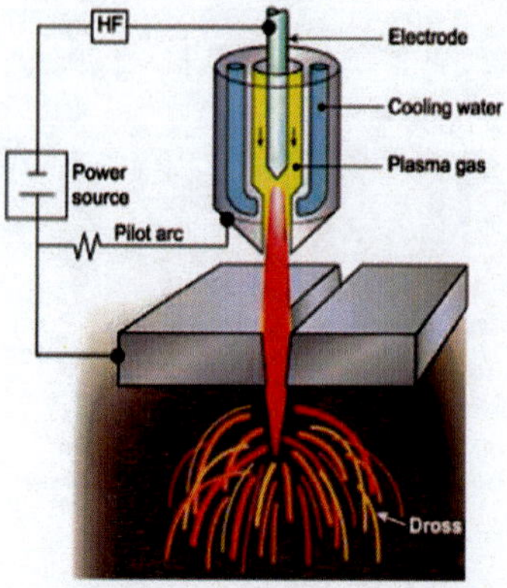

Fig. 2. CNC plasma torch diagram. Source: ASHM

Thus, it has been elastic collisions experience with each other continuously moving through the dry compressed air pressure generated a large amount of heat energy creating a hot plasma able to penetrate into the material and cut it.

Considering that the equipment and machines used in the industry need to have high duty cycles and better quality of air compressed for its performance, it is extremely important that industrial mechanics students get a better understanding of exactly why the air compressed suffers a constant humidification in the CNC plasma machine performance. Taking into consideration the fact of without the efficiency indicators to determine the processes capacity generates a competitive disadvantage within the production line, for this reason it is high-priority and high-benefit the implementation of a planning and control system for the machine production, which allows to improve the resources usage for the elaboration of the output product.

The primary objective of this research is based on demonstrating measurable behaviors of the dehumidification parameters of the air compressed produced by the air compressor, by exploration and descriptions of the machine working levels, based on the following specific aims:

- To carry out technical tests on different metals with the cutting performed by the CNC plasma machine to demonstrate a correct dehumidification of the air and allow a proper functioning of the screw compressor within the production line.
- To make a work planning with different records of the quantitative results to have a supporting source where it allows the chronological order of the research progress.
- To diagnose by interviewing the knowledge on the use of air compressed with technical specifications of the compressor for the equipment preservation.
- The idea of focusing on this area is due to the fact of empowering important professional fields, such as the appropriate handling of compressors, its maintenance and the performance quality of the CNC plasma to which it is feeding.

2 Methods and Resources

This research elaboration was carried out in different ways; at the beginning in an exploratory way, through field observations to analyze the real performance of the compressor at the moment of feeding the CNC plasma machine, based on the question:

What is the relevance of dry air during cutting? In this part we needed to see the external reactions that could have been generating the air dehumidification and the air outflow on the material to be cut.

Secondly, it is descriptive, where the fundamental objective was the data collection that allows a suitable experimentation quantitative analysis, as well as most relevant aspects to deepen in the research.

Finally, the last part of the research was developed from a causal research that helps us to differentiate and understand the causes variables the effect ones to determine the relevance of the air dehumidification from the knowledge that a group of people could have about the handling of the air compressed in the CNC plasma machine.

2.1 Testing

The study of the dehumidification parameters of the air compressed is a procedure that required several technical evaluations, in addition to the analysis of the reactions at different moments of its transformation. That is the reason why the methodology to achieve important and essential results, focused on the execution of:

A practical analysis on the influence of air compressed on the cutting performed by CNC plasma on probes of a variety of materials. Cutting analysis.

Having determined the performance and importance of the maintenance for the air compressed dehumidification based on the data obtained, it was proceeded to carry out practical tests together with the CNC plasma machine.

In the cutting performed by a CNC plasma, the main advantage is the high-quality cuts and very well-finished, as mentioned above the cutting quality is directly proportional to the quality of the air compressed, which means a previous successful dehumidification process. In this part of the research the testing was aimed at performing cross sections on probes of various materials such as black steel, stainless steel, aluminum, among other non-ferrous metals, taking into account some of the physical structural properties of the steels such as; hardness, toughness, heat and electrical conductivity.

The quality of the cut depends on the settings of the input parameters, furthermore, the distance from the torch to the workpiece, the cutting speed, the amperage, the gas pressure among others.

The cutting speed is defined which the torch moves in relation to the workpiece, and its value depends on the type pf material, thickness, amperage and type of consumables.

The cutting speeds, direction, distance from the work surface, consumables and plasma cutting technique in thickness of the materials, in addition several technical descriptions of the compressor were taken into account to avoid misalignment at the time of making the connections of the maintenance unit (Table 1).

Table 1. Minimum performance parameters.

Min. drive power	4,0 [HP]
Min. drive flow	13,8 [CFM]
Min. pressure	2 bar – 29PSI
Max. pressure	8 bar – 166PSI

Source: Porten Compresores

The quality of the air compressed depends on the least amount of humidity that the machine could generate at its outlet. In an air compressor, the effectiveness in performing in its different maintenance units is essential to meet a high level of dehumidification and therefore a high-quality CNC plasma cutting, that is why the cuts were made in two different ways, focused to check the real influence that this method has as described below (Table 2) (Fig. 3):

Table 2. Materials and dimensions of the samples.

Material	Dimensions [mm]	Thickness [mm]
Steel (ASTM A36)	150 × 100	10
Aluminum (ASTM 1200)	150 × 100	10
Stainless steel (AISI 304)	150 × 100	10
Galvanized steel (ASTM A653)	150 × 100	2

Own source.

Fig. 3. Unit maintenance switch off. Own source

- The first test in during this process was the running of several cuts with the CNC plasma with no prior connection of the maintenance unit, limiting the air dehumidification of the air compressed in the compressor and generating changes in the plasma quality and in its cutting (Fig. 4).

Fig. 4. Unit maintenance connection. Own source

- The second test consisted of making the same cross sections, in addition to the first test parameters including their dimensions and thicknesses. On this occasion with the availability of the unit connection maintenance which allowed a more successful dehumidification process thanks to its elements shown below (Fig. 5):

Fig. 5. FRL (unit maintenance). (a) air filter; (b) pressure regulator; (c) lubricator. Own source

The settings for cutting on the CNC plasma machine were also relevant for the behavior of the air compressed dehumidification, as several of these helped us to avoid any deformation in the physical structure of the materials used in the experimentation. Three important main factors were determined; power, shear rate and pressure.

The following charts describe the proposed parameters to differentiate the significance of air compressed in a CNC plasma cut on several materials, with and without the use of maintenance units (Tables 3, 4, 5 and 6).

Table 3. Parameters for the steel (ASTM A36)

Cutting	Amparage power [amp]	Cutting speed [m/min]	Compressor pressure [PSI]
Feed 1	85	400	96
Feed 2	85	2500	96
Feed 3	47	400	96

Own source

Table 4. Parameters for the aluminum (ASTM 1200)

Cutting	Amparage power [amp]	Cutting speed [m/min]	Compressor pressure [PSI]
Feed 1	85	400	96
Feed 2	85	2500	96
Feed 3	47	400	96

Own source

Table 5. Parameters for the stainless steel (AISI 304)

Cutting	Amparage power [amp]	Cutting speed [m/min]	Compressor pressure [PSI]
Feed 1	85	400	96
Feed 2	85	2500	96
Feed 3	47	400	96

Own source

Table 6. Parameters for the galvanized steel (A653)

Cutting	Amparage power [amp]	Cutting speed [m/min]	Compressor pressure [PSI]
Feed 1	55	3000	96
Feed 2	45	800	96

Own source

As shown in the tables, all the cuttings were made at a steady pressure by the compressor, which helped us to maintain the same uniformity in order to determine differences in all the probes, also with three different advances, varying both in power and cutting speed, but with a transverse direction in test 1 and 2, as shown in Fig. 6.

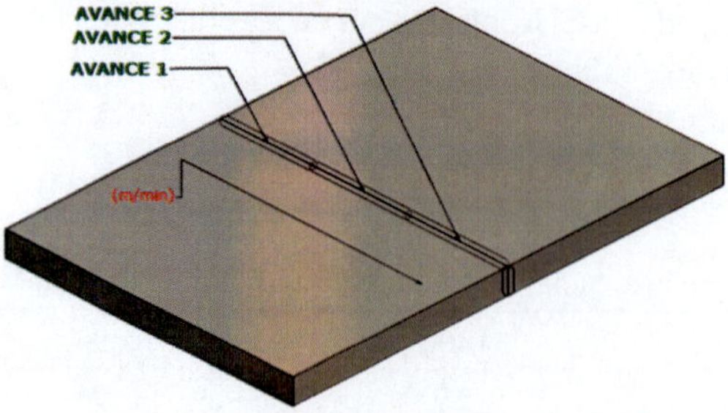

Fig. 6. CNC plasma cutting advancements. Own source

The systemic approach method identifies the input and output parameters to define the cutting process, the parameters are detailed. Torch-to-workpiece distance.

The torch – workpiece standoff distance is from the torch tip to the workpiece and it is highly critical for producing quality cuts.

This clearance increases as the arc force decreases and tends to spread and consequently more residue will form on the bottom edge of the plate. A narrow gap will reduce the lifetime of the nozzle tip, in many cases an arc is formed between the nozzle point and the metal, which can damage the tip.

Proper torch – workpiece height plays a significant role in obtaining high quality cuts, the gap significantly affects the bevel angle [1].

3 Results

For the results presentations it has been made tables, diagrams, which provide an effective method for the planning and control of the research.

Analysis of air dehumidification in CNC plasma. The results in this phase of the research are shown on the basis of established ranges, where the three previously chosen parameters; penetration, deformation and shear, may be compared on a numerical scale (Table 7, Figs. 7 and 8).

Table 7. Measurement parameters – results.

Penetration	
Successful Penetration	80 - 100[%]
Unsuccessful – Medium	5 - 80[%]
No penetration	0 - 5[%]
Cutting	
linear bevel	90°
Y-bevel	0° – 10°
V-bevel	0°- 10°
Deformation	
Largest amount - Burrs and Porosity	10 - 20[%]
Lesser amount - Burrs and Porosity	0 - 10[%]
Tears in the material	90 - 100[%]

Own source

Steel (A36).

Table 8. Cutting analysis of A36 steel.

Material	Cutting Quality					
	With dry air compressed			With humid air compressed		
BLACK STEEL	Penetration	Deformation	Cutting	Penetration	Deformation	Cutting
Feed 1	100%	10%	90°	100%	0%	5°
Feed 2	5%	15%	90°	80%	5%	90°
Feed 3	70%	10%	90°	80%	2%	5°

Own source

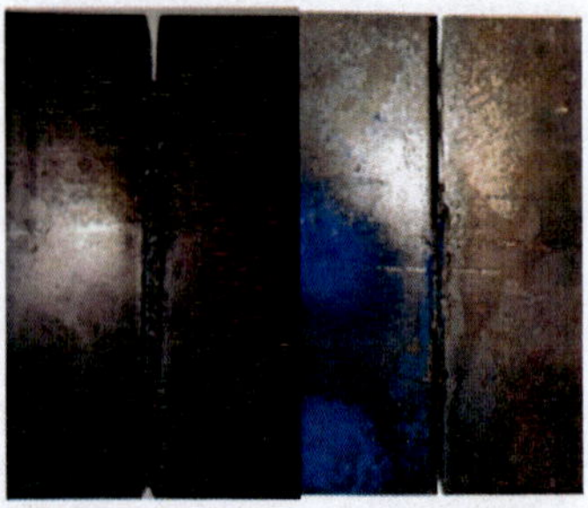

(a) (b)

Fig. 7. CNC plasma cutting on black steel. (a) with dehumidification; (b) without dehumidification. Own source. Aluminum (ASTM 1200).

Table 9. Cutting aluminium analysis. (ASTM 1200).

Material	Cutting Quality					
	With dry air compressed			With humid air compressed		
ALUMINUM	Penetration	Deformation	Cutting	Penetration	Deformation	Cutting
Feed 1	100%	0%	7°	100%	0%	10°
Feed 2	5%	15%	7°	100%	0%	90°
Feed 3	100%	0%	7°	5%	2%	90°

Own source

(a) (b)

Fig. 8. CNC plasma cutting of aluminum. (a) with dehumidification; (b) without dehumidification. Own source

Table 10. Cutting stainless steel analysis (AISI 304).

Material	Cutting Quality					
	With dry air compressed			With humid air compressed		
STAINLESS STEEL	Penetration	Deformation	Cutting	Penetration	Deformation	Cutting
Feed 1	100%	0%	90°	100%	0%	90°
Feed 2	5%	2%	90°	5%	5%	90°
Feed 3	80%	0%	90°	5%	5%	90°

Own source

(a) (b)

Fig. 9. CNC plasma cutting of stainless steel. (a) with dehumidification; (b) without dehumidification. Own source. Galvanized steel (A653).

Table 11. Cutting galvanized steel analysis (A653).

Material	Cutting Quality					
	With dry air compressed			With humid air compressed		
GALVANIZED STEEL	Penetration	Deformation	Cutting	Penetration	Deformation	Cutting
Feed 1	100%	0%	90°	100%	90%	3°
Feed 2	100%	0%	90°	100%	90%	3°

Own source

(a) (b)

Fig. 10. CNC plasma cutting of galvanized steel. (a) with dehumidification; (b) without dehumidification.

4 Discussion

As part of the cutting analysis, Table 8 showed that the penetration in the Black Steel with dry air compressed at a higher speed and intensity of 2500 [m/min] and 85 [amp] respectively, was insufficient, so that the CNC plasma barely grazed the material surface, meanwhile without the dehumidification process achieved 80% penetration, a higher percentage than the first one, however it was not satisfactory to completely overpass the material. In Steel A36 the deformations were very visible regardless of their speed and intensity range, since even with the dehumidification process the number of burrs and porosity did not decrease. The most remarkable difference was visualized in the cutting parameter, since the dry air compressed allowed a completely straight cut, unlike the cut without dehumidification, which, as the speed decreased, generated bevels in its frontal faces, particularly in feed rates 1 and 3.

In Table 9, the penetration in 2 advancements of aluminum was 100% with dry air compressed, something like the cutting with wet air compressed since it also obtained a rating of 100%, whereas in one advancement it did not penetrate in its entirety, due to the low amount of electricity. In the deformation with dry air compressed, the second advancement was qualified with 15% for high level of burrs and porosity, generated by the high current intensity and speed of the plasma machine, while in the cut with no dry air compressed, there was no deformation greater than 2%. The differences in the aluminum cuts are not significant, since being a light metal did not generate difficulties to overcome the material, although it generated bevels of lesser degrees in its edges.

In Stainless Steel (AISI 304) the use of dry air compressed to obtain a cut allowed us to visualize a great inconsistency in the cutting quality. It was demonstrated that in this metal, exceeding the limits of the cutting speed parameter caused the plasma failed to penetrate, although it hardly generated deformations in its edges in all the cutting advancements. The cut in Stainless Steel (AISI 304) quality made with humid air compressed, had the worst results of all the materials examined, since it barely managed to penetrate in the first progress, this due to the increase and decrease of its amperage and cutting speed. Despite the fact that the material penetration had drawbacks, it always had a directionality of 90° on the surface of the material in all the tests (Table 10).

Table 11 revealed a wide difference in the cutting quality of the CNC plasma in the Galvanized Steel, both with and no dehumidification process. Even though, the features of the probe were different from the previous materials and the parameters such as the amperage range were not higher than 55[amp] in their advancements, in order to prevent the melting of the same. The cut with dry air compressed allowed a total penetration, no porosities or burrs were formed and all its advancements were with a linear bevel, thus achieving a perfect cut. Performing a test with wet air compressed, caused tearing of the material due to the low amperage and high speeds in their advances as shown in Fig. 9, although there were no inconveniences to penetrate the material entirety, because of the material characteristics (Fig. 10).

The dry air compressed quality improves the quality features of the cut, avoiding reprocesses due to bevel cutting.

While research has been done on the use of different types of gases in the plasma cutting process, this research focused on the parameters of air as a gas but with a drying treatment prior to entering the plasma equipment.

Concerning the deformation of the material, it was found the fact of having dry air compressed does not have a directly influence. These results could be taken as a reference to further investigation of how much the dry air compressed influences the plasma cutting process.

5 Conclusions

The efficiency of a plasma machine depends on an effective air dehumidification process in the components of the unit maintenance; air filter, pressure regulator and air compressed lubricator. The lack of a unit maintenance at the compressor of air outlet almost always results in angular misalignment of the CNC plasma cutting torch due to lack of compressor pressure regulation causing bevels in its thicknesses.

In the analysis of the air compressed dehumidification, limiting this process in a plasma cutting, led several evident deformations in the physical appearance of most materials, such as the generation of porosity, burrs and material tearing in several advancements of the plasma machine.

In the unit maintenance, the condensed water in the filtering, must be purged before it reaches the maximum pressure level, thus preventing water from entering the CNC plasma torch and mix with the air and generate moisture avoiding damage the electrodes.

It is essential to take into account that dry air compressed helps and improves the cutting characteristics as well as productivity, contributing to improve both quality and production times. To avoid corrosion in the different elements of the plasma as well as to have a prolonged life of the consumables, it is required the use of dry air compressed.

References

1. Calupiña, C., Oña, D.: Mejoramiento de los parámetros de trabajo para una máquina de corte por plasma y oxiacetilénica Tipo CNC. EPN, Quito – Ecuador (2012)
2. Sánchez, L.: Mejora A La Metodología De Dimensionamiento De Redes De Suministro De Aire Comprimido En Operaciones Mineras De Gran Altitud. LIMA (2019)
3. Gonzales, B.: Proyecto de una planta auxiliar de suministro de aire comprimido para uso en una línea de montaje de automóvile. Tesis de pregrado, Universidad Politécnica de Valencia, Valencia (2016)
4. Diaz, N.F., Gutiérrez, M.: Estudio Del Proceso De Acondicionamiento De Aire Mediante Sales Disecantes En Aplicaciones HVAC/R. Scientia et Technica Año XV, **42**, pp. 69–74. ISSN 0122–1701 (2009)
5. Raffino, M.E.: Estado plasmático, 28 Junio 2020. [En línea]. Available: https://concepto.de/estado-plasmatico/
6. Deaza, D.: Cálculo Y Selección De Un Sistema De Tratamiento De Aire Comprimido Para La Red Neumática Del Laboratorio De Automatización, Bogota (2019)
7. Torres, E.: Diseño De Una Máquina Cnc De Corte Por Plasma Concontrol De Altura De La Antorcha, Ecuador (2020)
8. Gómez, W.: orte por Plasma con la Técnica de Control Numérico Computarizado, para Mejorar la Calidad de Corte de Metales Del I.E.S.T.P. Nueva Esperanza 2019. Tesis de pregrado, Universidad Cesar Vallejo, Lima (2019)
9. Suntaxi, M.: Diseño Y Construcción De Una Máquna Cnc Cortadora Por Plasma Para Planchas De Acero De Hasta 10mm De Espesor. Tesis de Pregrado, Universidad Internacional del Ecuador, Ecuador (2017)
10. Paredes, M.: Análisis Del Proceso De Corte Por Plasma En Planchasde Acero En La Empresa Atu Artículos De Acero S.A. Y Su Incidencia En La Productividad. Tesis de Pregrado, Universidad Tecnológica Indomérica, Quito (2017)
11. Bonilla, C.P.F.: Diseño De Un Desalinizador De Agua A Escala Piloto Por Humidificación-Deshumidificación De Aire Empleando Columnas De Burbujeo., Colombia (2016)
12. Serna, C.N.: Estudio Del Proceso De Acondicionamiento De Aire Mediante Sales Disecantes. Universidad Tecnológica de Pereira -Scientia Et Technica, Pereira, XV(42) (2009)
13. Ruiz Martínez, Á.: Análisis y mejora de una instalación de aire comprimido, Quito (2011)
14. Arcos, T.: Plasma: El cuatro estado de la materia. Editorial CSIC – CSIC Press, Bogota (2011)

15. Copco, A.: Qué es el aire comprimido Atlascopco. Atlas Copco., Buenos Aires (2021)
16. AIRE COMPRIMIDO: 26 Febrero 2010. [En línea]. Available: https://www.atlascopco.com/es-ec/compressors/wiki/compressed-air-articles/what-is-compressed-air
17. Cerdá Filui, L.M.: Automatismos Neumáticos e Hidráulicos. Paraninfo, Madrid (2018)
18. Pérez Bello, M.A.: Circuitos de fluidos. Suspensión y Dirección. Paraninfo, Madrid (2018)
19. Hyperterm: Manual del operador. Hyperterm Inc., Hanover (2018)

Low-Cost RTK System for Positioning Error Correction in Autonomous Vehicles

Alex Toapanta[1,2], Danny Zea[1,2], Cristian Tasiguano Pozo[1], and María Gabriela Vera[1](✉)

[1] Instituto Tecnológico Universitario Rumiñahui, Sangolquí, Ecuador
{alex.toapanta,mariagabriela.vera}@ister.edu.ec
[2] Universidad Politécnica de Madrid, Madrid, Spain

Abstract. State-of-the-art autonomous driving systems have been developed and researched extensively to solve traffic-related problems and to reach the concept of an autonomous and intelligent transport system. To achieve these objectives, using a highly precise positioning system, flexibility, mobility, scalability, and the ability to make corrections in real-time is essential. However, existing high-precision equipment is quite expensive for commercial use in autonomous vehicles, which is a significant limitation to the massive applicability of autonomous driving technology, which has the potential for an incredibly positive social impact. In this article, the problem of the cost of this equipment is addressed, and through a novel system using a low-cost GNSS (Global Navigation Satellite System) receiver, achieve real-time and high-precision positioning performance by error correction. In this research, RTKLIB software was used to implement a low-cost GNSS receiver positioning system through connection to the caster server of the Military Geographical Institute of Ecuador (IGM) for RTK (Real-Time Kinematic) correction, using the NTRIP protocol. The positioning system has been developed following the precision requirements of autonomous vehicle technology, evaluated, and kinematic tests were performed to test the low-cost receiver positioning performance in a real driving environment. The results obtained showed that the receiver satisfied the required level of precision (less than 0.1 m) and similar positioning has been achieved in rural, interurban, urban, and highway areas.

Keywords: GNSS · NTRIP · RTK network · Positioning · Autonomous vehicles

1 Introduction

Currently, the use of autonomous vehicles has become a trend to solve problems related to traffic, solutions in agriculture, industry, and especially in the current COVID-19 pandemic as a vital home delivery solution. The use of autonomous vehicles helps to improve road safety, reduce emissions, reduce traffic congestion, and especially avoid contact between people to reduce the spread of viruses [1]. Autonomous vehicles guide

M. Zambrano Vizuete et al. (Eds.): CI3 2021, LNNS 511, pp. 270–282, 2022.
https://doi.org/10.1007/978-3-031-11438-0_22

themselves, without any human intervention, through the use and convergence of different technologies such as sensing, decision-making, accident warning, lane keeping converged in sophisticated control systems [2]. However, a well-designed and implemented positioning system is necessary for an autonomous vehicle to achieve full automation while driving alone [3]. According to this requirement, the positioning system must guarantee high precision, flexibility, reliability, and scalability [4].

High-end global navigation satellite systems (GNSS) provide high precision, high reliability and they are the most realistic and viable solution to accurately guide autonomous vehicles on the roads. These systems have the possibility of using the technique of RTK (Real-Time Kinematic) positioning or satellite kinetic navigation in real-time, which constitutes a solution based on the use of phase measurements of navigators with GPS, GLONASS and/or Galileo signals, where a single reference station provides corrections in real-time, obtaining sub metric accuracy [5]. However, these high-performances, precision kits are too expensive for commercial use in autonomous vehicles. Therefore, the need arises to use a low-cost GNSS receiver in which different techniques can be applied to improve positioning accuracy such as Real Time Kinematics (RTK) and Precise Point Positioning (PPP) [6], through which a high positioning performance can be achieved with error levels in the sub metric order and in real-time. The use of low-cost receivers is a current and future trend for implementation in autonomous vehicles [7]. On the other hand, the use of GNSS information is not as common as would be expected in the field of Simultaneous Localization and Mapping (SLAM) due to the difficulty of incorporating data from GNSS into a SLAM framework due to characterization problems and modeling of the uncertainties inherent to GNSS systems [8].

For the development of these investigations, including the design and implementation of a low-cost location system, it is necessary to have an experimental platform that is equipped with the necessary sensors to be able to perform both conventional SLAM and GNSS positioning. The DTI research group has a conventional car equipped with the necessary sensors and actuators as shown in Fig. 1, and it will be used as an experimental platform for current and future research.

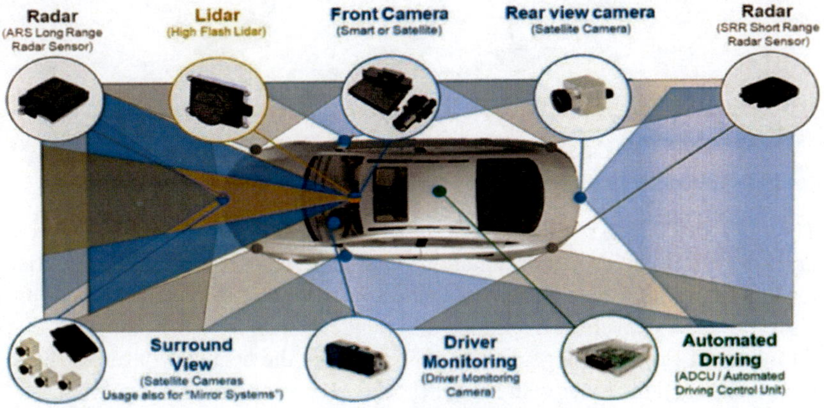

Fig. 1. Low cost GNSS receiver developed.

This article presents the design and implementation of a low-cost positioning system based on a satellite signal receiver and an embedded card, which have been placed in the car to integrate, evaluate and perform differential corrections using the protocol NTRIP (Differential Corrections Service provided by the Ecuador Military Geographical Institute), to improve the precision of the positioning of the experimental car and obtain errors in the order of centimeters including precision measurements compared to geo-referenced points through the use of a geodetic system.

The article is structured as follows: in Sect. 2, the characteristics of the low-cost system implemented are quickly presented, using a ZED-F9P u-blox receiver and a Raspberry Pi 3B; in Sect. 3, tests performed are described; in Sect. 4, the results obtained are discussed, and finally, in Sect. 5, the pertinent conclusions of the implemented system are made and future research is mentioned.

2 Implemented Low-Cost System Description

The RTK positioning system developed using a low-cost GNSS receiver (u-blox ZED-F9P) is based on the RTKLIB library and was implemented using Raspbian (Raspberry), using the architecture presented in Fig. 2.

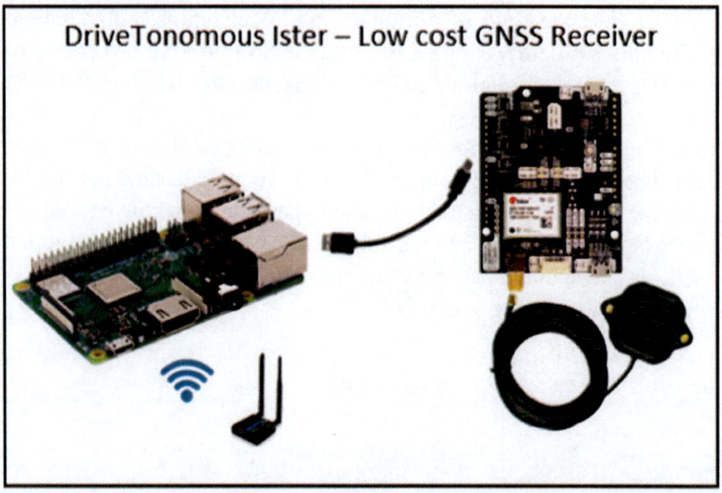

Fig. 2. Low cost GNSS receiver developed.

The ZED-F9P GNSS module is a simultaneous GNSS receiver capable of receiving and tracking multiple GNSS constellations and being able to perform differential corrections to obtain high precision. To be able to carry out differential corrections, it is necessary to use a pair of receivers that are identical from the hardware point of view, but by means of certain changes in the configuration, the behavior of one of them can be set as a base and the other with rover [9]. This receiver pair system provides an RTK solution by which precision values in the order of centimeters can be obtained. The ZED-F9P modules are simultaneous GNSS receivers and trackers of multiple GNSS constellations. The multi-band RF front-end architecture allows the main GNSS constellations (GPS, GLONASS, Galileo, and BeiDou) and the SBAS and QZSS satellites reception simultaneously. All satellites mentioned can be processed to provide an RTK navigation solution when used with correction data [10]. The QZSS system shares the same L1 and L2 frequency bands as GPS and can always be processed together with GPS. The differential corrections are sent from the base station located at the Military Geographical Institute, using the NTRIP (Network Transport RTCM Internet Protocol) protocol. The receiver, configured as a rover, enters RTK mode as soon as it starts receiving RTCM messages. In the event of losing the signal carrying the RTCM corrections (when the rover moves too far from the base or when there are obstacles that obscure the reception of the internet signal), the rover's receiver immediately goes into standalone mode, which means, it passes the mode in which a standard precision is obtained without corrections in GNSS positioning [11].

One of the most remarkable characteristics of the u-blox ZED-F9P receivers is that they allow access to the low-level information extracted from the satellites, with which it is possible to have pseudo ranges, the phase, and the Doppler frequency of the carrier. This available information could be used to implement multiple sensory fusion architectures to develop SLAM-type algorithms that can fuse both local sensors (Lidar, IMU, odometers, etc.), as well as low-level observations from satellites, achieving precise and robust positioning.

The low-cost positioning system implemented is based on a satellite signal receiver and an embedded card, which have been placed in the car to integrate, evaluate and perform differential corrections through the NTRIP protocol provided by the Military Geographical Institute of Ecuador and whose cost was approximately USD 300. Table 1 shows a comparative table in terms of the costs of the GNSS modules, where it is evident that the selected model has a lower cost concerning other equipment.

Table 1. GNSS module cost comparison.

GNSS Module Model	Features	Referential Cost
u-blox ZED-F9P 	The multi-band GNSS receiver offers centimeter-level accuracy in seconds. Concurrent reception of GPS, GLONASS, Galileo, and BeiDou. Multiband RTK with fast convergence times and reliable performance. The high update rate for highly dynamic applications. Centimeter precision in a small, energy-efficient module, and easy RTK integration.	USD 299,25 [12]
Trimble BD990 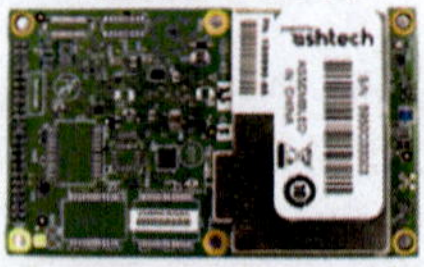	This receiver is used for a wide range of precise positioning and navigation. Applications include Unmanned Vehicles. The receiver offers centimeter-level accuracy based on the RTK carrier phase and submeter.	USD 2,592.00 [13]
Trimble BD982 	It is a single-board solution for precise heading and position. The product offers the latest GNSS signal support with multi-constellation RTK baselines between the two connected antennas and a remote base station. The Trimble BD982 GNSS supports GPS L1 / L2 / L5, GLONASS L1 / L2 / L3, and BeiDou B1, B2 signals.	USD 2,328.13 [14]

3 Methodology and Experimental Design

3.1 Experimental Environment

To carry out the tests exposed in this section, the area of the Rumiñahui Institute campus has been chosen, in which it is intended to carry out the current and future investigations of the DTI group (Drivetonomous Ister) concerning the autonomous driving research line. As can be seen in Fig. 3, in this area of the campus there are some cleared areas, areas with the presence of trees, and areas where nearby buildings can be found.

Fig. 3. Aerial view of the DTI group test area.

3.2 Base Station

The base receiver remains stationary, which is located at the Military Geographical Institute and uses a radio link transmission. The base receiver samples the data every second and transmits this unprocessed data along with its position over the communications link. The base receiver will receive the UTM coordinates of the point to be georeferenced, thanks to the radio frequency transmission that these devices have, guaranteeing more range and above all the security in the reception of coordinates.

3.3 Ntrip Server

The main Ntrip base server is located at the Military Geographical Institute, Quito, with the characteristics shown in Table 2.

Table 2. Main characteristics of the Ntrip IGM server.

BGK Ntrip Caster connection availability standard version	100 users per station (at the same time)
Data stream	RTCM 2.3 and RTCM 3.0
NTRIP version	1
Mount points (source ntrip)	REGME stations online connection via Internet

Source: [15]

In cases of possible crashes of the main Caster Ntrip IGM server (incidents, claims, risk factors, and vulnerabilities of a technical and/or social nature), the secondary Caster Ntrip backup server located in the ESPOCH Riobamba, assumes the flow of traffic of connections from Ntrip users and provides the real-time positioning service, until the main server is restored and vice versa.

Among the different modes allowed by the GNSS receiver used to function as a base station (Moving baseline, Attitude sensing, and Static mode), the static configuration is selected. In this selected mode, it is necessary for the receiver to know its own position exactly, to achieve the relative positioning of the rover with respect to the base with high precision. For the receptor, configured as a base, to emit differential corrections through NTRIP, it is necessary to configure the UART port specifying the input protocol none, and the output protocol as RTCM3 (Real-Time Correction Message 3).

3.4 Rover

It consists of an experimental research platform integrated with ROS and derived from a conventional Mazda 6 car, mechanically conditioned and fully instrumented with LIDAR, IMU, GNSS sensors, and a thermal imaging camera. Among the modifications that have been made to the car are an electromechanical system for the steering wheel, brake, and accelerator (using gears, DC motors, and servomotors). In the same way, a structure was incorporated to facilitate the location and placement of the necessary sensors and controllers, as well as the placement of the GPS and the GNSS antenna (magnetically added) responsible for capturing the information from the different satellites.

In addition, a low-cost positioning system is proposed that corrects the default positioning error of approximately 2 m (which in autonomous driving can be dangerous), using an NTRIP protocol by connecting to the caster server of the Military Geographical Institute. (IGM) through a Wifi protocol obtaining a coordinate correction of approximately 2 cm in RTK (Real-Time Kinematic). Next, the implementation of the low-cost positioning error correction system for autonomous driving is detailed.

The ZED-F9P GNSS receiver module was connected with the Raspberry Pi 3B together with a geodetic antenna and configured this system as a rover to listen to the corrections emitted by the base located at the Military Geographical Institute. In the low-cost receiver, it is used RTK positioning technique, receiving corrections through the Internet protocol (RTCM) (NTRIP), through which positioning information was obtained at 1 Hz. RTKLIB supports various positioning modes with GNSS for real-time processing and post-processing, in the present study, the kinematic model in RTKLIB was used for data processing.

The implemented system requires a configuration file containing input, processing, and output options. In terms of input options, the u-blox ZED-F9P module is configured as a portable input stream so the system can receive raw location data through the serial port. The system can recognize raw location data in a u-blox format and perform further calculations. The input stream is configured to connect to the server so that the system can receive RTCM3 format changes over NTRIP. The MAX-RTCMv3 mount point was used to reformat the obtained data.

In terms of processing options, the u-blox ZED-F9P module supports dual frequencies from the satellite system, so the frequencies are set to L1 and L2. A "direct filtering" type was used. Ionospheric calibration was based on the Krobsha model, the tropospheric correction on the Saastamoinen model, and the satellite ephemeris used was based on the transmission ephemeris. The maximum value of the time difference between the mobile terminal and the base station is set at 35 s, which is the time to support the delay of the Internet connection.

Regarding the output options, the solution format is set in latitude, longitude, and altitude. The system time is set to the GPS time. Statistics World Geodetic System (WGS), and the height of the ellipsoidal shape.

The positioning system developed and implemented in the Raspberry Pi was based on the open-source software RTKLIB. The experiments were carried out in Sangolquí - Ecuador, using a conventional car belonging to the Drivetonomous Ister (DTI) research group. There is an inertial navigation system (IMU) installed in the car which is the MTI – 610, with fairly high precision and used as a reference trajectory in the experiment.

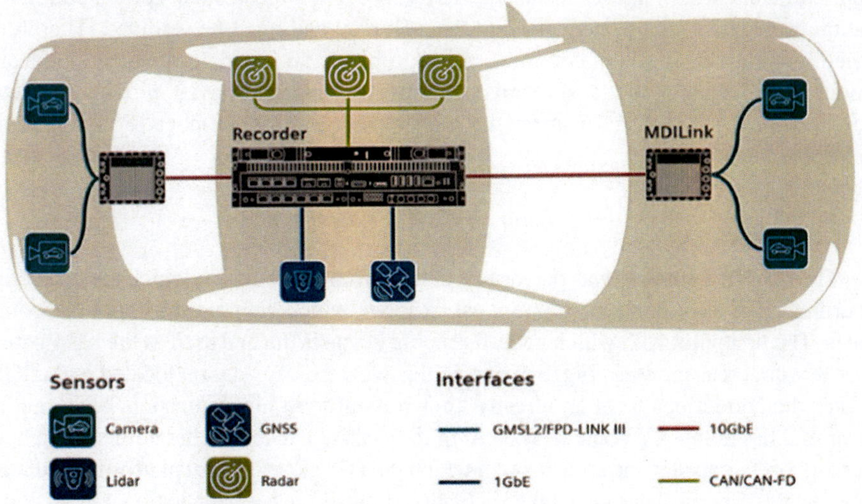

Fig. 4. Low cost system configured as a rover in the car.

As shown in Fig. 4, the DTI car is equipped with an ANN-MB-00-00 ublox antenna located on the top of the car, a ZED-F9P u-blox receiver, and a Raspberry pi 3B, which together form the low-cost receiver system and an MTI-610 Inertial Measurement Unit (IMU) from Xsens. The results of the kinematic test data were collected by the ublox antenna and transmitted to the low-cost receiver to generate the reference trajectory of the kinematic test.

4 Results and Discussion

The results obtained are discussed for certain combinations made between the low-cost receiver implemented that performs real-time positioning (ublox + antenna + Raspberry Pi 3) and the low-cost receiver that performs RTK positioning with correction (receiver of low cost + NTRIP).

First, the positioning data of the geodetic receiver and the implemented low-cost receiver were analyzed, evaluating the accuracy according to the requirements for autonomous vehicles. Finally, the performance of the low-cost receiver was evaluated in different areas, such as the analysis of the kinematic tests which mainly focused on the positioning performance of ublox + RTK (NTRIP) comparing it with other combinations and focusing on the requirements for autonomous vehicle applications.

In the same way, the accuracy and precision in east, north, and plane coordinates taken during the different tests have been considered. For each point taken, the error was calculated using Eq. 1. The mean error obtained using the ublox + RTCM (RT) system was approximately 0.5 m, which is why the precision of the system is approximately at the level of precision for the autonomous vehicle requirements. The error obtained when comparing it with the standard deviation of the ublox + RTCM system constituted approximately half in figures. Consequently, this event indicates that certain communication problems such as latency affect the accuracy and must be resolved. Therefore, by eliminating the effects of communication problems during real-time positioning and minimizing defects in low-cost receivers, more accuracy is achieved, proving that low-cost receivers can be used to meet the higher precision requirements of autonomous vehicle applications.

$$Error = \sqrt{\Delta E^2 + \Delta N^2} \quad (1)$$

The NTRIP technique was performed at five different points, of which the respective coordinates of each point were previously known, which were used as control coordinates. The first point with which georeferencing began is the one used as the base station for the work area, shown in Fig. 3. Thus, in this way, the base station located in the IGM makes the corrections from an already known point with high precision, obtaining an error of 2 cm in the XY plane, and 3 cm in the z-axis. The rest of the points taken have been geo-referenced in the area through which the DTI experimental platform circulates. Specifically, 5 points have been taken. The first is closest to the base station, the second is located further from the previous point, the following points have been placed at greater distances and an attempt has been made to take them from a more unfavorable position, that is, as far as possible from the station base and with the presence of buildings and trees around it that can make it difficult to see with satellites. The distance of the points between the base and the rover reached up to 30 km approximately. Subsequently, these received data have been processed to obtain the values of precision (standard deviation when averaging the measurements) and accuracy (error between the averaged position of the measurements, and the position of the measurement used as a base). The results obtained are presented in Table 3.

Table 3. Precision and accuracy values obtained.

Points	Precision (m)	Accuracy (m)
P1	0.0144	0.1031
P2	0.1161	0.8021
P3	0.1202	0.8256
P4	0.1356	0.8512
P5	0.1386	0.9504

According to the results obtained, it can be observed that the precision and accuracy values are within the optimal range for autonomous driving applications.

4.1 Evaluation of the GNSS-NTRIP System for Navigation

For the evaluation process of the NTRIP system in navigation, a trajectory has been tested on the DTI group's research platform at the back of one of the buildings that has been shown in Fig. 2. The trajectory carried out consists of two loops closed, which will serve to measure possible divergences in the trajectories when it ends at the same point at which it started.

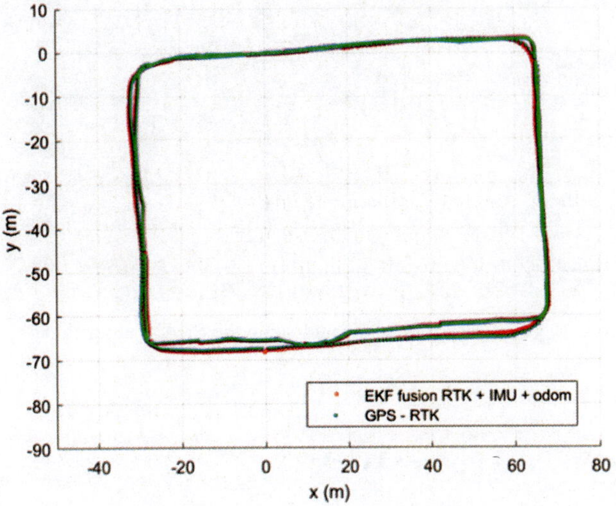

Fig. 5. Trajectory carried out through two closed loops behind the main building of the DTI Ister research area.

As shown in Fig. 5, the low-cost RTK NTRIP system implemented with the fusion of the generated trajectory with the vehicle odometry (IMU) (red colored path) has been compared to the previously performed trajectory. In addition, the low-cost system was also compared with the trajectory generated through the integration of a SLAM (Simultaneous Localization And Mapping) algorithm (green colored path). In Table 4, it is possible to see the position calculated in each loop and for each trajectory estimation system.

Table 4. Calculated positions for each implemented system.

	Closed loop 1		Closed loop 2	
	x (m)	y (m)	x (m)	y (m)
Odometry + IMU	−0.58	−0.11	−0.61	−0.05
SLAM Gmapping	0.24	−0.1	0.52	−3.59
RTK NTRIP System	−0.53	−0.9	−0.64	−0.07

In the case of a totally ideal system, these values should be zero, however, the values presented in Table 4 show small positioning errors for x and y. These small errors occur because the loop was made on marks previously placed on the ground and by manual conduction. In Figs. 5 and 6, the trajectories obtained for each system are presented.

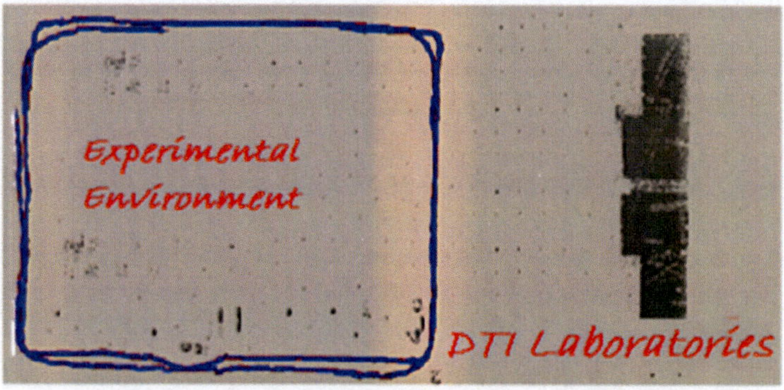

Fig. 6. Trajectory calculated using the SLAM algorithm.

5 Conclusions and Future Work

The results obtained with the low-cost receiver implemented are within the level of precision required for autonomous driving applications (0.5 m). However, on certain occasions, the performance of the low-cost system implemented was affected by the coverage of the internet signal, the correction latency, and the interruption of the satellite signal. By using other sensor systems such as an IMU and in conjunction with the car's odometry, the positioning system can compensate for the intervals in which the performance of the low-cost system implemented fails.

As future work, it is proposed to carry out several experiments that include multiple low-cost sensors and the use of possible multi-constellations. Similarly, the integration and low-level fusion of GNSS observables through SLAM algorithms would be important.

References

1. Liu, T., et al.: Hercules: An Autonomous Logistic Vehicle for Contact-less Goods Transportation During the COVID-19 Outbreak (2020)
2. Van Brummelen, J., O'Brien, M., Gruyer, D., Najjaran, H.: Autonomous vehicle perception: the technology of today and tomorrow. Transp. Res. Part C Emerg. Technol., **89**, 384–406 (2018). https://doi.org/10.1016/j.trc.2018.02.012
3. Artuñedo, A., Villagra, J., Godoy, J., Castillo, M.D.d.: Motion planning approach considering localization uncertainty. IEEE Trans. Veh. Technol. **69**(6), 5983–5994 (2020). https://doi.org/10.1109/TVT.2020.2985546
4. Maksimychev, E., Gafurov, S., Shapovalov, I.: Autonomous navigation in rural environment based on digital filters and LIDAR data (2020). https://doi.org/10.1109/NIR50484.2020.9290215
5. Sun, R., et al.: Improving GPS code phase positioning accuracy in urban environments using machine learning. IEEE Internet Things J. **8**(8), 7065–7078 (2021). https://doi.org/10.1109/JIOT.2020.3037074
6. Xue, X., Qin, H., Lu, H. High-precision time synchronization of kinematic navigation system using GNSS RTK differential carrier phase time transfer. Meas. J. Int. Meas. Confed. **176**, (2021). https://doi.org/10.1016/j.measurement.2021.109132
7. Broekman, A., Gräbe, P.J.: A low-cost, mobile real-time kinematic geolocation service for engineering and research applications. HardwareX **10**, e00203 (2021). https://doi.org/10.1016/j.ohx.2021.e00203
8. Liu, J., Tu, R., Zhang, R., Han, J., Fan, L., Lu, X.: Positioning bias of different frequency observations in double-differenced relative positioning. Adv. Sp. Res. **66**(6), 1321–1328 (2020). https://doi.org/10.1016/j.asr.2020.05.020
9. Kur, T., Liwosz, T., Kalarus, M.: The application of inter-satellite links connectivity schemes in various satellite navigation systems for orbit and clock corrections determination: simulation study. Acta Geod. Geoph. **56**(1), 1–28 (2020). https://doi.org/10.1007/s40328-020-00322-4
10. Hamza, V., Stopar, B., Sterle, O.: Testing the performance of multi-frequency low-cost GNSS receivers and antennas. Sensors **21**(6), 1–16. https://doi.org/10.3390/s21062029
11. Poteko, J., Eder, D., Noack, P.O.: Identifying operation modes of agricultural vehicles based on GNSS measurements. Comput. Electron. Agric. **185** (2021). https://doi.org/10.1016/j.compag.2021.106105
12. ALIEXPRESS: Módulo de ZED-F9P de alta precisión RTK UAV RTK, GNSS, placa de salida, RTCM, NMEA TOP906 Módulo de ZED-F9P de alta precisión RTK UAV RTK, GNSS, placa de salida, RTCM, NMEA TOP906 Módulo de ZED-F9P de alta precisión RTK UAV RTK, GNSS, placa de salida, RTC, 2021. https://es.aliexpress.com/item/4000849905931.html?spm=a2g0o.search0304.0.0.477a7457idl2YP&algo_pvid=d97c9c20-5f70-4e6d-83bb-d920a6ddcb1f&algo_exp_id=d97c9c20-5f70-4e6d-83bb-d920a6ddcb1f-4
13. ALIEXPRESS: Módulo receptor GNSS RTK, multifrecuencia, alta precisión, 100%, nuevo y Original, Trimble BD990 (2021). https://es.aliexpress.com/item/1005003103353602.html?spm=a2g0o.productlist.0.0.444b1272g52RDx&algo_pvid=d5fd49d5-7e31-447b-9723-fb9496ab4a9d&algo_exp_id=d5fd49d5-7e31-447b-9723-fb9496ab4a9d-48&pdp_ext_f=%7B%22sku_id%22%3A%2212000024100938820%22%7D

14. ALIEXPRESS: Receptor GNSS RTK, dispositivo de medición de posicionamiento de alta precisión, GPS, BDS, GLONASS, Galileo, módulo Trimble BD982 (2021). https://es.aliexpress.com/item/1005003141248019.html?spm=a2g0o.productlist.0.0.444b1272g52RDx&algo_pvid=47b803ba-dbd9-4353-bd79-2dd475f66e25&algo_exp_id=47b803ba-dbd9-4353-bd79-2dd475f66e25-26&pdp_ext_f=%7B%22sku_id%22%3A%221200002431 4241103%22%7D
15. Comisión Técnica de Radio para Servicios Marítimos: TCM 10403.3, Servicios diferenciales GNSS (Sistemas globales de navegación por satélite) (2020). https://rtcm.memberclicks.net/publications

Kinematic Control of a Vehicle on a Defined Trajectory Through a GPS Sensor and a Compass Sensor

Danny Zea[1,2], Alex Toapanta[2], María Gabriela Vera[2](✉), and Cristian Tasiguano Pozo[2]

[1] Escuela Politécnica de Chimborazo, Riobamba, Ecuador
danny.zea@espoch.edu.ec, dj.zea@alumnos.upm.es
[2] Instituto Tecnológico Universitario Rumiñahui, Riobamba, Ecuador
mariagabriela.vera@ister.edu.ec

Abstract. This research paper is focused in analyzes the behavior and performance of a control law that has been validated in a previous study in a three dimensions (3D) simulation software, now evaluated in a Mazda 6 vehicle with the objective that it travels a previously designed trajectory. This is possible, thanks to the use of a Global Positioning System (GPS) sensor as well as additional sensors that allow us to know the direct kinematics of the vehicle at each instant of time. The experimental tests were carried out with the use of a Raspberry pi 3 b + (small single-board computers) that is in charge of receiving the information from the sensors and then processed this information so that the control law allows the control actions sent to the actuators. Finally, results are presented to be considered for future research that allows the implementation of a fully autonomous driving system.

Keywords: Kinematic control · GPS sensor · Autonomous driving

1 Introduction

For decades, the Global Positioning System GPS has allowed the triangulation of countless objects on planet Earth, triggering the proliferation of various applications. Autonomous driving systems in the last decade have played an important role and have been one of the main research topics, especially they have focused on reliability when interacting with the environment, especially when there are various objects static and dynamic involved in real traffic [1]. One of the relevant points when it comes to reliability, are the sensors, and specifically, the GPS sensor that allows knowing the location at every moment, but is often affected by interference when traveling through closed places such as infrastructure of considerable heights that weaken the reception and can repeatedly lose the signal [2]; causing the vehicle to lose location information [3, 4]. The development of more robust GPS sensors with better techniques to improve their accuracy is booming, especially for autonomous driving applications. However, when

M. Zambrano Vizuete et al. (Eds.): CI3 2021, LNNS 511, pp. 283–292, 2022.
https://doi.org/10.1007/978-3-031-11438-0_23

the GPS signal is affected, the fusion of sensors such as a LIDAR, 3D vision camera, etc. comes into play to compensate for the disorientation of the vehicle and avoid accidents on the road [5, 6].

Among a variety of GPS sensors that exist today on the market that offer different features, most of them are limited when it comes to acquiring them due to their high economic value and making research difficult in research groups [7]. An alternative is to use 3D simulation software that allows shortening research times and through emulation validate proposals for applications in autonomous driving before taking them to the real world [8, 9]. In this article, the proposal developed in a previous investigation that was carried out in a 3D Webots simulation software is evaluated and will now be tested in a Mazda 6 vehicle that has an electromechanical system that allows an action to be taken on the vehicle. It is proposed a control law for tracking the vehicle on a previously defined desired path through a GPS sensor, then the performance of the GPS sensor is validated by analyzing its accuracy error when the vehicle travels on the path. It is important to note that the control law does not consider obstacles on the road nor does it allow the vehicle to interact with its environment, for this reason, the tests are carried out on a track for driving applications located at the Rumiñahui Institute, Province of Pichincha, Ecuador. In addition, the experimental tests are carried out with a driver who is inside the vehicle, to take control of the vehicle when an emergency arises and, above all, the driver has an emergency button that allows to turn off and disconnect the entire system.

The article is structured as follows: in Sect. 2 the controller design and control law are explained; in Sect. 3 the structure of the system is described with the description of each of the blocks; Sect. 4 shows the scenario where the experimental tests and the characteristics of the vehicle are carried out; in Sect. 5 he results of the behavior of the vehicle with the GPS sensor are presented, and finally, Sect. 6 presents the conclusions.

2 Kinematic Design of the Controller

The kinematics of the controller is based on the simplified kinematics model, which uses Acker-man steering on the front wheels and creates a center of rotation at some point along the axis of the rear wheels [10]. Figure 1 illustrates the simplified kinematic model.

The kinematic model configuration is defined by the following equation:

$$\begin{bmatrix} \dot{x} \\ \dot{y} \\ \dot{\phi} \end{bmatrix} = \begin{bmatrix} \cos(\phi) & -a\sin(\phi) \\ \sin(\phi) & a\cos(\phi) \\ 0 & 1 \end{bmatrix} \begin{bmatrix} \mu \\ \omega \end{bmatrix} \tag{1}$$

The reduced way of writing Eq. (2) is defined by:

$$\dot{\mathrm{h}}(t) = \mathbf{J}(\phi)\mathbf{v}(t) \tag{2}$$

Therefore, $\dot{\mathbf{h}}(t) = [\dot{x}\ \dot{y}]^T \in R^2$, represents the vector of Cartesian velocities at some point of interest defined by the distance$a$; $\mathbf{J}(\phi) \in R^{2x2}$ defines the Jacobian matrix and $\mathbf{v}(t) = [\mu\ \omega]^T$ s the control and maneuverability vector with μ and ω and as the

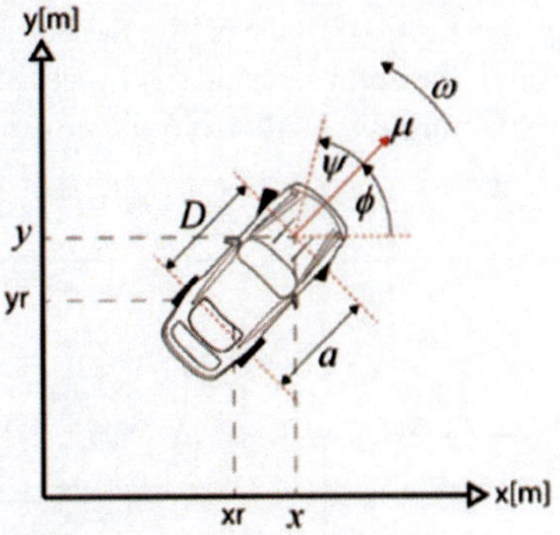

Fig. 1. Geometric model of a car-like robot.

angular and linear speed of the car-like vehicle respectively. The angle of the front wheels establishes the direction the vehicle takes and is defined by:

$$\psi = \tan^{-1}\left(\frac{D\omega}{\mu}\right) \tag{3}$$

The controller design starts from the simplified kinematic model defined in Eq. (2), where the control vector is defined by:

$$\mathbf{v_c}(t) = \mathbf{J}^{-1}(\phi)\dot{\mathbf{h}}(t) \tag{4}$$

Equation (4) defines the control law that allows the vehicle to adjust its movement to follow a previously configured trajectory. Therefore, $\mathbf{J}^{-1}(\phi)$ represents the inverse Jacobian matrix, $\dot{\mathbf{h}}(t) = \begin{bmatrix} \dot{x}_d \; \dot{y}_d \end{bmatrix}^T$ is the velocity vector and will now be defined by $\dot{\mathbf{h}}_\mathbf{d}(t) = \begin{bmatrix} \dot{x}_d \; \dot{y}_d \end{bmatrix}^T$ as the vector of desired velocities, then a new parameter is defined $\tilde{\mathbf{h}} = \mathbf{h_d} - \mathbf{h}$ which allows to quantify the position error, where $\mathbf{h_d}$ represents the desired position vector and $\mathbf{h}$ the current position of the vehicle. Finally, the control law is defined in Eq. (5), where $\mathbf{K_1}$ is a diagonal matrix that allows compensating for the position error and $\mathbf{K_2}$ is a diagonal matrix that allows fine-tuning the control actions.

$$\mathbf{v_c}(t) = \mathbf{J}^{-1}(\phi)\left(\dot{\mathbf{h}}_\mathbf{d}(t) + \mathbf{K_2}\tanh\left(\mathbf{K_2}^{-1}\mathbf{K_1}\tilde{\mathbf{h}}(t)\right)\right) \tag{5}$$

3 System Structure

The full-scale implementation of the proposal, where the control law is validated in the real world, is possible thanks to the configuration illustrated in Fig. 2, with the following blocks: i) Mazda 6, is a conventional vehicle that has been instrumented with an electromechanical system to take control of the steering wheel, accelerator and brake, it also has sensors such as a GPS and a compass that allow knowing its position and orientation, on the other hand, there are sensors for reading its speed and steering wheel position at each instant of time; ii) Raspberry pi 3 b+, is a low-cost single-board computer, being in charge of reading the input signals of the sensors that will be processed to perform control actions through the actuators and iii) Power stage, is the block in charge of isolating and amplifying the output signals from the computer to the actuators.

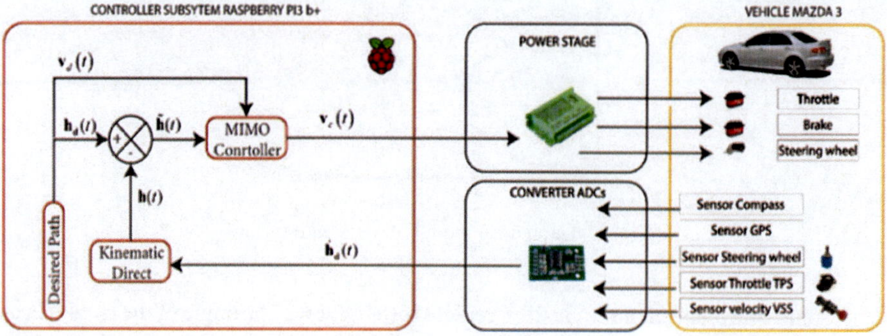

Fig. 2. System structure.

4 Scenario

An important point to be able to validate the proposal is the consideration that the vehicle does not have an obstacle avoidance system and it will not be able to interact with pedestrians or other vehicles, for which it will be validated in a suitable scenario. The experimentation area, Fig. 3, to tests autonomous driving is at the Rumiñahui Institute in the province of Pichincha. The dimensions of the stage are 177 m long and 60 m wide.

Fig. 3. Experimentation scenario.

4.1 Mazda 6 Vehicle

The vehicle used is a Mazda brand car, model 6, 2008, Fig. 4, which has an electromechanical system for the actuators and sensors. The control actions are sent to three servomotors that control the accelerator, brake, and steering wheel respectively. Signals are taken from four sensors that allow knowing the position of the steering wheel with a ten-turn potentiometer, the TPS sensor (Throttle Position Sensor) to know the throttle position, the VSS sensor (Vehicle Speed Sensor) to read the vehicle speed, the compass sensor to know the orientation of the vehicle and finally the GPS sensor for the vehicle location. The input signals are read through a digital and analog I / O module that allow parameterization and sending to a Raspberry Pi 3 b + to process these signals and then sent to a power module that allows controlling the servomotors.

Fig. 4. Mazda 6 vehicle.

5 Rsults and Discussion

After making some adjustments and tests of the controller in the Mazda 6 vehicle, the control law (5) is established with the following parameters: the desired route is defined by $\tilde{\mathbf{h}}_{\mathbf{d}}(t) = [1.5\text{t} + 30\ 10\sin(0.2\text{t} + 80)][m]$ with the respective transformation of the geodetic coordinates to Cartesian coordinates in order to facilitate the calculations in the controller, gain matrices $\mathbf{K_1} = diag[0.50\ 0.45]$ and $\mathbf{K_2} = diag[0.7\ 0.52]$. The initial starting points of the vehicle are: $\mathbf{p}(0) = [25\ -15\ -1.10]^T[m]$.

The sensors are installed in the trunk of the vehicle at the height of the axle of the rear wheels and from this point the control point is established $a = 2.725[m]$. The sampling time set for the controller on the Raspberry pi 3 b + is $100[s]$ with total vehicle travel time of $30[s]$. The characteristics of the Mazda 6 vehicle are detailed in Table 1:

Table 1. Mazda 6 vehicle main characteristics

Features, dimensions, weight and capacities	
Maximum speed	214 km/h9,9 s
Acceleration 0–100 km / h	
Number of doors	4
Length	4.735 mm
Width	1.795 mm
Height	1.440 mm
Battle	2.725 mm
Front track	1.550 mm
Rear track	1.550 mm
Weight	1.370 kg
Combustion Engine	
Purpose	Boost the vehicle
Fuel	Gasoline
Maximum power	147 hp/108 kW
Revolutions maximum power	6,500 rpm
Maximum torque	184 Nm
Revolutions at maximum torque	4,000 rpm
Situation	Cross forward
Number of cylinders	4
Cylinder arrangement	Online
Displacement	1,999 cm^3
Compression ratio	10 to 1
Transmision	
Traction	Lead
Gearbox	Automatic
Number of speeds	6

Table 2. Sensors equipped on the vehicle.

Position on the vehicle	Sensor	Model	Specifications/inputs	Values
Sensors slot rear	GPS	ProPak-V3TM	Horizontal position accuracy	0.35 m
			Resolution	0.05 m
			Speed Resolution	0.03 m/s
			Velocity maximum	515 m/s
Sensors slot rear	Compass	NEO-M8T	xAxis	True
			yAxis	True
			zAxis	True
			Resolution	5 milli-gauss
Vehicle steering wheel	Potentiometry	3590S-2-103L	Resistance Range:	10 kOhm
			Standard tolerance	±5%
			Linearity	0, 20%
			Power	2 Watts
			Number of turns	10
			Shaft Diameter	6, 35 mm

The characteristics of the sensors used in the Mazda 6 vehicle instrumentation are shown in Table 2.

The path of the vehicle on the previously desired trajectory is illustrated in Fig. 5 and shows the Cartesian positions of the vehicle $\tilde{\mathbf{h}}$ when traveling the desired path $\mathbf{h_d}$.

The control actions sent from Raspberry pi 3 b + to the actuators versus the dynamic behavior performed by the vehicle is illustrated in Fig. 6. In this figure, it is also shown three control actions, among which is the desired linear speed μ_c versus what the vehicle is dynamically performing μ, then there is the desired angular velocity ω_c versus what the vehicle is dynamically performing ω and finally there is the angle ψ that the front wheels should take versus the one that the vehicle is actually making ψ_c.

The errors are shown in Fig. 7 $\tilde{\mathbf{h}} = \left[\,\tilde{\mathbf{h}}_x\ \tilde{\mathbf{h}}_y\,\right]$ both latitude and longitude, once these values have been transformed to Cartesian coordinates. It can be seen that initially the error between the desired trajectory and the trajectory made by the vehicle in the case of the x- axis and the y- axis have a value of 2 and 1 $[m]$ respectively. At an initial point, the vehicle is positioned at a certain distance from the desired trajectory, and then at $4[s]$ it is observed that the error oscillates between $\pm 1.25[m]$. This result is mainly due to the accuracy error generated by the GPS sensor and, on the other hand, due to the dynamics of the MAZDA 6 vehicle.

In future research, experimental tests will be carried out with a new controller that, on the one hand, considers dynamic compensations and, on the other, takes into account avoiding obstacles in the vehicle's path.

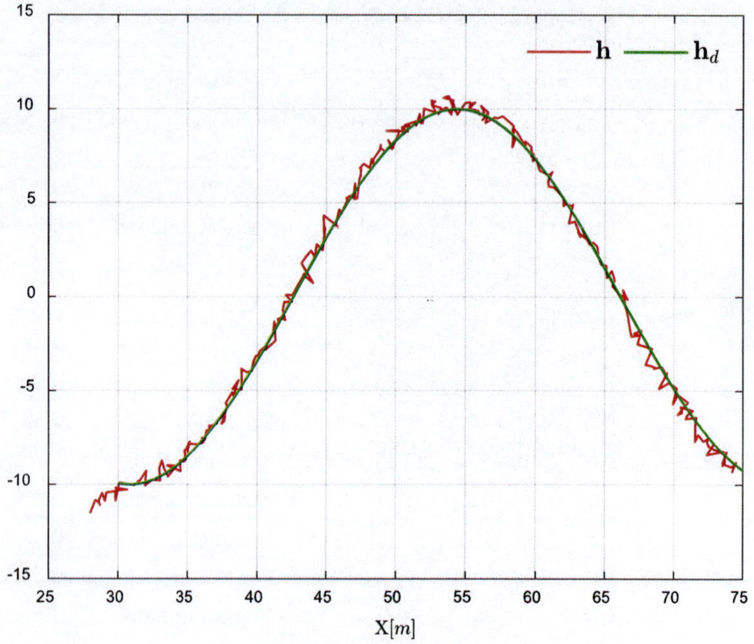

Fig. 5. Path definition

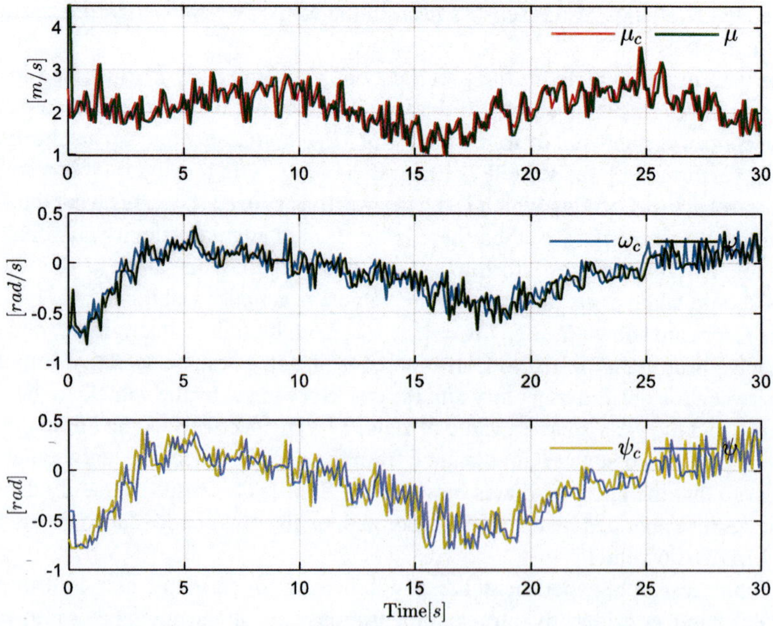

Fig. 6. Control actions from Raspberry pi 3 b + to the actuators versus the dynamic behavior performed by the vehicle.

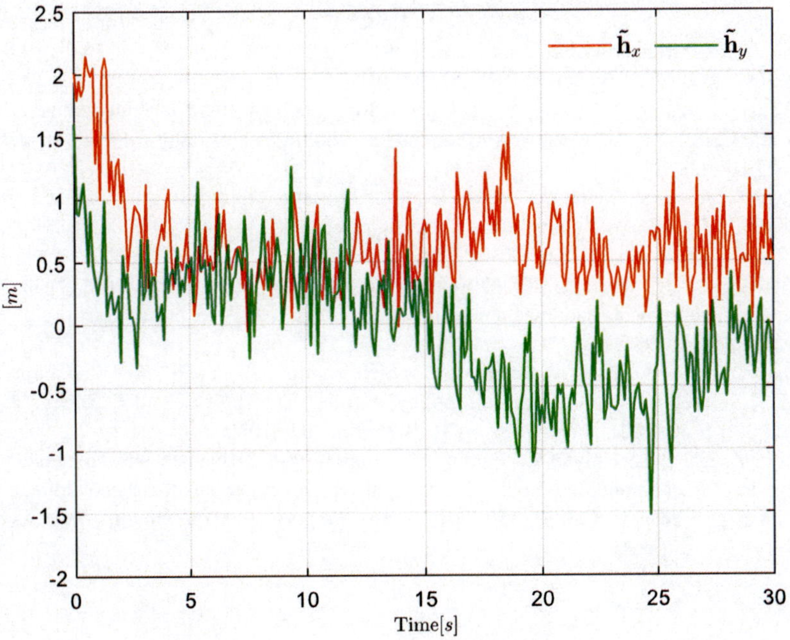

Fig. 7. Errors of both latitude and longitude

6 Conclusion

This article is the continuation of the proposal that has been previously validated in a 3D simulation program and is now experimentally tested in a Mazda 6 vehicle instrumented with an electromechanical system that allows the guidance of a vehicle on a previously designed trajectory for autonomous driving applications. The errors between the vehicle position and the desired trajectory over time range between 1.25 m of accuracy.

References

1. Bersani, M., et al.: An integrated algorithm for ego-vehicle and obstacles state estimation for autonomous driving. Rob. Auton. Syst. **139**, 1–16 (2021). https://doi.org/10.1016/j.robot.2020.103662
2. Atoui, H., Sename, O., Alcala, V., Puig, V.: parameter varying approach for a combined (kinematic + dynamic) model of autonomous vehicles. IFAC-PapersOnLine **53**(2), 15071–15076 (2020). https://doi.org/10.1016/j.ifacol.2020.12.2028
3. Kummerle, R., Hahnel, D., Dolgov, D., Thrun, S., Wolfram, B.: Autonomous driving in a multi-level parking structure. Robot. Autom. 3395–3400, 2009. https://doi.org/10.1109/ROBOT.2009.5152365
4. Yu, Z., Hu, Y., Huang, J.: GPS/INS/Odometer/DR integrated navigation system aided with vehicular dynamic characteristics for autonomous vehicle application. IFAC-PapersOnLine **51**(31), 936–942 (2018). https://doi.org/10.1016/j.ifacol.2018.10.060

5. Castaño, F., Beruvides, G., Villalonga, A., Haber, R.: Self-tuning method for increased obstacle detection reliability based on internet of things LiDAR sensor models. Sensors **8**, 1–16 (2018). https://doi.org/10.3390/s18051508
6. Wang, L., Zhang, Y., Wang, J.: Map-based localization method for autonomous vehicles using 3D-LIDAR. IFAC-PapersOnLine **50**(1), 276–281 (2017). https://doi.org/10.1016/j.ifacol.2017.08.046
7. Park, G., Choi, S., Hyun, D., Lee, J.: Integrated observer approach using in-vehicle sensors and GPS for vehicle state estimation. Mechatronics **50**, 134–147 (2018). https://doi.org/10.1016/j.mechatronics.2018.02.004
8. Tuncali, C.E., Fainekos, G., Prokhorov, D., Ito, H.: Requirements-driven test generation for autonomous vehicles with machine learning components **2**, 1–16. https://arxiv.org/pdf/1908.01094.pdf (2019)
9. Di, X., Shi, R.: A survey on autonomous vehicle control in the era of mixed-autonomy: from physics-based to AI-guided driving policy learning. Transp. Res. Part C Emerg. Technol. **125**(40), 1–40 (2021). https://doi.org/10.1016/j.trc.2021.103008
10. Ziye, Z., Haiou, L., Huiyan, C., Jiaming, H., Hongming, G.: Kinematics-aware model predictive control for autonomous high-speed tracked vehicles under the off-road conditions. Mech. Syst. Signal Process. **123**, 33–350 (2019). https://doi.org/10.1016/j.ymssp.2019.01.005

Energy and Environment

Constant Voltage Battery Charger Energized from an MPPT Photovoltaic System

Javier Rojas(✉), Carlos Lucero, and Iliana Merchán

Salesian Polytechnic University, Quito, Ecuador
jrojasu@ups.edu.ec

Abstract. Maximum power point tracking (MPPT) enables extracting the maximum power that a photovoltaic panel is capable of delivering regardless of the change in solar irradiance or ambient temperature throughout the day, maximizing its efficiency; it is carried out by means of the control of a DC-DC converter to modify the voltage delivered to the load, such that the variation in the current keeps the power constant. In this work it is designed and simulated an MPPT with an SEPIC converter whose pulse width control signal is determined by the Perturb & Observe (P&O) algorithm to identify the voltage that produces the maximum power, it is available at the output of a BUCK converter controlled by a PID that enables energizing loads with constant voltage despite of the variations produced by the P&O algorithm. The will be energized from a 30 W panel and will charge 9 V batteries at a constant charge voltage, that are commonly used in the robotics club of the Universidad Politécnica Salesiana, located in Quito, Ecuador. Matlab/Simulink is used for simulation considering scenarios of variable irradiance levels and also load variations. The results show a good transient response of the SEPIC converter with the MPPT control to extract the maximum power, moreover the PID c ontroller enables attenuating the ripple obtaining a constant output with a ripple of 2.5%, and does not generate overshoot.

Keywords: Battery · Converter · Photovoltaic · MPPT · PID · Subsystem

1 Introduction

The concern generated by climate changes produced by the greenhouse effect and the search for solutions that contribute to reduce its impact, have promoted research and utilization of renewable energy sources that diminish the use of fossil fuels [1]. Taking this into account it has been increased the use of photovoltaic panels, which are capable of generating electric energy from solar irradiance. A photovoltaic panel delivers a DC voltage at a level that varies according to solar irradiance, ambient temperature and position with respect to the sun, and consequently the instantaneous voltage and power vary throughout the day which may generate problems when the voltage raises more than what is tolerated by the load, or is very low such that the load does not operate correctly; besides, it does not make the most of the energy generated.

In order to prevent these problems it may be inserted a DC-DC converter between the panel and the load, if it is a step-down converter the voltage may be substantially

M. Zambrano Vizuete et al. (Eds.): CI3 2021, LNNS 511, pp. 295–306, 2022.
https://doi.org/10.1007/978-3-031-11438-0_24

reduced at particulars hours of the day, while if it is a step-up converter it stops operating when the irradiance level is very high, and consequently the most convenient option is to use a DC-DC converter with step-down/step-up features such as the Cuk, Sepic or Zeta topologies, so that the voltage delivered may be regulated by means of the pulse width of the control signal according to the requirement of the load as well as the level of solar irradiance [1]. The power vs. voltage characteristic of a photovoltaic panel, for any level of irradiance, shows a maximum point at a particular value of voltage, such that if it operates at that point, through a DC-DC converter, it is extracted the maximum power from the panel, and therefore, if the operating point is modified as the irradiance changes so that throughout the day it operates at the maximum, it is maximized the efficiency of the photovoltaic system [2].

There are strategies, such as the Perturb & Observe (P&O) and the Incremental Conductance (IC) algorithms, to identify the voltage of maximum power. In P&O, small perturbations are performed in the control of the converter, to increase or decrease the voltage delivered to the load; if there is an increment in the power when the voltage is increased, the same perturbation (increase) is performed again, on the contrary it is used the opposite perturbation (decrease), thus enabling to find the maximum operating point, since the perturbation is constantly repeated the operating point will remain oscillating around the maximum. The IC strategy performs the same perturbations than the P&O, but instead of monitoring the power change it is determined the slope of the dP/dV curve, the perturbation is performed until the slope is zero [3]. Both strategies require few measurement parameters and are relatively easy to implement, P&O depends on the magnitude of the perturbation, such that, as the perturbation is larger, the maximum is identified faster but the oscillations around this point will be larger; IC has the same issue but determining the value of zero in the slope is complicated due to the presence of noise or failures in the measurements, and also due to the estimation in the calculations of dP and dV [4].

There is much research activity to maximize the efficiency of a photovoltaic system, in [5] it is analyzed the behavior of four DC-DC converters in the implementation of MPPT with P&O and with a fuzzy algorithm, by means of Matlab-Simulink simulations obtaining that the SEPIC converter enables to obtain increased power at the output of the photovoltaic system, [6] compares the behavior with the Boost and SEPIC converters with the IC algorithm, simulated in Matlab-Simulink and demonstrating that the SEPIC converter has a more stable behavior in the presence of variations in solar irradiance and ambient temperature. In the work [3] it is implemented an MPPT with a SEPIC converter with a controller to improve the response of the MPPT algorithm, by means of Matlab-Simulink simulations it is evaluated a nonlinear integral backstepping (IBS) controller, a PID, and one based on fuzzy logic, in their results the IBS controller reduces the magnitude of the oscillations around the point of maximum power. In [7] it is implemented the MPPT with a SEPIC converter using a modified algorithm that reduces the difference between the actual voltage and current with calculated reference values, the results obtained indicate that it reaches the maximum in less time and reduces the complexity. The research conducted in [8] compares the performance of the ZETA and SEPIC converters with ZVS to reduce losses, it uses the P&O and IC algorithms and it focuses on low power applications, the evaluation is carried out by means of

Matlab simulations and experimentally obtaining that when using ZVS with P&O there is a larger increment and in [9] it is implemented a photovoltaic system with a SEPIC, and the P&O and IC algorithms are evaluated determining that both show sensitivity problems to changes in atmospheric conditions and require variations in the regulated and non-fixed perturbations.

In this paper, Matlab-Simulink are used to carry out the simulation of a photovoltaic system with a SEPIC converter controlled by the P&O algorithm to obtain the point of maximum power in the presence of variations of solar irradiance. The design of the system considers the use of a 30W photovoltaic panel, and it is intended for charging 9 V LiPo batteries that require constant charge voltage, and thus it is inserted a BUCK converter controlled by a PID. The paper is organized as follows, Sect. 2 presents the methodology used to design and simulate the photovoltaic system, Sect. 3 details the tests carried out by means of simulations to evaluate the operation of the system and Sect. 4 exposes the conclusions derived from the results.

2 Methodology

2.1 Design of the Photovoltaic System

The system seeks to take advantage of the electric energy generated by a 30 W photovoltaic panel to charge 9 V batteries at a constant charge voltage. The design of the system is carried out according to the scheme presented in Fig. 1 the interaction of each of the subsystems enables delivering a constant voltage to the battery and transferring the maximum power from the photovoltaic panel, despite of the changes in the level of solar radiation.

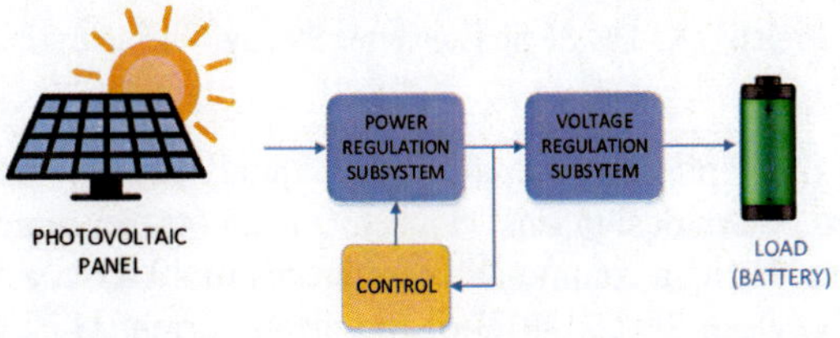

Fig. 1. Block diagram of the photovoltaic module.

Power Regulation Subsystem. It is in charge of managing the power delivered by the photovoltaic panel with the MPPT controller, acting as a variable power source, and consequently it is constituted by a DC-DC converter that receives as input the voltage generated by the photovoltaic panel and delivers at its output the voltage determined by the MPPT controller.

For the design of the converter, the input voltage is between 10 V and 18 V, according to measurements at the photovoltaic panel, carried out during the operation hours; the output voltage is 15 V, according to the results obtained with Matlab-Simulink simulations using the technical data of the photovoltaic panel. It is used a step-down/step-up DC-DC converter with SEPIC topology, since it does not reverse the voltage polarity, and moreover it provides the best stability in the presence of variations of climatic conditions and enables extracting increased power.

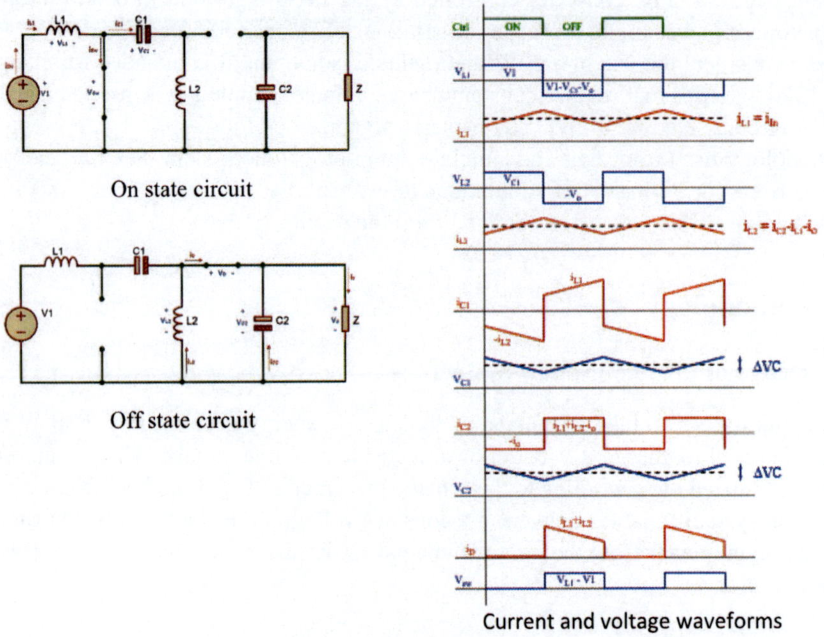

Fig. 2. DC-DC Sepic Converter Steady state analysis.

Figure 2 shows the Sepic converter steady state analysis based on its on and off state circuits, it is used to obtain design Eqs. (1), (2), (3) and (4), for dimensioning inductors and capacitors. Considering a commutation frequency of 40 KHz, a ripple of the current in inductors (Δi) between 20 and 40% of the mean current [11]; and a ripple of the voltage in capacitors (ΔV) of 1% of its mean voltage, it is determined that L1 and L2 are of 270 μH, the capacitor C1 of 10 μF and the capacitor C2 of 220 μF.

$$L_1 \geq \frac{V_1 K}{\Delta i_{L1} f_{sw}} \tag{1}$$

$$L_2 \geq \frac{V_1 K}{\Delta i_{L2} f_{sw}} \tag{2}$$

$$C_1 \geq \frac{i_o K}{\Delta V_{C1} f_{sw}} \tag{3}$$

$$C_2 \geq \frac{i_o K}{\Delta V_{C2} f_{sw}} \tag{4}$$

Voltage Regulation Subsystem. It is in charge of delivering a constant output voltage according to the charge requirements of the battery, from the voltage of the SEPIC, i.e., it has an approximate input voltage of 15 V, requires a constant voltage of 8.4 V to charge a LiPo battery.It is used a step-down DC-DC converter with BUCK topology in closed-loop with a PID controller to compensate the voltage variation produced by the MPPT, and also improves the transient response to load variations as the battery charges.

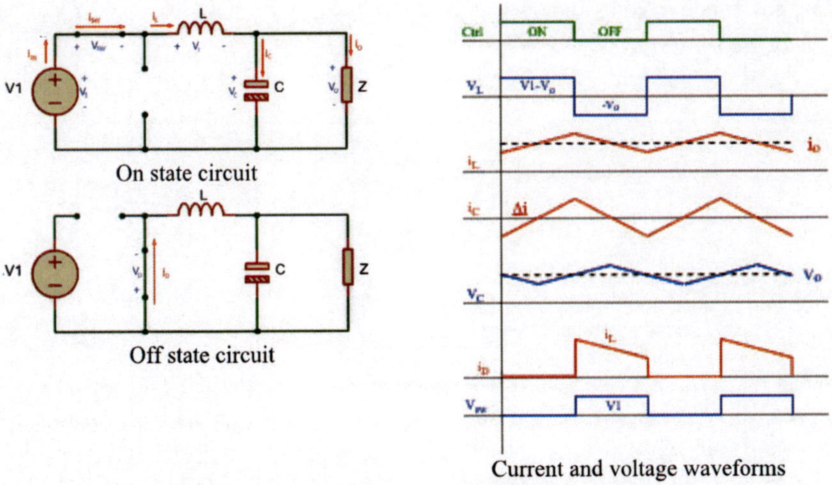

Fig. 3. DC-DC buck converter steady state analysis.

Like the Sepic converter design showed previously, the design of the Buck converter is carried out according to the steady state analysis showed in Fig. 3 with which Eqs. (5) and (6) are obtained. The design parameters for this converter with respect to frequency and ripple are similar to the ones already mentioned for the SEPIC converter, obtaining an inductor L of 75 μH and a capacitor C of 47 μF.

$$L \geq \frac{(V_{in} - V_o)KT}{2\Delta i} \tag{5}$$

$$C \geq \frac{T\Delta i}{8\Delta V_C} \tag{6}$$

Measurement and Control Subsystem. It is in charge of monitoring the variables of the system and generating the control signals required by the power switches of the power regulation and voltage regulation subsystems implementing the control algorithms.

MPPT Controller with P&O. It is in charge of implementing the P&O algorithm according to the flow diagram shown in Fig. 4, where the power of the operating point is measured through voltage and current in the Sepic converter output then it is compared with a previous power value to determine if a duty cycle perturbation is needed, this perturbation could be forward, backward or none depending of the comparison.

The process is repeated in a loop to guarantee that the operating point correspond with the maximum power point; the variables used in the flow chart are:

- Duty, is the pulse width of the frequency signal.
- duty_pre, stores the last value obtained for duty.
- Delta, modifies the value of duty according to the voltage conditions.
- Vpv, voltage delivered by the photovoltaic module.
- Vpre, stores the last value obtained for Vpv.
- Ppv, power delivered by the photovoltaic module.
- Ppre, stores the last value obtained for Ppv.

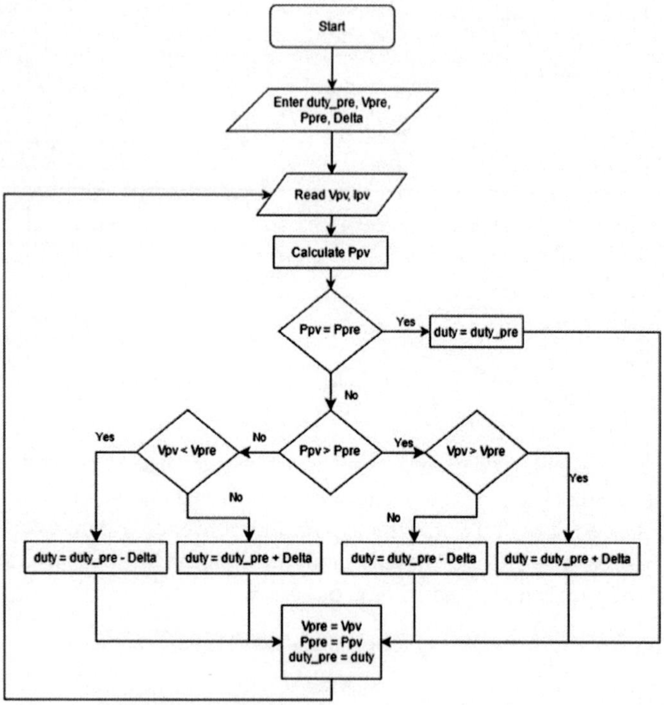

Fig. 4. Flow diagram for the P&O algorithm.

Voltage Controller. The Sepic converter allows to extract the maximum power available from the solar panel however it gives variable output voltage because of the P&O algorithm, so it produces a variable voltage in the Buck output. The Li-Po battery that is intended to be charged by the system, requires constant voltage, so the Buck converter operates in closed loop and with a controller to have a better transient response.

The design of the controller is carried out from the small-signal model of the Buck converter with the discrete-time transfer function of the converter that is indicated in

Eq. (7).

$$G_{(Z)} = \frac{0,05237\,z + 0,05158}{z^2 - 1,949\,z + 0,9557} \tag{7}$$

To establish the design criteria for the controller, the dynamic response of the converter for a unit-step signal is evaluated through simulation in Matlab Simulink, the converter has a maximum overshoot of 44.4% and a settling time of 1.4 ms, as it can be observed in Fig. 5a.

A PID controller reduces the overshoot value and improves the settling time thus improving the transient response and makes the system immune to load or input voltage variations. The design of the PID is carried out using the "PID Tuner" Matlab tool, obtaining the Proportional, Integral and Derivative constants as 0.2436, 1717.7908 and 7.8928e-6 respectively. Figure 5b shows the response of the system with the PID controller to a unit-step signal, where it may be observed a transient response which is better than the one shown in Fig. 5a with a maximum overshoot of 11.11% and a settling time of 0.3 ms.

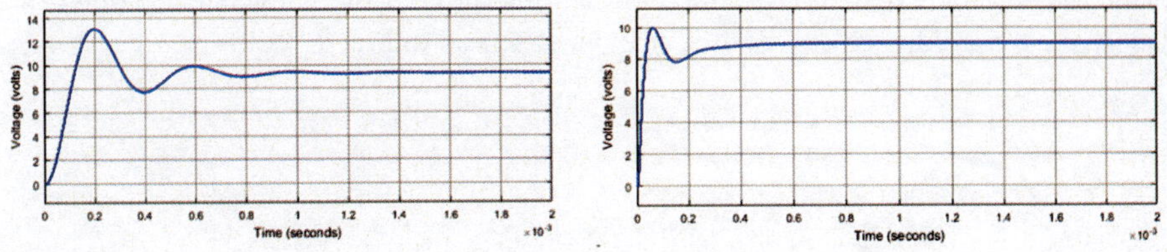

a b

Fig. 5. Unit-step response of the subsystem, a) Buck converter in open loop; b) Buck converter in closed loop with a PID controller.

The simulation scheme implemented in Matlab-Simulink can be observed in Fig. 6, in which all the designed subsystems have been integrated. The MPPT controller is implemented with the "Matlab Function" tool and the PID controller with the "Discrete PID Controller" tool; both controllers combine with the "PWM Generator" blocks to generate the trigger signal for each converter. The converters are integrated by a controlled voltage source to avoid coupling effects and include diodes that make the power flow unidirectional.

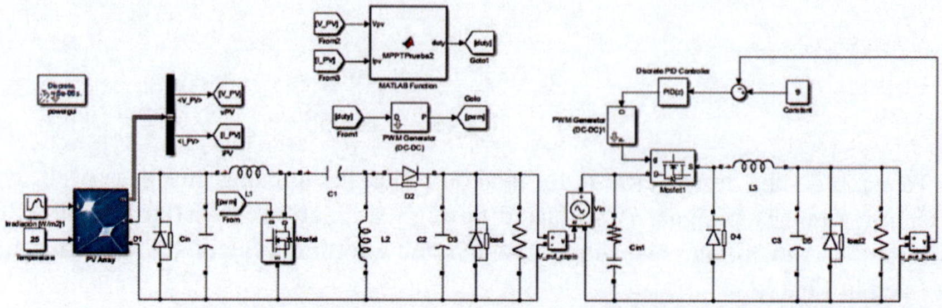

Fig. 6. Simulation scheme of the photovoltaic system implemented in Matlab Simulink.

3 Results

3.1 P&O Algorithm Simulation

This test simulates the SEPIC converter isolated from the rest of the system, using for the solar panel the solar irradiance values obtained at two different days at the automatic meteorological station installed in the backyards of the Universidad Politécnica Salesiana, the irradiance curves used may be observed in Fig. 7.

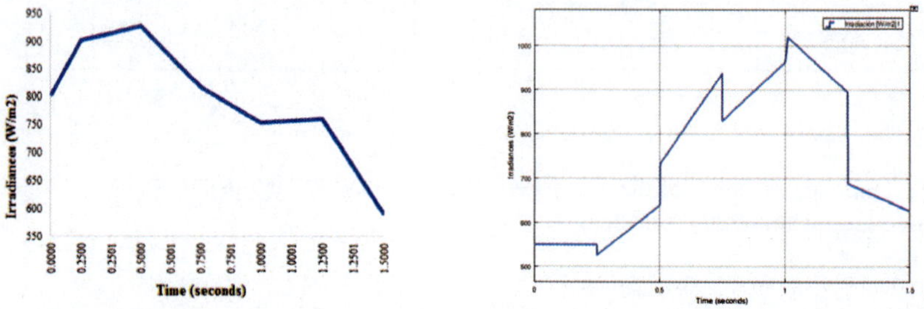

a b

Fig. 7. Irradiance curve of the automatic meteorological station on days a) August 1st, 2020, b) August 2nd, 2020.

The results obtained in the simulation may be observed in Fig. 8, where it is identified the change in the voltage level of the photovoltaic panel with the values of solar irradiance, while at the output of the SEPIC, under the action of the MPPT algorithm, the levels that produce the maximum power are obtained between 10 and 15 V.

It is observed in Fig. 8b that the power delivered by the SEPIC is almost equal to the power delivered by the photovoltaic panel, the difference is because power switches with conduction losses were used in the simulation, with these results it may be appreciated that the MPPT algorithm enables delivering maximum power, with the highest point being 30 W and the lowest one being 10 W.

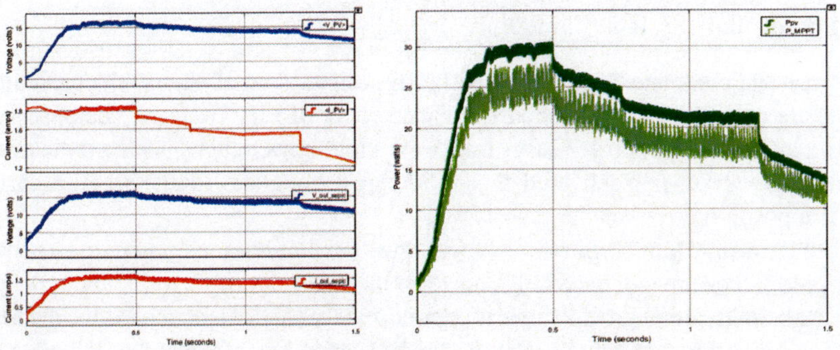

a b

Fig. 8. Response to a variation in solar irradiance. a) from the top, Voltage of the photovoltaic panel, Current of the photovoltaic panel, Output voltage of the Sepic converter and Output current of the Sepic converter. b) Power curve at the input of the photovoltaic panel and at the output of the Sepic converter.

3.2 Buck Converter with PID Controller Simulation

It is evaluated the response of the Buck converter when the SEPIC delivers a voltage as the one indicated in Fig. 9a, for this purpose the BUCK converter is simulated with an input voltage controlled from a function block, the ripple of the voltage is 14% with voltages larger than 10 V.

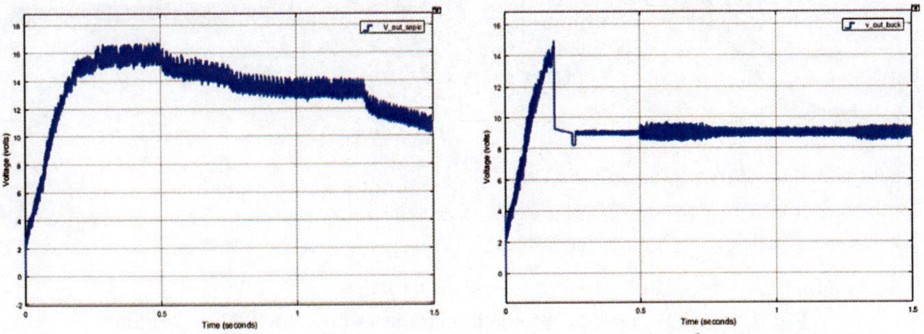

a b

Fig. 9. a) Input voltage for the test. b) Response of the Buck with PID

The output voltage of the BUCK converter is observed in Fig. 9b, it is seen that the operation is consistent with the configuration because there is a fast charge stage that delivers a voltage of up to 14 V followed by a charge stage at a constant voltage of 8.4 V.

The PID enables the existence of a minimum overshoot when the stage change is performed and then keeps the ripple in less than 3%, it may be also observed a settling time smaller than 0.01 s when going to the constant voltage charge stage and a rise time of 0.2 s.

3.3 Photovoltaic System Simulation

In this test the complete system showed in Fig. 6 is evaluated, the results presented in Fig. 10 are obtained using the solar irradiance curve of Fig. 7b. It may be observed in Fig. 10a that the output power curve tracks the input power curve, which indicates that maximum power is being delivered, thus validating the operation of the MPPT algorithm, there is a position error smaller than 10%.

The "PV array" Matlab-Simulink block shows oscillations thus generating a behavior with ripple at the output of the SEPIC, however the P&O algorithm recalculates the pulse width at all times, although this ripple is approximately 15% of the mean value the SEPIC is always delivering maximum power. It may be seen in Fig. 10b that the BUCK reaches 8.4 V in 0.2 s, and then continues with the initial fast charge stage and thus the voltage increases, according to the load, to 12 V, this lasts approximately 0.6 s and afterwards, it goes to the stage of absorbing charge at a constant voltage of approximately 8.4 V enabling that the battery is charged correctly.

The PID enables the existence of no overshoot in the fast charge stage, as well as at the instant of change to the absorption stage, and also controls the ripple which is maintained in 2.5% of the mean value even though the ripple delivered in the SEPIC is greater.

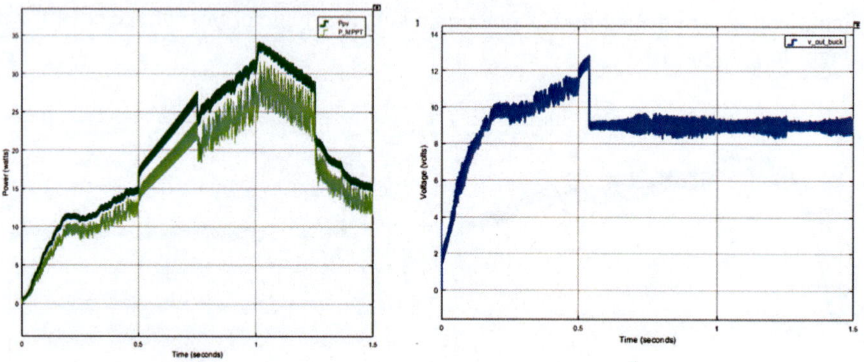

a b

Fig. 10. Power curve photovoltaic module vs Power MPPT algorithm

4 Conclusions

The photovoltaic system designed and simulated enables maximizing the efficiency of a photovoltaic panel to charge LiPo batteries with constant voltage, it comprises a fast charge stage and afterwards regulates the voltage at a constant value such that the useful life of the battery is not reduced. The tests of each converter separately validate the design of the converter and its corresponding controller, and when integrated into a single system the transient operating conditions as well as the stability of the system are maintained.

The PID controller maintains the ripple of the charge voltage in 2.5% and does not generate overshoot, thus attenuating the characteristic ripple due to the P&O algorithm which is 15%, and the variation in the voltage delivered by the solar panel. The operation obtained enables concluding that for this application it is not necessary that the SEPIC operates in closed loop because with the values used the obtained MPPT has no problems, and besides the controller in the BUCK improves the total response of the circuit, simplifying the implementation of the system.

References

1. Panaiotov, P., Goranov, G.: Development of a solar MPPT controller with AI for control SEPIC converter. In: 2019 IEEE XXVIII International Scientific Conference Electronics (ET), pp. 1-4. Sozopol, Bulgaria (2019). https://doi.org/10.1109/ET.2019.8878606
2. Suyanto, S., Mohammad, L., Setiadi, I.C., Roekmono, R.: Analysis and evaluation performance of MPPT algorithms: perturb & observe (P&O), firefly, and flower pollination (FPA) in smart microgrid solar panel systems. In: 2019 International Conference on Technologies and Policies in Electric Power & Energy, pp. 1–6. Yogyakarta, Indonesia (2019). https://doi.org/10.1109/IEEECONF48524.2019.9102532
3. Shirazi, R.A., Ahmad, I., Arsalan, M., Liaquat, M.: Integral backstepping based MPPT controller for photo-voltaic system using SEPIC converter. In: 2019 7th International Conference on Control, Mechatronics and Automation (ICCMA), pp. 62–67. Delft, Netherlands (2019). https://doi.org/10.1109/ICCMA46720.2019.8988614
4. Falin, J.: Texas instruments. [Online]. Available: http://www.ti.com/lit/an/slyt309/slyt309.pdf (2008). Accessed 30 Jan 2020
5. Zouirech, S., Zerouali, M., Ougli, A.E., Tidhaf, B.: The impact of the type of converter and the algorithm of the control on the production of maximum power by a photovoltaic system. In: 2019 7th International Renewable and Sustainable Energy Conference (IRSEC), Agadir, Morocco, pp. 1-5 (2019). https://doi.org/10.1109/IRSEC48032.2019.9078204
6. Srivastava, A.K., Kumar, S., Pandey, A.S., Mukherjee, D., Behera, R.R.: Performance comparison Of PV module using INC MPPT with boost & SEPIC converter. In: 2019 International Journal of Scientific & Technology Research, vol. 8, Issue 11 (2019)
7. Kumari, N., Kumar, S.S., Laxmi, V.: Modified MPPT topology: a comparative analysis. In: 2019 IEEE 16th India Council International Conference (INDICON), pp. 1-4. Rajkot, India (2019). https://doi.org/10.1109/INDICON47234.2019.9029042
8. Raj, A., Arya, S.R., Gupta, J.: Solar PV array-based DC–DC converter with MPPT for low power applications. Renewable Energy Focus, **34**, 109–119 (2020). ISSN 1755–0084. https://doi.org/10.1016/j.ref.2020.05.003
9. Majstorović, M., Mršević, D., Đurić, B., Milešević, M., Stević, Z., Despotović, Ž.V.: Implementation of MPPT methods with SEPIC converter. In: 2020 19th International Symposium INFOTEH-JAHORINA (INFOTEH), East Sarajevo, Bosnia and Herzegovina, pp. 1–6 (2020). https://doi.org/10.1109/INFOTEH48170.2020.9066296
10. Subirón, E.M.: Desarrollo de una Aplicación en Matlab Para la Evaluación de Algoritmos MPPT y GMPPT. 2016, Universitat Politécnica de Catalunya, Barcelona (2016)
11. Muñoz, S.C.G.: Análisis de Convertidores de Potencia DC-DC con Software Libre OPENMODELICA. Biblioteca de la Escuela Politécnica del Ejército, Quito (2012)
12. Ruiz, P.L: Diseño de una Instalación Solar Fotovoltaica para el Suministro de Energía Eléctrica de una Vivienda Aislada (2015)

13. Guachamin, J.F.V.: Diseño y Simulación de la Optimización en la Generación de Energía en una Central Fotovoltaica Mediante Conversores DC/DC y la Técnica de Control Mppt, Quito (2018)
14. Durango, J.J.M., Ordoñez, J.J., Machado, L.F.M.: Universidad Tecnológica de Pereira. 01 03 2017. [Online]. Accessed 16 Feb 2020

Comparison of Mechanical Losses in a 1100 cc Gasoline Engine Under the Morse and Thermodynamic Methods

Klever Tumbaco Casa[1](✉), Abrahan Jorque Rea[2], Daniel Casaliglia Gordon[2], and Christian Tupiza Quimbiulco[2]

[1] Escuela Politécnica Nacional, Quito 170517, Ecuador
klever.tumbaco@epn.edu.ec
[2] Instituto Superior Universitario Central Técnico, Quito 170138, Ecuador

Abstract. Mechanical losses in a vehicle can increase or decrease depending on different factors such as the type of fuel, atmospheric pressure, lubricant viscosity coefficient, among others; these in turn have a direct relationship with engine efficiency, so it is important to monitor them; there are several methods to evaluate mechanical losses; in the present investigation they were determined by applying the Morse and thermodynamic methods in a gasoline internal combustion engine. For this purpose, a 1,100 cc Kia Picanto vehicle was used; this study was carried out in the Laboratory of the Technology Transfer Center for Training and Research in Vehicle Emissions Control (CCICEV) of the National Polytechnic School (EPN), which is located in the city of Quito, which is at 2810 m above sea level and has an atmospheric pressure of 0.71 atm. The results obtained in this study are based on the initial operating conditions of the engine, such as temperature and pressure in the intake manifold. The data obtained from the mechanical losses with the Morse and thermodynamic methods show a variation of 1.1–3.9 hp.

Keywords: Mechanical losses · Morse method · Thermodynamic method

1 Introduction

A vehicle with a gasoline internal combustion engine uses at most 30% of the fuel efficiency [1]. The low efficiency is due to the existence of energy losses [2, 3]. The energy losses that occur in the vehicle are directly related to the high rates of pollution into the atmosphere and especially affect large cities, which have a significant growth in the vehicle fleet. Quito is one of the cities in Ecuador that has been affected by the excess of pollutants emitted into the atmosphere [4]. The altitude of this city is 2810 m above sea level [5]. The municipality of Quito, in order to reduce these high pollution rates, has implemented measures such as the "pico y placa" [6] and other measures to rationalize the use of vehicles [7].

There are several studies that attempt to reduce pollutant emissions by means of lower fuel consumption caused by using engine lubricants with lower viscosity [8], another study where the tribological properties of engine oils are improved by using

M. Zambrano Vizuete et al. (Eds.): CI3 2021, LNNS 511, pp. 307–322, 2022.
https://doi.org/10.1007/978-3-031-11438-0_25

nano particulate additives in commercial oils such as 5w30 [9]. Vehicles with spark ignition engines (MEP) distribute the energy losses in the engine [10], in the transmission system [11] and those caused by the control elements. Of the total losses that occur in the vehicle, the highest percentage are those in the internal combustion engine, such as friction, pumping and auxiliary element losses [12–14].

It should be emphasized that mechanical friction losses are those generated between the moving parts of the engine and are the highest percentage [15], since they are in the range of (45–65%), the main elements considered for the analysis are those generated in the pistons (40–75%), in the bearings (20–40%) and in the valve system (3–30%) [3]. Pumping losses are around (15–30%) [3] and those caused by auxiliary elements drive (20–25%) respectively [14, 16].

Considering that, in order to determine the mechanical losses of the motor, the use of mathematical models is very complex, in this project we have tried to calculate them by means of two methods that are highly manageable: the Morse method and the indicator or thermodynamic diagram. It should be emphasized that knowing the results of the mechanical losses in a vehicle is very useful when you want to make a sampling of the technical or mechanical condition of an engine and know if the use of a particular fuel or lubricating oil is adequate.

2 Equipment and Methodology

2.1 Equipment

In order to obtain the necessary parameters, it is essential to use technological equipment such as: the automotive scanner, the LPS3000 power test bench in which torque and power tests are performed, and a vehicle with a spark ignition engine with electronic injection system.

M, Vehicle. The vehicle used for the study is a Kia Picanto of four cylinders in line, 1086 cc of displacement, of the year 2008, with 140200 km of travel, of particular use as a summary of the main technical data can be seen in Table 1.

Table 1. Technical specifications of the test vehicle.

Description	Specifications	Description	Specifications
Version	Picanto 1.1 Active	Stroke	77 mm
Fuel	Gasoline	Compression ratio	10
Maximum power	85 cv	Power supply	Indirect injection
Maximum power revolutions	5500 rpm	Maximum torque	97 NM
Cylinder bore	67 mm	Maximum torque revolutions	2800 rpm

Automotive Scanner. It allows to monitor the temperature parameters and the inlet pressure to the intake manifold, when connected through the 16-pin DLC port. Figure 1 shows the vehicle and the type of scanner used.

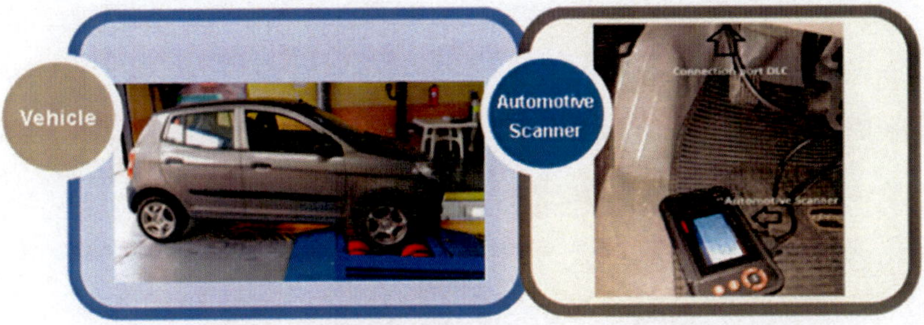

Fig. 1. Test vehicle and automotive scanner.

Power Dynamometer LPS3000. This equipment allows to obtain the data of torque and power of the motor, the elements that conform it are specified in Fig. 2.

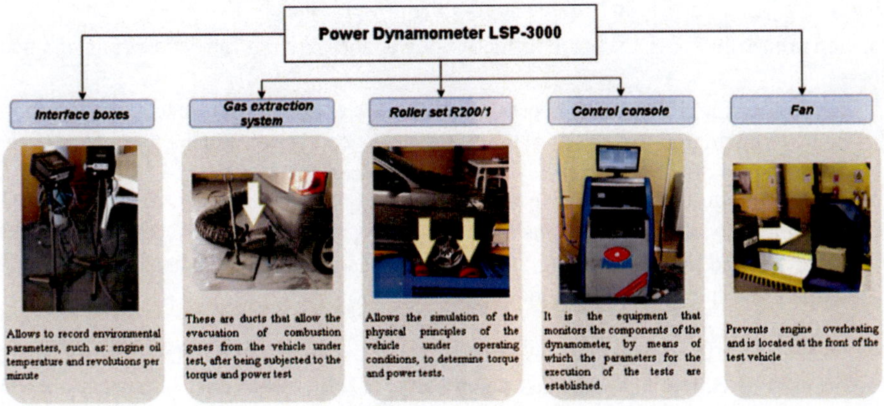

Fig. 2. Dynamometer.

2.2 Methods

Morse Method. It consists of suspending the ignition or interrupting the fuel flow in one of the cylinders, to measure the effective power of the engine (Ne) in that condition, this activity is repeated in each cylinder.

For a four-cylinder engine the following equations are presented [12]:
Spark cut in cylinder one:

$$Ne^{(I)} = Ne^{(2)} + Ne^{(3)} + Ne^{(4)} - Npm^{(1)} \tag{1}$$

Spark cut-off on cylinder two:

$$Ne^{(II)} = Ne^{(1)} + Ne^{(3)} + Ne^{(4)} - Npm^{(2)} \tag{2}$$

Spark cut-off on cylinder three:

$$Ne^{(III)} = Ne^{(1)} + Ne^{(2)} + Ne^{(4)} - Npm^{(3)} \tag{3}$$

Spark cut-off on cylinder four:

$$Ne^{(IV)} = Ne^{(1)} + Ne^{(2)} + Ne^{(3)} - Npm^{(4)} \tag{4}$$

The sum of the four equations indicated above is made, and the following expression is obtained.

$$\sum_{I=I}^{IV} Ne^{(i)} = 3Ne - Npm \tag{5}$$

where: $\left(\sum_{I=I}^{IV}\right)$ corresponds to the sum of the power measured when eliminating combustion in each cylinder, (Ne) is the effective power and (Npm) is the power of mechanical losses.

Before starting the test, it is necessary to start the engine until it reaches its optimum operating temperature; then, each of the cylinders is deactivated and the torque and speed data are taken to obtain the power delivered by the engine [12].

Thermodynamic Method. It consists of obtaining the pressure vs. volume diagram, based on real and initial operating values; in this case, since it is a gasoline engine, the OTTO thermodynamic cycle diagram is used, as shown in Fig. 3.

To determine the mechanical losses (N_{pm}) by means of this method it is indispensable to know the effective and indicated parameters, of which the most important are: the effective power (N_e) and the indicated (N_i) [3].

$$N_i = N_e + N_{pm} \tag{6}$$

Effective Parameters. These values are obtained at the motor output shaft and can be measured by means of technological equipment. In this case, the effective torque and power values are obtained by means of the power dynamometer.

Indicated Parameter. These parameters can be found by means of the thermodynamic diagram shown in Fig. 3.

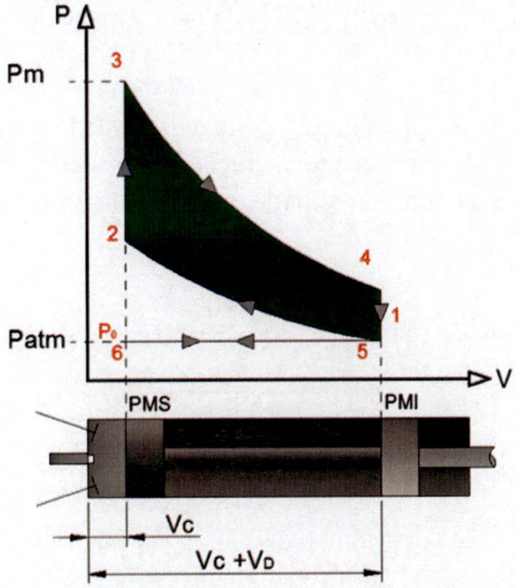

Fig. 3. Pressure vs. volume diagram of the OTTO cycle.

In the thermodynamic analysis of the vehicle it is indispensable to know several parameters, which are represented in the thermodynamic diagram of the Otto cycle in four fundamental points. The points of the OTTO thermo-dynamic cycle and the mathematical expressions to be used at each point are detailed below.

Point 1. The mathematical expressions needed to obtain all the data at this point are as follows [17]:

$$r_c = \frac{V_c + V_D}{V_c} \tag{7}$$

$$V_1 = V_c + V_D \tag{8}$$

$$m = \frac{P_1 \cdot V_1}{R \cdot T_1} \tag{9}$$

where:r_c is the compression ratio, V_c is the combustion chamber volume, V_D is the cylinder volume, m is the mass, P_1 is the atmospheric pressure, R is the ideal gas constant [18], T_1 is the temperature at the inlet of the intake manifold and V_1 is the volume at point 1 of the thermodynamic cycle.

Point 2. The following expressions should be used in this point:

$$P_2 = P_1 r_c^k \tag{10}$$

$$T_2 = T_1 r_c^{k-1} \tag{11}$$

$$V_2 = \frac{mRT_2}{P_2} \tag{12}$$

At this point it is indispensable to know relation of specific heats ratio of specific heats.

Point 3. At this point one must know Q_{in} is the heat added to the system, m_f mass of fuel for one cycle, Q_{HV} calorific value of the fuel [19], η_c fuel efficiency, m_m mass of the mixture, c_v Specific heat at constant volume, m_a the mass of air and (λ) lambda factor [20].

$$Q_{in} = m_f Q_{HV} \eta_c = m_m c_v (T_3 - T_2) \tag{13}$$

$$m_f = \frac{m_m}{14,7 * \lambda + 1} \tag{14}$$

$$m_a = m_m - m_f \tag{15}$$

Knowing that $V_3 = V_1$, the value of the volume at point three is obtained from the first point.

$$P_3 = P_2\left(\frac{T_3}{T_2}\right) \tag{16}$$

Point 4. Finally, the expressions that allow us to know the data of the fourth point are shown below.

$$P_4 = P_3(1/r_c)^k \tag{17}$$

$$T_4 = T_3(1/r_c)^{k-1} \tag{18}$$

$$V_4 = \frac{mRT_4}{P_4} \tag{19}$$

Net Work (W_{net}). To obtain the net work in the system, the following expressions are applied, where the work that enters the system is related to the work of the system (W_{1-2}) and leaving (W_{3-4}) the system.

$$W_{3-4} = mR(T_4 - T_3)/(1 - k) \tag{20}$$

$$W_{1-2} = mR(T_2 - T_1)/(1 - k) \tag{21}$$

$$W_{net} = W_{1-2} + W_{3-4} \tag{22}$$

Indicated Power. This value is obtained by means of Eq. 23, where N is the engine rpm and n is the number of cylinders of the engine.

$$N_i = \frac{W_{net} \cdot N}{n} \tag{23}$$

Table 2. Vehicle data.

Description	Values	Units
Number of cylinders	4	
Compression Ratio	0,001086	m^3
Maximum Power	10	
RPM (max)	65	Cv
Compression Ratio	5500	

Table 3. Universal constants.

Description Units	Abbreviation	Values	Units
Atmospheric pressure	P.adm	71,77	kPa
Ideal gas constant	R	0,287	kJ/kgK
Ratio of specific heats	k	1,35	
Calorific value of the fuel	QHV	44300	kJ/kg
Specific heat at constant volume	Cv	0,821	kJ/kgk
Density of air Quito		0,832	kg/ m^3
Fuel Efficiency	η_c	0,95	

3 Results

3.1 Initial Data

Table 2 shows the values obtained from the technical data sheet of the vehicle manufacturer's manual, which are necessary to develop the respective calculations.

The constants required for the calculations are shown in Table 3.

Table 4 shows the values of the lambda factor recorded by means of the scanner.

Table 4. Lambda factor

Rpm	1011	1760	2509	3258	4009	4758	5500
Lambda	1,01	1,001	1	1	0,99	1	1

3.2 Results of the Thermodynamic Method

Results at Point 1. The results correspond to the data in point 1, the pressure values correspond to those obtained at the inlet of the intake manifold and are obtained by means of the automotive scanner. Table 5 shows the values of the initial pressures at the intake manifold inlet.

Table 5. Initial pressures.

rpm	1011	1760	2509	3258	4009	4758	5500
P1(kPa)	73	72	71	71	70	70	70

The intake manifold inlet temperature values obtained from the scanner can be seen in Table 6.

Table 6. Intake manifold inlet temperatura.

rpm	°C	K
1011	51	324,15
1760	52	325,15
2509	51	324,15
3258	52	325,15
4009	53	326,15
4758	50	323,15
5500	51	324,15

The initial volume at point 1 can be visualized in Table 7 and to obtain this result, Eqs. 7 and 8 must be applied, as well as the specifications of the manufacturer's manual specified in Table 1.

Table 7. Volume data in item 1.

VD (m3)	Vc (m3)	V1(m3)
0,0002715	3,02E-05	0,000301667

The results of mass of the mixture at point 1 can be seen in Table 8, these values are obtained by applying Eq. 8.

Point 2. Table 9 shows the results of pressure, temperature and volume, which were obtained in point 2.

Table 8. Mass values.

rpm	1011	1760	2509	3258	4009	4758	5500
mm(kg)	2,37E-04	2,33E-04	2,30E-04	2,30E-04	2,26E-04	2,28E-04	2,27E-04

Table 9. Results in point 2 (P, V and T)

Rpm	P2 (kPa)	T2 (°C)	V2 (m3)
1011	1634,266431	725,6814571	3,01667E-05
1760	1611,87922	727,9201782	3,01667E-05
2509	1589,492008	725,6814571	3,01667E-05
3258	1589,492008	727,9201782	3,01667E-05
4009	1567,104797	730,1588993	3,01667E-05
4758	1567,104797	723,4427359	3,01667E-05
5500	1567,104797	725,6814571	3,01667E-05

Point 3. The results of input Q_{in}, temperature, pressure, fuel mass m_f and air mass m_a, can be seen in Table 10.

Table 10. Results in item 3.

rpm	mf (kg)	ma (kg)	Qin (kj)	T3 (K)	P3 (kpa)
1011	1,49374E-05	0,00022178	0,66172711	4130,65319	9302,41194
1760	1,48111E-05	0,00021794	0,65613287	4161,55792	9215,19823
2509	1,46642E-05	0,00021556	0,64962365	4162,53414	9117,38158
3258	1,46191E-05	0,0002149	0,64762573	4164,77286	9094,22952
4009	1,45048E-05	0,00021318	0,64256289	4199,49517	9013,17376
4758	1,45024E-05	0,00021248	0,64245599	4160,29542	9011,93499
5500	1,44577E-05	−1,4458E-05	0,64047402	4162,53414	8988,96775

Point 4. The results obtained at this point are shown in Table 11, where the pressure, temperature and volume data are given.

Table 11. Results in item 4

rpm	P4 (kPa)	T4 (K)	V4 (m3)
1011	415,5234781	1845,095006	0,000301667
1760	411,6277849	1858,899643	0,000301667
2509	407,2584753	1859,335701	0,000301667
3258	406,2243112	1860,335701	0,000301667
4009	402,6036832	1875,84559	0,000301667
4758	402,5483493	1858,335701	0,000301667
5500	401,5224405	1859,335701	0,000301667

Net Work. The result of the network at different rpm are obtained by means of Eqs. 20, 21 and 22; these values can be seen in Table 12.

Table 12. Net work

rpm	W3-4 (kj)	W1-2 (kj)	Wneto (kj)
1011	0,44363765	−0,077939154	0,365698496
1760	0,439478376	−0,076871495	0,362606881
2509	0,43481344	−0,075803835	0,359009605
3258	0,433709305	−0,075803835	0,35790547
4009	0,429843707	−0,074736175	0,355107531
4758	0,429784629	−0,074736175	0,355048454
5500	0,428689307	−0,074736175	0,353953132

Mechanical Losses. Table 13 shows the results of the indicated power N_i, these values are obtained by applying Eq. 23; the effective power N_e, are values generated by the power dynamometer LSP2000; finally the mechanical losses are obtained by means of Eq. 6.

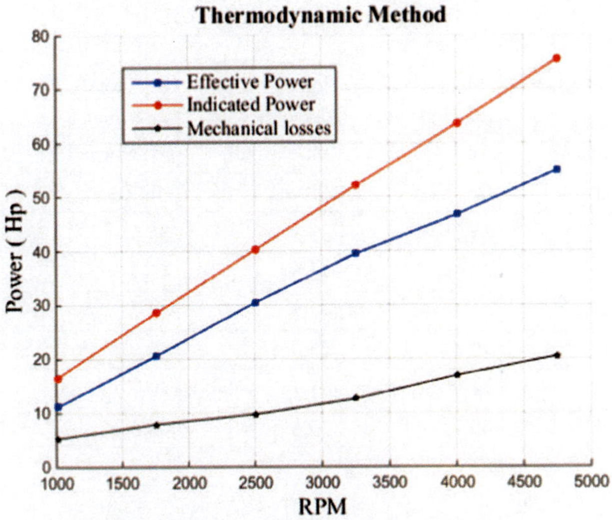

Fig. 4. Powers of the thermodynamic method.

Figure 4 shows the trend of the three types of power. These have an ascending slope as the engine rpm increases.

Table 13. Mechanical losses (Thermodynamic Method)

rpm	Power Effective (Hp)	Power Indicated (Hp)	Losses Mechanical (Hp)
1011	11,2	16,52653672	5,326536725
1760	20,6	28,52700857	7,927008565
2509	30,5	40,26375297	9,76375297
3258	39,5	52,12270416	12,62270416
4009	46,7	63,63608637	16,93608637
4758	55	75,51262823	20,51262823

3.3 Morse Method Results

The torque and power results can be seen in Tables 14 and 15.

Table 14. Torque test results.

n	All Cylinders	Injector 1 disconnected	Injector 2 disconnected	Injector 3 disconnected	Injector 4 disconnected
	M. Normal	M. Normal	M. Normal	M. Normal	M. Normal
(rpm)	(lbf.ft)	(lbf.ft)	(lbf.ft)	(lbf.ft)	(lbf.ft)
1011	58,1	35,7	38,5	35,5	25,9
1760	61,6	40,5	42,5	41,7	27,3
2509	63,8	41,3	42,5	44,2	46,7
3258	63,7	42,4	42,6	43	45,1
4009	61,2	41,9	41,7	40,7	43
4758	60,7	40,2	39,8	40,6	40,8
5495	52,5				

Table 15. Power test results.

n	All Cylinders	Injector 1 disconnected	Injector 2 disconnected	Injector 3 disconnected	Injector 4 disconnected
	P. Normal	P. Normal	P. Normal	P. Normal	P. Normal
(rpm)	(HP)	(HP)	(HP)	(HP)	(HP)
1011	11,2	6,8	7,4	6,8	4,9
1760	20,6	13,5	14,2	13,9	9,1
2509	30,5	19,6	20,2	21,1	22,3
3258	39,5	26,2	26,3	26,6	27,9
4009	46,7	31,8	31,7	30,9	32,7
4758	55	36,3	36	36,7	36,9
5495	54,9				

With the data obtained, corresponding to the torque and power values of the operating engine, with the different spark gaps in the different cylinders, Eq. 5 is applied and the values of the mechanical losses are obtained, this value can be visualized in Table 16.

Figure 5 shows the graphical representation of the power and mechanical losses generated by this method.

Table 16. Mechanical losses (Morse Method)

[rpm]	Sum of powers [hp]	P normal [hp]	P. mechanics [hp]
1007,25	25,9	33,6	7,7
1749,25	50,7	61,8	11,1
2503	83,2	91,5	8,3
3248,5	107	118,5	11,5
3994,75	127,1	140,1	13
4741,5	145,9	165	19,1

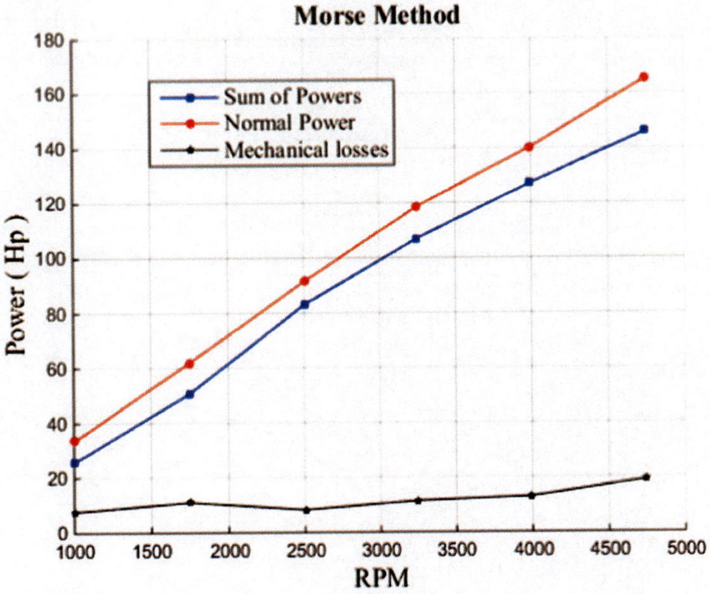

Fig. 5. Mechanical losses (Morse Method).

4 Discussion

Table 17 shows the summary of the mechanical losses obtained by the two methods, with a minimum difference of 1.12 hp and a maximum of 3.93 hp, as previously determined in the calculation process, the thermo-dynamic method is purely theoretical in that it assumes values from the manufacturer's technical data sheet, universal constants, mathematical formulas and values measured with the automotive scanner, while the Morse method is more real, which is based on the parameters that the dynamometer directly dates on the output shafts.

After obtaining the different data of the mechanical losses of the research, it is established that with the Morse method there is a prolonged variation of the data, mainly in the range of 1000 to 2500 rpm see Fig. 6, which could be mentioned that there is

Table 17. Summary of mechanical losses with Morse and Thermodynamic methods.

N°	rpm	Morse Method (Hp)	Thermodynamic Method (Hp)	Difference
1	1011	7,7	5,33	2,373
2	1760	11,1	7,93	3,173
3	2509	8,3	9,76	1,464
4	3258	11,5	12,62	1,123
5	4009	13	16,94	3,936
6	4758	19,1	20,51	1,413
			min	1,123
			máx	3,936
			Variation (máx.-min)	2,813

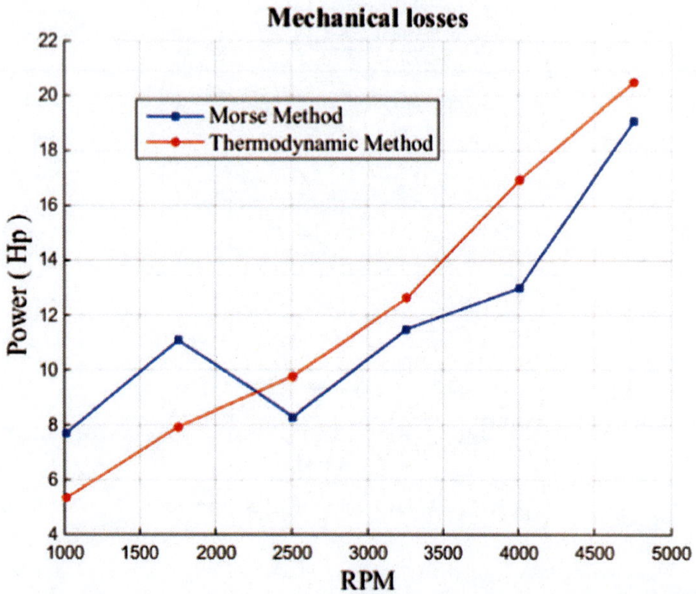

Fig. 6. Summary of mechanical losses obtained with the Morse and Thermodynamic methods.

a greater tolerance in the veracity of a result, this can happen because of two main factors, the first one due to the environmental conditions in which the experimentation is performed, such as the height above sea level and the atmospheric pressure.

On the other hand, with the thermodynamic method it is observed that there is an increasing tendency of the mechanical losses in a direct way with the revolutions per minute of the engine, it can be mentioned that the data collection carried out with this method is reliable because they were carried out in real operating conditions of the engine measured with adequate techno-logical instrumentation. This method consists in obtaining simultaneously the pressure-volume diagram, from which the indicated power can be obtained.

The effective power of the engine obtained from the dynamometer can be used to calculate the effective power, the difference of these parameters represents the mechanical losses.

5 Conclusions

With the data collection it can be seen that with the Morse method there is a maximum of 19.1 hp in mechanical losses, while with the thermodynamic diagram method a result of approximately 20.5 hp is obtained, this represents a range of 3% difference between the two methods.

It is worth mentioning that the calculations made to obtain the mechanical losses in the two methods applied in this research present a great difference in their methodology due to the handling of different parameters and measuring equipment. Thus, as a result of the graph it can be noted that the indicator diagram in the range of 1000 to 5000 rpm presents an increasing trend, i.e. the mechanical losses increase as a function of the revolutions per minute, this is presented by the mathematical model of the Otto cycle. While the results of the Morse method date up to a maximum of 3.9 hp difference in mechanical losses at low revolutions assuming real operating conditions, where it is concluded that there is greater engine wear.

As a comparative summary between the two methods used to determine mechanical losses, it can be mentioned that, in order to achieve accuracy and reliability of reading and analysis, it is advisable to keep all the boundary conditions associated with each of the measurement methods within limits.

In the Morse method, the test bench to measure torque and power, as well as equipment that allows mechanical or electronic deactivation to modify the fuel injection or ignition parameters is necessary, avoiding a variation of more than 5% of the value of the coolant and oil temperature.

The indicator diagram method would be more reliable if the data acquisition is done with technological equipment of high sampling frequency, visualizing the torque and rotational speed to calculate the average effective pressure and the total mechanical losses.

The research carried out in the city of Quito, establishes that the higher the atmospheric pressure, the higher the mechanical losses of the internal combustion engines, applied directly to the process of load renewal or also known as mechanical losses.

As for the cost, the value is higher with the Morse method, that is why fleets of vehicles try to avoid this methodology to determine the mechanical losses, on the other hand for the indicator diagram method the cost is well below the previous one being the most used for this analysis and the pumping losses, considering the above comments.

References

1. Thomas, J.: Drive cycle powertrain efficiencies and trends derived from EPA vehicle dynamometer results. SAE Int J. Passeng. Cars – Mech. Syst. **7**(4), 1374–1384 (2014). https://doi.org/10.4271/2014-01-2562

2. . Milan, K.: Fuel agnostic zero emission engine technology. Creat. Futur. [Online]. Available: https://contest.techbriefs.com/2020/entries/automotive-transportation/10795 (2020)
3. Payri, F., Olmeda, P., Martin, J., Carreño, R.: A new tool to perform global energy balances in DI diesel engines. SAE Int. J. Engines **7**(1), 43–59 (2014). https://doi.org/10.4271/2014-01-0665
4. OA(CODIGO ORGÁNICO DEL AMBIENTE): Codigo Organico Del Ambiente. Regist. Of. Supl. **983**, 1–92. [Online]. Available: http://gobiernoabierto.quito.gob.ec/Archivos/Transparencia/2017/07julio/A2/ANEXOS/PROCU_CODIGO_ORGANICO_ADMINISTRATIVO.pdf (2017)
5. Llanes Cedeño, E.A., Rocha-Hoyos, J.C., Peralta Zurita, D.B., Leguísamo Milla, J.C.: Evaluación de emisiones de gases en un vehículo liviano a gasolina en condiciones de altura. Caso de estudio Quito, Ecuador. Enfoque UTE **9**(2), 149–158 (2018). https://doi.org/10.29019/enfoqueute.v9n2.201
6. Polivio, A., Coyago, R.: Análisis de la aplicación del pico y placa en la ciudad de Quito **2**(6), 136–142 (2017)
7. Mosquera, R.: Racionalización del uso del automóvil en quito: un análisis de incentivos económico (2010)
8. Tormos Martínez, B., Miró Mezquita, G., Pérez Gutiérrez, T., De Diego Pardo, J.: Aceites de motor de baja viscosidad: ahorro de combustible y ensayos en condiciones reales. Dyna (Spain) (2016)
9. Davis, D., Shah, A.F., Panigrahi, B.B., Singh, S.: Effect of Cr2AlC nanolamella addition on tribological properties of 5W–30 engine oil. Appl. Surf. Sci. **493**, 1098–1105 (2019). https://doi.org/10.1016/j.apsusc.2019.07.097
10. Heywood, J.B.: Internal combustion engine fundamentals (1988)
11. Antoni, G.: On the mechanical friction losses occurring in automotive differential gearboxes. Sci. World J. **2014** (2014). https://doi.org/10.1155/2014/523281
12. Payri, F., Desantes, J.: Motores de Dombustión Interna Alternativos. Editorial Reverte, España (2011)
13. Wong, V.W., Tung, S.C.: Overview of automotive engine friction and reduction trends–effects of surface, material, and lubricant-additive technologies. Friction **4**(1), 1–28 (2016). https://doi.org/10.1007/s40544-016-0107-9
14. Taraza, D., Henein, N.: SAE technical friction losses in multi-cylinder diesel engines (724) (2018)
15. Jia, B., Mikalsen, R., Smallbone, A., Roskilly, A.P.: A study and comparison of frictional losses in free-piston engine and crankshaft engines. Appl. Therm. Eng. **140**(April), 217–224 (2018). https://doi.org/10.1016/j.applthermaleng.2018.05.018
16. Tormos, B., Martín, J., Carreño, R., Ramírez, L.: A general model to evaluate mechanical losses and auxiliary energy consumption in reciprocating internal combustion engines. Tribol. Int. **123**, 161–179 (2018). https://doi.org/10.1016/j.triboint.2018.03.007
17. Ehsani, M., Gao, Y., Emadi, A.: Internal combustion engines. Mod. Electr. Hybrid Electr. Fuel Cell Veh. 1–20 (2006). https://doi.org/10.1201/9781420054002-3
18. Çengel, Y.A., Boles, M.A.: Termodinámica, 8a. ed (2015)
19. Baglione, M., Duty, M., Pannone, G., Corp, D.: Vehicle system energy analysis methodology and tool for determining vehicle subsystem. SAE Techinical (724), 9 (2007)
20. Pulkrabek, W.W.: Engineering fundamentals of the internal combustion engine. J. Eng. Gas Turbines Power **126**(1), 198 (2004). https://doi.org/10.1115/1.1669459

Comparison of Photovoltaic Energy Production Potential Between Different Regions of Ecuador and the Incidence in Implants of Photovoltaic Energy Production Plants

Omar Fernando Mejía Mendía[1](✉), Robert Anthony Toala Constante[1], Harry Ricardo Arias Realpe[1], and Wellington Isaac Maliza Cruz[2]

[1] Instituto Superior Universitario Central Técnico, Quito, Ecuador
omar1914@hotmail.es
[2] Instituto Superior Universitario Bolivariano, Guayaquil, Ecuador

Abstract. This current research paper has studied the centralized photovoltaic energy potential production in three provinces of the Ecuadorian highland through comparative analysis of incidents characteristics for the photovoltaic generation (irradiation, temperature, humidity, land variation and the electricity grid connectivity) added to the analysis by emulation and simulation of the potential production in a large scale simulated plus losses. The results allow us to identify to Imbabura as the province with the best incidents characteristics performance, higher useful potential energy produced per year and a total grid connection energy of 15449 MWh with a percentage loss of system of 12.2% to the first year of operation.

Keywords: Centralized photovoltaic systems · Solar energy · Power plant · Climate resource · Emulation · Simulation · PVsyst software · Design · Photovoltaic losses

1 Introduction

The current energy matrix of the country is mainly made up of renewable energies, the water resource predominated in this aspect while that another large percentage of total production is based on non-renewable sources, causing more pollution due to CO_2 emissions, leaving only a small percentage to unconventional energy sources as photovoltaic and wind. Ecuador is a country with great potential for photovoltaic solar energy, even so, it has not been used properly, generating the challenge of an energy matrix that considers increasing the generation of electricity from unconventional sources, mainly photovoltaic solar.

Velasco and others [20] conducted the study to contribute to the dissemination about the state of the art in implementation of photovoltaic systems connected to the grid in Imbabura through the presentation of a summary of the fundamentals of solar technology, its basic configuration and energy saving opportunities of energy saving. The results

M. Zambrano Vizuete et al. (Eds.): CI3 2021, LNNS 511, pp. 323–338, 2022.
https://doi.org/10.1007/978-3-031-11438-0_26

showed an almost costly monthly generated energy, an annual generation of 10627,59 kWh and a reduction of 9 tons. of CO_2 generated by the burning of fossil fuels.

Carvajal and others [3] carried out the study to determine the appropriate areas for the location of power plants, generated from renewable wind and photovoltaic sources in places with an energy deficit, as a result of the AHP hierarchy, the sectors of preference in order are had for the photovoltaic potential: Cangahua, Zapotillo, Olmedo, Guangaje and for the PROMETHEE methodology the net flow of ranking of photovoltaic potential is: Cangahua, Zapotillo, Guangaje and Olmedo.

This research paper mainly seeks to solve the problem of pollution by CO_2 emissions caused by the burning of fossil fuels in non-renewable electricity generation sources by means of the implementation of centralized photovoltaic energy projects, in addition to the uncertainty generated by the use of the solar resource in the country based on its incident's characteristics and effects on a photovoltaic energy production plant, mainly in the studied areas of the Ecuadorian highlands.

2 Methods

In the main search to reduce CO_2 emissions and to have a clean energy matrix worldwide, solar energy is an inexhaustible source for the production of electricity free of pollution to avoid the negative environmental impacts in the planet. Opting for the generation of electricity from solar resources is increasingly promoted in the world, with China and Europe where this type of project has been most installed [18], to the end of 2015, there was an estimate that 1% of the total electricity production in the world was from solar sources, being around 227 GW [6]. Ecuador, for its location, has a great potential for solar energy, studied in this section, which has not yet been used.

2.1 Production of Photovoltaic Solar Energy in the Country

According to the Atlas of the Ecuadorian Electricity Sector 2018 [1] in the country there is a total of 8676.89 MW of installed nominal power, but 40% of this energy is from non-renewable sources, which prioritizes the use of renewable energy. In Ecuador, according to the CONELEC solar map [5] the sectors with the most global solar irradiation during the year are the provinces of Carchi, Imbabura and Loja.

The approximate average value of global solar radiation in the country is 4575 Wh/m^2/day that giving a solar energy potential of 456 TWh per year, this in electrical energy, translates into 312 GW, values that are, in reference, about 15 times country's hydroelectric potential [18]. Even with this great potential it's estimated that only 00.36% of the total installed power in MW, of the country, is for photovoltaic solar energy sources [11]. What makes it vital to analyze the possibilities of photovoltaic electricity generation that has the country.

Solar Irradiation, Climate and Electrical Characteristics in the Province of Carchi, Imbabura and Loja

Carchi located in the extreme north of the country is with a high potential in solar energy. Carchi has a nominal installed power of 4.82 MW only by renewable sources

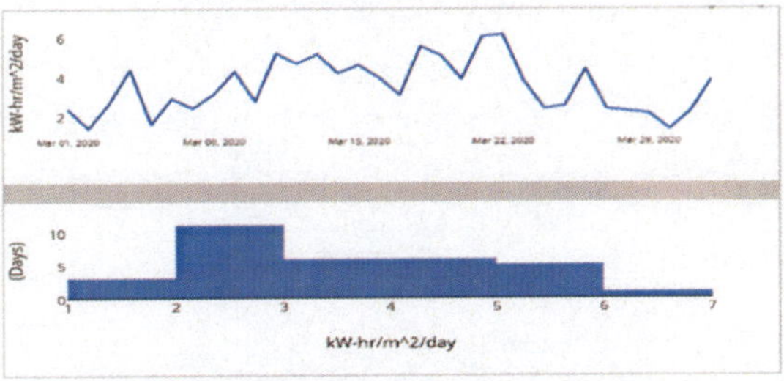

Fig. 1. Monthly solar resource in Carchi.

and an overall production of 24838.17 MWh, its main generator is the San Miguel de Car hydroelectric plant with 2.95 MW (Fig. 1).

Located at the southern limit of Carchi is the province of Imbabura, its global irradiation is superior to the other provinces of the country. The main power generators in Imbabura are the Selva Alegre Thermal Power Plant with 29.28 MW and the Ambi Hydroelectric Power Plant with 8 MW, Imbabura has two photovoltaic projects, the Salinas and Paragachi power plants, together have a nominal power of 4 MW (Fig. 2).

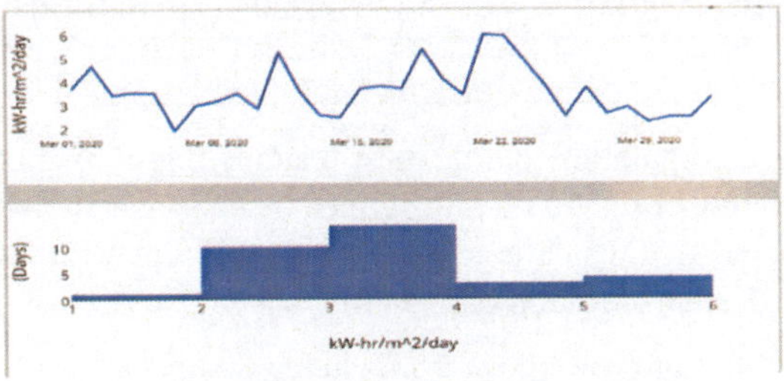

Fig. 2. Monthly solar resource in Imbabura.

Loja is located in the extreme south of the country; its solar irradiation is just below Imbabura with considerable values. Loja has a nominal installed power of 42.22 MW of which 19.73 MW are for non-renewable energy, its main generator is the Catamayo thermal power plant with 19.74 MW. It is currently the province with the most alternative generation projects with 4 photovoltaic plants and the main wind generator in the country (Fig. 3, Table 1).

The average solar irradiation levels are shown with better characteristics in the province of Imbabura, as well as the monthly behavior, the average temperature and humidity levels are adequate, the variation of terrain and its geographical location is feasible for this type of project, the results show Imbabura as the best province in the study carried out. For this reason and, in order to analyze the incidence in a deep photovoltaic plant, the study is expanded in this province.

With the climatic characteristics and solar irradiation in Imbabura, a prospective project is proposed, in the first instance a study of the solar resource of the site is prepared to know its potential for generation, an optimal study of the plant is carried out

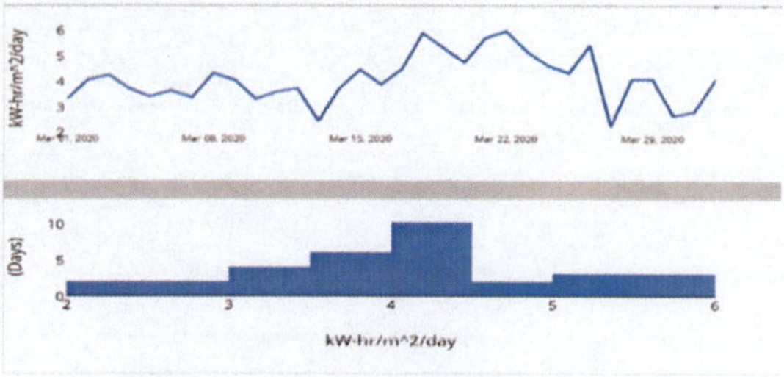

Fig. 3. Monthly solar resource in Loja.

Table 1. Characteristics of photovoltaic analysis in Carchi, Imbabura and Loja

Characteristics	Carchi	Imbabura	Loja
Irradiación global [Wh/m^2/day]	4200	5100	4861
Temperature [°C]	15	16	14
Humidity [%]	70.87	63	69.62
Connectivity in the electrical grid (MW)	4.82	108.73	42.22
Heigh variation (terrain) [MSNM]	1200–4729	1100–4700	594–2390

to select a suitable configuration and the best features that maximize its operation, in addition to its operating costs. To do this, it must be chosen a site with suitable terrain and horizon characteristics which is located north of Ibarra with latitude: 0.4519 and longitude: −78.1198 at an elevation of 2634 m.

2.2 Photovoltaic Power Plants

In a photovoltaic plant there are a large number of panels arranged in such way to offer an adequate performance, in principle, a photovoltaic installation according to [7] is composed of: modules solar photovoltaic, inverters and blocks power, wiring and electrical protections, transformers and substations, structure of bracket, solar tracking system and systems auxiliaries. To consider the use of solar resource the large-scale design of a centralized photovoltaic plant, the model to be analyzed using the PVsyst software.

2.3 Large-Scale Photovoltaic Plant Prospect Interconnected to the Grid

As an initial part of a project, is considered the irradiation characteristics of the site, the geographical location, modules and inverters, as well as their orientation at the site in question, are taken into account all these factors in PVSYST.

Climate and Geographic Resource of the Site. Data from weather in climate resource is used from [10] and [12].

Table 2. Weather data in the ideal place to analyze

Characteristics	Average per year
Global horizontal irradiation [kW/m^2/day]	5.39
Diffuse horizontal irradiation [kW/m^2/day]	2.16
Temperature [°C]	14.9
Wind speed [m/s]	2.7
Relative humidity [%]	76.7

The climate resource of the site is described in Table 2, with an annual average, these values have a variability of 6.8% per year. To define the orientation of the modules the solar path is analyzed at the point described which we will call "Hacienda la Unión".

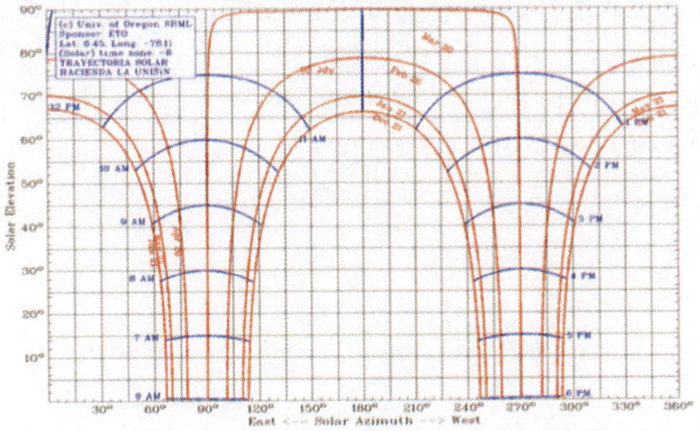

Fig. 4. Solar trajectory at the study site.

System Orientation

The location, orientation and inclination of the panels is vital for the use of the solar resource, Fig. 4 shows the path that the sun has for the winter solstice, its route varies depending on the season of the year. The fixed orientation of the panels for an average annual is estimated with an Azimuth of 0° relative to the south and an inclination of 15° (Fig. 5).

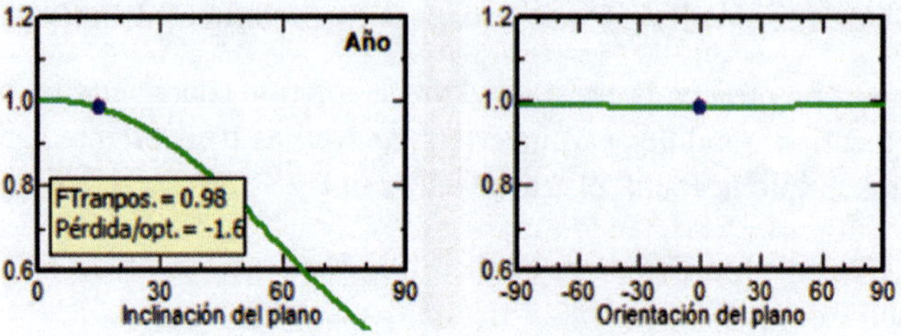

Fig. 5. Orientation optimization with respect to the year.

System Features

The designed system is targeted as a prospect for this site and other locations in the ideal region that can cover the features of land and available area. For optimal plant design, the data in Table 3 is taken (Fig. 6).

Table 3. System characteristics.

FV MODULES:	JKM 350PP-72-DV Jinkosolar
Rated power	350 Wp
Num. of modules	26208
Total power	9173 Kwp
INVERTER:	SINACON PV1045 Siemens
Rated power	1045 Kw ac
Num. of units	8
Total power	8360 Kw ac
Num. of strings in series	26
Num. de strings (parallel)	1008

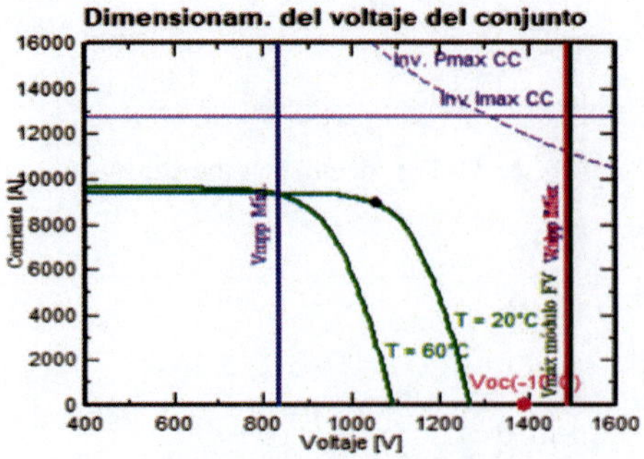

Fig. 6. V-I sizing of the PV set.

The system is divided into 8 subsets of 126 strings, each with 26 modules in series, a design with a nominal power ratio of 1.10 and an overload loss of 0.6%. With the characteristics described in Table 3 the dimensioning is achieved with a nominal power of the FV assembly of 9173 kWp (Fig. 7).

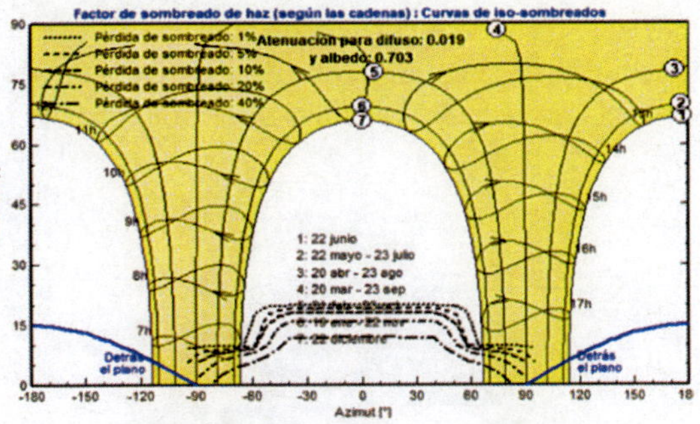

Fig. 7. Shading factor in module strings.

The configuration of the photovoltaic array is arranged with 63 sheds that are transposed in rows towards the Azimuth, each set is composed of 2 rows and 208 columns with the portrait-like orientation, and has a free horizon, the shadows in the set are according to the modules chains and the area they cover is 51164 m^2.

Loss Factors in the PV Array

The PV plant under constant operating conditions is subject to losses caused by environmental, electrical or mechanical factors that will establish a percentage reduction in energy production, the factors to be taken into account are described in Figs. 8 and 15. Account is taken of the loss on the entire system throughout its operation.

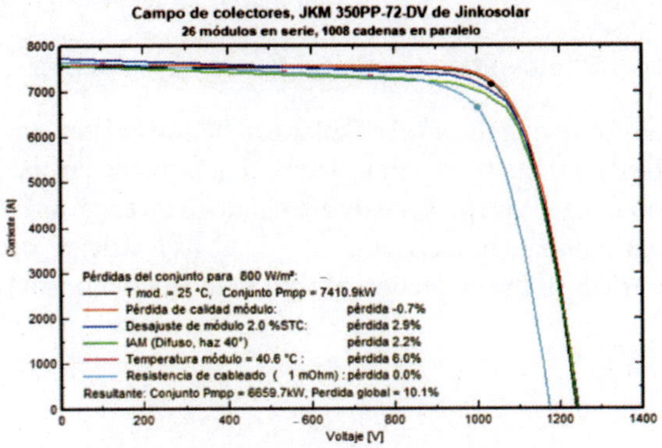

Fig. 8. Loss factors in PV array.

3 Results

3.1 Characteristic Comparison of Suitable Sites

Solar irradiation, terrain analysis and its possible location in the electricity grid [17] as well as climatic factors are taken into account in a photovoltaic plant. The PV system modules ensure a correct operation in STC conditions (temperature of 25° and an irradiation of 1000 Wh/m^2) [8]. The humidity considers a greater structural deterioration of the system, so these data are taken [13] and they are compared in Fig. 10.

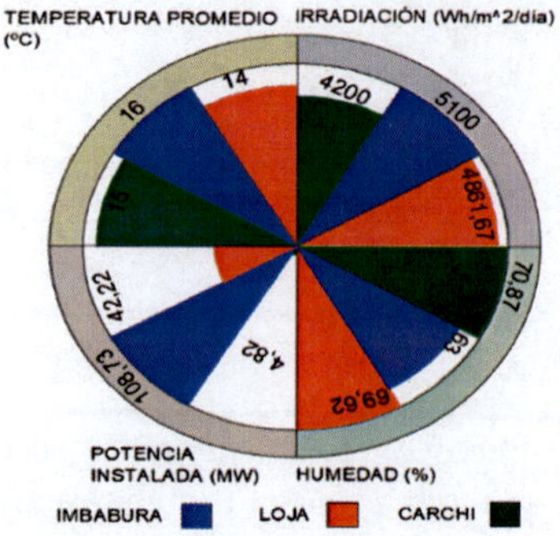

Fig. 9. Comparison of incident characteristics in suitable sites.

For all the exposed, in view of the similar climatic characteristics and the most important details, Fig. 9 is mainly analyzed in relation to the above, the province of Imbabura is considered to have better incident characteristics for the installation of centralized photovoltaic plants, due to its solar irradiation superior to the other provinces, an ambient temperature that approaches the ideal, adequate humidity, its location and expansion in the power grid, in addition to its geographical location.

3.2 Comparison of Photovoltaic Electrical Energy Production

By characteristically knowing the level of solar irradiation existing in each study area, in order to analyze the electricity production, a midpoint is defined in the 3 study provinces, in which the current voltage and power data are obtained by experimentation using Lucas Nulle's photovoltaic equipment that is in the I.S.U.C.T electricity career. The following characteristics are used for the collection of data with irradiation and equal conditions.

Table 4. Experimentation data on photovoltaic equipment.

Data by province	° h	° inc	° Sol	Conditions
Imbabura, Lat: 0.44, Long: −78.75	66°	23.1°	0°	-Polycrystalline panel of 10 w, 26 v y 650 mA -500 w halogen light -Irradiation characteristics of each province
	Azimuth = 183°			
Carchi, Lat: 0.53, Long: −77.82	67°	23.1°	0°	
	Azimuth = 183°			
Loja, Lat: −3.91, Long: −70.45	71°	23.1°	0°	
	Azimuth = 183°			

The main production effects in a cell are related to the irradiance and solar path, to practical purposes the panel emulator is located with the data in Table 4 for each case, the behavior of the panel is analyzed with the analogue digital multimeter supported by the laboratory assistant Labsoft, using a load of 1 kΩ the results are shown in Fig. 10, where Imbabura shows a slightly higher power than Loja and Carchi.

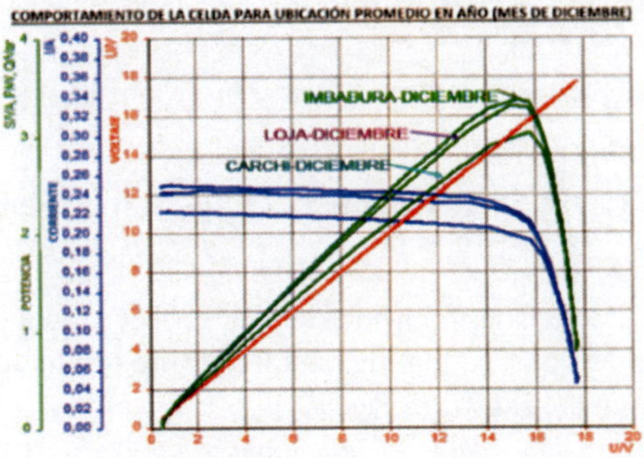

Fig. 10. Potential behavior of test panel for the 3 provinces.

Later, using PVSyst software we compare a photovoltaic energy production for the 3 provinces for a maximum average potential, we observe that the energy curve produced is very similar, for an average of electricity produced per year Imbabura at the given site presents a higher value. Loja has higher losses while Carchi and Imbabura maintain similar parameters (Fig. 11).

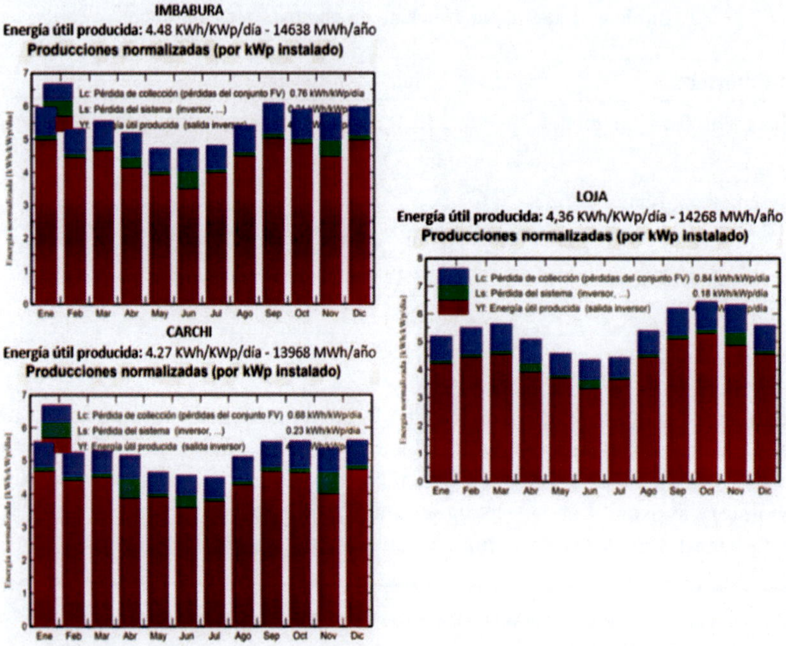

Fig. 11. Comparison of generation curves by year in the study provinces.

3.3 Incidence Analysis of Solar Potential in Photovoltaic Plant in the Province of Imbabura

The incident of the characteristics for a photovoltaic energy production in the experiment is analyzed based on the solar path at the study site, the results for the test panel with the characteristics given are shown in Fig. 12.

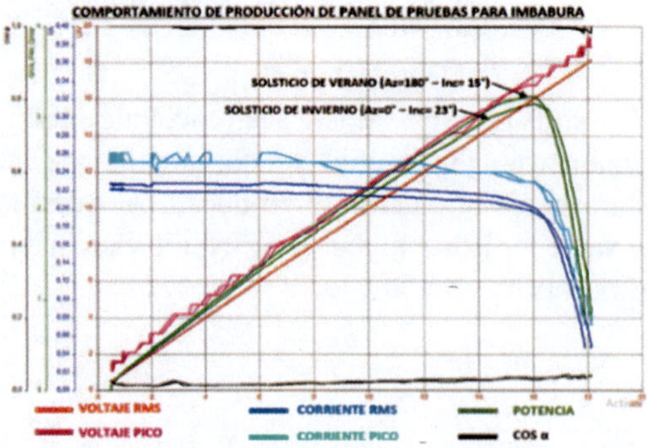

Fig. 12. Behavior of the test cell in production for the 2 seasons of the year in Imbabura.

Using PVsyst the comparative study of the 2 possible cases that can be presented in the photovoltaic installation at the site selected for the most suitable province based on the designs made, the one on the winter solstice with an azimuth at 0° and an inclination of 23° and the second on the summer solstice with an azimuth at 180° and an inclination of 15°, under the same conditions. Table 5 shows the comparison of annual data and its main results for the 2 cases.

Table 5. Comparison of production results.

Characteristics	Winter	Summer
Global horizontal irradiation [kW/m^2]	1970	1970
Diffuse horizontal irradiation [kW/m^2]	812.10	812.10
Room temperature [°C]	15.14	15.14
Global incident irradiation [kW/m^2]	1861.5	1935.7
Effective global [kW/m^2]	1718.1	1809.2
Energy at the output of the set [MWh]	14455	15214
Energy injected into the grid [MWh]	13782	14487
Performance ratio [%]	0.807	0.816

It's observed that a fixed summer configuration delivers to the grid 705 MWh more than a winter configuration, considering that in summer according to the solar path enter the months of April, May, June, July, August and September and for winter the remaining 6 months. Figures 13 and 14 show the energy produced by each installation and the behavior of the system, identifying its losses in the VF set and system.

The nominal power of the system is 9MWp, can be identified in winter production in the months of October to March which is considerably higher than the rest of months, in this configuration for an annual average, a standard output of 4.11 kWh/kWp/day is obtained and, in those months, the system loss are more attenuated.

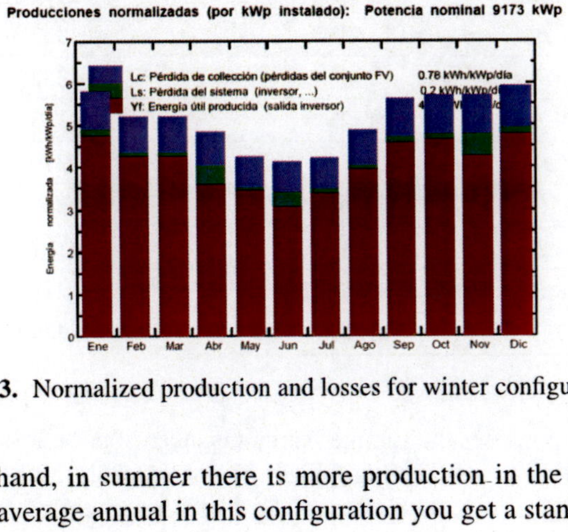

Fig. 13. Normalized production and losses for winter configuration.

On the other hand, in summer there is more production in the months from May to September, on average annual in this configuration you get a standard production of

4.33 kWh/kWp/day and in the same way the loss in these months is a lot softer than the other semester.

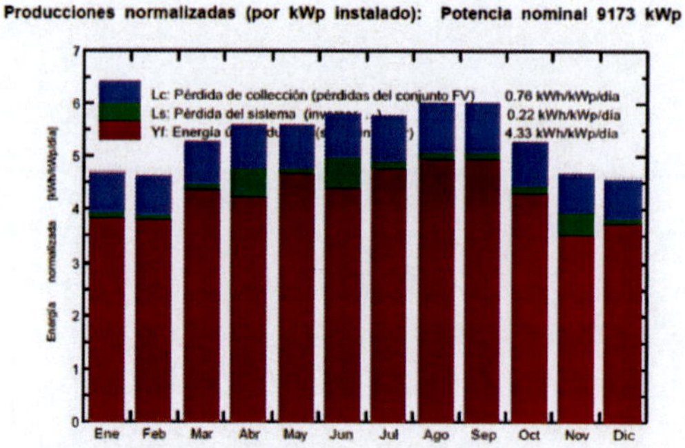

Fig. 14. Normalized production and losses for summer configuration.

To maximize the annual production at the study site, both configurations of both semesters are used to obtain the data in Table 6, where a considerable increase in the energy injected into the grid of 962 MWh is seen.

Table 6. Standardized production with configuration variant for summer and winter solstice.

	Effective Global. [kW/m^2]	Effective energy output. [MWh]	Energy injected into the grid [MWh]
April	156.8	1317	1173
May	163.8	1367	1329
June	162.9	1373	1219
July	167.2	1405	1367
August	174.8	1457	1417
September	168.6	1405	1367
October	163.7	1369	1332
November	159.3	1325	1185
December	170.0	1416	1376
January	167.6	1402	1364
February	134.2	1131	1101
March	148.6	1252	1219
Year	**1937.5**	**16219**	**15449**

For simulation purposes the summer variant is taken and the loss balance is shown. Horizontal irradiation in the receiving plane is considerably high to achieve MPPT, shading factors reduce to 1809 kW/m^2 of final irradiation collected in the plane, the

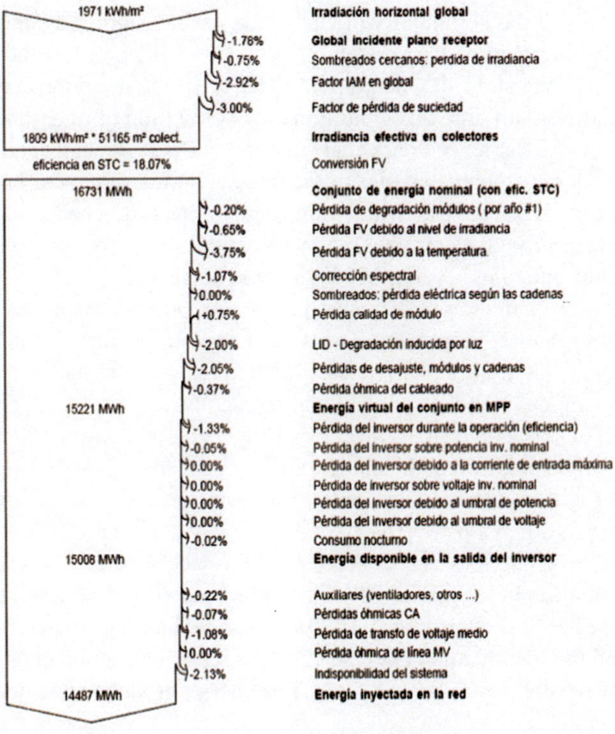

Fig. 15. Loss diagram of the photovoltaic system in summer variant.

panel gives us 18.07% efficiency under STC conditions for photovoltaic conversion which translates into 16731 MWh of rated system power. The loss in the photovoltaic array and D.C connections, the inverters and A.C connections and the loss generated by auxiliary systems and M.T.'s system, reduce the power of grid connection to 14487 MWh, for an optimal use of the solar potential on the site, the ratio of performance of the photovoltaic plant is 81.6% in annual average, the main loss in the system to take into account after the collection reach a value of 12.2% for the first year of operation.

4 Discussion

Results achieved in this work reflect an important and useful source of information for the appropriate use of solar resource in Ecuador, as well as an excellent prospectus in the country for both evaluation in the energy future giving way to a matrix of renewable energies as their expansion in larger and more robust projects, and complementing external evaluation analyses or loss in the following stages which involves such a project.

These results are related to the study supported by Velasco & Cabrera [21], which points out the great potential of distributed photovoltaic solar energy that the country has, focusing on the most suitable province that is Imbabura, where it finds almost constant energy throughout the year, however they are not related to the study it maintains Carvajal & Rodrigo [3], where based on a study focused on marginal areas and eroded soils, have as a result the community of Cangahua in the canton Cayambe as the most ideal for establishing photovoltaic energy projects, also, the present study is related to the results of Peralta et al. [15], who point out and stress the importance of the analysis of climatic variations and time that condition the exploitation of renewable energies.

The revised and experienced data analyzing the climatic, technical and useful electrical energy production characteristics of the three potential provinces indicate that the province of Imbabura has the best irradiation characteristics and behavior, temperature, humidity, expansion in the electrical grid and location, the power of generation in this province is higher due to its level of irradiation and solar path.

The impact of this photovoltaic power production potential is tested in simulation for a large-scale prospective project. The design and study carried out in the PVsyst software indicates that in the study site the global horizontal irradiation reaches a value of 1970 kW/m^2 annually and a final ambient temperature of 15.14 °C, for a given test configuration of 26208 panels of 350 Wp providing a nominal power of 9 MW, different energy values can be obtained at the output of the set depending on the orientation used, for a winter variant the effective irradiation in the plane is 1718.1 kW/m^2, and for a summer variant of 1809.2 kW/m^2, to have a more linear generation curve and maximize the potential of the plant, the 2 variants are used for each time of the year, In this way, an effective irradiation in the plane of 1937.5 kW/m^2 per year is achieved to obtain a net injection power of 15.4 GWh per year. This power could be even greater; however, it has to be considered the loss factors in the collector plane, in the set of panels, inverters and wiring, transformers and auxiliary systems. The energy produced could supply around 5000 families.

It should be considered that the present study does not take into account economic details, to complement it, is recommended to carry out an economic evaluation, as well as studies of interconnection analyzing the flow of load, short circuit and harmonics between the main ones. The study is optimal to take advantage of and encourage the exploitation of the solar resource and the implementation of large-scale centralized photovoltaic projects, generating the disuse of non-renewable sources, thus reducing in the short and long term the burning of fossil fuels and the polluting emissions of CO_2.

5 Conclusions

The incident characteristics in the potential of photovoltaic energy production, irradition and solar trajectory show Imbabura as the most suitable province for the study and implementation of large-scale photovoltaic projects, the study of the real characteristic curve in the 3 provinces demonstrates a very similar behavior, as well as an almost equal power, Imbabura maintains the best parameters, followed by Loja and Carchi.

The 3 provinces of study have optimal characteristics for the development of photovoltaic projects, as a result of the analysis of the useful energy produced it shows a behavior of greater incidence in the province of Imbabura besides an adequate yield, the difference between the 3 provinces is due to the average of solar irradiation and its climatic characteristics incident that reduce loss of collection and in the system delivering a greater specific production (KWh/KWp/day).

The loss balance is a key factor for the exploitation of the solar resource and thus achieve greater electricity production, the results show 3 stages of loss power reduction, in the photovoltaic assembly for the collection of solar energy, shows the importance of adequate orientation, design and efficiency in the panels, then in the system corresponding to an appropriate structure, conditions, dimensioning and connections of panels and inverters, is in this process to obtain a power of injection to network in which there is more percentage of main loss to take into account, and those associated with the coupling to the external network and continuous operation added to the projection of useful life.

In order to obtain the maximum potential of a photovoltaic installation in the country, it is important to study the solar trajectory during the year, from there the use of the solar monitoring system is recommended as the resource can be better used and the production of photovoltaic energy can be improved, minimizing the costs of the system and making great use of the solar resource for 12 months, almost constant production is achieved throughout the year.

The incidence of the solar resource in an electricity production is variable and considers the requirement that the user assigns to the system, the climate resource in the study sites can vary 6.8% each year for the implementation of a photovoltaic plant, the feasibility of the system and its applications as a prospective plant described in the study are a potential reinforcement for a clean energy matrix in Imbabura and in the country in general, as a photovoltaic plant requires little maintenance, its durability and reliability is high, and the solar resource is infinite unlike fossil fuels.

Taking advantage of the solar resource for the production of electric power is the next step to take to have a totally clean electric sustainability, the feasibility of implementing photovoltaic systems grows with the powerful solar resource in many areas of the country and based on accurate studies can increase the cost-benefit ratio to implement new systems with a high percentage of performance, by opting for a global energy matrix change at an enormous pace.

References

1. ARCONEL: Atlas del Sector Eléctrico Ecuatoriano 2018. Quito – Ecuador (2019)
2. Álvarez, O., Montaño, T., Quentin, E., Maldonado, J., Solano, J.: La radiación solar global en la región sur del Ecuador. Reanálisis de la nubosidad diurna (2013)
3. Carvajal, P., Rodrigo, J.: Determinación de zonas óptimas para ubicación de energías alternativas (eólica y fotovoltaica) en las provincias de la region Sierra del Ecuador mediante evaluación multicriterio. (Thesis). ESPE. Quito – Ecuador (2020)
4. Clima promedio en Loja, Ecuador, durante todo el año – Weather Spark, https://es.weatherspark.com/y/19339/Clima-promedio-en-Loja-Ecuador-durante-todo-el-año. Last accessed 29 April 2020
5. Conelec, C.: Atlas Solar del Ecuador con Fines de Generación Eléctrica. Quito - Ecuador (2008)
6. Peláez, M., Espinoza, J.: Energías Renovables en el Ecuador: Situación Actual, Tendencias y Perspectivas. 1st edn. Cuenca – Ecuador (2015)
7. Domínguez, J., Rosendo, J.: Mantenimiento y explotación de una planta fotovoltaica (Thesis). Universidad de Sevilla, Sevilla (2016)
8. Farhat, M., Barambones, Ó., Ramons, J.A., Durán, E., Andújar, J.M.: Diseño e implementación de un Sistema de Control estable basado en Lógica Borrosa para optimizer el rendimiento de un Sistema de Generación Fotovoltaico. Ibero-American J. Autom. Industr. Inf. **12**(4), 176–487 (2015)

9. Herrera, G.: Plan de Desarrollo y Ordenamiento Territorial de la Provincia del Carchi. Carchi (2015)
10. Intro – Meteonorm: https://meteonorm.com/. Last accessed 24 Aug 2020
11. La energía solar se abre terreno en Ecuador: https://www.eluniverso.com/tendencias/2017/06/10/nota/6222868/energia-solar-se-abre-terreno-ecuador. Last accessed 20 March 2020
12. NASA POWER – Prediction of worldwide energy resources, https://power.larc.nasa.gov/. Last accessed 25 Aug 2020
13. National and Local Weather Radar: Daily Forecast Hurricane and information from The Weather Channel and weather.com, https://weather.com/. Last accessed 9 Sep 2020
14. Ortega, S., Jaramillo, J.: Diseño de sistemas para el aprovechamiento de ERNC en la provición de energía a una Vivienda de campo en la provincial de Loja. (Thesis). UTPL, Loja (2014)
15. Peralta, J., Lopez, Á., Barriga, A., Sosa, I., Delgado, E.: Análisis estadístico de la información meteorológica para la explotación de energías renovables en el Ecuador. In: Conference Paper (2), pp. 9–9 (2013)
16. PVsyst (Version 7.0) [Software]: Windows. Wegenéve, Francia (2020)
17. Rodríguez, M., Martín, R., Perles, M.: Las plantas fotovoltaicas en el paisaje. Tipificación de impactos y directrices de integración paisajística. Nimbus **25**(26), 129–154 (2015)
18. Muños, J., Rojas, M., Barreto, C.: Incentivo a la generación distribuida en el Ecuador. Ingenius **19**, 60–68 (2018)
19. Situación geográfica: https://www.otavalo.gob.ec/otavalo/situacion-geografica.html. Last accessed 7 May 2020
20. UO SRML: programa de gráficos solares, http://solardat.uoregon.edu/SunChartProgram.php. Last accessed 05 Aug 2020
21. Velasco, G., Cabrera, E.: Generación solar fotovoltaica dentro del esquema de generación distribuida para la provincial de Imbabura (2009)

Characterization of Biodegradable Nonwoven with Coconut Shell Fibers “Cocos nucifera” and Natural Latex for Application in Floating Root Hydroponic Crops

Willlam Ricardo Esparza Encalada(✉), Pablo Giovany Ayala Pineda, Wilson Adrián Herrera Villarreal, and Luis Adalberto Chamorro Ortega

Universidad Técnica del Norte, Av. 17 de julio 15-21, Ibarra, Ecuador
wresparza@utn.edu.ec

Abstract. The aim of this research was to produce 100% biodegradable non-woven with natural coconut fiber (CF) “Cocos nucifera” employing natural latex (NL) and characterize it to apply as a flotation medium for plant support in floating root hydroponic crops. The procedure employed was by mean of chemical impregnation of natural latex (NL) to form the non-woven, applying two impregnation ratios 1:1 and 1:2, for weights (W) (M1) 636.16 (M2) 825.89 (M3) 1238.84 (M4) 1540.18 g/m^2 (CV = 38.38) obtaining 4 samples dried at 37 °C. Parameters of tensile strength (TS) (CV = 62.23), elongation (E) (CV = 21.01), vertical moisture absorption (MAV) (CV = 0), drying (DT) (CV = 27.42) and evaporated water time (EWT) (CV = 40.94) were tested. The data provided a reliability of 95% ($p > 0.05$) using the Past 4 statistical software. sample 4 (M4) has the maximum tensile strength (TS) 105.52 N, sample one (M1) obtained a longer drying time 262 min at 37 °C and no sample presents relevant vertical absorption (MA). To be applied in Hydroponic crops, sample 3 (M3) is considered the best because it has significant characteristics such as 49.76 N of tensile strength (TS) and 206 min of drying (DT), with total dry weight 1238.84 g/m^2 of non-woven of coconut fiber (CF) required for its application.

Keywords: Nonwoven · Coconut fiber · Natural latex · Hydroponic · Biodegradable

1 Introduction

This study is about the elaboration of the nonwoven using fibers from the coconut shell “Cocos nucifera”. The coconut tree is described as a monoecious palm with a single trunk frequently inclined, 10 to 20 m high and 0.5 m thick at the base and tapering toward the top. Coconut fiber has the potential as a raw material for the development of non-woven, that is, the cross-linking of fibers to form a mattress with characteristics of resistance and porosity to be applied in hydroponic crops. For the formation of non-wovens based on coconut fiber, there are various processes and methods such as plasma

M. Zambrano Vizuete et al. (Eds.): CI3 2021, LNNS 511, pp. 339–350, 2022.
https://doi.org/10.1007/978-3-031-11438-0_27

treatment on the sorption capacity of treated and untreated ones, using coconut fibers known as Nonwoven Coco mats. The fibers were first molded by hand, then washed with boiling water, dried and finally treated with an argon plasma jet, showing that the plasma treatment gives Coco fibers an increase in hydrophilicity and surface roughness, which allows them better sorption properties [11], Furthermore, all these properties are asymmetric by nature and positively skewed, the diameter of a fiber varies along its length with a tapered shape at both ends, the three physical characteristics, length, diameter, linear density, and are correlated with flexural stiffness [15]. Recently, the addition of natural fibers to high-strength concrete (HSC) has been of great interest in the field of building materials. Compared to man-made fibers, natural fibers are cheap and available locally. Among all natural coconut fibers, the fibers have the highest known tenacity [1], The attributes of coconut, improved flexural and compressive strength with an increase in coconut fiber, with the interconnection between the natural fiber composite (NFC) and an epoxy matrix, when treated with alkali, show attributes of traction, bending higher compression, impact resistance and toughness, modifying the fiber surface occurred and improved fiber-adhesion of the matrix [7]. The uses of coconut fibers provide an improvement in the physical and thermal properties of MDP panels, although the increase in the amount of coconut fibers reduces the mechanical properties, using; 11% and 7% urea-formaldehyde adhesive; and a specific pressing cycle of 3.92 MPa at 160 °C with a pressing time of 8 min [10]. The incorporation of coconut coir into the myofibrillar protein (MP) gel can improve its properties and the effects of coconut coir on water, texture, geological behavior, and microstructural properties of the MP gel system, and An increase in gel strength was induced by increasing coconut fiber concentrations [12]. In addition to the acoustic absorption and properties of coconut fiber with and without reinforced polyurethane (PU) holes for acoustic applications with materials using coconut fiber as reinforcement and PU as matrix, the results reveal that it shows a good agreement in the value of the absorption coefficient which is 0.75 between the frequencies 3500 to 4000 Hz [19], another form was developed with a polypropylene (PP) composite with 30% coconut fibers without additives with the help of a rotating twin-screw extruder. in which samples were obtained over molded with PP and PP-coconut fiber composite, to evaluate the adhesive strength of different materials. The results indicated that the adhesive force increased significantly with a greater overlap area. injection temperature and flow were the factors that most contributed to strengthening the bonds of different materials [13]. In addition, derivatives of coconut shell residues were prepared by means of a hydrothermal treatment, by a foaming process that allowed the addition of carbon (up to seven parts per hundred of rubber) to natural rubber with a high content of ammonia-latex. The composite foams had open pore structures with good dispersion of the functionalized carbon [6]. Likewise, polymeric compounds were made with coconut fibers (CF), polyamide 1010 (PA1010) was melt-extruded with 20% by weight of CF, and then it was shaped into pieces by injection molding, increasing the interactions between polymer-fiber improving toughness by adding two chemically modified vegetable oils, linseed oil and given epoxy [14]. Taking advantage of the properties of resistance to sea water, it was applied to measure the degree of carbonation of Coco. Fiber-reinforced mortars that were prepared from Portland cement (PC) and calcium sulfoaluminate cement (CSA cement). Accelerated carbonation was carried out at 4% CO_2

concentration, a temperature of 20 °C and 65% relative humidity (RH) in 12 weeks [2], There are countless applications of coconut fiber, such as in the field for oil spills from the petrochemical industries in marine areas to clean up affectations from oil pollutants that use waste Coco fibers (Cocos nucife-ra L.) pre solvent treated with a "green" protic ionic liquid for oil remediation in saline water [9], and evaluates the characteristics of the surface of the coir fiber and its delimitation with the polyester matrix after being treated with limestone water, showing that the fiber surface tends to be clean, rough and grooved. The highest surface energy was obtained at 40.74 mN/m during immersion in limestone water for 8 h. The highest value of the shear strength of the interface between fiber and the matrix is 3.80 MPa during 8 h of immersion [16]. With all these properties and characteristics of resistance that coconut fiber has to water, oil, polymers applied by different methods, it was applied for the formation of non-woven to be used in the cultivation of soybean plants and can use nitrogen derived from the biological fixation of N2, the use of plants inoculated with Rhizobium can eliminate or diminish the need for fertilization with mineral nitrogen in hydroponic systems, Inoculated plants were produced efficiently by rooting cuttings of seedlings stems in a medium containing a Rhizobium inoculant [5]. Based on the efficiency of the use of water and nutrients and the growth time of the seedlings, produced hydroponically presented a faster growth of both the shoot and root systems. There was a reduction of 12 days to reach the transplant point [17]. In addition, the impact of the application of the compost extract and urea fertilizer forms on the development of kailan plants in the hydroponic wick method can be observed, and it could satisfy the nutrient requirements of the plants [4].

1.1 Methodology

The methods used are based on methodologies according to the standards for the analysis of the samples obtained from the coconut fiber non-woven (CF) and their characteristics with respect to tensile strength (TR), elongation (EG), absorption (AB) and moisture retention (MR), the following is taken as a reference:

•Standard ISO 13934–2 (2014): Tensile strength and elongation.
•Standard AACTT 197 (2003): Vertical moisture absorption.
•Standard AACTT 199 (2013): Drying time of textiles: Moisture analyzer method.

The process was based on the formation of the coconut fiber non-woven, especially focused on the appropriate density parameters to support the plants for hydroponic crops, and the natural polymer (NL) with the latex concentration (NL), which gives structural stability, support and takes advantage of its buoyancy. Using methods in its formation, it was developed with percentages ranging up to 90.45% of natural latex (NL) dissolved with water in relation to the weight of coconut fiber non-woven (CF). Applied on latex crosslinked films and coatings made from 25–90 wt. of % EBM/GBM and 10–75 wt.% of HOSBM. Demonstrating that the characteristics of the resulting biobased latex materials are determined by the nature and proportion of the aliphatic oil residues of HOSBM and the aromatic fragments of EBM/GBM in the macromolecular structure [3].

1.2 Materials and Equipment

The materials used in the formation of the nonwoven with coconut fibers are:

- Natural latex
- Coconut fiber
- Water
- Mold of (28 cm × 32 cm)
- Oven
- Dynamometer
- Balance

1.3 Process

In the selection of the samples, two densities 558 g/m^2 ± 10 and 1116 g/m^2 ± 10 were taken into account; With a dimension of 28 × 32 cm, it corresponds to 50 g ± 1 and 100 g ± 1, respectively, the latex attachment ratios were 1:1 and 1:2.

For the 4 samples, 2 variables were selected, one corresponding to the density and the other to the attached relationship. For this, a mold (oven grill) with dimensions (28 cm × 32 cm) was used for the placement of the fibers and the latex solution for subsequent curing.

- A first test was made with two (a-b) coconut fiber (CF) nonwovens with an approximate weight of 50 g ± 1 each, with a surface area of (28 cm × 32 cm). For these first two rafts, a 1:2 ratio solution (158 g of latex in 316 mL of water) was prepared, a ratio (1:1) was added with respect to the weight of the sample (100%) and procu-trying to completely soak the fiber with this solution.
- The second test was made with two (a-b) nonwovens of coconut fiber (CF), the weight was increased to 50 g each, on a surface of (28 cm × 32 cm) and with the same solution a ratio (1:2); that is (200%) with respect to the weight of the sample.
- The third test was made two (a-b) nonwovens of coconut fiber (CF) with a weight of 100 g each, with impregnation ratio (1:1); that is (100%) with respect to the sample weight. Another 1:2 ratio solution was used (200 g of latex in 400 ml of water). In this way, nonwovens with a lower percentage of impregnated latex were achieved.
- The fourth test was made equally as the previous two samples (a-b) nonwovens of coconut fiber (CF) with a weight of 100 g each and impregnation ratio (1:2), with respect to the weight of the sample (200%). In this case, it was the heaviest with the highest percentage of attached for analysis.

For this non-woven formation process, a method of direct cold application of a water solution with natural latex (NL) was used on the coconut fibers (CF), to then proceed to a curing (drying), as detailed in Table 1.

Table 1. Formulation for the solution to be impregnated in the coconut fiber nonwoven

Substances	Weight (G)	Volume (Ml)	Relationship
Water		400	2
Natural latex	200		1

After dissolving in the ratios of the compounds detailed in Table 1, the solution was applied to the coconut fiber nonwoven (CF). It was realized by means of the steps represented (see Fig. 1).

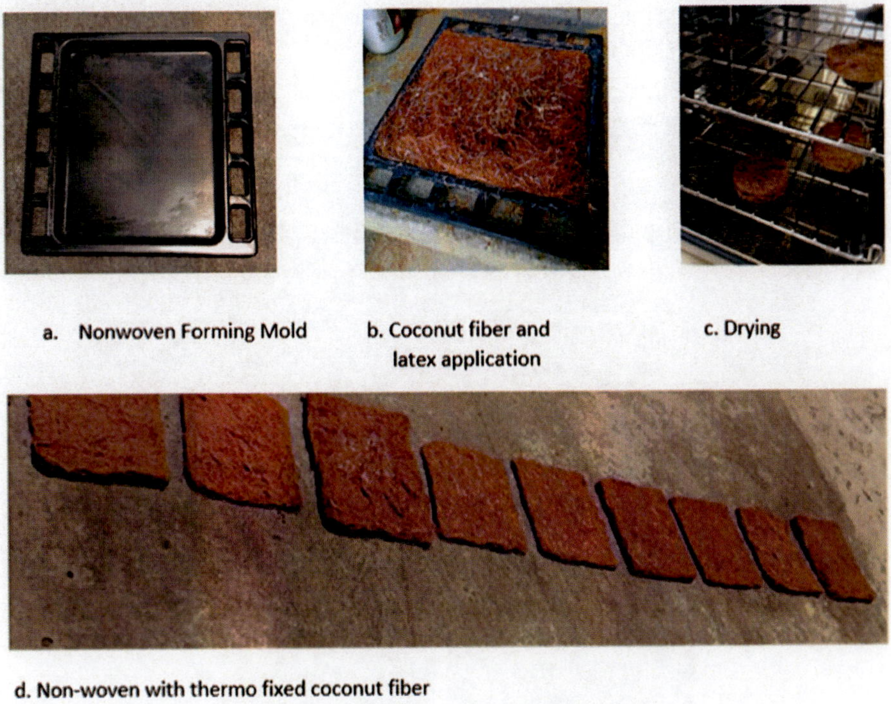

Fig. 1. Steps for shaping non-woven from coconut fiber and natural latex

Next, it is poured into a 100 mL container of water and 50 g of natural latex (NL) is added and mixed until a homogeneous mixture is obtained, and the dissolution ratio is established due to its difficulty in using latex (NL) in its natural state of preservation, due to its short-term handling and its rapid drying in contact with the environment. For this reason, it was necessary to dissolve it in water and in the drying process all the excess moisture evaporated, the structure remaining own latex (NL), taking into account the pickup (PU), the dissolution weight (DW) and the drying time (DT) indicated in Table 2.

Table 2. Parameters applied to non-woven with coconut fiber (CF) and natural latex (NL)

Show (a-b)	Dimensions (cm)	Weight (g)	Pick up (%)	Dissolution weight (g)	Drying time (min)
M1	28 × 32	50	100	50	120
M2	28 × 32	50	200	100	120
M3	28 × 32	100	100	100	150
M4	28 × 32	100	200	200	150

Developed the non-woven samples with coconut fiber (CF) and natural latex (NL) with the parameters indicated in Table 2, the tests were carried out to determine the vertical humidity (VM), tensile strength and elongation (TS), drying time (DT) and evaporated water (EW) indicated (see Fig. 2).

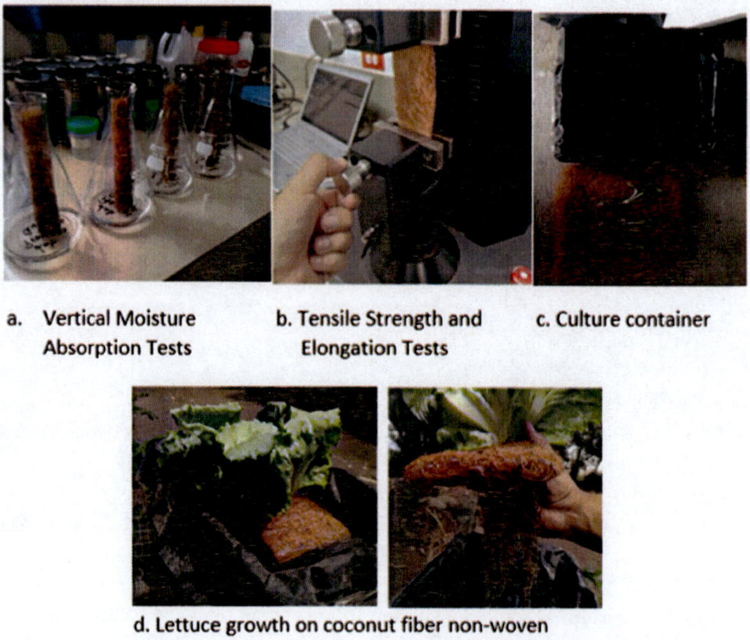

Fig. 2. Laboratory tests of vertical humidity (VM), tensile strength (TS) and elongation (E), drying time (DT) and evaporated water (EW)

2 Results and Discussion

With the data obtained from the tests carried out in the formation of the non-woven applying coconut fiber (CF) and natural latex (NL), they were carried out using the average values with respect to their density, tensile strength, elongation (elongation), moisture absorption, drying time and evaporated water, as indicated in Table 3.

Table 3. Results data of the laboratory tests of the nonwoven with coconut fiber (CF) and natural latex (NL)

Show	weight (g)	Pick up (%)	Dissolution weight (g)	Drying time NT (min)	Weight (gr/m 2)	Tensile strength (N)	Elongation (mm)
M1	50	100	50	120	636,16	17,62	16,65
M2	50	200	100	120	825,89	60,89	10,95
M3	100	100	100	150	1238,84	49,67	17,53
M4	100	200	200	150	1540,18	105,52	13,17

The data obtained were analyzed using the Past 4 statistical program and the data were validated in the parameters of: Weight (W), pick up (PK), dissolution weight (SW), non-woven drying time (NTDT), weight per area (WA), tensile strength (TS), elongation (E), moisture absorption (MA), drying time (DT) and evaporated water (EW), through the analysis of variance and the norm test. quality of the data, finding that the data distributions are within normal with a reliability of 95% ($p > 0.05$) and there are no significant differences between their data, as detailed in Table 4.

Analyzing the variation of the parameters, it is found that the weight (W) in (g/m^2) of the samples used have a high variation, finding that most of the data are in the 75th percentile, and by analyzing the method Box plot, it is observed that the sample (M1) increased the drying time to 262 min applying an impregnation ratio (1: 2) of coconut fiber (CF) and natural latex (NL) (CV = 55.92) (100g and 200g) respectively higher than all the samples analyzed, in addition the samples (M1–M3) present an increase in their elongation (E) of 16.65 and 17.53 mm (CV = 21.01) respectively, this is due to Because, the processability and mechanical properties of natural rubber depend to a great extent on its molecular weight (MW) and molecular weight distribution (MWD) [18] improving the elongation (E) while the other parameters remain stable and the variables analyzed of the samples indicate that the samples do not have significant differences ($p > 0.05$), being located in the third quartile in relation to all the samples analyzed as observed (see Fig. 3).

According to the variation graph it can be observed, as the variations are presented, there are variations in their properties in elongation (E) 16.35 and 13.17 mm (CV = 21.01) and drying time (DT) in the samples (M1–M4) 262 and 162 min (CV = 27.42) respectively and there are no significant differences ($p > 0.05$), this change presented is due to the increase in the weight of the latex solution (NL) applied to the coconut fiber non-woven (CF), that is, as the concentration of latex (NT) increases, the elongation and maximum drying time (DT) increases in 262 min with a pick up (PK) of 100%, exhibiting a better viscosity increasing effect after heat treatment and stronger ability to reduce fluid loss at high temperature [8] obtaining a correlation similar to the sample (M3) with a smaller difference in drying time (DT) 208 min, as shown (see Fig. 4).

Table 4. Normality and reliability of the data

	weight (g)	Weight (g/m2)	Pick up (%)	Dissolution weight (g)	Drying time NT (min)	Tensile strength (N)	Elongation (mm)	Drying time (min)	Evaporated water (min)
N	4	4	4	4	4	4	4	4	4
Shapiro-Wilk W	0,7286	0,9579	0,7286	0,8949	0,7286	0,978	0,9243	0,9919	0,8891
p(normal)	0,02386	0,7658	0,02386	0,4064	0,02386	0,8901	0,5612	0,9668	0,379
Anderson-Darling A	0,576	0,2071	0,576	0,361	0,576	0,2043	0,257	0,165	0,3101
p(normal)	0,04675	0,6607	0,04675	0,2343	0,04675	0,6733	0,4948	0,8375	0,3399
p(Monte Carlo)	0,043	0,8039	0,0448	0,2684	0,0423	0,8095	0,6116	0,9568	0,4291
Lilliefors L	0,3068	0,2176	0,3068	0,3287	0,3068	0,223	0,2509	0,154	0,2769
p(normal)	0,2139	0,7309	0,2139	0,1388	0,2139	0,6969	0,5129	10,06	0,3558
p(Monte Carlo)	0,1843	0,7739	0,187	0,1258	0,1877	0,7368	0,5476	0,9787	0,3623
Jarque-Bera JB	0,6667	0,4173	0,6667	0,4193	0,6667	0,236	0,4724	0,287	0,4883
p(normal)	0,7165	0,8117	0,7165	0,8109	0,7165	0,8887	0,7896	0,8663	0,7834
p(Monte Carlo)	0,1534	0,6435	0,1576	0,652	0,1503	0,9221	0,5087	0,8534	0,452

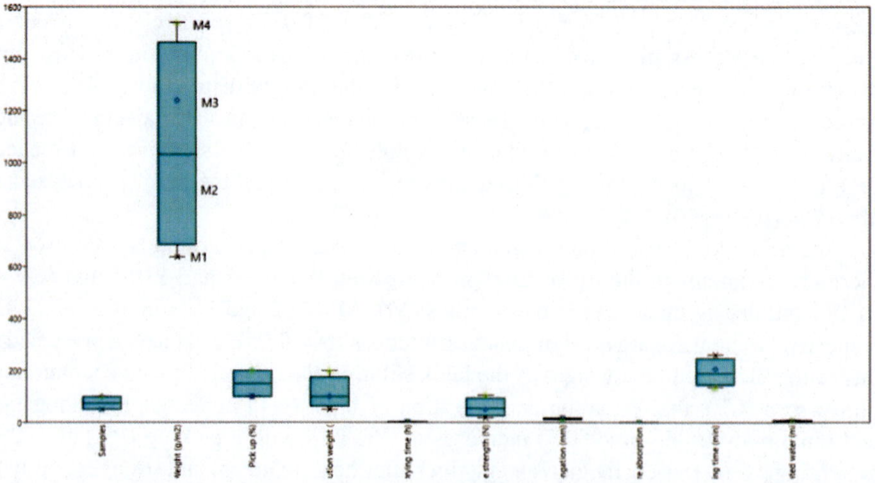

Fig. 3. Analysis of data variation, Bar chat/plot method.

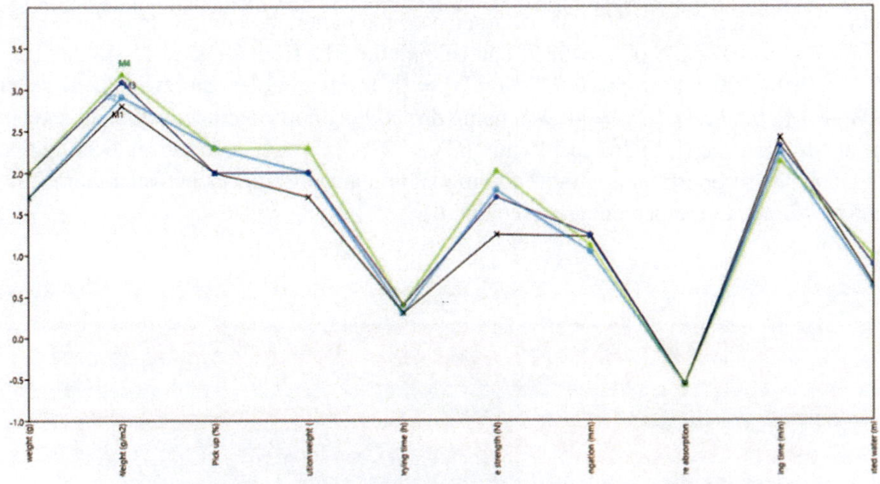

Fig. 4. Variation of the parameters of the samples M1, M2, M3, M4.

While in the clustering analysis it was found that the two samples (M3–M4) are very similar in relation to the rest of the samples, due to their similar concentrations of weight (W) 100 g (CV = 38.49), natural latex (NL) 100 g and 200 g respectively (CV = 55.92), and Pick up 100% and 200% (PK) (CV = 38.49). No significant differences were found between the means ($p > 0.05$), but yes, finding a difference with the sample (M2) and significantly a large difference with the sample (M1), which is the lowest in concentrations of (PK) and drying time (DT) (CV = 12.8) as indicated (see Fig. 5).

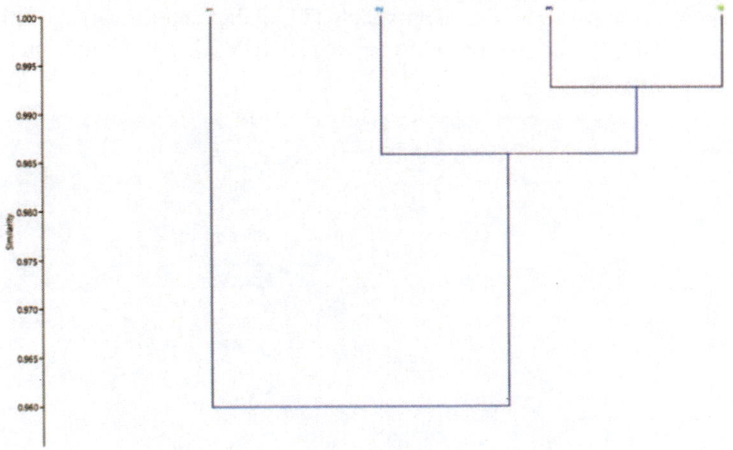

Fig. 5. Similarity between samples M1, M2, M3, M4.

Regarding the evolution of the samples, it can be observed in Figs. 6 and 7 that, the higher the weight (W) g/m^2 (CV = 38.38) of the non-woven coconut fiber (CF) 1540.18 g with natural latex (NL) (CV = 38.38), increases the parameters of tensile strength (TS) from 17.52 to 105.52 N (CV = 62.23) and elongation (E) from 10.95 to 17, 53 mm (CV = 21.01), moisture absorption 0.27% (CV = 0) in all samples and evaporated water (EW) 4.1 to 9.57 min (CV = 40.94), being directly proportional between them, except for the drying time (DT) 262 to 136 min (CV = 27.42) which decreases indicating that it is inversely proportional, without finding significant differences between them ($p > 0.05$) as shown in the percentages (see Fig. 6).

Fig. 6. Percentages of the variables applied to the samples (M1, M2, M3, M4).

Using the Matrix graph, the relationship between the four samples (M1, M2, M3, M4) is described, with a high value of the weight (W) of the sample (M4) in relation to the sample (M1) 1540, 18 and 636.16 g/m^2 respectively (CV = 38.38) with a pick up (PK)

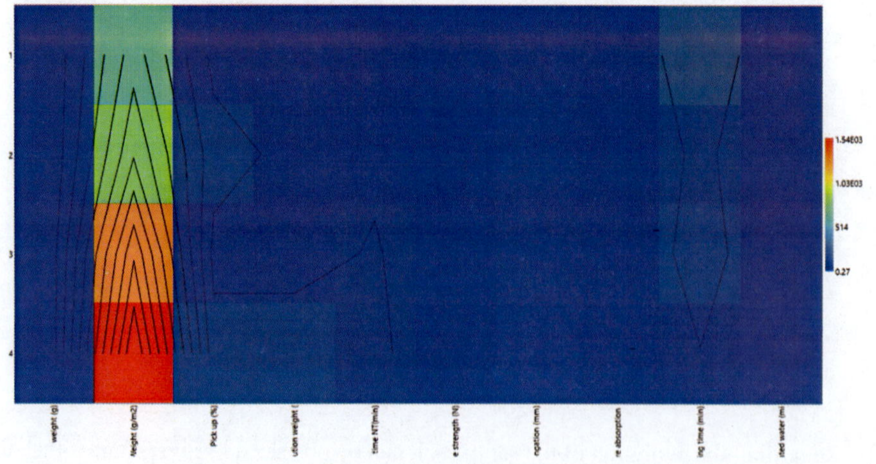

Fig. 7. Matrix graph of similarity of the variables.

between 100 and 200% (CV = 38.49), finding similarity of the variables with deviations within normal in all the applied parameters and there are no significant differences between them ($p > 0.05$) detailed (see Fig. 7).

3 Conclusions

When developing non-woven using coconut fiber (CF) with natural latex (NL) for application in hydroponic crops, results were obtained indicating that by increasing the weight (W) in g/m^2 applying pick up (PK) 200% natural latex (NL) and water ratio (1:2), increased elongation (E), tensile strength (TS), drying time (DT), evaporated water time (EWT) when thermosetting the latex with the non-woven, reaching the conclusion that the sample (M3) contains the best properties and is the best in the parameters analyzed.

References

1. Ahmad, W., et al.: Effect of coconut fiber length and content on properties of high strength concrete. Materials (Basel) **13**(5), 1075 (2020). https://doi.org/10.3390/ma13051075
2. Bui, H., Boutouil, M., Levacher, D., Sebaibi, N.: Evaluation of the influence of accelerated carbonation on the microstructure and mechanical characteristics of coconut fibre-reinforced cementitious matrix. J. Buil. Eng. **39**, 102269 (2021). https://doi.org/10.1016/j.jobe.2021.102269
3. Demchuk, Z., Mora, A.-S., Choudhary, S., Caillol, S., Voronov, A.: Biobased latexes from natural oil derivatives. Ind. Crops Prod. **162**, 113237 (2021). https://doi.org/10.1016/j.indcrop.2021.113237
4. Fitriani, T., Pangaribuan, D.H., Niswati, A., Yusnaini, S.: Improving nitrogen fertilizer efficiency with the addition of compost extracts to kailan (Brassica oleracea L.) plants with wick hydroponic cultivation. SAINS TANAH – J. Soil Sci. Agroclimatology **17**(2), 122 (2020). https://doi.org/10.20961/stjssa.v17i2.41370
5. Hata, N., Futamura, H.: Production of soybean plants for hydroponic cultivation from seedling cuttings in a medium containing r hizobium inoculum depending on various concentrations of nutrient solution and different nitrogen sources. J. Hortic. Res. **28**(2), 71–82 (2021). https://doi.org/10.2478/johr-2020-0015
6. Kettum, W., et al.: Enhanced adsorptive composite foams for copper (II) removal utilising bio-renewable polyisoprene-functionalised carbon derived from coconut shell waste. Sci. Rep. **11**(1), 1459 (2021). https://doi.org/10.1038/s41598-020-80789-x
7. Kumar, S.S., Raja, V.M.: Processing and determination of mechanical properties of Prosopis juliflora bark, banana and coconut fiber reinforced hybrid bio composites for an engineering field. Compos. Sci. Technol. **208**, 108695 (2021). https://doi.org/10.1016/j.compscitech.2021.108695
8. Lei, M., et al.: Synthesis and characterization of high-temperature self-crosslinking polymer latexes and their application in water-based drilling fluid. Powder Technol. **389**, 392–405 (2021). https://doi.org/10.1016/j.powtec.2021.05.045
9. Maia Cardoso, C.K., Mattedi, S., de Carvalho Lima Lobato, A.K., Andrade Moreira, Í.T.: Remediation of petroleum contaminated saline water using value-added adsorbents derived from waste coconut fibres. Chemosphere **279**, 130562 (2021). https://doi.org/10.1016/j.chemosphere.2021.130562

10. Narciso, C.R.P., Reis, A.H.S., Mendes, J.F., Nogueira, N.D., Mendes, R.F.: Potential for the use of coconut husk in the production of medium density particleboard. Waste Biomass Valor. **12**(3), 1647–1658 (2020). https://doi.org/10.1007/s12649-020-01099-x
11. de Oliveira, D.M., Cioffi, M.O.H., de Carvalho Benini, K.C.C., Voorwald, H.J.C.: Effects of plasma treatment on the sorption properties of coconut fibers. Procedia Eng. **200**, 357–364 (2017). https://doi.org/10.1016/j.proeng.2017.07.050
12. Pereira, J., et al.: Insight into the effects of coconut kernel fiber on the functional and microstructural properties of myofibrillar protein gel system. LWT **138**, 110745 (2021). https://doi.org/10.1016/j.lwt.2020.110745
13. Pisanu, L., Santiago, L.C., Barbosa, J.D.V., Beal, V.E., Nascimento, M.L.F.: Effect of the process parameters on the adhesive strength of dissimilar polymers obtained by multicomponent injection molding. Polymers **13**(7), 1039 (2021). https://doi.org/10.3390/polym13071039
14. Quiles-Carrillo, L., et al.: Development of compatibilized polyamide 1010/coconut fibers composites by reactive extrusion with modified linseed oil and multi-functional petroleum derived compatibilizers. Fibers Polym. **22**(3), 728–744 (2021). https://doi.org/10.1007/s12221-021-0024-z
15. Sengupta, S., et al.: Stochastic analysis of major physical properties of coconut fibre. Indian J. Fibre Text. Res. **39**(1), 14–23 (2014)
16. Sutrisno, S., Soenoko, R., Irawan, Y.S., Widodo, T.D.: Effect of coconut fiber treatment with limestone water media on the fiber surface, wettability, and interface shear strength. East.-Eur. J. Enterp. Technol. **1**(6 (109)), 48–56 (2021). https://doi.org/10.15587/1729-4061.2021.217730
17. da Silva Toyosumi, I., da Silva, T.S.M., Melo, D.M., de Azevedo Neto, A.D., Soares, T.M., Filho, M.A.C.: Optimization of banana plantlets acclimatization by hydroponic cultivation. Sci. Agric. **78**, e20200165 (2021). https://doi.org/10.1590/1678-992x-2020-0165
18. Xin, S., et al.: Comparative analysis of latex transcriptomes reveals the potential mechanisms underlying rubber molecular weight variations between the Hevea brasiliensis clones RRIM600 and Reyan7-33-97. BMC Plant Biol. **21**(1), 244 (2021). https://doi.org/10.1186/s12870-021-03022-5
19. Yuhazri, M.Y., et al.: Coconut fiber panel towards acoustical performance. Int. J. Mech. Mechatronics Eng. **19**(2), 11–24 (2019)

Analysis of Quality vs Price for Selection of Alternative Energy Sources

Salguero Alison[1], Villegas Pablo[1], Campaña Cristina[1](✉), Narváez Joyce[1], and Héctor Chancay[2]

[1] Instituto Superior Universitario Central Técnico, Quito, Ecuador
ccampana@istct.edu.ec

[2] Instituto Superior Universitario Bolivariano, Guayaquil, Ecuador

Abstract. This research makes an introduction to photovoltaic energy, wind energy and its components and performs an analysis of the advantages and disadvantages of using an alternative energy source. It also performs an analysis of the parameters in quality and price when choosing an alternative energy source and took into account the analysis of two types of energy: solar and wind. The analysis is carried out between different brands of solar panels and terrestrial wind turbines establishing a technical quality criterion such as energy cost, useful life, guarantee, initial investment, operation and maintenance cost, comparing it with statistical graphs to obtain the best option. When choosing a brand, the user or the client must take into account the quality, the price and the technical characteristics to choose the most suitable panel or wind turbine. For the presentation of the results, radial graphs are used for a better comparison of the options.

Keywords: Self-sustaining energy · Renewable energy · Generation · Efficiency · Cost · Quality

1 Introduction

Renewable energy is known as the inexhaustible natural sources that are capable of generating by themselves, they are friendly to the environment, due to their abundance of energy they can replace current energies by being self-sustaining.

1.1 Photovoltaic Energy and Its Components

The Solar Photovoltaic system has been developed as a technology that generates direct current with the help of semiconductor components, when these are reached by the solar rays, the light activates the solar cell to generate electrical power (power measured in watts or kilowatts); when the sun goes down the electricity disappears. Solar cells do not need to be charged like batteries. Some solar cells have been in operation on the ground or in space for 30 years [2].

M. Zambrano Vizuete et al. (Eds.): CI3 2021, LNNS 511, pp. 351–364, 2022.
https://doi.org/10.1007/978-3-031-11438-0_28

The main elements of the photovoltaic system for isolated houses are:

Photovoltaic Generator. It transforms the energy originating from the sun into a direct current. They are usually installed on roofs, so they are usually fixed support structures.
Investor. It transforms the DC (direct current) into alternating current AC (alternating current) useful for consumption. In some cases its use is not necessary because the equipment used works in direct current.
Batteries. The current is not used at the same time it is generated, so it is feasible to use an accumulation system to store the energy. Batteries are also a provision measure for those days with unfavorable weather conditions.
Charge Regulator. It is an electronic device that controls the input of excess electricity to the battery (overcharging) and also prevents over-discharges.
Loads or Consumptions. They are the elements (lamps, appliances) that, as their name indicates, consume the electrical energy produced.
The installation of these systems must have adequate electrical protection:
Grounding. Protection against direct and indirect contacts, against short circuits, overloads and over voltages [2] (Fig. 1).

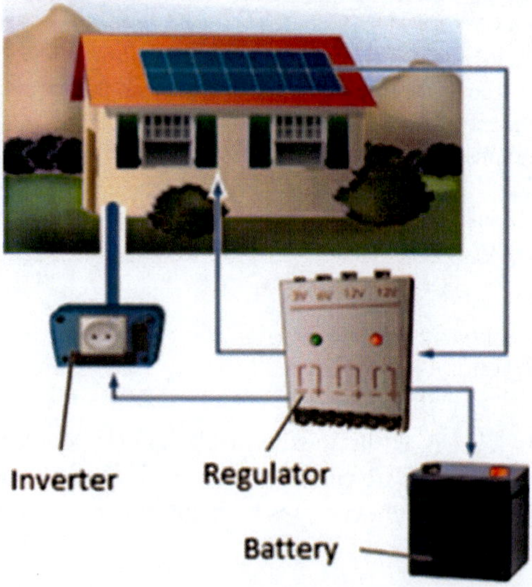

Fig. 1. Components of the photovoltaic solar power plant [1]

1.2 Wind Energy and Its Components

It is the energy obtained from the wind, it is the transformation of kinetic energy to electrical energy generated by the effect of air currents.

The wind turbine is designed at different heights based on the wind regime of the site. Currently, turbines can be found that exceed 100 m in height.

Wind Turbine Components

Base. In the lower part (base or ground), the low voltage boards and the high voltage switching sections are usually placed. Sometimes these electrical cabinets or panels are installed outside the wind turbine in prefabricated facilities. In some machines, the soil of the base or ground can be entered into the pit, it is a place of difficult access where the high voltage conductors that are responsible for transmitting the current to the substation are placed.

Tower or Tube. On the base, a tube or tower is placed, which is normally metallic, it can be made of concrete or lattice. Some turbines have a service lift that is used to transfer personnel to different working heights and stairs for ascending on foot.

Important Components that make up the Nacelle. The multiplier is a gear system that converts energy from low speed to high speed comes from the wind reaching the generator [4]. The generator transforms the kinetic energy of rotation originating from the multiplier into electrical energy. It has a rotor and a stator. The wind turbine rotor is located at the front of the nacelle. The rotor assembly is made up of the blades and the hub [4] (Fig. 2).

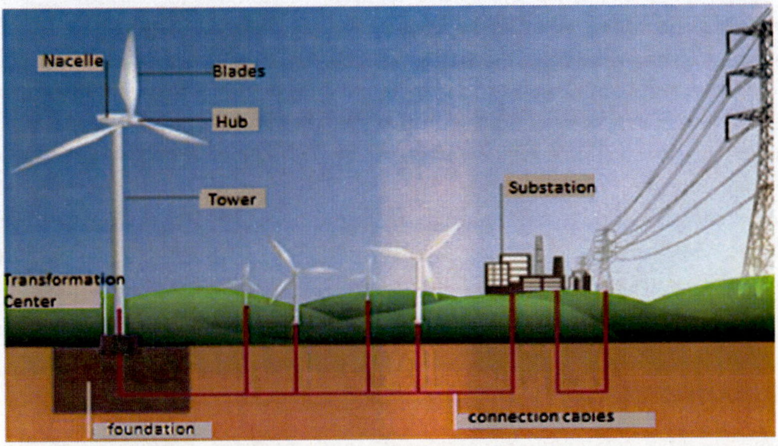

Fig. 2. Components of the onshore wind power plant [3]

1.3 Energy Natural Factors to Take into Account for the Selection of Photovoltaic and Wind Energy Production Sources

There are factors that determine the criteria for the selection of sources:

Global Warming. Volatile gases from products such as paints and other aerosol mechanisms give way to the greenhouse effect, thus causing global warming. CO_2 is a clear indicator of the combustion that is generated by the excessive use of fossil fuels.

Acidification. These two elements (nitrogen and sulfur dioxide) present in combustion cause the well-known acid rain, due to the combination of these elements in the air, it affects ecosystems and the health of the beings that interact in the environment.
Eutrophication. The indiscriminate burning of fossil fuels plays a very important role since they produce nitrogen oxides, which provide an increase in nutrients that adhere to water or soil. This causes an increase in algae in rivers or lakes resulting in a decrease in other species.
Photovoltaic Energy. Consider the geographical location where the system is going to be implanted, for which it is induced to verify the solar atlas of Ecuador. Efficient irradiation on the photovoltaic panel. Verify the space available for the placement of your photovoltaic panels and their components [5].
Wind Power. Consider the geographical location where the system will be implemented, for which it is urged to verify the eolic atlas of Ecuador. Efficient wind force of the wind over the blades. The distribution of the wind where the turbine is installed, therefore, a place with a good potential for wind energy must be chosen, because the wind speed changes second by second [5].

1.4 Advantages and Disadvantages of Using Alternative Energy Sources

Advantage. Solar and wind energy are relatively clean and inexhaustible, avoid the consumption of fossil fuels and contribute to climate change.

In the electricity generation stage, it does not emit polluting gases that deteriorate the atmosphere, problems of pollution due to transport, extraction and fossil combustion are ruled out, which favors the non-pollution of the atmosphere, water and soil.

Solar energy is available almost anywhere on the planet, especially if it is used on a small scale.

These energies can be used in areas far from electrical networks or with difficult access, they can be installed in rural sectors creating jobs, helping the growth and equality of services of each population, since it is not essential to connect a network of an electrical system.

It is own and universal, it exists anywhere in the world.

Disadvantages. Solar and wind energy have a high investment cost.

Wind energy needs ample spaces for its implementation and operation due to its large generating machines, which is detrimental to the cultivation of land to be able to operate in this sector.

Wind farms cause an aesthetic impact and alter the landscape in the surrounding areas.

Birds can crash into the blades, killing them; however, these accidents can be avoided by painting the blades in light colors.

At the end of its life cycle, it can cause an impact on the environment due to photovoltaic waste.

Visual impact on the landscape by solar capture screens.

They have a high initial investment cost, although if they are compared with traditional energies and their costs due to damage to the environment are evaluated, they may be at the same or less cost [6].

2 State of Art

The state of the art is presented below.

Guaman, in his work "I study the arrangement of solar cells in 10w photovoltaic panels to analyze their energy conversion efficiency", prepared by CONELEC says that Ecuador is a country with highly varied topographic characteristics of great climatic diversity and inimitable conditions that It gives it a high potential for renewable and clean energy. He mentions that one of the renewable energies that Ecuador has is solar energy, as a good option for the use of the raw material that is solar radiation.

He focused on the study of cell orientation configurations in photovoltaic panels to analyze their energy conversion efficiency.

By investigating the conversion efficiency of photovoltaic solar panels, Guaman selecting a panel for the respective analysis, calculating the electrical energy produced and evaluating the effects of the different configurations of the photovoltaic panels.

The results obtained through the tests carried out on different days to obtain better efficiency applying different configurations exposed to the analysis (trapezoidal, concave, convex, triangular, zigzag, etc.), obtained that the most efficient of them is the zigzag configuration, since that the solar rays impact most of its projected area, thus obtaining an accumulation of energy in a shorter time compared to the others. The evaluation that the zigzag configuration underwent resulted in an average storage efficiency of 9.06% in a time of two hours between 9:30 and 11:30 with a temperature of 21 °C, with a wind speed of 2.5 m/s and a solar radiation of 458.14 W/m^2, in relation to the other configurations that were subjected to the same situations [7].

Gonzalez, Jurado, and Granados, in their work "Analysis of Efficiency and degradation of a photovoltaic system interconnected to the electricity grid" affirm that most of the installations of photovoltaic systems interconnected to the electricity grid drive to examine the factors that directly alter or indirectly to its efficiency and wear. At present it is accessible and partially easy to acquire a photovoltaic system interconnected to the electricity grid. In some moments without following step by step the applicable regulations on photovoltaic systems or those pertinent to electrical installations.

Therefore, tests and calculation methods are analyzed to demonstrate the quality of a photovoltaic system in relation to its efficiency.

The research examined the factors and their impact when installing an Interconnected Photovoltaic System (SFVI), its possible effects on efficiency, decay or aging, in addition to its possible early deficiency.

They determined that for the design of photovoltaic systems, efficiency must be taken into account, as well as the performance of each of the components that make it up to produce the required level of electrical energy generation. According to the previous table, they determined that the useful life of a SFV (Photovoltaic System) can be reduced from 20 years to only 6 years, accelerating the aging and degradation of the system. Therefore, it is recommended that the installation of a photovoltaic system be installed by trained personnel with experience in electrical installations.

In addition, to ensure the useful life and efficiency of the photovoltaic system it is necessary to have the best quality equipment due to the amount used in this type of installation [8].

Escandon, in his work "Analysis, specification and development of operating procedures for the management of wind energy in Ecuador" investigated the factors to promote the development of wind technology and analyzed technical, financial and economic aspects of wind technology and other renewable energies to participate in the national energy matrix, energy security, environmental protection and economic sustainability are essential.

He performed a sensitivity analysis considering the local environment to identify which parameters are more sensitive from the financial point of view of wind projects. He also establishes minimum conditions for the investment of this technology in Ecuador.

This study carried out an introductory analysis in which the state of renewable energies is described, to analyze the production, regulations and production of energy costs, among other things the review of the components of a wind generator, wind resource, environmental law, international, relative to the connection of wind farms, compared with the requirements for connection to the grid in Ecuador.

Generation will originate from hydroelectric, photovoltaic and wind power to take the first steps to disseminate research nationwide on key renewable technologies.

It establishes the economic factors involved and determines which ones should be prioritized to structure and promote this technology in Ecuador [9].

Alvarez and Targarona, according to their work "Wind generation using different types of generators considering their impact on the power system" refers to renewable alternative energy sources, one of the most used is wind from the point of view of costs and generally your building permits are more easily obtained.

They concluded that, despite its advantages, in the use of wind there are also disadvantages from a technical point of view, the energy efficiency of generators and the electrical impact that wind farms cause on the power system, particularly on the quality of the product.

Given this proposed drawback, the analytical study of the operating schemes of onshore wind turbines is considered of fundamental importance, in order to conclude which are the ones that produce the greatest energy transfer, improving the use of the wind and considering its impact from the electrical point of view in addition to its quality and efficiency when it comes to producing long-term energy [10].

3 Development

3.1 Technical Quality Criteria

Next, we present the description of the quality criteria.

Efficiency. It refers to the relationship between the output energy and the continuous energy at the primary source. It indicates how much energy can be obtained after energy conversion [5].

Usefull Lifetime. Solar installations have a useful life of more than 30 years and close to 40 years in full activity. Onshore wind installations have a useful life of more than 25 years depending on the speed of the wind [11].

3.2 Price Criteria

Initial Investment. Corresponds to the cost of provision, operation; These costs depend on taxes, on the promotion of the system or the importation of the equipment, and in general on the locality in which they are installed. The criterion of evaluating only technologies may be insufficient for decision-making, since there are alternatives such as photovoltaic, solar thermal, and even wind that can become massive and require greater investment, while mini-hydroelectric, tidal, biomass or biogas technologies are conditioned to few plants, normally requiring less investment [5].

Operation and Maintenance. Cost this is the most important criterion for comparing various alternatives in the technological field. It represents the cost of the different equipment, the installation, the construction of communication networks and the engineering services used for the start-up of the project.

The operating prices (personnel, products or services) are taken into consideration throughout their useful life for a better knowledge of the income and expenses that I must generate and especially if the project is profitable.

The monetary value of an energy unit obtained must be considered [5] (Table 1).

Table 1. Energy prices for unconventional plants [12]

Energy prices for non-conventional power plants		
Prices	Aeolian	Photovoltaic
(USD/kWh) Territory Continental	0.0939	0.05204
USD/kWh) Territory Insular de Galápagos	0.01221	0.05724

Energy Cost. It depends a lot on the country where the study is carried out, not only due to the quality of the wind and photovoltaic resources of the geographical areas where the plants are built.

The analysis is required Net Present Value (NPV), Internal Rate of Return (IRR), this analysis can also be carried out through Capital Amortization Tables, which will be used to know the status of the investor's indebtedness.

The kWh cost is between USD 0.08–0.09, according to data from the Quito Electric Company (EEQ), however, the Ecuadorian state through CONECEL regulation 009/06 makes a reference to the costs of energy from renewable sources [13].

Quality and Price Criteria by Type of Alternative Energy Source Solar Photovoltaic and Onshore Wind

Panel GMA SOLAR 60P Series GMA250-60. It presents the new generation of solar modules, providing more energy through the use of high-efficiency solar cells. Its outstanding power performance helps our customers maximize return on investment especially in grid-tie applications.

All solar modules are manufactured in accordance with UL and IEC standards. Each module undergoes a strict quality check through an EL testing process. The EL tester works as an X-ray scanner that immediately exposes all possible hidden defects in the solar panel. In this way, a clear image is obtained on the glass of the proper functioning of the solar panels.

Our solar modules are designed for convenience. Our frame allows installation in multiple configurations, making installation easy, safe and fast even in the harshest climates. Drain holes prevent frame breakage in freezing temperatures.

The cost of the panel including IVA is $ 591.36 [14] (Table 2).

Table 2. GMA SOLAR 60P Series GMA250-60 technical quality criteria

Panel GMA SOLAR 60P series GMA250-60	
1 Efficiency	
Output power	250 W
Tolerance	0–5%
2 Usefull lifetime	
Warranty	12 años
Power output guarantee	25 años
3 Price criteria	
Initial investment (panel)	$4730880
Operation cost and maintenance	$45191
4 Energy cost	
Energy cost	0,05204 $/kWh

Yingli Solar YGE 60 Cell 40 mm SERIES Tipo YL250P-29b/250 W

It is a vertically integrated manufacturer of solar photovoltaic modules.

Ingots, wafers, cells and modules are manufactured in their own facilities. This ensures that material and production quality are closely controlled, offering leading product durability and sustainable performance backed by a 25-year limited power warranty.

Robust, corrosion resistant aluminum frame independently tested to withstand 2.4 kPa wind and 5.4 kPa snow loads ensuring stable mechanical life in the modules. Prepared packaging to protect modules during transport and minimize waste during installation. Modules independently tested to ensure compliance with certificates and regulatory standards. 10-year limited product warranty. The cost of the panel including Iva is $ 269.92 [15] (Table 3).

Aerogenerador 1500 kW/GOLDWING Modelo: GW70/1500

Power: 1500 kW

Rotor diameter: 70.3 m

Class: IEC Ia/IIa

Manufacturer: LM Gasfíter

Table 3. YINGLI SOLAR panel technical quality criteria

Yingli solar YGE 60 cell 40 mm SERIES type YL250P-29b/250W	
1 Efficiency	
Output power	250 W
Tolerance	0–5%
2 Usefull lifetime	
Warranty	12 años
Power output guarantee	25 años
3 Price criteria	
Initial investment (panel)	$2159360
Operation cost and maintenance	$19456
4 Energy cost	
Energy cost	0,05204 $/kWh

This company provides wind farm operation and management services in addition to the entire wind turbine manufacturing and installation process.

Wind turbines are characterized by the electric generator.

All Goldwind generators have a permanent magnet direct drive (PMDD) generator, making Goldwind the world's largest generator of such generators [16].

In Ecuador, hydro energy is the most used, but in Loja and in Galapagos.

Mainly this technology combines the European Onshore and the American AG 4.0 which are located in three wind farms in Loja (Villonaco), Galapagos (San Cristóbal Island (Cerro Tropezón) and Baltra Island).

Its construction was carried out between the highest points of the Villonaco hill, at an approximate height of 2,720 m above sea level. It has 11 wind turbines, each one with a maximum height of 100 m high; and an interpretation center, designed to welcome visitors who wish to learn about the development of wind energy in Ecuador (Table 4).

Aerogenerador AW-1500

Model: ACCIONA Wind power

Potential: 1500 kW

Rotor diameter: 70 m

Class: IEC Ia

Manufacturer: Acciona (Espagne)

The El Cabrito eólico farm is located in a mountain range near Tarifa, in southern Spain.

ACCIONA inaugurated a battery-powered wind energy storage plant in the municipality of Barásoain (Navarra). It is the first plant of its kind integrated into a wind farm connected to the grid in Spain. The application of electrical storage systems with batteries linked to wind farms and solar plants is a field with great growth potential due to the strong development of both renewable energies at a global level and the cheaper battery technology and the improvement of its efficiency.

Table 4. Technical quality criteria of the GOLDWING wind turbine

Wind turbine 1500 kW/GOLDWING model: GW70/1500	
1 Efficiency	
Output power	1500 Kw
Swept area	3.88 m^2
2 Shelf life	
Years warranty	10
Years power output warranty	25 years
3 Price criteria	
Final project investment	$24366874.66
Operation and maintenance cost	2% to 3% of the initial cost
4 Energy cost	
Energy cost	0,0939 $/kWh

Replacement of 90 wind turbines, with a nominal capacity of 330 kW, with 12 more efficient ones, with the aim of optimizing the park's production and reducing operating costs, consisting of eight N100/3000 wind turbines and four AW70/1500 wind turbines [17] (Table 5).

Table 5. ACCIONA wind turbine technical quality criteria

Wind turbine 1500 kW/GOLDWING model: GW70/1500	
1 Efficiency	
Output Power	1500 Kw
Swept area	3.88 m^2
2 Shelf life	
Years warranty	10
Years power output warranty	25 years
3 Price criteria	
Final project investment	$25,183.363.51
Operation and maintenance cost	2% to 3% of the initial cost
4 Energy cost	
Energy cost	0.0733 $/kWh

4 Analysis of Results

4.1 Solar Energy

By analyzing the quality, efficiency and guarantee it will be shown that it is better to obtain a result greater than 8 out of 10. To see the results in a better way, a radial graph was made between the panels studied previously, resulting in the following. Output energy for Eco Green is 10 because the output energy is higher than Yingli Solar. Tolerance in this case Yingli Solar is 10 better than Eco Green due to its wide tolerance of the panel. The two panels are similar in that they give a 10-year warranty against manufacturing defects. Output power guarantee according to the graph we can say that they are the same because the two fully guarantee the power for correct operation. Initial investment in this case we have the Eco Green Panel that is 9 therefore this panel is better due to its lower investment while the Yingli Solar needs more investment. Cost of operation and maintenance the Eco Green panel needs lower investment costs for its operation and maintenance compared to the Yingli Solar panel. Energy cost for the two panel cases is given a value of $ 0.4003 kWh (Fig. 3).

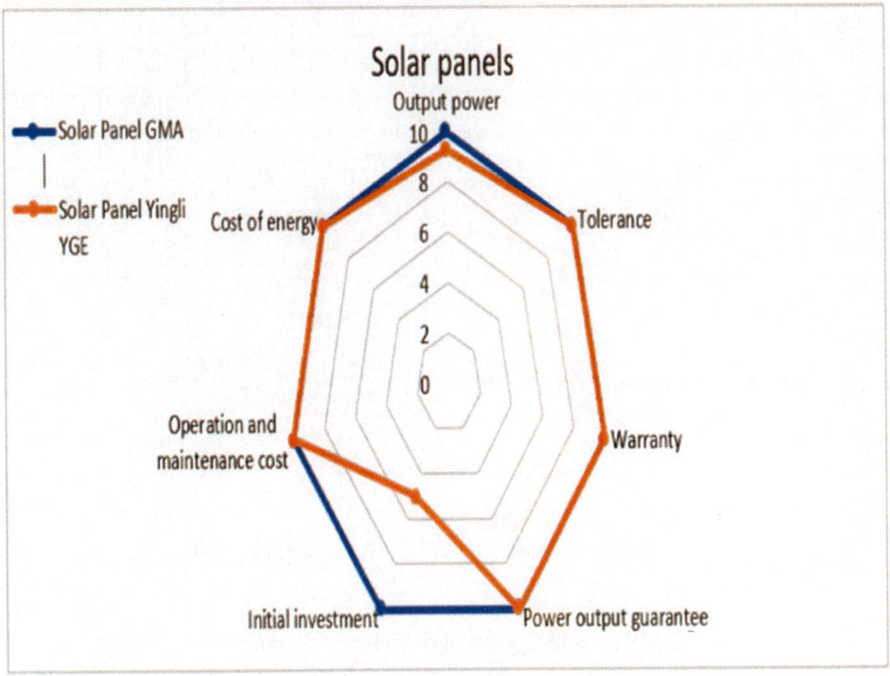

Fig. 3. Radial graph of the comparison of solar panels

4.2 Wind Power

By analyzing the quality, efficiency and guarantee it will be shown that it is better to obtain a result greater than 8 out of 10. A radial graph was made where it can be seen that the Goldwing wind turbine is the same as the Goldwing wind turbine, when observing the output energy they have the best quality in kW/h (kilowatt/hour). It can be seen that the swept area is 10, for the Goldwing ECU wind turbine, unlike the Goldwing SPA, which is 6 due to its difference in square meters. When observing the guarantee and the power output guarantee, it was possible to examine that they have a quality of 10/10 because it has a guarantee of 10 to 25 years. When analyzing prices, costs and investment, you should choose a wind turbine that when comparing is less than 8 out of 10. It can be seen that the Goldwing ECUA wind turbine has a higher initial investment cost than the Goldwing SPA wind turbine. It can be seen that the operation and maintenance cost are the same because in the Goldwing ECU and Golding SPA wind turbine, the cost is 2 to 3% of the total investment. The cost of Energy is lower in the Goldwing ECU wind turbine, it can be seen that the price of energy is higher in the Goldwing SPA wind turbine due to the difference in states, as well as that in Spain the implementation of wind energy has been since 1982 (Fig. 4).

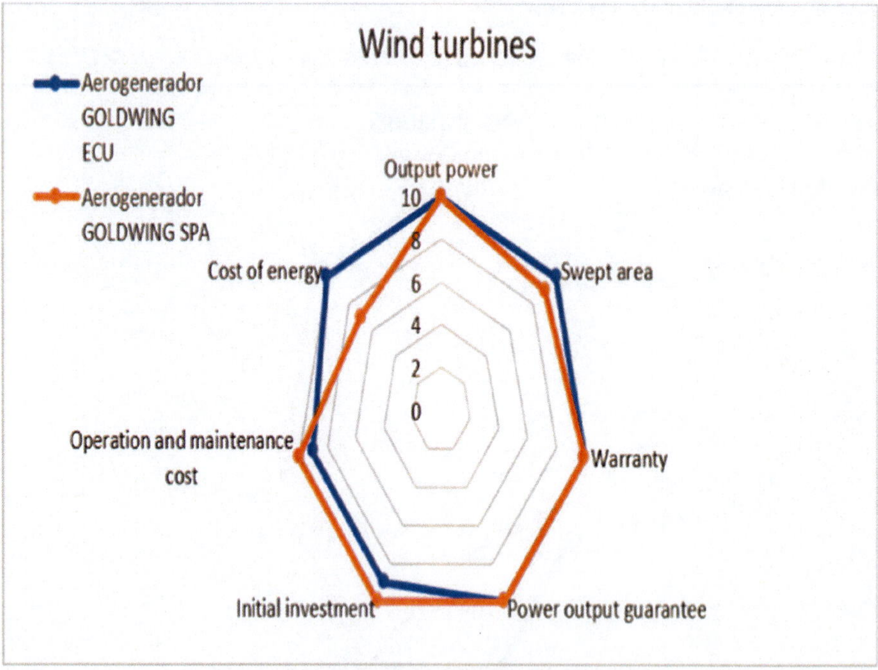

Fig. 4. Radial graph of onshore wind turbines

5 Discussions

The most common panels on the market are those of pure monocrystalline silicon, pure polycrystalline silicon and those of amorphous silicon, which are easy to acquire, therefore for our research we use those of pure monocrystalline silicon [7].

In this research it was found that panels made of monocrystalline pure silicon tend to absorb radiation better and withstand less overheating. The useful life of an interconnected photovoltaic system can be limited from 20 years to only 6 years, accelerating the aging and degeneration of the photovoltaic system.

It is highly advisable that the installation of a photovoltaic system be carried out, installed by trained personnel with experience in installations already mentioned above [8].

In this investigation, it was possible to deduce that to avoid the degradation of the photovoltaic system, the technical data of the manufacturer must be taken into account and both operating personnel and our technical personnel be duly trained for said installation to avoid future damage to our system.

Escandón, investigated the technical requirements necessary for wind generators to operate in the Ecuadorian network, based on existing local procedures, also delves into the financial analysis of a wind installation in the Ecuadorian environment, determining what type of factors are most important for public or private developers invest in this technology [9].

It coincides with the investigation because to acquire an onshore wind generator, the technical and economic characteristics must be taken into account in addition to the place that is suitable for the onshore wind system to operate without problems in the future.

In Ecuador, the wind potential is not known in detail, it is expected that, with the development of the wind atlas, promotion will increase, as well as local and international interest for its promotion [10].

This research is essential for the renewable alternative sources previously studied such as terrestrial wind energy, photovoltaic solar power to predominate in the place they want to implement due to their quality and price parameters so that interest increases in future projects.

6 Conclusions and Recommendations

In this research, the quality and price criteria that must be qualified before purchasing and installing solar panels and onshore wind generators were determined. For solar panels and wind generators, quality criteria were considered, such as their efficiency in addition to their useful life, prices and their energy cost in the market, which will allow choosing the most appropriate option.

This research was of great help because of the great field that includes renewable energies as the sources of energy investigated, because when acquiring a panel or a wind turbine we have to take into account the price without neglecting the quality that will determine the years that our system will be up and running and encourage the country to implement renewable sources as they are self-sustaining energies.

Before implementing a photovoltaic system or an onshore wind farm, the price should be analyzed taking into account its efficiency, quality and guarantee so that it has a long-term useful life and effectively.

According to the research carried out, it is concluded that the Yingli Solar panel is the most efficient thanks to the fact that it has a greater tolerance in output energy than the Eco Green Energy panel.

According to the research carried out, it is concluded that the Made generator is better in terms of quality, efficiency and useful life, unlike the Goldwing wind turbine, which has a lower cost but its efficiency is very low.

This research was developed to learn more about renewable energies where different brands were compared to find out the characteristics where a direct relationship was observed in both prices and quality where technology advances over time, renewing their systems.

It is recommended for a future investigation that in addition to investigating the quality and price. It is recommended for a future investigation that in addition to investigating the quality and price parameters, the geography of the place is investigated, to know what type of renewable energy is best. Since in a place with little solar radiation it would not be suitable to install photovoltaic panels, or a place with low wind speed would not be suitable to install onshore wind turbines.

References

1. Díaz, T., Carmona, G.: Instalaciones Solares Fotovoltaicas. McGrawHill, España (2010)
2. Colegio Oficial de Ingenieros de Telecomunicación: Energía Solar Fotovoltaica 2E5C69A6. España (2002)
3. Ruiz, J.: Energía Eólica, pp. 1–18. Sevilla (2006)
4. Enair E30 Wind Turbine. www.enair.es. Accessed 01 Apr 2021
5. Barragán, E., Zalamea, E., Vanegas, P.: Factores que influyen en la selección de energías renovables en la ciudad. EURE **45**(134), 1–19 (2019)
6. Garzón, C.: Evaluación de alternativas de generación de electricidad desde el punto de vista de su impacto ambiental, para sectores no conectados a redes elétricas. Universidad Técnica Cotopaxi, Ecuador (2010)
7. Guamán, J.: Estudio de la disposición de celdas solares en paneles fotovoltaicos de 10W para analizar su eficiencia de conversión de energía. Repo. Uta. Edu. Ec. **593**(3), 130 (2014)
8. González, P., Jurado, F., Granados, D.: Análisis de eficiencia y degradación de un sistema fotovoltaico interconectado a la red eléctrica. Revista Iberoamericana de Ciencias, pp. 1–22, United States of America, Texas (2018)
9. Escandón, A.: Análisis, especificación y desarrollo de procedimientos de operación para la gestión de la energía eólica en el Ecuador. Universidad de Cuenca **1**, 222 (2012). Ecuador
10. Álvarez, J., Targarona, J.: Generación eólica empleando distintos tipos de generadores considerando su impacto en el sistema de potencia. DYNA **78**(169), 95–104 (2011). Colombia
11. Añazco, A.: Guía para la preparación de anteproyectos de energía solar fotovoltaica. EPN, p. 153. Ecuador (2009)
12. CONELEC: Regulación Conelec 006-09. Ecuador (2009)
13. Ricaurte, P.: Análisis de las energías renovables no convencionales (ERNC), situación actual y costos de producción con énfasis en los tipos de generación más comunes: Energía Solar Fotovoltaica, Energía Hidroeléctrica, Energía Eólica y Biomasa. Universidad: San Francisco de Quito **1**, 21–79 (2015). Ecuador
14. GMA Company: Panel GMA 60-P Series 225-260W. Canada (2016)
15. Yingli: Yingli Solar. China (2014)
16. Goldwind: The Wind Power, GW70/1500. China (2020)
17. David, A., Barrera, S., Sevilla, D.: Análisis Coste Beneficio Aplicado a un proyecto de un parque eólico. Departamento de organización industrial y gestión de empresas: Bibing, pp. 1–161. España (2012)

Comparative Analysis of Plasma Cutting Parameters Between K100 and W300 Steels

F. Gabriel Auz C[1](✉), Diego W. Chillagano[1], G. Alexander Paucar[1], and Rogelio Chou Rodriguez[2]

[1] Instituto Superior Universitario Central Técnico, Quito, Ecuador
gabriel_auz@hotmail.com
[2] Instituto Superior Universitario Bolivariano, Guayaquil, Ecuador

Abstract. This paper analyzes the new technologies that are used today for the metal-working industry, which are based on the comparative analysis of Plasma arc cutting (PAC) plasma cutting processes between K100 (a Galvanized Steel edge rail) and ASTM A36 steels. Initially, an analysis of the properties that are presented by each material (such as physical, chemical, mechanical) was carried out, with this it was decided to also compare the PAC plasma arc cutting process and the OAG oxy-acetylene cutting process. In order to carry out the analysis, in the testing, process 2 specimens of 8 × 100 × 300 mm were manufactured, of each material, which were subjected to longitudinal cuts, in order to obtain data about the best cut with each technology, for each material. Verified and compared the variables operating time, type of slag, finish of the cut, it was confirmed that the plasma-cutting machine provides better results for use in the industry.

Keywords: K100 steel · ASTM A36 steel · PAC · OAG

1 Introduction

The research is based on the study and comparison of plasma cutting parameters between K100 and ASTM A36 steels, to reduce costs in the use of machines and determine the appropriate material to be used.

K100 steel, according to "AISI-SAE D3", is from the D family, group 3 with a high content of carbon and chromium, it is used in cold work processes. While ASTM A36 steel is an alloy with a yield point of 250 MPa, used in hot work, it offers high impact resistance, with excellent hot tensile properties, and in the air, it offers exceptional hardening, admits water-cooling.

The interest of the research is to achieve the optimal cutting parameters for each material and the appropriate cutting process, be it (PAC) "Plasma arc cutting" or (OAW) "Oxy-acetylene cutting". Using the Hyperterm Powermax 30 XP machine, the best configuration is determined at the time of cutting on the machine in the two selected materials, at the same time a comparison is made to determine which process offers the best result.

M. Zambrano Vizuete et al. (Eds.): CI3 2021, LNNS 511, pp. 365–376, 2022.
https://doi.org/10.1007/978-3-031-11438-0_29

2 Methodology

To carry out this investigation, the physical, chemical and mechanical characteristics of each of the materials are identified, and then proceed to the manufacture of specimens, verifying which are the optimal cutting parameters in each material in the face of the factors that affect it, such as hardness, tensile strength, elongation and maximum strength. The standardized specimen cutting processes are carried out in each of the machines, with the results obtained it is determined which process is suitable for each material, proceeding to carry out cost analyzes.

2.1 Selection of Materials

The selection of materials is made according to the physical and chemical properties and their use in the metalworking industry. The selected materials are K100 steel and ASTM A36 steel, whose properties are indicated in Tables 1 and 2 respectively.

Table 1. General properties of K100 steel (Source: [1])

Chemical composition			
C	Si	Mn	Cr
2.00	0.25	0.35	11.50
Physical properties			
Thermal expansion		10E-6/K	
Thermal conductivity		25 W/mK	
Melting temperature		[1450–1510] C	
Density		7700 kg/m^3	
Resistivity		0.55 Ω mm^2/m	
Mechanical properties			
Modulus of elasticity		[190–210] GPa	
Poisson's modulus		[0.27–0.3]	
Tensile strength		[650–880] Mpa	
Elongation		[8–25]%	
Fatigue resistance		275 Mpa	
Elastic limit		[350–650] MPa	

Table 2. General properties of ASTM A-3 steel. (Source: [2])

Chemical composition				
C	Si	Mn	Cr	P
0.26	0.40	0.80–1.20	0.05	0.04
Physical properties				
Melting temperature			1020 C	
Density			7860 kg/m^3	
Mechanical properties				
Modulus of elasticity			[190–210] GPa	
Poisson's modulus			0.3	
Tensile strength			400 MPa	
Elongation			20%	
Fatigue resistance			275 MPa	
Elastic limit			[220–250] MPa	

2.2 Machine Selection

The machine selected to perform the shear test is the PLASMA Hypertherm Powermax 30 XP model (Fig. 1). This plasma cutter features a two-in-one design with great cutting power for thick metal (up to 16 mm with 125 mm/min cutting speed), 50% faster achieving higher cutting speeds and twice the life of consumables with 70% more efficient.

The Air T30 hand torch, designed to withstand demanding conditions, comes installed in the Powermax30 plasma system, with a set of 4.5 m cables and hoses and a work clamp [3, 4]. Table 3 lists the machine specifications.

Fig. 1. PLASMA Hypertherm Powermax 30 XP (Source: [3]).

2.3 Plasma Cutting Procedure (PAC)

Plasma cutting is characterized by being a process that uses high temperatures, in a high-speed gas jet called plasma, which conducts electricity from the plasma torch to

Table 3. Hypertherm Powermax 30 Specifications. (Source: [3])

Input voltage	120–240 V, 1 F, 50/60 Hz
Input current at 3.8 kW	120–240 V, 1 F, 25.5–18,8 A
Input current at 3.8 kW	15–30 A
Nominal output voltage	125 VCD
Duty cycle at 40 ºC	35%–240 V, 20%–120 V
Power source type	Inverter-IGBT
Engine requirement	5.5 kW, 30 A output

the work material, where the high plasma flow quickly blows into the molten metal, breaking the material [4]. The plasma cutting process usually reaches temperatures of more than 20000 °C, which enables it to cut any type of conductive metal, optimal finishes, iron and steel, shipyards and heavy machinery, leaving high quality jobs [5].

Its advantages are a fast-cutting speed and high efficiency, high cutting precision, quality and finishes, higher productivity (depending on the material), cheaper in thicknesses less than 25 mm. Its disadvantages are permanent deformations in materials with a thickness of less than 5 mm, a high cost when cutting thicknesses of more than 30 mm and the production of polluting gases.

2.4 Oxygen Gas Cutting Procedure

With flame cutting, metals can be sectioned by their local and continuous combustion in the presence of a jet of oxygen. Under normal conditions, at room temperature, steel undergoes an oxidation process, which is slow and non-combustible, due to the proportion of oxygen in the atmosphere, approximately 21%, however, under an atmosphere of oxygen (greater than 88%) and at a temperature that reaches the combustion temperature of steel (870 ºC approx.), it becomes fuel.

For there to be oxy-cutting, the metal must be heated (oxidized) under a suitable atmosphere (projection of pure oxygen), thereby achieving its violent burning, giving rise to oxy-cutting, the fuel is Fe, the oxidizing agent is O_2 and the initiating agent is flame from the torch. Good application is achieved in carbon steels and low alloy steels [6, 7].

2.5 Preparation of Specimens

To perform the shear test with the selected materials (K100 and ASTM A36), it is necessary to have the surface to perform the test, thus determining the dimensional characteristics and the number of specimens to be used, in this case four, two for each material.

Dimensions of the Specimens. The dimensions of the specimens to perform the shear test by the PAC and oxy-cutting processes are 8 × 100 × 300 mm, as indicated in Fig. 2.

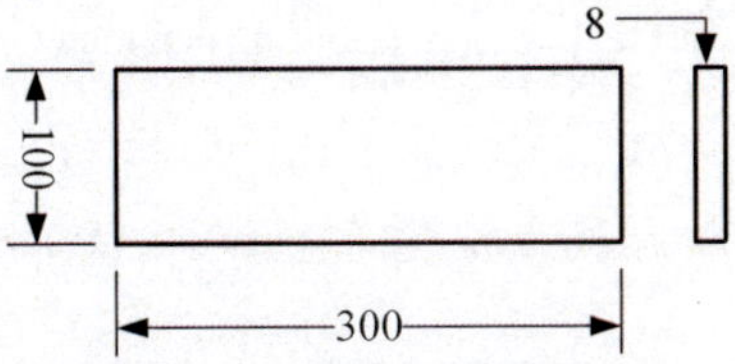

Fig. 2. Specimen dimensions.

Marking of the Specimens. The K100 and ASTM A36 specimens are marked (Fig. 3), according to the process, so that the type of test to be carried out can be distinguished. Taking into account the 4 test tubes, the distinction of A-001 and A-002 was made for the ASTM A36 steel, as in the material K100, the distinction of K-001 and K-002 was made, the test tubes 001 will be made the PAC process while at 002 the OAW process.

Probeta acero K100 Probeta acero ASTM A36

Fig. 3. Specimen marking.

The cuts are made symmetrically, making two per specimen, identified as K-001-1 and A-001-1 (Fig. 4).

K-001-1 A-001-1

Fig. 4. Specimen sectioning.

2.6 Machine Cutting Parameters

The appropriate parameters are established for each machine to perform the corresponding cut of the first 2 plates of the K100 and ASTM A36 steels and the 2 second plates of the steels, as indicated in Table 4.

Table 4. Machine cutting parameter

	2 first plates		2 second plates	
Cut parameter	Plasma	OAW	Plasma	OAW
Air pressure (psi)	100	–	110	–
Config. machine	Thick thickness	–	Thick thickness	–
Voltage	230–400	–	230–400	–
Nozzle diameter (mm)	–	0.8	–	1
Torch O_2 pressure (psi)	–	25–35	–	25–35
C_2H_2 pressure (psi)	–	6	–	6

2.7 PAC Cutting Test

In the PAC cutting process (Fig. 5), the selected cutting parameter was 100 psi air pressure for the two specimens (K-001 and A-001). Which, after changing the cutting parameters in the PAC process, were cut into 100 × 100 mm sections (K-001-1, A-001-1) and into 50 × 100 mm sections, with an air pressure of 110 psi for other results, basis of comparison.

Fig. 5. PAC process at 110 psi.

2.8 OAW Cutting Test

For this test, 100 mm cuts were made vertically using the OAW process to obtain three equal parts of 100 × 100 mm, Fig. 6.

Fig. 6. OAW cut test.

The parameters used to perform the cut in the two materials were, the pressure of the oxygen tank 30 psi and 10 psi for the liquefied gas tank [8] (Fig. 7).

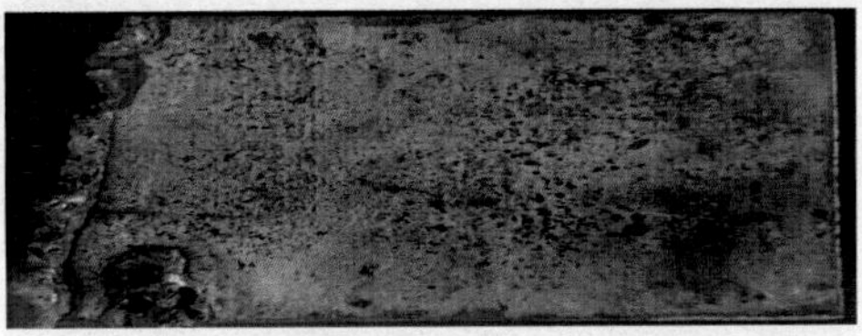

Fig. 7. OAW cut.

3 Results

Once the cuts were finished, a comparative analysis of the cutting parameters was carried out for K100 and ASTM A36 steels, as well as a comparison between the PAC and OAW process, to determine the best cutting process. The results will show the differences when working with different cutting parameters in the PAC process, by varying the air pressure, with a pressure of 100 psi in the K-001 and A-001 specimens, while in the K-001 specimens -1 and A-001-1 a pressure of 110 psi was used.

3.1 Comparison of K100 and ASTM A36 Steel Properties

Based on the data of the properties, indicated in Tables 1 and 2, we have:

Tensile Strength. K100 steel has higher tensile strength when stretched than ASTM A36 so it will resist more traction before breaking.

Elastic Limit. The elastic limit of K100 steel is higher than that of ASTM A36, which is why it will withstand more stress without suffering permanent deformation.

Density. In this case, the density of ASTM A36 steel is higher, approaching the average density of steel, which is 7850 kg/m^3, while that of K100 steel remains below average, making it a little lighter [9].

Carbon Percentage. K100 steel has a higher percentage of carbon making its tensile strength greater, steels that do not exceed 0.25% carbon can be cut without reserve, while at a higher carbon content the steel develops a high resistance to the action of oxygen making the cutting process difficult. In this case, ASTM A36 steel will be easy to cut, while K100 steel will be more complicated.

Table 5. Operation time for cutting K100 steel at 100 psi and 110 psi in 200 mm

Specimen	Operating time (min) 100 psi	Operating time (min) 110 psi
K-001	1.2	–
K-001-1	–	0.85
Total	1.2	0.85

3.2 K100 Steel

First, the cutting time will be analyzed, which involves working at 100 psi and 110 psi in a 200 mm cut, as shown in Table 5.

Cutting time at 110 psi is shorter due to more air pressure, allowing you to cut faster.

The second point to analyze is the type of slag that is produced at different air pressure. As seen in Fig. 8, the increase in air pressure from 100 psi to 110 psi has a slight influence on the increase in slag, determining that at 100 psi the slag is less, therefore, it is advisable to work at that pressure.

Acero K100 escoria producida a 100 psi   Acero K100 escoria producida a 110 psi

Fig. 8. Slag according to working pressure, K100.

The third point to analyze is the finish of the cut, to visualize which cut surface presents fewer defects in relation to other pressure [10, 11] (Fig. 9).

Corte K100 a 100 psi   Corte K100 a 110 psi

Fig. 9. Finishing according to working pressure, K100.

Cutting at 100 psi exhibits less surface deformation, while 110 psi exhibits more slag deposits creating a worse finish.

3.3 ASTM A36 Steel

For the ASTM A36 steel, the same points treated with the K100 will be analyzed, in a cut of 200 mm at 100 psi and 110 psi [14, 15], in Table 6, the operating times are indicated.

Table 6. Operating time for cutting steel ASTM A36 at 100 psi and 110 psi in 200 mm.

Specimen	Operating time (min) 100 psi	Operating time (min) 110 psi
A-001	1.3	–
A-001-1	–	0.98
Total	1.3	0.98

Run time at 110 psi is less than 100 psi, saving time in the cutting process. Regarding the production of slag in the cutting process, in Fig. 10, it can be seen that ASTM A36 steel, cut at 110 psi, produces a lot of slag while, at 100 psi, a cleaner cut is obtained.

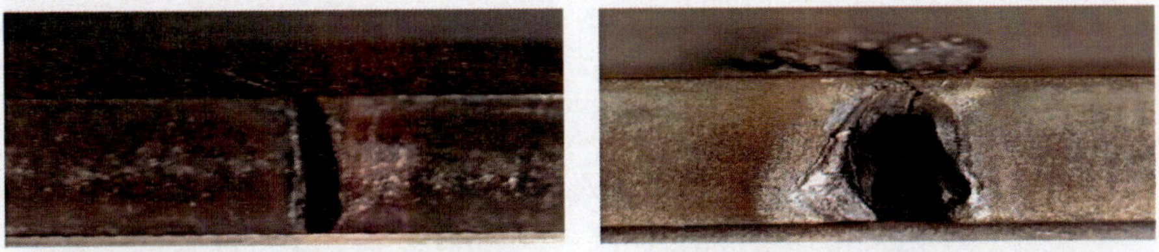

ASTM A36 slag produced at 100 psi   ASTM A36 slag produced at 110 psi

Fig. 10. Slag according to working pressure ASTM A36.

As the last point, it is determined which parameter leaves a better finish in the PAC cutting of ASTM A36 steel (Fig. 11).

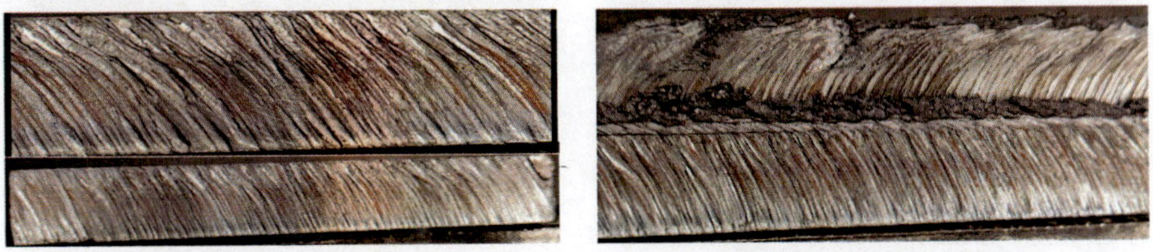

ASTM A36 cut at 100 psi   ASTM A36 cut at 110 psi

Fig. 11. Finishing according to working pressure ASTM A36.

It is observed that the cut at 100 psi is cleaner, while at 110 psi the cut presents deformations and excess slag, which determines that the best way to work is at 100 psi.

3.4 PAC vs OAW Process Comparison

A comparison will be made, equivalent to that made on the materials, treating the same analysis points, applying it to the PAC and OAW processes.

For this comparison, it works at a pressure of 100 psi in the PAC cutting process in the test tubes K-001 and A-001, while for the OAW process at a pressure of the oxygen tank of 30 psi and of the liquefied gas of 10 psi on specimens K-002 and A-002. Machine setup time and cutting operation time are counted first, as shown in Table 7.

Table 7. Machine preparation time and cutting operation at 200 mm (PAC and OAW)

PAC		OAW	
Machine preparation time (min): 5.06		Machine preparation time (min): 3.15	
Specimen	Operating time (min)	Specimen	Operating time (min)
A-001	1.2	A-002	2.35
K-001	1.3	K-002	2.40
Total	2.5	Total	4.75

Due to its carbon content, K100 steel is more difficult to cut by the OAW process, so it would be necessary to use acetylene to reach a higher temperature and not liquefied gas.

In terms of machine setup time, the OAW process is the fastest while the PAC process has an advantage in operating time, making the PAC process more profitable in the long term. The next point of comparison is that of slag production between the PAC and OAW process, Fig. 12.

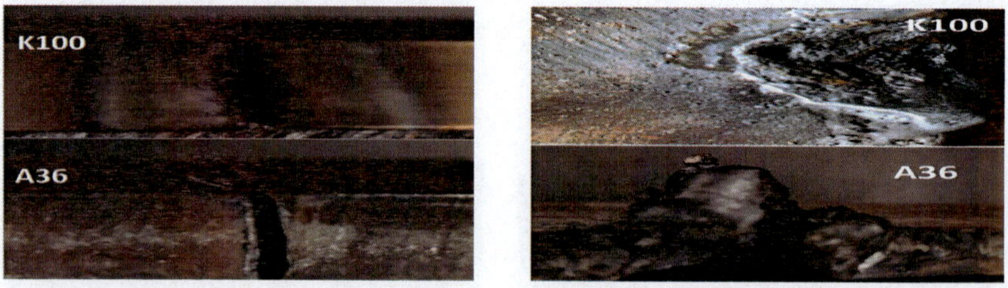

K100 and ASTM A36 steel slag PAC process

K100 and ASTM A36 steel slag OAW process

Fig. 12. Slag according to PAC and OAW process.

The slag produced by the PAC process is minimal, while that produced by the OAW process is abundant and requires a grinding process to remove it. As a last point of comparison, the type of cut that leaves a better finish at the end of the process is analyzed. In Fig. 13 shows the finishes.

K100 and ASTM A36 steel finish
PAC process

ASTM A36 steel finish
OAW process

Fig. 13. Finishing according to PAC and OAW process.

As can be seen, the final cut finish of the PAC process leaves a clean surface without much deformation, while the OAW process leaves a rough surface with a lot of deformation, which is why the PAC process has an advantage at this point.

3.5 Analysis of Results

In the PAC process, when changing the cutting parameter from 100 psi to 110 psi, changes in the cutting material occur, increasing the production of slag in both materials, especially in ASTM A36 steel.

As for the cutting time at 110 psi, it is 27% faster in both materials, therefore, working at that pressure decreases the cutting operation time, but increases the production of slag.

Finally, for the cut finish on both materials, at 100 psi a cleaner cut is observed, with less deformation than at 110 psi. In reference to the cutting comparison between PAC and OAW processes, in machine setup time, OAW cutting relative to PAC is 37% faster.

In the OAW cutting operation, the K100 steel could not be cut (Fig. 14), so it would be necessary to use acetylene to reach a higher temperature and not liquefied gas. Therefore, the cutting time of K100 steel should be a little longer than that of ASTM A36, therefore, a time of 2.40 min is assumed, in that case the PAC cutting process is 53% faster.

Fig. 14. Failed OAW cut, K100 with liquefied gas.

In the OAW process, the slag production is significantly higher, with the PAC process being the most efficient, while in the cut finish the PAC process is the cleanest. The PAC cutting process is the best option to choose, since in all the aspects that have been compared, it offers advantages over the OAW.

4 Conclusions

The PAC plasma cutting tests, carried out on the two materials K100 and ASTM A36, show an advantage over the oxyacetylene OAW, indicating greater efficiency when cutting the materials, reducing the cutting time, cost of using the machine, and surface finish cutting material, while cutting (OAW) leaves impurities on the surface, in addition to increasing use times and cutting time.

The cost of using the plasma cutting machine (PAC), decreases with respect to the use that it is given and the cuts it makes, it does not present an excessive effort and it is not necessary to acquire additional tools. With the plasma cutting machine (PAC), a more homogeneous and precise cut can be established without making a greater effort compared to other cutting processes, because the machine has the characteristics to be maneuverable, in this way to use in any type of field. Plasma cutting (PAC) represents the new technology for metalworking industries, as well as small-medium industrial workshops that can reduce operating costs and provide better service to external customers.

References

1. Böhler Voestalpine: https://www.acerosbohler.com/es/products/k100/.uthor. Accessed 10 May 2021
2. Material mundial: https://www.materialmundial.com/acero-astm-a36-propiedades-ficha-tecnica-estructural/. Accessed 05 Feb 2021
3. Hypertherm: https://www.hypertherm.com/es/hypertherm/powermax/powermax30-xp/. Accessed 05 May 2021
4. Calupiña, C., Oña, D.: Mejoramiento de los parámetros de trabajo para una máquina de corte por plasma y oxiacetilénica tipo CNC - 4000 marca Hugong Welder, EPN, Quito (2012)
5. Mano Mano Como elegir un cortador por plasma. https://consejos.manomano.es/como-elegir-un-cortador-por-plasma-n3507. Accessed 05 May 2021
6. Cuesta, E., Rico, J., Valiño, G., Villanueva, A.: El corte de chapa mediante oxicorte. MetalUnivers **5**(2), 50–65 (2002)
7. Aguilar, J.: Soldadura oxido/acetilénico. http://biblio3.url.edu.gt/Libros/2013/pro_ma/22.pdf (2013)
8. G. Auz VID_20210206_115559: https://drive.google.com/file/d/11-41HZ19OIblpxDvY700Iwk64A5uEINi/view. Accessed 09 Feb 2021
9. Böhler Voelstalpine One Step A Head. https://www.voestalpine.com/highperformancemetals/ecuador/es/bohler/. Accessed 09 May 2021
10. G. Auz Corte por plasma acero K100. https://drive.google.com/file/d/1_EDjTC2kJuBCWfupFq9bpIbC1V2zELyC/view. Accessed 09 Feb 2021
11. G. Auz Corte por plasma acero K100 a 110 psi. https://drive.google.com/file/d/15oKl_luYs40acRxGqRXKnsSWm93aC1kK/view. Accessed 09 Feb 2021
12. Hypertherm Sistema de plasma Powermax30 XP. https://www.hypertherm.com/es/hypertherm/powermax/powermax30-xp/. Accessed 15 Jan 2021
13. G. Auz Ensamble de consumibles plasma. https://drive.google.com/file/d/1jjrvfCBa8aTs2obWwGNOsZBIIVXYAJti/view. Accessed 20 Apr 2021
14. G. Auz Corte por plasma ASTM A-36. https://drive.google.com/file/d/1zWccTeDdjZp_vLuRLk3M0_GBBdgLr21u/view. Accessed 09 Feb 2021
15. G. Auz Corte por plasma acero A-36 a 110 psi. https://drive.google.com/file/d/1LRBCP5ql8J9uCKYdFaiMmJhF4G-rjyfO/view. Accessed 09 Feb 2021

Efficiency of the Box-Type Solar Cooker in the Cooking of Andean Tubers in Puno

Antonio Holguino-Huarza[1](✉) and Jose Quiñonez-Choquecota[1,2]

[1] Universidad Nacional Del Altiplano de Puno, Floral Avenue 1153, P.O. Box 291, Puno 21001, Peru
aholguino@unap.edu.pe

[2] National University of Moquegua, Ilo 18611, Peru

Abstract. This paper details the research results on the efficiency of a box-type solar cooker made of wood, with walls and floors covered with air-containing partitions and a double-glazed area where the solar energy accesses. This cooker has been tested for cooking existing tubers in the highland region of Puno, where the average annual solar radiation is 2500 kWh/m^2. To evaluate the performance of the designed and constructed solar cooker, the first and second merit factor criteria have been considered, which relate to the optical efficiency of the energy loss through the cooker components, as well as the efficiency of the transfer of energy to the tubers during cooking; obtaining average values of 9.14 and 50.49 °Cm^2/W respectively, which are within the suggested values for a good cooking process. Concerning the percentage, energy and exergetic efficiencies are 29.91 and 28.13 respectively, which show an adequate transfer of energy in the cooker during the cooking process of tubers. Baking is more convenient than cooking it immersed in water; since the cooking time and temperature for native potatoes and oca follow the Arrhenius equation for tuber cooking.

Keywords: Energy efficiency · Solar cooker · Tuber cooking

1 Introduction

The Puno region is the high Andean region of Peru, located at 15° south latitude in western South America, where the global horizontal irradiation has an annual value of greater than 2200 kWh/m^2 and the normal direct irradiation has an average value of 2500 kWh/m^2 [1]. The overall thermal performance of solar cookers depends on the solar radiation potential of the region [2]. About 20% of the Peruvian population, especially in the rural sector, still uses biomass to cook their food [3], and the process of cooking food is mainly done by women and older adults; therefore, the probability of increasing respiratory infections has risen by 12% [4].

However, passive solar energy is an alternative to cooking food, which is a clean and free source, whose use has been carried out since ancient civilizations, where food processing and preservation played an essential role in nutritional, aesthetic and spiritual applications [5].

M. Zambrano Vizuete et al. (Eds.): CI3 2021, LNNS 511, pp. 377–389, 2022.
https://doi.org/10.1007/978-3-031-11438-0_30

The existing tubers in the South American Andes are of diverse varieties such as potatoes, oca, olluco and mashua [6], being foods that allow the subsistence of the rural population.

The performance of the solar cooker depends on several factors, among them are: the metal absorption platform must be painted matte black [2], the space containing the cooking platform must have thermal insulation, and the heat concentrator is made of phase change materials such as organic oils [7], these materials have a composition and characteristics suitable for the use applied to food cooking processes [27], and do not distort the food during the heat transfer process through the absorb plate. The use of solar cookers reduces the emission of pollutants and does not require the use of fuels during the food cooking process [16], which implies that food cooked in solar cookers is ambientally friendly.

According to Guidara [9], box-type cookers are feasible because this type of solar cookers has less convective heat loss, and their construction allows many possibilities for improvement. The heat loss by conduction through the walls and floor is reduced because the materials used for their construction, such as wood, glass wool and other materials, have low thermal conductivity.

In this research work, the efficiency of a box-type solar cooker is evaluated experimentally; for this purpose, the coefficient of energy efficiency of the cooker was determined, the physical characteristics or merit tests of the cooker that depend on the optical efficiency of the access area of solar radiation, the accumulation and retention of energy with the materials used.

2 Methods and Procedures

2.1 Description of the Box-Type Solar Cooker

In the design of the experimental solar cooker, the following parts have been considered:

- A double-walled trapezoidal box with an inner volume of 14.7×10^{-3} m^3.
- The sides of the inner wall and the floor are covered with air partition between wooden sheets in an area of 0.25 m^2.
- Its outer wall, trapezoidal sides and rectangular floor are made of wood in an area of 0.51 m^2.
- The access area, where the solar radiation enters the cooker, is separated with double glass sheets; the internal region has a horizontal tempered glass sheet of 0.075 m^2, the outer part is repaired with a single glass sheet inclined at 27° with respect to the horizontal of 0.10 m^2 of area.
- A black painted metal box, whose sidewall has a perimeter of 0.76 m, with a height of 0.16 m; a door of 0.054 m^2 placed at the back of the cooker.
- The floor of the metal box is the platform painted matte black, while the sidewalls, including the door, are lined with polished aluminum plates in an area of 0.17 m^2.
- The reflector systems are constituted by an external movable mirror of 0.12 m^2, an internal fixed mirror at the rear upper part of the 0.042 m^2 area and the polished aluminum surface placed on the sidewalls of the metal box.

One of the most essential components of a solar cooker is the thermal insulation that prevents heat transfer between the cooker and the ambient [10]. Therefore, in the design and construction of the solar cooker, the accumulated amount of energy from the solar radiation reflected by the reflectors, which enters the metal box through the double-glazing zone, was considered. The dual glazing zone reduces energy losses due to the greenhouse effect in the walls and floor due to the low thermal conductivity of the air that occupies the partitions.

On the other hand, solar radiation passing through the double-glazing zone transmits this radiation to the furnace area and is absorbed by a metallic surface with high specific heat and painted matte black to maximize absorption of the radiated energy [2].

The physical properties of the materials used in the construction of the cooker are shown in Table 1.

Table 1. Geometric and physical properties of the materials used in the construction.

Material	Description	Value
Absorber plate	Specific heat c_{plate} (kJ/kgK)	0.502 [11]
Glass	Thermal conductivity k_{glass} (W/mK)	0.780 [11]
Air	Thermal conductivity k_{air} (W/mK)	0.026 [11]
Wood	Thermal conductivity k_{wood} (W/mK)	0.274 [8]

2.2 Measuring Instruments Used

The recorded data of temperature and direct solar radiation as a function of time were collected in LabQuest 2. Temperatures were measured with a Wide-Range Temperature Probe. For the measurement of solar radiation, the Pyranometer Tech T - Apogee Instruments, which records electromagnetic radiation including near-infrared, visible and ultraviolet radiation, was used, recording the measurements in units of watts per square meter. The graphs were prepared in SigmaPlot 10, while the descriptive statistics were performed with InfoStat.

2.3 Temperature and Solar Radiation Measurements

The simultaneous temperature and solar radiation measurement tests were carried out in Alto Puno, located at latitude −15.814357 and longitude −70.031849, during 4 days, from May 28 to June 21, 2021, the experimental test data have been recorded from 9:00 to 14:30 h. In the test, the incidence area has been reoriented every 20 min, and the moving reflector allowed the reflected rays to strike the platform.

2.4 Thermal Efficiency

The energy balance in the solar cooker under stationary conditions is given by

$$E_{inp} = E_{out} + E_{lost} \tag{1}$$

where E_{lost} represents the energy lost through the walls, floor and double-glazing area; E_{inp} is the solar radiation energy entering the cooker platform through the double-glazing area and is determined by [12].

$$E_{inp} = I_{rad}A \tag{2}$$

where I_{rad} is the solar radiation intensity, and A is the collection area for solar radiation.

The output energy, which is the energy of sensible heat gain in a food heating or cooking process, is determined by [12].

$$E_{out} = \frac{m_w c_w (T_{fw} - T_{iw})}{t} \tag{3}$$

where m_{w} and c_w are the mass and specific heat of water respectively, T_{iw} and T_{fw} are initial and final temperatures of water, and t is the time interval in which the temperature changes. Heating tests are done with water because food has high water content.

With the input and output energy, the thermal efficiency of the cooker is determined, given by

$$\eta = \frac{E_{out}}{E_{inp}} \tag{4}$$

Mullick [13] indicates that the efficiency of a box-type solar cooker can be evaluated with merit tests, which characterize the solar cooker, independent of climatic variables.

The first merit factor relates the optical efficiency to the heat loss factor. This value of merit, according to Mullick [13], is given as

$$F_1 = \frac{T_p - T_{amb}}{I_{rad}} \tag{5}$$

where T_p is the equilibrium temperature of the absorb plate, I_{rad} represents the direct radiation on a horizontal surface, and T_{amb} is the ambient temperature at the moment when the stabilization temperature is reached.

When the solar cooker is in operation, its efficiency can be evaluated with the second merit factor. A high value of the second merit factor represents good heat exchange, good optical efficiency and low heat loss inside the cooker. This factor, according to Mullick [13], is given by

$$F_2 = \frac{F_1 m_w c_w}{At} \ln\left(\frac{1 - \frac{1}{F_1}\left(\frac{T_{w1} - T_{amb}}{H}\right)}{1 - \frac{1}{F_1}\left(\frac{T_{w2} - T_{amb}}{H}\right)}\right) \tag{6}$$

where t is the time measured for sensible heating of water between two known temperatures, $m_w c_w$ is the heat capacity of the water in the container, A is the area of energy collection in the solar cooker.

Another parameter that evaluates the efficiency of the solar cooker is the exergy, which allows finding the maximum energy degraded or used in the cooker [14]. According to Herez [10], the exergy efficiency is given by

$$\psi = \frac{Ex_{out}}{Ex_{inp}} = \frac{\frac{m_w c_w}{t}\left[\left(T_{fw} - T_{iw}\right) - T_o \ln\left(\frac{T_{fw}}{T_{iw}}\right)\right]}{I_{rad} A\left(1 - \frac{4T_{amb}}{3T_s}\right)} \tag{7}$$

where T_o is the outside temperature of the cooker, T_{amb} is the ambient temperature, $T_s = 6000$ K is the sun temperature [14], I_t is the instantaneous total radiation.

In a comparison between energy and exergy efficiencies, according to Pandey [15], energy efficiency can only account for the amount of energy transferred; however, exergy efficiency represents the quality of energy transferred.

2.5 Cooking of Native Tubers

For the cooking of tubers to take place, the temperature gradient inside the cooker increases progressively, so this tuber cooking process has an Arrhenius-type behavior. According to Ebersviller [16], the cooking time of a tuber is given by

$$t_c = Be^{-\frac{\phi}{RT}} \tag{8}$$

where ϕ and B characterize the tubers and represent the activation energy and the chemical reaction respectively, $R = 8,314$ J/mol K the universal gas constant, and T is the cooking temperature.

To evaluate the efficiency of the box-type solar cooker during the food cooking process, native tubers commonly used in food have been used; due to two transcendental factors such as the importance of the cooking temperature below the boiling point of water, because of this is that food cooked in solar cookers retain their nutritional value. According to Lennox [6], the native tubers of the altiplano region of Puno, have an important nutritional value in the diet; whose categorization by high nutritional value and use in the diet of the population of this area, in order of importance are: mashua, native potatoes, oca and melloco [17]. On the other hand, the cooking of tubers causes the reduction of polyphenols; this significant difference between the total content of polyphenols in the order of importance is raw potato, steamed potato, cooked in water and microwave [18].

3 Data Collection and Experimental Analysis

The experimental test has been carried out between May and June 2021, with intermittent solar radiation that characterizes the highland region of Puno. This test corresponds to the heating processes of the box-type solar cooker without and with load and the tuber cooking processes (see Fig. 1).

On June 2 and 18, data were collected on the evolution of the parameters that allowed evaluating the efficiency of the solar cooker in the process of cooking tubers and heating water. For this data collection process, we started with the exposure of the cooker to the ambient with the energy access area exposed so that the box and the metallic platform reach the thermal equilibrium with the ambient. For the thermal evaluation of the loaded cooker, a metal container containing 1.11 L of water was placed on the platform from the beginning.

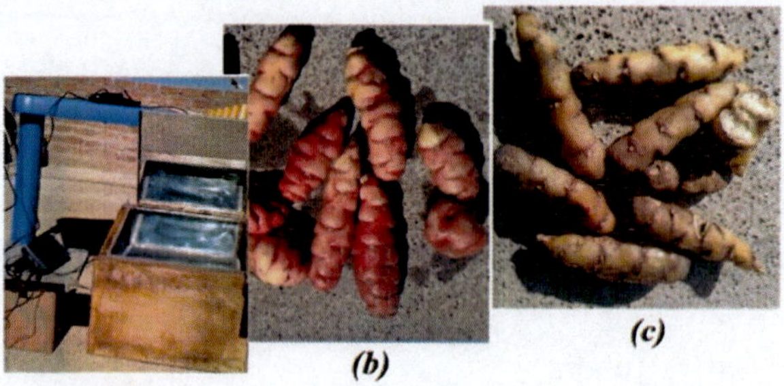

Fig. 1. Photo of: (a) Box type solar cooker installed, (b) Raw tuber, (c) Baked tuber.

During the days of experimental data recording, there was always cloud cover on some days more than on others, due to which the direct solar radiation had an intermittent variation (see Fig. 2).

The Fig. 2a shows the heating of the absorb plate without load, performed on a partially cloudy day, while Fig. 2b shows the evolution of the heating of the platform with the load on a partially clear day. The performance parameters such as solar radiation, the evolution of the temperatures of the metal platform without and with load and the temperature of the environment, which allow determining the parameters of the merit tests, energy and exergy efficiencies, are shown in Table 2. Also, the effective area of solar radiation capture of 0.076 m^2, the average external surface temperature of the cooker of 27.5 ± 1.24 °C has been determined by direct measurements.

Table 2. Solar radiation, evolution temperatures of the metallic platform, and ambient.

Magnitude	Under no-load	Under load (water)
Initial temperature T_i (°C)	19.2	29.5
Final temperature T_f (°C)	104.1	73.9
Ambient temperature T_{amb} (°C)	12.59 ± 2.08	26.66 ± 3.43
Outside temperature of the cooker T_o (°C)	–	27.5 ± 1.24
Incident solar radiation I (W/m^2)	1001.91 ± 25.58	873.2 ± 156.62
Sun temperature T_s (K)	6000 [14]	

The cooking of tubers (see Fig. 3) was done by two different processes; the first by heat transfer through water and the second by direct contact of the tuber with the absorb plate or baking process. During both cooking processes, the temperatures of both the unloaded platform and the platform with water remain almost stable, as shown in Fig. 3a and Fig. 3b for the cooking of native huayro potatoes and oca (Oxalis tuberosa), respectively. However, solar radiation is intermittent due to the cloudy conditions that interfere with direct solar radiation (see Fig. 3).

The cooking time of potato immersed in water has been 159 min, the mean water temperature was 81.64 ± 2.92 °C, which has evolved from 70.4 to 85.7 °C, under the mean solar irradiation of 940.30 ± 138.41 W/m^2, in an ambient whose mean temperature was 17.28 ± 0.86 °C (see Fig. 3a). Regarding oca baking (see Fig. 3b), the mean temperature of the platform was 81.49 ± 0.89 °C, it has evolved from 79.40 to 82.8 °C, with mean solar irradiation of 985.92 ± 58.79 W/m^2, with the mean ambient temperature of 16.6 ± 1.34 °C, this process has been carried out in a time of 126 min. In each tuber cooking process, 420 g were used. The tubers used are the native huayro potato and oca; these tubers are native to the high zones of the Andean Mountain range, which are used to feed the people of the Puno.

The solar cooker was used at two times different temperatures inside the cooker to determine the cooking time through the baking process, both for the native potato and oca (see Fig. 4). In the procedure, the data recorded provide statistics on the average cooking time and temperature of these two Andean tubers.

4 Results and Discussion

With the data obtained from the processes of solar cooker heating (see Fig. 2) and tuber cooking (see Fig. 3), the performance parameters of the cooker were determined; the first merit factor was determined with Eq. (5) for the heating of the cooker without load, while the second merit factor was determined with Eq. (6) for the cooker with load or tuber immersed in water. The energy and exergy efficiency coefficients were also determined, using Eqs. (4) and (7), respectively. These results are shown in Table 3.

The efficiency of the cooker is related to the optimal use of solar radiation for heating the platform with minimal heat loss to the outside. According to Mullick [13], the value of the first merit factor greater than 5% represents good optical efficiency and a low heat loss factor. For the solar cooker evaluated, the first merit factor has a 9.135 ± 0.025% value and is greater than the suggested 5%.

For a solar cooker with phase change materials and heat accumulator, the average value of the first merit factor determined by Coccia [19] was 0.19 °Cm^2/W in an ambient whose average temperature was 28.2 °C and for average solar irradiation of 884.23 W/m^2. While this merit factor determined for the evaluated cooker has been lower because it does not have a heat accumulator system, and the outside temperature was higher.

While the second factor of merit represents the heat exchange ratio of the platform with the water or food to be cooked and the optical efficiency of the components of the solar cooker. Coccia [19] claims, in the process of heating 1.5 L of water, the average temperature of the ambient was 26.6 °C and the average solar irradiation of 803.9 W/m^2; the value of the second factor of merit has been 0.052. Meanwhile, during the evaluation process of the solar cooker with load, 1.11 L of water was used, obtaining the value for the second factor of merit 0.5049, which is 9.7 times higher than that obtained by Coccia

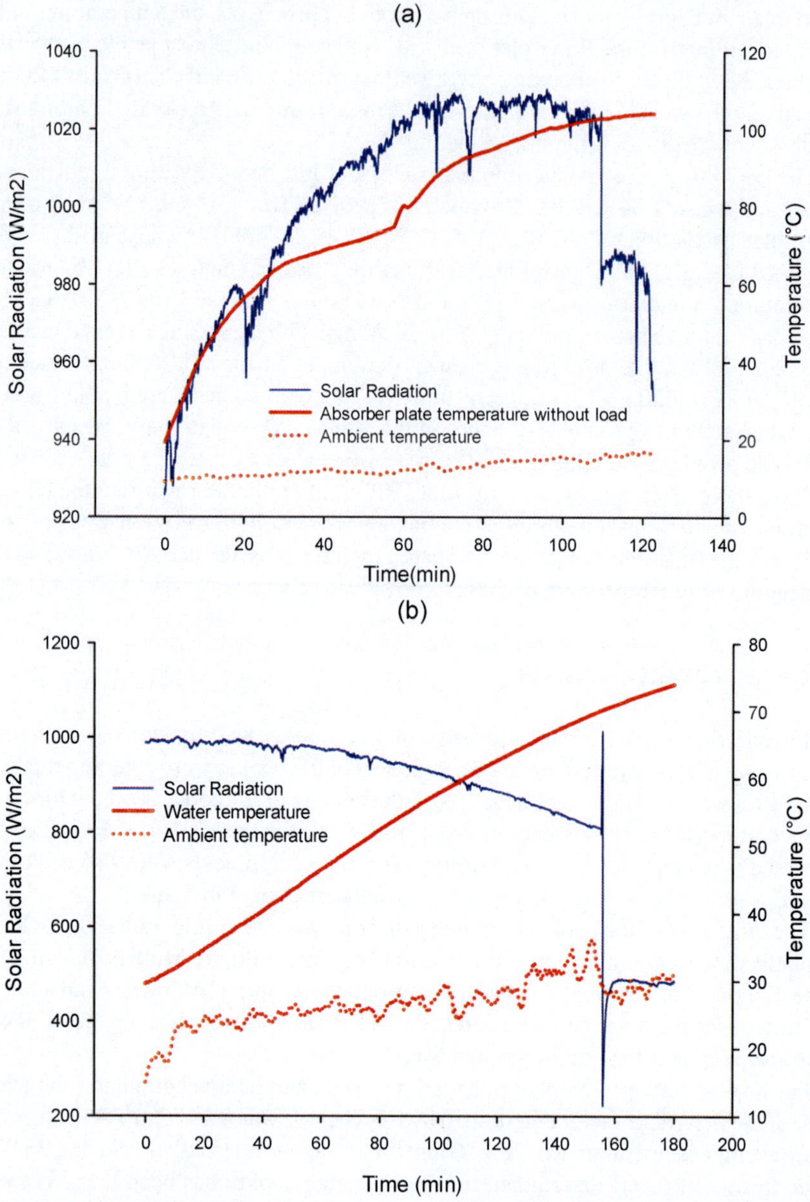

Fig. 2. Heating Process: (a) Absorb plate without load; (b) Absorber plate with load: water and tuber.

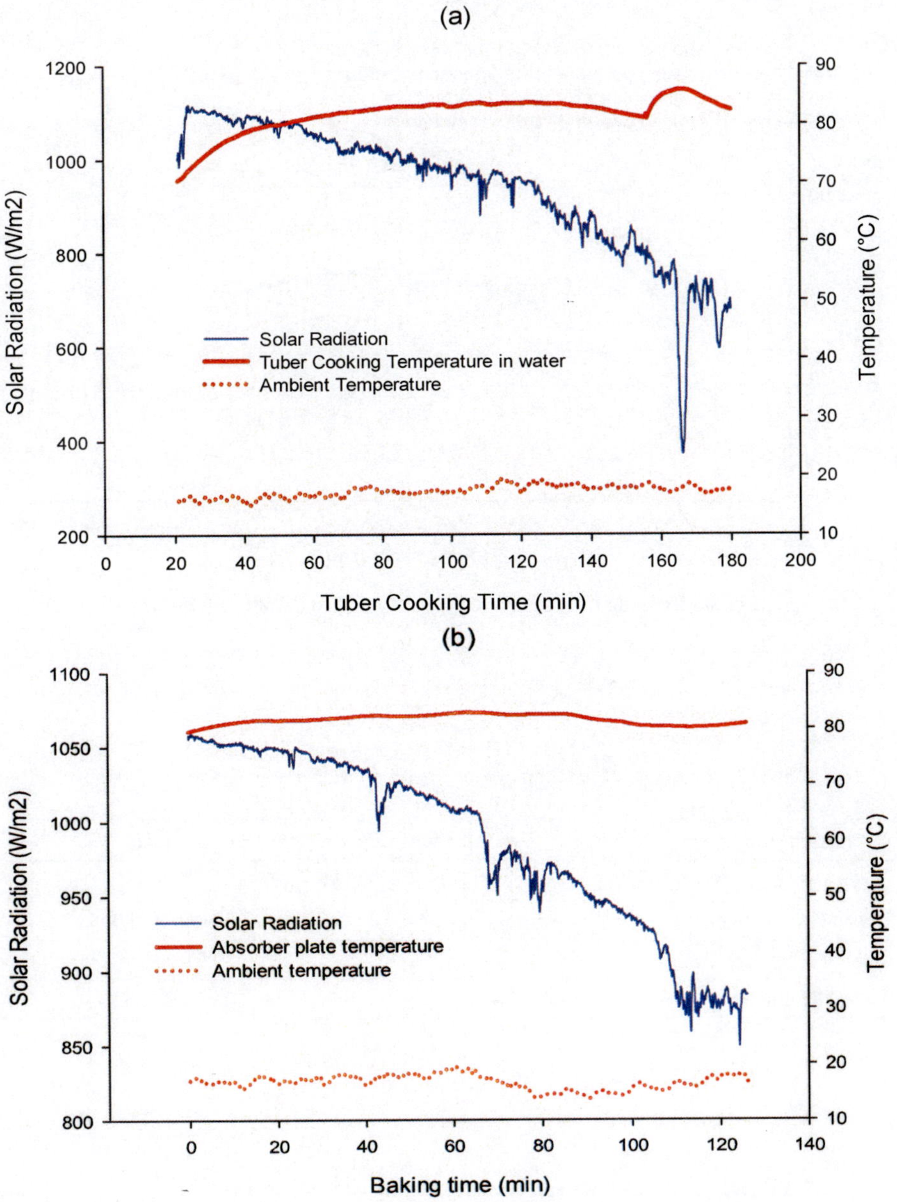

Fig. 3. Tuber cooking processes: (a) Cooking submerged in water; (b) Baked on the absorber plate.

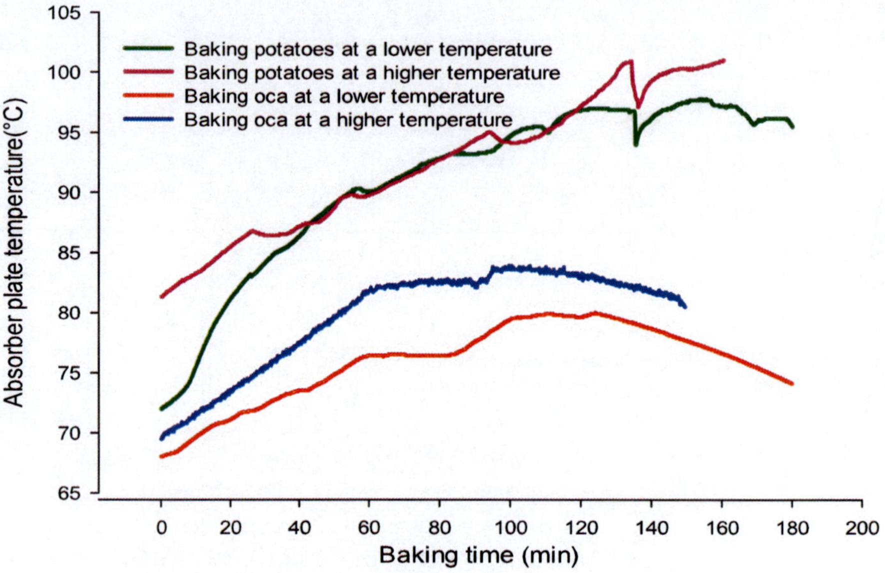

Fig. 4. Cooking time of native potatoes and oca (Oxalis tuberosa).

Table 3. Parameters for evaluating the thermal efficiency of the solar cooker.

Evaluation parameters	Under no-load	Under load
First merit factor F_1(°Cm^2/W)	0.09135 ± 0.00025	–
Second merit factor F_2	–	0.5049 ± 0.0058
Energy efficiency $\eta(\%)$	–	29.91 ± 2.89
Exergetic efficiency $\psi(\%)$	–	28.13 ± 1.52

Table 4. Time intervals and average cooking temperatures of tubers used.

Parameters	Native potato (huayro)		Oca (Oxalis tuberosa)	
	High temp.	Low temp.	High temp.	Low temp.
Average cooking temp. (°C)	92.27 ± 5.62	91.07 ± 6.87	79.77 ± 4.17	75.88 ± 3.2
Cooking time (min)	160.4	180	149.8	178.2

[19]; therefore, the solar cooker evaluated in the ambient conditions of the Puno region, can heat 9.7 times more water used.

According to Saxena [20], the value for the first factor of merit obtained was 0.13 m^2 °C/W, in an environment with an average temperature of 34 °C, with average solar irradiation of 630 W/m^2; while the value of the second factor of merit was 0.44 and the estimated thermal efficiency was 37.1%.

Another parameter that allows evaluating the performance of the solar cooker is the thermal efficiency determined by Eq. (4). For the sample data of the experimental tests, this value was 29.91 ± 2.89% (see Table 3).

According to Nayak [21], for an efficient solar cooker, the values of the first and second merit factors should have an average value of 0.123 m^2 °C/W and 0.44, respectively. According to the value of the first factor of merit, improvements need to be made in the optical system and energy accumulation of the evaluated solar cooker; for the second factor of merit, it is higher than suggested and indicates that there is an adequate heat exchange with the food during the cooking process.

Concerning the obtained value of energy and exergy efficiency of 29.91 ± 2.89 and 28.13 ± 1.52% respectively (see Table 3), the value of energy efficiency is within the range, and the value of exergy efficiency is above the value suggested by Herez [10]; who, from the performance parameters of the solar cooker, determined an energy efficiency that varies between 20–30% and an exergy efficiency that varies between 4–6%. According to Pandey [15], the average exergy efficiency is higher when more water is heated. This parameter is higher for parabolic solar cookers compared to box-type solar cookers.

In cooking the tuber submerged in water, half of the polyphenols contained are lost; these polyphenols are antioxidant compounds, which eliminate radicals in the organisms of consumers [18]. The antioxidants of tuber, as in potato, are significantly reduced by cooking by baking at high temperature and microwave [22]. Therefore, to preserve polyphenols, other antioxidants, and vitamin C, tubers should be cooked in the water below boiling temperature or baked at temperatures below the boiling point of the water contained in the tubers. During the solar cooking evaluation, the cooking process of the tubers was carried out at an average temperature of 81.64 °C (see Fig. 3a), which is below the boiling point. In the city of Puno, the boiling temperature corresponds to a mean value of 87.67 °C [23]. According to Jayanty [24], leaching is an important factor affecting the loss of vitamin C in boiling cooking. Leaching can be reduced by cooking potatoes with the skin on, limiting the surface area of the pulp exposed to water. There is an additional advantage of cooking potatoes with the peel on, in that most of the phenolic compounds migrate to the peel during the cooking process [25].

For cooking native potato immersed in water, the averages of cooking temperature and cooking time are 81.64 ± 2.92 °C and 159 min respectively; these values are congruent to the cooking temperature of 80 °C and a cooking time of 121 min, determined by Chatelain [26] which agrees with the Arrhenius equation. The values of both parameters are relatively higher because these data have been recorded in a heating process where the temperature was slightly rising (see Fig. 3a). Regarding oca baking, the mean values of temperature and cooking time were 81.49 ± 0.89 °C and 126 min respectively (see Fig. 3b), which are similar to the cooking of potato immersed in water, where the temperature is also similar of the platform is rising during the cooking process.

In the tuber cooking processes, the efficiency of the box-type solar cooker allows inferring those native potatoes require more cooking time and temperature than oca (see Table 4). It is also observed that in the cooking process for both the native huayro potato and oca, the increase in cooking temperature allows for a decrease in cooking time (see Fig. 4).

5 Conclusions

The values of the merit factors reflect that the optical efficiency of the box-type solar cooker is acceptable since the value of the first merit factor is higher than the suggested 5%, while, according to the value obtained for the second merit factor, there is an efficient energy exchange with the food during the cooking process.

According to the value determined for exergy, the quality of energy used in the tuber cooking process is appropriate since it agrees with the second factor of merit calculated.

The cooking temperature constantly rises during the process, whether by baking or exposure to water, because the tubers begin at room temperature, which is lower than the temperature of the water or platform, and must reach thermal equilibrium over time.

References

1. Prăvălie, R., Patriche, C., Bandoc, G.: Spatial assessment of solar energy potential at global scale. A geographical approach. J. Clean. Prod. **209**, 692–721 (2019)
2. Cuce, P.M.: Box type solar cookers with sensible thermal energy storage medium: a comparative experimental investigation and thermodynamic analysis. Sol. Energy **166**, 432–440 (2018)
3. Zafra-Tanaka, J.H., Tenorio-Mucha, J., Villarreal-Zegarra, D., Carrillo-Larco, R., Bernabe-Ortiz, A.: Cancer-related mortality in Peru: trends from 2003 to 2016. PLoS ONE **15**(2), 1–13 (2020)
4. Canales Gutiérrez, Á., Belizario Quispe, G., Calatayud Mendoza, A.P., Chui Betancur, H.N., Huaquisto Ramos, E.: Thermal comfort and the risk of respiratory infections in older adults in the Peruvian highlands. Revista Espanola de Geriatria y Gerontologia **56**(1), 24–28 (2021)
5. Barba, F.J., et al.: Solar radiation as a prospective energy source for green and economic processes in the food industry: from waste biomass valorization to dehydration, cooking, and baking. J. Clean. Prod. **220**, 1121–1130 (2019)
6. Lennox, E.: Double exposure to climate change and globalization in a Peruvian highland community. Soc. Nat. Resour. **28**(7), 781–796 (2015)
7. Nkhonjera, L., Bello-Ochende, T., John, G., King'ondu, C.K.: A review of thermal energy storage designs, heat storage materials and cooking performance of solar cookers with heat storage. Renew. Sustain. Energy Rev. **75**, 157–167 (2017)
8. Holguino, A., Olivera, L., Escobar, K.: Thermal comfort in an adobe room with heat storage system in the Andes of Peru. Revista de Investigación Altoandinas **20**(3), 289–300 (2018)
9. Guidara, Z., Souissi, M., Morgenstern, A., Maalej, A.: Thermal performance of a solar box cooker with outer reflectors: numerical study and experimental investigation. Sol. Energy **158**, 347–359 (2017)

10. Herez, A., Ramadan, M., Khaled, M.: Review on solar cooker systems: economic and environmental study for different Lebanese scenarios. Renew. Sustain. Energy Rev. **81**, 421–432 (2018)
11. Yunus, A.C., Afshin, J.G.: Heat and mass transfer. Fundamentals and applications. 4th edn. McGraw-Hill, Mexico, D.F. S.A. (2011)
12. Ozturk, H.H.: Comparison of energy and exergy efficiency for solar box and parabolic cookers. J. Energy Eng. **133**(1), 53–62 (2007)
13. Mullick, S.C., Kandpal, T.C., Saxena, A.K.: Thermal test procedure for box-type solar cookers. Sol. Energy **39**(4), 353–360 (1987)
14. Panwar, N.L.: Thermal modeling, energy and exergy analysis of animal feed solar cooker. J. Renew. Sustain. Energy **5**(4), 043105 (2013)
15. Pandey, A.K., Tyagi, V.V., Park, S.R., Tyagi, S.K.: Comparative experimental study of solar cookers using exergy analysis. J. Therm. Anal. Calorim. **109**(1), 425–431 (2012)
16. Ebersviller, S.M., Jetter, J.J.: Evaluation of performance of household solar cookers. Sol. Energy **208**, 166–172 (2020)
17. Santillán, L.B., Moreno, M.J.P., Martínez-Monzó, J., García-Segovia, P.: Functional properties of tubers native to the Andean region of Chimborazo (Ecuador): a revision. Revista Espanola de Nutricion Comunitaria **22**(4), 28–33 (2016)
18. Laib, I., Barkat, M.: Optimization of conditions for extraction of polyphenols and the determination of the impact of cooking on total polyphenolic, antioxidant, and anticholinesterase activities of potato. Foods **7**(3), 36 (2018)
19. Coccia, G., Aquilanti, A., Tomassetti, S., Comodi, G., Di Nicola, G.: Design, realization, and tests of a portable solar box cooker coupled with an erythritol-based PCM thermal energy storage. Sol. Energy **201**(3), 530–540 (2020)
20. Saxena, A., Karakilcik, M.: Performance evaluation of a solar cooker with low-cost heat storage material. Int. J. Sustain. Green Energy **6**(4), 57–63 (2017)
21. Nayak, J., Sahoo, S.S., Swain, R.K., Mishra, A.: Construction of box type solar cooker and its adaptability to industrialized zone. Mater. Today Proc. **4**(14), 12565–12570 (2017)
22. Perla, V., Holm, D.G., Jayanty, S.S.: Effects of cooking methods on polyphenols, pigments and antioxidant activity in potato tubers. LWT-Food Sci. Technol. **45**(2), 161–171 (2012)
23. Gibson, B., Pamidi, S., Kim, J.: FSU presidential scholars program superconductors at varying temperatures (2016). https://diginole.lib.fsu.edu/islandora/object/fsu:291066/datastream/PDF/view
24. Jayanty, S.S., Diganta, K., Raven, B.: Effects of cooking methods on nutritional content in potato tubers. Am. J. Potato Res. **96**(2), 183–194 (2019)
25. Ezekiel, R., Singh, N., Sharma, S., Kaur, A.: Beneficial phytochemicals in potato - a review. Food Res. Int. **50**(2), 487–496 (2013)
26. Chatelain, T., et al.: Solar cooking potential in Switzerland: nodal modelling and optimization. Sol. Energy **194**(6), 788–803 (2019)
27. Reyes-Cueva, E., Martínez-Gómez, J., Delgado Yánez, M.: Phase change materials. Material selection based on better thermal properties: a literature review. In: Botto-Tobar, M., Zambrano Vizuete, M., Díaz Cadena, A. (eds.) CI3 2020. AISC, vol. 1277, pp. 450–463. Springer, Cham (2021). https://doi.org/10.1007/978-3-030-60467-7_37

Security

Electromagnetic Fields Estimation of the Santa Rosa-Totoras 230 kV Transmission Line

M. F. Velasco(✉), X. A. Proaño, W. P. Guamán, and G. N. Pesántez

Universidad Técnica de Cotopaxi, Latacunga, Ecuador
maria.velasco4345@utc.edu.ec

Abstract. This article develops the calculation and estimation of the Electric and Magnetic Fields (EMF) of the Santa Rosa-Totoras 230 kV line in the Cotopaxi Province. First, the state of the art of works associated with the study of the electric field, magnetic field, and EMF in electric power systems was reviewed. Then, the transmission line route was designed using PLS_CADD software and applying the calculation methodology used by Electric Power Research Institute (EPRI), the EMF levels emitted by the line under study were determined and compared with the results obtained from the EMF calculation module of PLS_CADD.

Finally, a statistical analysis was performed to determine if the EMF emissions of the Santa Rosa-Totoras line are within the limits established by the International Commission on Non-Ionizing Radiation Protection (ICNIRP), as well as the most representative values of electric and magnetic fields present within the transmission line's right of way.

Keywords: Electric field · Magnetic field · Transmission lines · Power systems

1 Introduction

High voltage lines generate around them variable low-frequency electric and magnetic fields (50 Hz in Europe, 60 Hz in the USA), the levels of which must be carefully measures to ensure that they are below the limits set by national and international regulations. For a given voltage, the electric field strength may to vary from one line to another depending on the characteristics of the line, the number of circuits, and their geometrical arrangement. The magnetic field on a line also changes according to the current intensity circulating as a function of the power demand [1].

EMF, generated by high-voltage transmission lines, are elements of great importance that power companies carefully consider during line design and maintenance. As a protection measure for the inhabitants near the transmission lines, the energy companies establish right of way, as well as safety measures to be adopted during maintenance works.

For many years, most of the research in electrical engineering has focused on the distribution of EMFs in electrical power system equipment and their possible effects on health. Notable among the studies, carried out were the reviews of the United Kingdom's National Radiological Protection Board (1992 and 1994), whose delegation investigated

M. Zambrano Vizuete et al. (Eds.): CI3 2021, LNNS 511, pp. 393–405, 2022.
https://doi.org/10.1007/978-3-031-11438-0_31

the hazards that EMFs can cause. Based on the results obtained, in 1996 the WHO, through the International Commission on Non-Ionizing Radiation Protection (ICNIRP), developed EMF exposure limits, taking into account the biological effects that may affect health [2].

Then, in 1998, the report prepared by the U.S. National Academy of Sciences and the review by the International Commission on Non-Ionizing Radiation Protection (1998), determined that emissions associated with high voltage power lines comply with permissible limits, concluding that there is no evidence of biological harm to humans.

In 2015, one of the first studies on EMF in electrical power systems was developed in Ecuador, which concluded with the publication of the book entitled "Determination of electromagnetic fields in electrical systems. Application to real cases". The work was based on the measurement of magnetic field and electric field variables in more than 400 points distributed inside and outside the facilities of different electrical substations in the city of Quito, including distribution feeders. This study verified that the maximum field values for occupational and population exposure are well below the limits proposed by international standards [3].

A more recent study was carried out in early 2020 in the Loja city, where *Empresa Eléctrica Regional del Sur* (EERSSA) measured EMF levels associated with its power systems, applying the criteria established in *Libro VI anexo 10 del Texto Unificado de la Legislación Secundaria del Ministerio del Ambiente (TULSMA)*. The results showed that both the electric and magnetic field levels are within the limits established by the regulations [4].

2 Electric and Magnetic Fields in Transmission Lines

2.1 Electric Field

The magnitude and direction of the force exerted on an electric charge are defined as electric field, which can vary depending on the characteristics of the line, i.e., number of conductors per phase, conductor diameter, number of circuits, the geometrical arrangement of the conductor, among others [5].

The unit of measurement of the electric field is the volt per meter (V/m). According to the EPRI red book [6], the electric field at a point in space can be calculated from the charges. The charges are calculated by means of the voltage V_k, applied to each conductor, k, the calculation of these charges is given by:

$$[Q] = [P]^{-1} \cdot [V] \tag{1}$$

where:

[Q] = is the array of the line chargues, (C/m)

$[P]^{-1}$ = is the array of Maxwell potential coefficients, (m/F)

[V] = is the array of the conductor voltages, (V)

Each charge of the conductor contributes to the electric field, which is calculated by summing the contributions of all the charges. Figure 1 refers to the electric field, $\overline{\boldsymbol{E}}_k$, at point **M** caused by the load of the line, $\tilde{\boldsymbol{Q}}_k$, on the conductor, $\boldsymbol{k}$, is equal to the vectorial

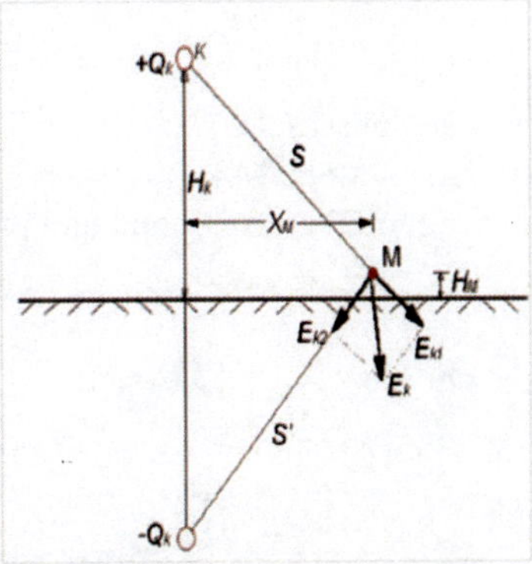

Fig. 1. Calculation of the electric field of a linear charge [6].

sum of the fields, $\overline{E}_{k1}$, due to, $\widetilde{Q}_k$, and $\overline{E}_{k2}$ generated by the image, $-\widetilde{Q}_k$ of $\widetilde{Q}_k$, on the ground [6]. Now, H_k, it represents the height of the conductor to the ground, X_M, it is the horizontal distance that exists from the measuring point M, and the conductor k, also H_M, it is the height from the ground to the measuring point M. The magnitude of the horizontal and vertical components of the electric field caused by the charges on the conductor k, $\widetilde{E}_{kx}$ and $\widetilde{E}_{ky}$ [7], is given by the equation.

$$\widetilde{E}_{kx} = \frac{\left(\widetilde{Q}_{rk} + j\widetilde{Q}_{ik}\right)}{2\pi \cdot \varepsilon} \cdot \left[\frac{X_M}{X^2{}_M + (H_k - H_M)^2} - \frac{X_M}{X^2{}_M + (H_k + H_M)^2}\right] \tag{2}$$

$$\widetilde{E}_{ky} = \frac{\left(\widetilde{Q}_{rk} + j\widetilde{Q}_{ik}\right)}{2\pi \cdot \varepsilon} \cdot \left[\frac{H_M - H_k}{X^2{}_M + (H_k - H_M)^2} - \frac{H_M + H_k}{X^2{}_M + (H_k + H_M)^2}\right] \tag{3}$$

The r.m.s. value of the electric field is given by:

$$E_{rms} = \sqrt{E^2{}_{rx} + E^2{}_{ix} + E^2{}_{ry} + E^2{}_{iy}} \tag{4}$$

2.2 Magnetic Field

The magnitude and direction of the force exerted on a moving electric charge be defined as a magnetic field. The international unit of magnetic flux density is the tesla (T) [8]. As in the electric field, the EPRI book [6] proposes the equation to calculate magnetic field in a double-circuit transmission line with a vertical configuration (Fig. 2).

$$B = \frac{2 \cdot \sqrt{3} \cdot P_v \cdot (I_1 + I_2)}{R^2} \tag{5}$$

where:

B = Magnetic Field, [μT]
P_v = Vertical distance between phases, [m]
I_1; I_2= Currents flowing in the system, [A]
R = horizontal distance between measuring point and phase

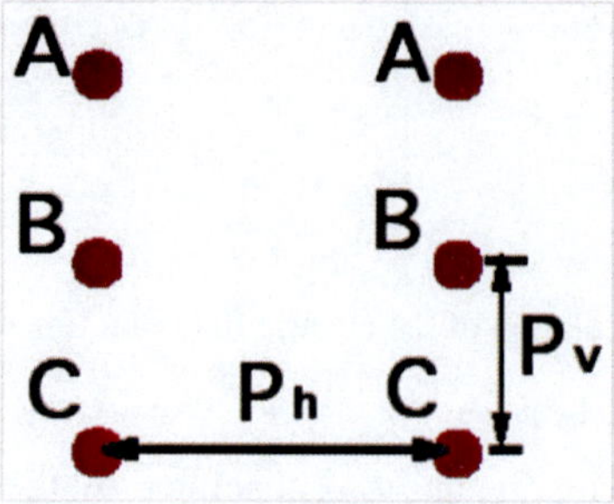

Fig. 2. Configuration of a double three-phase circuit [6].

3 Methodology

The simulation of the 230 kV transmission line section belonging to the province of Cotopaxi was carried out in two stages: the towers design in PLS_TOWER and route design with PLS_CADD. The transmission line under study interconnects the provinces of Pichincha and Tungurahua, with a length of 110.09 km, of which 60.5 km of this line borders the province of Cotopaxi. The transmission line crosses altitude zones that vary between 3500 and 2900 m above sea level. The Santa Rosa-Totoras line consists of two circuits A and B, with a single ACSR conductor (1x1113 MCM) of the Bluejay type. The shielding is provided by an OPGW conductor and another of the Steel type. Once the transmission line was simulated, the electromagnetic fields associated with the operation of the line are determined using the EMF calculation module of PLS_CADD. Then, applying the methodology of the EPRI red book, the analytical calculation of the electric and magnetic fields was performed to compare the results.

3.1 Tools

Metal Structures. In the Santa Rosa-Totoras transmission line, there are the following types of structures:

- **Suspension Structures:** Support the conductors, resist vertical loads, these towers do not support longitudinal stresses.
- **Terminal Structures:** Support the pull of the conductors on one side only, used at the beginning and end of the line.
- **Angular Structures:** Used at the vertices of the lines for changes of direction.

Insulation Criteria. For the selection of the insulation, it is necessary to know the separation between phases, phase, and ground and the number of insulators. In this work, was reviewed the selection of the number of insulators and the shielding angle.

Insulator String. The number of insulators per string, according to the contamination conditions in the project area, is given by:

$$D_c = \mathrm{V_{Max}} \cdot \mathrm{D_{cm}} \cdot \frac{1}{\delta} \tag{6}$$

$$\mathrm{N}_i = \frac{D_c}{D_i} \tag{7}$$

where:
D_c = Creepage distance, [mm]
$\mathrm{V_{Máx}}$ = Maximum system voltage (Phase to ground voltage)
$\mathrm{D_{cm}}$ = Minimum creepage distance, [kV/mm]
$\mathrm{N_i}$ = Number of insulators per string
D_i = Insulator creepage distance, [mm]
δ = Relative air density

Also, to comply with the atmospheric overvoltage evaluation, the critical flashover voltage (**CFO**) is calculated by the following relationship:

$$\mathrm{CFO_L} = \frac{\mathrm{LIWL_0}}{(1 - 3\sigma) \cdot \delta} \tag{8}$$

$$\mathrm{N}_i = \frac{\mathrm{CFO_L}}{\mathrm{CFO_i}} \tag{9}$$

where:
$\mathrm{CFO_L}$ = Critical flashover voltage at line
$\mathrm{CFO_i}$ = Critical flashover voltage at insulator
$\mathrm{LIWL_0}$ = Lightning impulse withstand voltage
σ = standard deviation
δ = Relative air density
$\mathrm{N_i}$ = Number of insulators per string

Shielding Angle. The calculation of the shield angle is given by BULLETIN 1724E-200 [9]. The shielding angle of a transmission line is found with the following equation:

$$\propto = \tan^{-1}\left(\frac{\text{Opposite}}{\text{Adjacent}}\right) \tag{10}$$

Tower Cross Arm Measure. The length of the transmission tower cross arm is related to the length of the insulator string, with the altitude of the area, the voltage level, all these factors influence the size that this element will have.

Horizontal Clearance. According to NESC-C2-2017 [10], the horizontal clearance of support is given by (Table 1):

Table 1. Horizontal clearance regulated by NESC-C2_107 [10].

Voltage level	Horizontal clearance (mm)
50 kV > x < 814 kV	d = 715 + 10 (This additional is given for each kV above 50 kV.)

Table 2. Vertical clearance regulated by NESC-C2_107 [10].

Voltage level	Vertical clearance (mm)
50 kV > x < 814 kV	d = 580 + 10 (This additional is given for each kV above 50 kV.)

Vertical Clearance. Clearance in any direction from line conductors at or near support to supports, and to vertical conductors (Table 2).

For vertical and horizontal clearances, it must have complied with that when voltages exceed 50 kV, the additional clearance will be increased by 3% for every 300 m above 1000 m above sea level [10].

Mechanical Loads on Towers. Transmission line towers must have sufficient mechanical resistance to withstand the loads that will act on them, without permanent deformation of their elements:

- Normal loads
- Exceptional loads
- Transverse loads

Safety Factors for Overloading. The loads must be applied in the mechanical design of the transmission towers but considering the safety factors for overloading [11] (Table 3):

Table 3. Overload factors [11].

Vertical loads	1,4
Vertical overload	1,2
Uplift load	1,5
Load angle	1,4
Longitudinal overload	1,2
Longitudinal unbalance	1,2
Terminal	1,2

3.2 Electromagnetic Field Emission Limits

The reference levels for exposure to the electric and magnetic field, for the general public and occupationally exposed personnel, are established in Table A4.1 of the Official Register No. 41 Book VI, Annex 10 of the Unified Text of Secondary Legislation, Environment (TULSMA). This standard is based on the International Commission on Non-Ionizing Radiation Protection (ICNIRP), 1998 [12] (Table 4).

Table 4. Reference levels for exposure to EMF [12].

Exposition	Intensity electric field (V/m)	Magnetic flux density (μT)
General public	4 167	83
Occupational exposure	8 333	417

4 Results

4.1 Simulation of Transmission Towers

Before simulating the electromagnetic fields in the transmission lines, it was necessary to model towers, taking into account design parameters given in the ASCE 10 standard that are applied by the PLS_TOWER software. Then, three types of structures with the following characteristics are modeled (Table 5):

Table 5. Design characteristics of transmission towers

	Towers		
	Terminal tower (DT1 + 0)	Angle tower (DA1 + 0)	Suspension tower (DS1 + 0)
Function	Terminal	Angular	Suspension
Material	Steel lattice	Steel lattice	Steel lattice
Voltage	230 kV	230 kV	230 kV
Height	45 m	45 m	46,25 m

The towers are double-circuit, horizontal, conical type, with connections for the guard wire, square (angular) and triangular (terminal and suspension) cross arms and insulators (thirteen), connected at each vertex of the cross arm (Fig. 3).

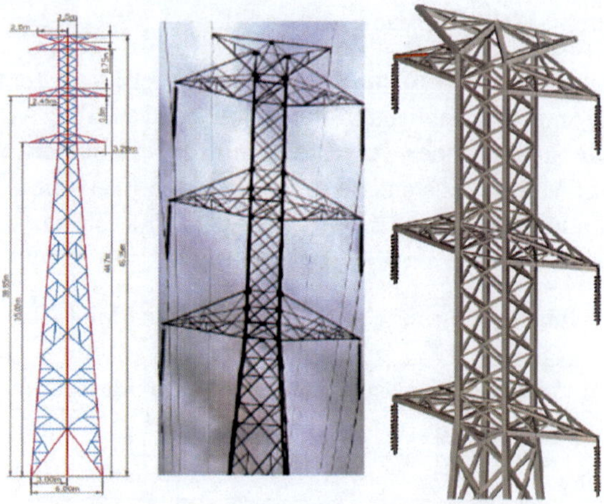

Fig. 3. Modeled suspension tower.

4.2 Santa Rosa-Totoras Transmission Line Route Simulation (Cotopaxi Section)

Taking as reference the data of Sect. 3, and using the PLS_CADD software, the simulation of the line route is obtained, with a length of 60.5 km, this project has 160 structures between terminal, angles, and suspension towers, Fig. 4 shows the transmission line implemented in Google Earth, with the georeferenced support structures.

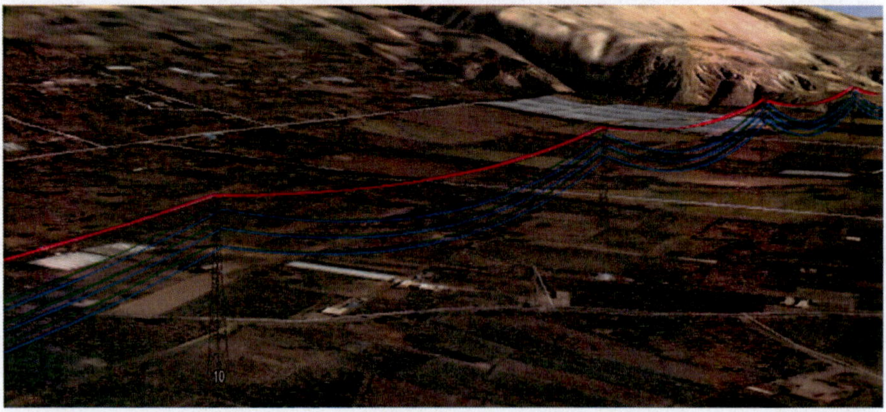

Fig. 4. 3D view of the transmission line stringing.

4.3 Simulation of Electromagnetic Fields

The calculations of the electromagnetic fields determined by PLS_CADD are based on the methodology of the EPRI Red Book [6]. The following factors must be taken into account for the simulation:

- The nominal voltage of the system, in this case, 230 kV.
- Current flowing through the conductor at 75, in this case, 878.3 A.
- The voltage and current for the guard wires are 0.
- Phase angles; (0°; 120°; −120°), for a, b and c respectively.
- Height of the measuring point (taking as reference the flat ground).
- Value in meters of the right of way (from the center of the tower to the left and right).
- Emission limits of the magnetic and electric field according to Table 6 [9].

Simulation Results. According to IEEE Std 664 [13], the electric and magnetic fields are measured for each structure, at mid-span and a height of one meter above ground level. The results provided by the EMF calculation module of PLS_CADD are: The EMF behavior within the right of way, being a double circuit line, the results will be the same towards +15 m and −15 m, having a reflection effect that creates a perfect parabola in both cases (Figs. 5 and 6).

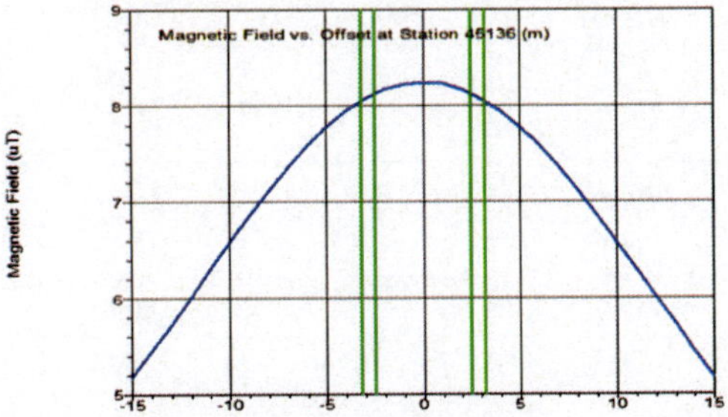

Fig. 5. Magnetic field graph.

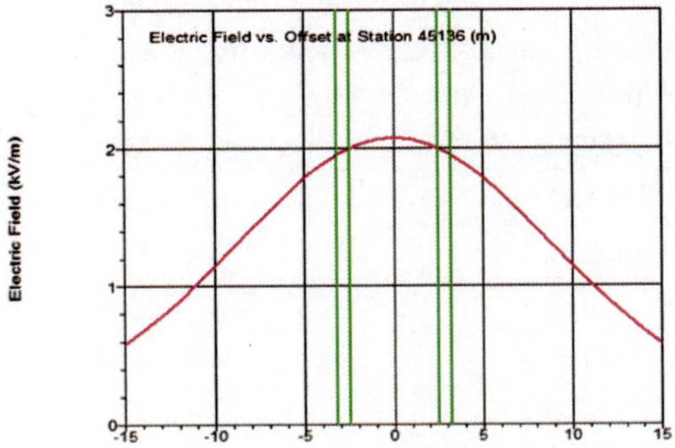

Fig. 6. Electric field graph.

The Table 6 shows the maximum emission values of the electric and magnetic fields of the 159 spans, thus verifying that the results obtained are within the emission limits regulated by ICNIRP.

Table 6. Maximum emission values of electric and magnetic fields.

Maximum emission values		
Distance	Magnetic field (uT)	Electric field (kV)
0	13,428	3,393
1	13,397	3,365
2	13,299	3,281
3	13,123	3,142
4	12,859	2,953
5	12,502	2,72
6	12,058	2,456
7	11,541	2,176
8	10,97	1,892
9	10,367	1,617
10	9,753	1,361
11	9,142	1,128
12	8,549	0,943
13	7,981	0,813
14	7,444	0,696
15	6,941	0,593

Table 7. EMF emission values with a 90% probability.

Weibull distribution at 90%		
Distance	Magnetic field (uT)	Electric field (kV/m)
0	9,8980	2,5018
1	9,8512	2,4370
2	9,8083	2,4506
3	9,6693	2,3481
4	9,4969	2,2078
5	9,2658	2,1050
6	8,9720	1,9459
7	8,6642	1,7767
8	8,3656	1,6076
9	8,0110	1,4309

(*continued*)

Table 7. (*continued*)

Weibull distribution at 90%		
Distance	Magnetic field (uT)	Electric field (kV/m)
10	7,5912	1,2531
11	7,1770	1,0879
12	6,8385	0,9280
13	6,4878	0,7896
14	6,1374	0,6754
15	5,7727	0,5927

Weibull Distribution. Using the Weibull distribution, the probability density function is obtained, which specifies the 90% relative probability that the electric and magnetic field results are below the simulated values (Table 7).

Finally, the variation per meter of the electric and magnetic field for a span is analyzed to show that the electric field varies to a greater extent than the magnetic field, as the measuring point approaches the center of the tower, the emission values are reduced because of the distance between the structure and the measuring point increases (Table 8).

Table 8. EMF emission values form 0 m to 15 m.

Electromagnetic fields		
Distance	Magnetic field (uT)	Electric field (kV)
0	1,692	6,428
1	1,682	6,414
2	1,654	6,372
3	1,607	6,301
4	1,544	6,204
5	1,467	6,082
6	1,378	5,937
7	1,281	5,772
8	1,178	5,59
9	1,073	5,395
10	0,967	5,19
11	0,865	4,978

(*continued*)

Table 8. (*continued*)

Electromagnetic fields		
Distance	Magnetic field (uT)	Electric field (kV)
12	0,766	4,764
13	0,673	4,549
14	0,587	4,337
15	0,507	4,128

With these data we calculate that the decrease for:

$$\% \text{ Electric Field} = \frac{(0,507 - 1,692)}{0,507} \cdot 100 = -233\% \tag{11}$$

$$\% \text{ Magnetic Field} = \frac{(4,128 - 6,428)}{4,128} \cdot 100 = -55,7\% \tag{12}$$

According to the results, it is evident that the magnetic field for structure N° 1 has a lower percentage of decrease than the electric field, which represents, approximately four times lower than the result of the electric field.

5 Conclusions

The standards and guidelines prescribe measurements at 1 m height. However, there is very little difference between the estimated values for the measurement point from 0 m to 1 m, in this case, the electric field changes by 0.501% and the magnetic field by 8.29%, showing that the magnetic field is more intense than the closer it is to the conductor. When analyzing the variation per meter of the electric and magnetic field for each span, it is evident that the electric field varies to a greater extent than the magnetic field; as the measurement point approaches the center of the tower, the emission values are reduced because the distance between the structure and the measurement point increases. In the case studied, an electric field variation of 1.185 (kV/m) was identified, which represents 233.73% of the highest emission value. While for the magnetic field the variation percentage is 55.72%. The simulation of the Santa Rosa-Totoras line (Cotopaxi section), allows establishing that the estimated electric and magnetic field values do not exceed the electromagnetic field emission limits established in both IEEE Std 664 and ICNIRP, which establish limits of 4.167 (kV/m) and 83 (uT), for the general public and for occasionally exposed personnel is 8.33 (kV/m) and 416.7 (uT) of electric field and magnetic field, respectively.

References

1. Castaño Lara, S., Gómez Ross, J., Gallego Real, A.: Campos Electromagnéticos generados por las líneas eléctricas de alta tensión. Efectos posibles la sobre el salud y medio ambiente. CIEMAT, Madrid (1997)

2. World Health Organization (OMS): Campos Electromágneticos. OMS (2016)
3. Morales, J.A., Gavela, P.: Determinación de Campos Electromagnéticos en Sistemas Eléctricos. Universidad Politécnica Salesiana **1**, 151 (2015)
4. Pineda Ochoa, F.E., Pineda Delgado, F.X.: Campos Eléctricos y Magnéticos (CEM) producidos por el Sistema Eléctrico de Potencia (SEP) de la Empresa Eléctrica Regional del Sur (EERSSA). ECUACIER **93**, 1–6 (2020)
5. de España, R.E.: Resumen sobre los campos eléctricos y magnéticos generados por las instalaciones eléctricas de alta tensión. Red Eléctrica España **1**, 1–8 (2010)
6. Electric Power Research Institute: AC Transmission Line Reference Book–200 kV and Above. Third Edition. EPRI **3**, 1–1074 (2005)
7. Morales, J., Gavela, P.: Determinación de Campos Electromágneticos en Sistemas Eléctricos. Quito (2015)
8. Pineda Ochoa, F., Pined Delgado, X.: Campos Eléctricos y Magnéticos (CEM) producidos por el Sistema Elécrico de Potencia (SEP) de la Empresa Eléctrica Regional del Sur (EERSSA). ECUACIER **93**, 1–6 (2020)
9. U.S.D. of A.R.U.S.E.S. Division: Bulletin 1724E-200 Design Manual for High Voltage Transmission Lines. No. May (2009)
10. A.S.C.C2-2017: 2017 National Electrical Safety Code (NESC-C2-2017) (2017)
11. M. de E. y E. Renovable: Diseño Electromecánico Linea De Subtransmisión a 69 kV. pp. 0–28
12. Tribunal Constitucional: Libro VI: De la Calidad Ambiental del Texto Unificado de Legislación Ambiental Secundaria del Ministerio del Ambiente. Regist. of. vol. VI, pp. 1–79 (2007)
13. D. Committee: IEEE Standard Procedures for Measurement of Power Frequency Electric and Magnetic Fields From AC Power Lines. vol. 644 (1994)

Anomaly Detection Method in Computer Systems by Means of Machine Learning

Henry Luis Mezones Santana[1](✉), Tatiana Elizabeth Cobeña Macias[2], and Mauricio Alexander Quimiz Moreira[3]

[1] Instituto de Postgrado, Universidad Tecnica de Manabí, Portoviejo, Ecuador
hmezones8602@utm.edu.ec
[2] Facultad de Ciencias Informáticas, Universidad Tecnica de Manabí, Portoviejo, Ecuador
[3] Dirección de TICS, Universidad Técnica de Manabí, Portoviejo, Ecuador

Abstract. The target of this study was to obtain a method of intruders' detection that offers alternative support for the protection of the computer systems, being necessary to obtain information derived from the LOGS of a host, classified by a Websockets across a computer system and perforations, for his respective evaluation, it was necessary to form the environment to compare the methods relating to the set of information preprocessed, to create the method and his respective putting on the verge of the model, to evaluate it and to implement it as support. The not supervised method was formulated to identify anomalies when the information is not labeled or classified. There were used skills of mining of information of automatic learning developed to implement a system of intruders' detection with the effective support of statistical specializing hardware and of mining of information: Orange and Python; Libraries for the manipulation and preparation of information: Pycaret, Pandas and Sklearn; Dash for the implementation. The different methods were evaluated establishing the comparative one based on such metric ones of classification as; Counterfoil of confusion, accuracy, precision, and sensibility. The quasi-experimental results of the new method provide few false positives allowing major valuations of detection on having used automatic learning, allowing the effectiveness of helping in the environment of application as support consuming few resources and contributing to the decision making.

Keywords: Unsupervised method · Intrusion detection · Websocket · Machine Learning

1 Introductions

The world of the technologies of the information is transforming, from what it needs updates you continue and practically instantaneous. With the development of the computer networks, the increase of the services that are big enough and the need to support the reliability, integrity, and availability of the transmitted information do that the safety of the computer systems is more important, since on the other hand, the attacks to the system increase turning into a serious problem [1], all this because the management of

M. Zambrano Vizuete et al. (Eds.): CI3 2021, LNNS 511, pp. 406–415, 2022.
https://doi.org/10.1007/978-3-031-11438-0_32

the information in digitized and electronic formats is growing rapidly, which leads to storing big volumes of information of diverse sources [2].

One of the first applications in safety and intruders' detection that has allowed to resist partially this evil is the Firewall and the systems of intruders' detection (IDS); who think that they are a tool more than it can use for the improvement of the safety of the systems of Information [3].

Automatic learning can be defined as the work realized by an artificial system on a set of information, suitable for learning the rules for yes alone from the information of training. The use of automatic learning has gained importance due to its efficacy in the evaluation of safety. This has allowed the inclusion of this type of learning in the majority of the applications of recognition of bosses and mining of information, optimizing considerably the effects obtained with the traditional methods in different tasks as the detection of anomalies. Also, at the time of developing a method for the detection of anomalies there exist several challenges to bear in mind because the environment of the anomalies is dynamic. Therefore, it is necessary to adapt an algorithm of learning automatic and to evaluate it with metric of performance [4], a characteristic that is present in the skills.

This growth owes to a great extent to the detection of specific vulnerabilities in computer systems and networks, which implies a major danger. For this reason, it is necessary to study other possible methods to detect anomalies in the system. Before this serious situation, more and more institutions invest in solutions of software and hardware to minimize the impact caused by the proliferation of local and remote threats. The detection of anomalies is a technology of mining information with a wide fan of applications, centered on social security, such as analysis of computer and social networks, analysis of bank deals, analysis of information of sensors, etc. [5]. In a computer network, the unusual bosses of behavior might mean that a host is sending information sensitive to not authorized destination.

At present, there exist several studies, like that the lectures of unusual behavior of a sensor of a spaceship might mean an error in some component of the ship (Fujimaki). Even when one works with medical images, a sudden change in the intensity of the pixels in unexpected places can indicate the presence of malignant tumors [5]. The different behaviors in the information of the deals of the credit cards might indicate the theft of the identity or the credit card [6]. As for the use of skills of mining of information applied to the financial sector, the study presented by [7], tells the atypical detection of the fraud with credit cards with skills that unusual bosses detect across the algorithm KNN. The skill detects the information of entry, calculates the most nearby neighbor and the punctuation of similarity, determining if the information has a suspicion of fraud. Another investigation presents an extract of the different remarks of detection of atypical values, being these: the approach based on the statistics, the approach based on the distance, and the approach based on the thickness [8]. The authors present a discussion related to the detection of atypical values, the methods used to group the set of information, conclude with the algorithm kmean as the most popular to group a set of information. Also, other studies use skills of mining information, statistical methods, or both. For the detection of anomalies, the skills of the knearest neighbor (KNN, Isolation

Forest) have been applied commonly together with others to find unusual patterns during the behavior of the information or to improve the yield of the process.

In the study of [9], Isolation Forest can detect anomalous sensors in time of execution, allowing the exclusion or cross-check of this sensor on the part of the agent of the environment. The related studies also use the statistician chisquare, since it is the case of [10] whose method the statistician implements for the detection of intrusions in a computer network to segment it in normal and abnormal traffic. The study includes a method that uses the skill of the machine of vectors of support (SVM). All these bosses that do not continue the desirable behavior are treated as anomalies and his detection allows to prepare new attacks and malfunctions. In this investigation of [11], they stress on the general vision of the system of detection of intrusive methods and skills where they showed that the safety of the systems and the networks at present keeps on being the principal worry of the investigators and system managers.

For the exposed previous thing, these studies are very important for the developed one of this investigation, since it was needed to analyze the problems on the different theoretical essentials based in machine learning. In this sense, it proposes to use two stages, the first one to use the algorithm Isolation Forest, which is a method not supervised to identify atypical values when the information is not labeled, that is to say, does not know the classification of the remarks, the second one to use supervised methods when the byline is already classified to determine if this information is real.

2 Materials and Methods

The methodology of this study has developed in six different phases that consist of multiple tasks: phase of comprehension of the business, the definition of the targets of the mining of information, comprehension of the information, cross-check of the quality of the information, preparation of the information, shaped, evaluation and implementation. He adapted himself according to the project methodology of mining of information CRISP-DM (Cross Industry Standard Process for Data Mining), which provides a normalized description of the cycle of life of a standard project of analysis of information [12]. This agile methodology of the science of the information gives solutions of predictive analysis and intelligent applications of efficient form.

As skills the mining of information was applied with automatic learning developed to implement a system of intruders' detection of an effective way, the same one it splits into two sections, skills of machine learning and not supervised and supervised methods, where it is structured by levels of abstraction of the methodology CRISP-DM, the first two levels include all the processes of mining of information that are present in a project of these characteristics. The third level alludes to those proper tasks of a specific typology of investigation and the fourth and last level, groups the set of actions, decisions, and results on the project of development of Information.

The specializing hardware used for the statistics and the mining of information consisted in: Orange and Python; Libraries for the manipulation and preparation of information: Pycaret, Pandas and Sklearn; Dash for the implementation; Node js for WebSocket and teams computational. Isolation Forest formulated to identify anomalies when the information is not classified; that is to say, there is not known the real categorization (normal abnormal). It was elected this model because it groups these not common remarks and it is not known if they are anomalous or normal. For it, one proceeded to create a file.CSV extracted from the records of a computer.

3 Results and Discussion

In this paragraph of the document there appear detailed the processes and results obtained following the methodology used and detailed previously. The reliability of these results appears in each of his phases, which have been established as skills for the discovery of the hidden studied knowledge. The different phases of methodology are described to offer solutions of analysis predictive and intelligently of efficient form, and the following results appear:

In the Table 1 there are described the information that were obtained by the tool that developed websocket, which were transformed so that the method could interpret them. The form in which we divide information is important because there is information that we will not use during the process of shaped and that will use in the end to validate our results by means of the simulation of real information. The information that we use for the shaped one are subdivided by us to evaluate two stages, that of training and that of test. Therefore, Fig. 1 appears in the following illustration (Table 2 and 3).

Table 1. Description of data obtained

IP	User	Date	Method	Code answer	Size object
14476103165	0	30042021141111	2	200	0
162214148112	0	30042021141110	1	200	8165
162214148112	0	30042021141122	2	200	8602
14476103165	0	30042021141422	2	200	0
Data =	**30524**	**Rows =**	**6**		

Table 2. Anomaly prediction data

IP	User	Date	Method	Code answer	Size object	Anomaly	Anomaly_Score
1399945136	0	4052021161913	1	200	8165	0	−0.152243
1615318020	0	23052021113030	2	200	413	0	−0.216191
1678675216	0	7052021085824	2	200	238	0	−0.213977
19052205125	0	16052021051817	1	304	0	0	−0.023594
4219222476	0	17052021090942	2	200	221	0	−0.238237

With an initial fact of 30524 lines and 6 columns, as soon as the information was initiated with the bookstore mentioned previously, this information transforms from 30524

Table 3. Comparison of supervised methods

Model	Accuracy	AUC	Recall	Prec.	F1	Kappa	MCC	TT (Sec)
Random forest classifier	0.9979	0.998	0.971	0.987	0.979	0.9785	0.978	1.11
Extra trees classifier	0.9977	0.998	0.970	0.984	0.977	0.9761	0.976	0.84
Light gradient boosting machine	0.9976	0.999	0.967	0.985	0.976	0.9751	0.975	0.20
Decision tree classifier	0.9969	0.982	0.967	0.971	0.969	0.9675	0.967	0.05
Gradient boosting classifier	0.9944	0.999	0.921	0.968	0.944	0.9411	0.941	1.19
Ada boost classifier	0.9894	0.998	0.834	0.951	0.888	0.8832	0.885	0.52
Ridge classifier	0.9647	0.000	0.314	0.980	0.475	0.4623	0.544	0.04
K neighbors classifier	0.9644	0.871	0.487	0.728	0.581	0.5638	0.577	0.29
Linear discriminant analysis	0.9521	0.979	0.748	0.524	0.615	0.5906	0.602	0.09
Logistic regression	0.9490	0.500	0.000	0.000	0.000	0.0000	0.000	0.07
Naive bayes	0.9275	0.681	0.265	0.279	0.271	0.2335	0.233	0.04
SVM - linear kernel	0.8471	0.000	0.194	0.397	0.121	0.0908	0.125	0.35
Quadratic discriminant analysis	0.1408	0.000	0.900	0.045	0.087	0.000	0.000	0.05

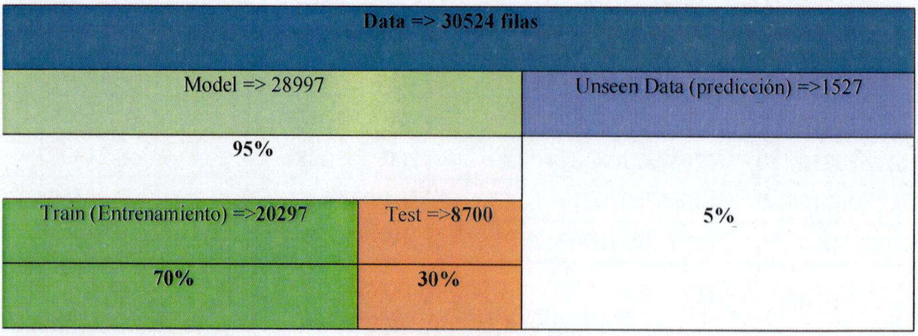

Fig. 1. Data division.

lines to 19 columns as there appears in the Fig. 2, which orders us for the characteristics that influence the detection of anomalies, representing characteristics that impel the prophecy higher, whereas the blue color, they impel to the lowest value.

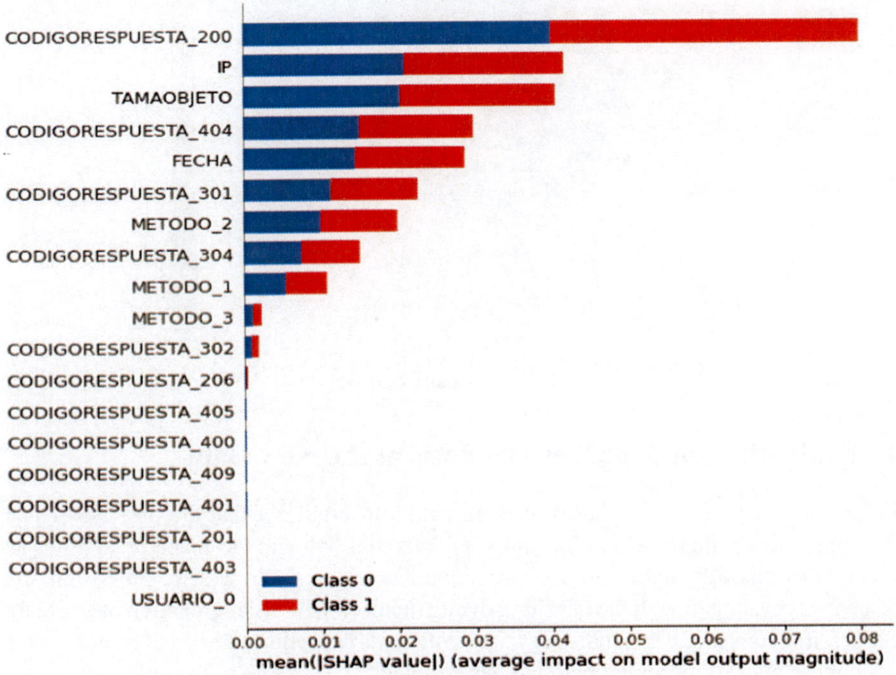

Fig. 2. Data preformatted by the Pycaret library

The parameters that appear next are established for the creation of the method of the forest of isolation:

Iforest (Contamination = auto, Max_features = 1.0, Max_samples = 'auto', N_estimators = 100, N_jobs = −1, Random_state = 123, Verbose = 0)

With the help of the bookstore Pycaret, we establish the parameters of initialization for the shortcoming of the algorithm isolation forest with automatic contamination that simply controls the threshold for the function of decision when a fact can be an anomalous observation. As soon as the creation of the method was established, the possible anomalies are identified and printing the new information labeled by the classes (Anomaly – Anomalous Punctuation), these are added after be awarded that the remarks with atypical values in one or more of his variables separate of the rest with the major facility, those remarks with the minor distance I mediate they should be the most atypical, grouping her for normal and anomalous class represented for (1–0) as 2 appears in the table.

Figure 3 is the graphic representation of the classes of abnormal and normal information, where the blue ones are 0 and the yellow ones are 1, showing the above-mentioned a different pattern from others.

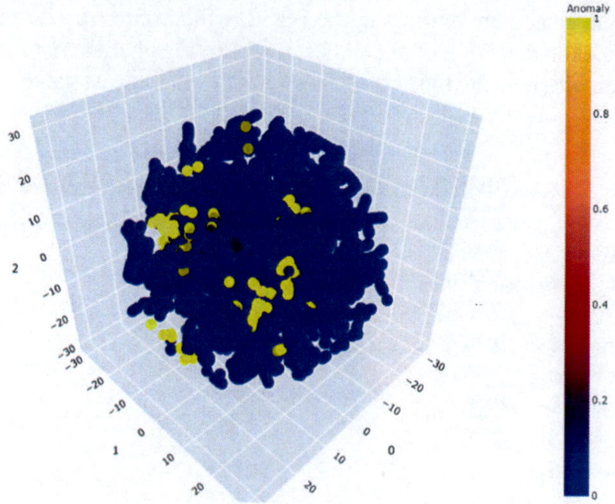

Fig. 3. Anomaly outliers

4 Evaluating the Model on the Basis of the New Data

Before breaking down the phases, it is important to highlight that the purpose of this research is to obtain a method for intrusion detection, which is to make predictions as accurate as possible and to be used as support in a computer system. To do this, its behavior is evaluated with the data already restructured with its aggregation of anomaly classes, it is convenient to handle supervised methods to predict and classify future cases since these algorithms are trained with labeled data.

To use a supervised method, 13 methods were compared and they were evaluated by means of the metric ones of evaluation of a model of automatic learning, like 3 appears in the Table. Leaving as result two almost perfect models, one with a major precision of 0,9879 and accuracy of 0,9979 for Random Forest, the second Extra Tree with a precision of 0,9842 and accuracy of 0,9977, it is necessary to mention that both algorithms are of classification, with which we will evaluate the information of the forest of predicted isolation.

Comparison of the Extra method three vs Random Forest: This comparison was realized between these two methods after refining them, for his evaluation with the metric ones of yield accuracy, (recall) sensibility, precision and punctuation f1, establishing a better result of the chance forest as it shows in the Table 4.

The compared methods were evaluated according to their metrics once the confusion matrix Table 5 was detected, where we can see the real positives, false positives, false negatives and real positives described by the most relevant methods according to our set of information.

4.1 Implementation and Prediction

Comparing the information of the methods from the training, one determined that the best classifier for the set of information is the random forest, both in the stage of training

Table 4. Comparison of methods

	Evaluation metrics				
	Exactitud	Recall	Prec.	F1	Time
Random forest	0.9977	0.9633	0.9906	0.9767	1.119
Extra tree	0.9972	0.9541	0.9905	0.9720	0.847

Table 5. Confusion matrix.

	Evaluation metrics			
	Positive true	False positives	False negatives	True negatives
Random forest	8716	5	10	427
Extra tree	8717	4	10	427

Table 6. Random forest comparation

Random forest					
	Evaluation metrics				
	Accuracy	Recall	Prec.	F1	Time
Random forest initial	0.9977	0.9633	0.9906	0.9767	1.119
Random forest prediction	0.9984	0.9771	0.9884	0.9827	0.847
Random forest	0.9889	09998	0.9999	0.9989	

and in that of test and prophecy to these we add the deployment. These results are corroborated by a study [13], where one affirms that using Random Forest Borderline SMOTE there have been detected almost all the present and anomalies almost all the normal elements. These results [14], indicate that for a correct interpretation the open bosses must be exhibited, as well as the possibility of visualizing them; in addition to using the knowledge discovered then to make use of the created model. For the problems of high dimension that contain a big number of irrelevant attributes, isolation forest can reach a high yield of detection rapidly with an additional selector of attributes; whereas a method based on the distance has a yield of detection or needs very much more time, we demonstrate this in the Table 6 establishing the comparative one of the method created from his tuning, prophecy, and deployment; these are valued with metric of evaluation of machine learning who corroborates with the study of [4], which realized his investigation focused in "Model of detection of intrusions in systems of network, realizing selection of characteristics with FDR and training and classification with SOM", being the target to evaluate the efficiency of a model of detection of intrusions of network proposed, using metric of sensibility and specificity, utilizing a process of simulation that uses the byline - set NSL-KDD the latter as his byline.

5 Conclusions

According to the bibliographic study, the application of data mining allows covering a wide variety of environments such as medicine, education, engineering, among others, to find optimal solutions to problems such as the detection of anomalies in computer systems through machine learning. During this research, an analysis of how logs act within a host was carried out, where some were identified, but only one was chosen, that of the Apache service, information that has been useful to train a machine learning method, which allowed the following conclusions to be reached:

Using the pycaret library enabled efficient modeling, advanced preprocessing and assembly, accelerating the experiment cycle exponentially and making it productive.

The implementation of the method of detection of anomalies is based on the levels of attraction of the methodology CRISP-DM, which has big relevancy on having analyzed and interpret the structure of the information, followed by the shaped one, evaluation and implementation, allowing the securing of a new method of detention of anomalies.

The investigator could state, that the methods of existing detection of anomalies in the investigation fit to the general target that is the use of the method as support for the hosts, this based on his metric ones obtained by his high yield, on having used supervised method Random Forest and not supervised Isolation Forest that is used as criterion to identify anomalies on values to typical of his variables that separate of the rest with major facility,

The raised result shows the effectiveness of the implementation of the method of detection of anomalies inside environments in computer systems supporting the decision making of contingencies that can persecutor the service offered to the host.

6 Recommendations

To give continuity to this investigation, it is recommended for future works, to raise a hybrid system that identifies attacks based on atypical, capable values of comparing connections in real-time with a database of signatures, and detecting new attacks by means of the use of combined algorithms of learning supervised and not supervised, allowing the evaluation of the traffic of the host and of the network of distributed form.

References

1. Rivero Perez, J.: Tecnicas de aprendizaje automatico para la deteccion de intrusos en redes de computadoras. Revista Cubana de Ciencias Informaticas, Universidad de las Ciencias Informaticas. **8**(4), 52–73 (2014)
2. Kuna, H., et al.: Procedimiento de explotacion de informacion para la identificacion de campos anomalos en base de datos alfanumericos. Revista Latinoamericana de Ingenieria. **3**, 102–106 (2013)
3. Martínez, G., et al.: Sistema de detección de intrusos en redes corporativas. Scientia et technica **22**(1), 60–68 (2017)
4. Hoz De la, D., et al: Modelo de detección de intrusiones en sistema de red, realizando selección de características con FDR, entrenamiento y clasificación con SOM Universidad de LaCosta. Inge Cuc (2012)

5. Lopez-Avila, L., Acosta-Mendoza, N.: Detección de anomalías basada en aprendizaje profundo: Revisión. Revista Cubana de Ciencias Informáticas **3**(3), 107–123 (2019)
6. Save, P., et al.: A novel idea for credit card fraud detection using decision tree. Int. J. Comput. Appl. **161**(13), 6–9 (2017)
7. Malini, N., Pushpa, M.: Machine learning review. In: Third International Conference on Advances in Electrical, Electronics, Information, Communication and Bio-Informatics, pp. 255–258. IEEE (2017)
8. Orellana, M., Cedillo, P.: Outlier detection with data mining techniques and statistical methods. In: International Conference on Information Systems and Computer Science (INCISCOS), pp. 51–56 (2019)
9. Bhatti, M., et al.: Outlier detection in indoor localization and Internet of Things (IoT) using machine learning. J. Commun. Netw. **22**(3), 236–243 (2020)
10. Thaseen, I., et al.: Modelo de detección de intrusiones que utiliza la fusión de la selección de características de chi-cuadrado y SVM de clases múltiples. J. King Saud Univ.Comput. Inf. Sci. **29**(4), 462–472 (2017)
11. Jakic, P., et al.: The overview of intrusion detection system methods and techniques. In: Rinteza 2019-International Scientific Conference on Information Technology and Data Related Research, vol. 29, no. 4, pp. 462–472 (2017)
12. Rodriguez, L., Garcia, L.: Adaptation to a methodology of data mining for applying to unsupervised problems type attribute-value. Revista Universidad y Sociedad **8**(4), 42–52 (2016)
13. Zamora, J.: Comparativa y análisis de algoritmos de aprendizaje automático para la predicción del tipo predominante de cubierta arbórea, Master Dissertation (2018)
14. Ballesteros, H.: Mineria de datos, RECIMUNDO: Revista cientifica de la investigacion y el conocimiento, vol. 2, no. 1, pp. 339–349 (2018)

Software

Mobile Application to Monitor Body Mass Index and Heart Rate in the Pandemic Stage Covid-19

J. A. Mocha-Bonilla(✉), Sánchez-Guerrero Javier, Flores Gabriela Rosita, and Núñez Ramírez Judith

Facultad de Ciencias Humanas y de La Educación, Unidad de Investigación y Desarrollo FCHE, Carrera PAFD, Universidad Técnica de Ambato, Ambato, Ecuador
{ja.mocha,jsanchez,rg.flores,judithdnunezr}@uta.edu.ec

Abstract. During the year 2020, with the arrival of the Covid-19 pandemic, the confinement and health emergency gave rise to the application of mobile technology in the areas of health, numerous scientific literature gives account of the support provided by the "mHealth" in health and personal care, Thus, the group of researchers developed a free mobile application to monitor the body mass index and heart rate of a group of people of limited economic resources, a quantitative methodology was applied in the study sample to collect numerical data, the app designed has a base to record user data; It was possible to obtain data on body mass index and heart rate during six months of the pandemic, and reliable results were obtained for the control, prevention and personal care of the participants.

Keywords: Mobile application · BMI · Heart rate · Pandemic · Pulse monitor

1 Introduction

Due to the health emergency of the Covid-19 pandemic, today technology is essential for life, because in it we find a number of mobile applications; regardless of what we are looking for, we will always find mobile applications (App), which are available on all virtual platforms for Android and iOS operating systems [1]; these applications have little weight for storage in the memory of the phone, but have great importance for life; thanks to them we find tips on how to improve fitness, exercise the body and various applications related to personal care and health [2]. The incorporation of information and communication technology (ICT) in everyday life has changed lifestyles and made us more technological. More and more young people have a smartphone with various applications (App) which are downloaded for free or paid, however, the practicality given at home will educate, raise awareness and manage a responsible use of digital tools, and promotes a proper application of APPS in daily life [3].

In the current times, because of the Covid-19 pandemic, the design of mobile applications Apps helps in health care, since for reasons of sars-cov-2 medical equipment greatly increase costs, similarly health professionals, to mitigate the effects of contagion among the population, implement some safety standards through health application related to Covid-19 [4], apply emerging measures such as disease prevention, as well

M. Zambrano Vizuete et al. (Eds.): CI3 2021, LNNS 511, pp. 419–430, 2022.
https://doi.org/10.1007/978-3-031-11438-0_33

as telephone medical consultations and telemedicine. A report provided by the World Health Organization WHO mentions the use of mobile health or "mHealth" defined as the use of mobile devices as well as wireless technology for the benefit of health and medical monitoring [5]; the development of mobile applications for personal use helps to improve fitness and personal health [6], there are heart rate monitors and applications that provide information on calorie consumption, so that people can access detailed information on the amount of food they need every day to be able to perform their activities, promoting systematic physical activity pro-grams and healthy eating [7].

More than a year ago, due to the Covid-19 pandemic, technological tools were used to collect data on vital signs in patients, such records are easily accessible on mobile devices by users. Thus, two specific needs arise for low-income families, the control of the so-called Body Mass Index (BMI) and also the monitoring of Heart Rate (HR); The Body Mass Index is an indicator to control the weight of the person and to know at what level he/she is (underweight, normal, overweight and obese) parameters that allow the person to maintain control over food and improve health levels; as for the heart rate, it is essential to control the heartbeat, i.e. the number of contractions or beats that the heart makes in a minute. With the situation of the Covid-19 health emergency, the implementation of wireless technology is motivated, as wired devices in medicine have a high cost, which is why we are considering carrying out a study to develop a mobile application to obtain Body Mass Index and Heart Rate values.

2 State of the Art

Various studies point to the regular use of mobile devices through apps for personal care, with which the use reaches a greater number of users, as smartphones provide the software packages to download and use the various applications. Mobile devices help to obtain personal data to record and maintain personal control and care in the individual's health [8]. Several researches agree that self-care has a multilateral point of view, as it covers a wide range of functions [9] by highlighting psychosocial aspects, health, care, personal well-being and physical condition; thanks to technological advances today we have the ability to perform various functions with the support and use of mobile apps, the new technological era and confinement, motivates the use of technologies, because thanks to them we can obtain valid and reliable data that are recorded on smartphones [10]. Nowadays, technological applications have great versatility, the existing mobile phones and computers are used for prevention and health care.

There are countless areas of the Apps, as some help to: count steps, calories burned, prepare exercise routines, simulate virtual landscapes for running, heart rate control, among others; the most important use is that they facilitate with information and calculation, by providing real data to promote the quality of life of people [11].

When reviewing several investigations that present as main objective to develop APP for care in the pandemic, we have the case of applications for food delivery [12], necessary for the promotion of physical activity, which today for reasons of the pandemic worldwide is applied, as it is carried out together with physical education professionals, the so-called virtual physical activity, nutrition and healthy physical activity at home; because with the support of technology and mobile applications, actions are recorded day by day through the app, and develop physical activity procedures to promote self-care [13]. At present, due to the pandemic, many people carry out physical activity at home, the reason is to take care of themselves and each member of the family group so as not to contract the virus; a situation for which mobile devices are currently widely used, we agree with [14] who mentions that the applications do not detect the virus, but they are a prevention tool against the spread of the virus, as people feel comfortable when having to use an application on their personal phone due to the confinement; the author [15] affirms that mobile devices are a powerful motivational resource for health, especially when performing personalized physical activity, as the apps present a large number of motivational components to improve personal health, highlighting the online application of metal care with the support of mobile applications [16].

As an example of applications related to our study we can mention the Fitbit app, which presents a device for physical activity monitors by means of heart rate, the purpose is to improve physical characteristics in individuals, it also has clinical use to improve health in people; we also have FibriCheck a smartphone application of re-mote mobile health based on demand (mHealth) that measures heart rate [17]. An-other example of Apps that today play a very important role in all areas, specifically in the field of health closely linked to body weight; being used by people with obesity, who want to improve their lifestyle and use mobile applications that contribute to treatment [18]. New applications provide exercise routines that should be performed daily, are fundamental in the control and diet that the person should consume to improve health; in such a way we agree with [19] who mentions that mobile applications have to be easy to access and use. Regarding mobile applications related to the Covid-19 pandemic, there are many high-quality applications, however, the useful-ness of the functions is limiting medical care [20]. Therefore, it should always be kept in mind that a person can gain, maintain or lose weight, as well as heart rates vary depending on the type of activity performed, therefore, the purpose is to have a low-cost application created by the group of researchers so that the person can obtain personal data of their Body Mass Index and Heart Rate for prevention and personal health care.

3 Methodology

The proposal to develop the device was born as a necessity in the pandemic, since the health system is collapsed, in the same way the measures of social distancing and confinement gave way to the application of mobile technology. In year 2020 the research team designed the mobile application; however, it was not possible to apply and record data for fear of contagion. The mobile application designed allows the calculation of data such as body mass index, step count, calorie consumption per day and heart rate [21], during the month of July 2020 the group of researchers decided to conduct the field work, and respected the biosafety standards issued by the Government of Ecuador to apply and record data with the designed APP.

The study was executed using a quantitative approach, since numerical data are presented that relate to the body mass index, which was performed in two moments with an interval of six months; the heart rate was taken in three moments after the performance of physical activity agreeing with [22].

3.1 Subjects

The participants of our study were considered according to the health emergency present in Ecuador, to avoid the spread of Covid-19 three families were taken from the city of Ambato, belonging to the neighborhood of La Pradera, people close to the home of one of the researchers, this sample represents a total of 20 people to test the functions of the application designed for the calculation of BMI and HR. The most relevant data and results are shown below.

3.2 Instruments and Design Phases

The present research work was developed through 2 phases, which are detailed below.

Phase 1. Development of the application to calculate BMI. The first phase was designed using MIT App Inventor because the operating system was planned for Android type phones, it is here where the graphic interface of the design is found in which it presents the labels, buttons, icons and logos, it was done using the free tool known as Mockflow. The main screen contains in the upper part registration and data view buttons, while in the lower part the access buttons for the users are presented.

The access screen to the Body Mass Index calculation shows the fields to enter the data for the calculation of the Body Mass Index; where it is necessary to enter user data such as: weight, height, age and gender, once these data are entered, click on the calculate button and the Body Mass Index percentage will be obtained, which is calculated based on the data provided by the person (Fig. 1).

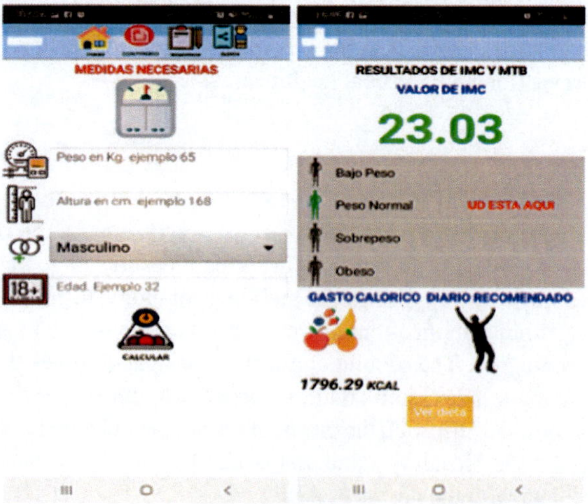

Fig. 1. Data entry and body mass index calculation buttons.

Phase 2. Development of the heart rate monitor. For the design of the sensor a friendly and easy to use programming was used, where all the buttons and elements that make up the interface are placed. The heart rate calculation function was designed to receive data through a sensor that is integrated into an Arduino board, at the top is the connection button to the bluetooth module, once the connection is initiated the application will receive the pulse data of the individual.

3.3 Components

The SEN-11574 sensor is a device composed of a green LED light; it was used to obtain the data of the heart rate in the person; for its application 4 mA of 5 V current consumption is needed; subsequently the data is transmitted and uses the wireless component through the Bluetooth HC-05, which presents great versatility due to the size, the most important thing is the range and the transmission of the heart rate data, only a supply voltage of 3.6 V–6 V is required.

The pulsations of the person are transmitted to the ARDUINO UNO board, which is an electronic board that has free hardware and software, has a USB connector for programming and a power connector, this device is designed to work with a voltage of 5 V, however, for our case a 9–V battery was used to turn on and off the monitor. Finally, we have the SHIELD BOARD which is an additional component that is placed on the Arduino board and allows to expand the capacity of the board as it was assembled by connecting pins.

The "eHealth-UTA" device allows taking the heart rate and uses the technique of photoplethysmography using light spikes that help the accuracy in the measurement of HR [23], which is applied by a non-invasive light source to the person on the index finger, and the heart rate data is obtained in one minute (Fig. 2).

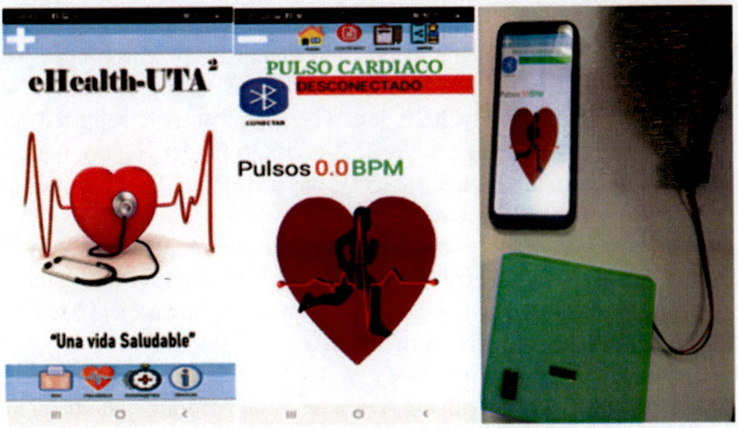

Fig. 2. Data entry buttons and heart rate calculation.

4 Results

In the first instance, the members of each family were socialized on the use of the application, the correct way to download and install the App, designed for devices with Android system; once it was installed, the data entry functions for the calculation of the body mass index were explained, registering weight, height, age and gen-der, which provides the BMI; subsequently, the index finger must be brought close to the sensor for 15 s to obtain the heart rate.

Due to the pandemic situation, it was applied to three families with 7 to 8 family members.

4.1 Results of the Body Mass Index Calculation

Table 1. Results of the body mass index in family 1 recorded with the application.

Subject	Age	Gender	Level	WEIGHT kg Take 1 PANDEMIC July 2020	BMI Take 1	WEIGHT kg Take 2 PANDEMIC January 2021	BMI Take 2	Level
1	37	Male	Normal	59,85	24,59	59,40	24,41	Normal
2	35	Female	Underweight	41,85	16,35	46,35	18,11	Normal
3	17	Male	Normal	60,75	24,34	61,20	24,52	Normal
4	16	Female	Normal	50,85	20,11	50,85	20,11	Normal
5	68	Male	Overweight	67,50	26,37	67,50	26,37	Overweight

Table 1 shows the results of the calculation of the Body Mass Index, which are recorded based on the data entered by the members of family 1; of the five family members in the first intake in July 2020, it can be seen that male subject 1 has a BMI of 24.59 (normal), female subject 2 has a BMI of 16.35 (underweight), male subject 3 has a BMI of 24.34 (normal), female subject 4 has a BMI of 20.11 (normal) and male subject 5 has a BMI of 26.37 (overweight).

While in the second sampling after six months in the month of January 2021, the most relevant data could be verified thanks to the application, registering that person 2 of (underweight) rose to 18.11 obtaining a BMI (normal), while subject 5 remained with a BMI of 26.37 that is to say with (overweight).

Table 2 shows the results of the Body Mass Index calculation based on the data entered by the members of family 2; out of a total of eight family members; in the first intake during the month of July 2020 it can be observed that male subject 1 has a BMI of 27.04 (overweight), female subject 2 has a BMI of 23.83 (normal), male subject 3 has a BMI of 21.05 (normal), female subject 4 has a BMI of 21, 60 (normal), male subject 5 has a BMI of 22.60 (normal), female subject 6 has a BMI of 19.80 (normal), male

Table 2. Results of the body mass index in family 2 recorded with the application.

Subject	Age	Gender	Level	WEIGHT kg Take 1 PANDEMIC July 2020	BMI Take 1	WEIGHT kg Take 2 PANDEMIC January 2021	BMI Take 2	Level
1	44	Male	Overweight	61,65	27,04	62,10	27,24	Overweight
2	40	Female	Normal	55,80	23,83	56,70	24,22	Normal
3	23	Male	Normal	43,65	21,05	43,65	21,05	Normal
4	22	Female	Normal	48,60	21,60	48,60	21,60	Normal
5	81	Male	Normal	49,50	22,60	49,50	22,60	Normal
6	79	Female	Normal	46,35	19,80	47,25	20,18	Normal
7	39	Male	Overweight	71,10	26,12	70,20	25,79	Overweight
8	28	Female	Normal	53,55	20,66	53,55	20,66	Normal

subject 7 has a BMI of 26.12 (overweight) and female subject 8 has a BMI of 20.66 (normal).

In the second sampling, after six months in January 2021, the most relevant data, based on the data from the application, showed that person 1, who was overweight, increased to 27.24, continuing with a BMI of (over-weight), while subject 7 lowered his BMI to 25.79 but continues with (overweight).

Table 3. Body mass index results in family 3 recorded with the application.

Subject	Age	Gender	Level	WEIGHT kg Take 1 PANDEMIC July 2020	BMI Take 1	WEIGHT kg Take 2 PANDEMIC January 2021	BMI Take 2	Level
1	39	Male	Overweight	72,90	27,10	71,10	26,44	Overweight
2	37	Female	Normal	56,25	20,91	56,25	20,91	Normal
3	71	Male	Normal	53,10	23,92	53,55	24,12	Normal
4	68	Female	Normal	48,60	19,72	48,60	19,72	Normal
5	20	Male	Normal	52,20	22,59	52,20	22,59	Normal
6	19	Female	Normal	51,75	22,11	51,75	22,11	Normal
7	16	Male	Normal	47,25	20,45	49,5	21,42	Normal

Table 3 shows the results of the Body Mass Index calculation based on the data entered by the members of the third family; out of a total of seven family members, the first measurement in the month of July 2020 shows that male subject 1 has a BMI of 27.10 (overweight), female subject 2 has a BMI of 20.91 (normal), male subject 3 has a BMI of 23, 92 (normal), female subject 4 has a BMI of 19.72 (normal) and male subject 5 has a BMI of 22.59 (normal), female subject 6 has a BMI of 22.11 (normal) and female subject 7 has a BMI of 20.45 (normal). Finally, during the second intake, after six months in January 2021, the most relevant data recorded in the application showed that person 1 who obtained overweight dropped to 26.44 continuing with a BMI of (overweight), while the other subjects are in a normal weight.

4.2 Results of the Heart Rate Calculation

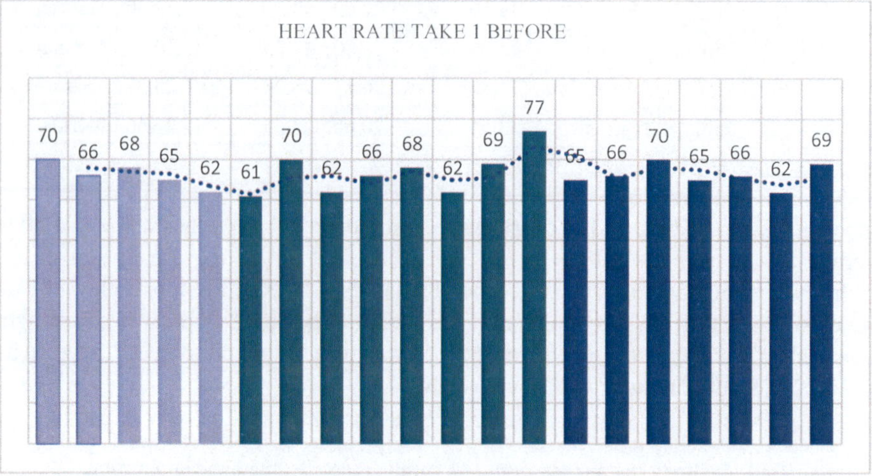

Fig. 3. Results of heart rate before physical activity.

The heart rates of the study subjects can be observed, which were taken with the monitor designed (Fig. 3).

In the first instance, a recording of the carotid pulse was made in 15 s (15 $\times$ 4) to contrast the values provided by the application, and gave similar values, therefore, the data recorded with the support of the application designed are presented. In family 1 with 5 members, it can be seen that the heart rate taken at rest (before physical activity) fluctuates with a maximum value of 70 bpm and a minimum value of 62 bpm, in family 2 with 8 members the maximum value was 77 bpm and a minimum value of 61 bpm, and in family 3 with 7 members the maximum heart rate was 70 bpm and a minimum value of 62 bpm (Fig. 4).

The heart rate data of the three families taken with the monitor designed by the research team are shown. In the first instance, the type of exercise that the members of the families should perform at home due to the Covid-19 pandemic situation was socialized.

The exercise consisted of a repetition of 15 vertical jumps [24]; the data recorded by means of the application designed show: in family 1 with 5 members it can be observed

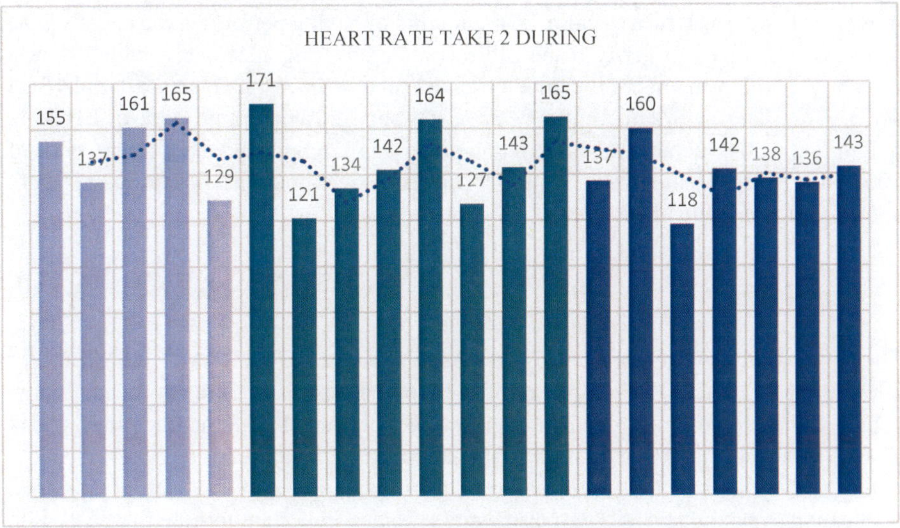

Fig. 4. Results of heart rate during physical activity.

that the heart rate taken (during physical activity) fluctuates with a maximum value of 165 bpm and a minimum value of 129 bpm; family 2 with 8 members the maximum value was 171 bpm and a minimum value of 121 bpm, while in family 3 composed of 7 members the maximum heart rate was 160 bpm and a minimum value of 118 bpm.

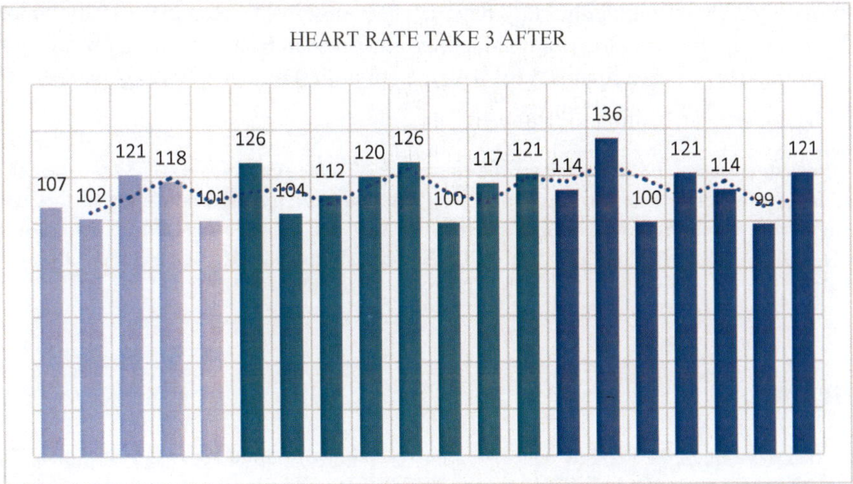

Fig. 5. Results of heart rate after physical activity.

The Fig. 5 shows the heart rates of the study subjects, which were taken the designed heart rate monitor. First of all, it was socialized to the participants that after finishing the physical exercise, the intensity should be lowered and gradually reach the recovery or what is known as return to calm, so after 3 min of having performed the exercise, the heart rate was taken to verify if the subjects are recovering adequately, which shows: in family 1 with 5 members it can be observed that the heart rate taken (after physical activity) oscillates with a maximum value of 121 bpm and a minimum value of 101 bpm, in family 2 with 8 members the maximum value was 126 bpm and a minimum value of 100 bpm, and finally in family 3 with 7 members the maximum heart rate was 136 bpm and a minimum value of 99 bpm.

5 Conclusions

The application designed presents great versatility since the use of MIT APP Inventor allowed the design and development of the App because it is a free software of low complexity in its programming code; with which the data of both the body mass index and the heart rate can be saved in the application's own database.

Regarding the calculation of the body mass index, it was noted that the designed application is easily accessible to users, and it is also important to note that, based on the data entered, participants obtain their BMI result based on the WHO classification in addition to the daily caloric expenditure, which leads to program a diet based on the result obtained [25].

The heart rate monitor that was designed is a prototype that allowed us to obtain the first results of the heart rate of the subjects in real time, however, it should be noted that the monitor designed cannot be compared with any existing heart rate monitor on the market for the control and care of personal health, especially in these times of pandemic since an erroneous data can lead to an erroneous diagnosis.

Finally, due to the pandemic stage it has not been possible to expand the study sample, which opens up the possibility of conducting future research and presenting results with a greater number of participants, [26] with the possibility of incorporating the heart rate sensor through bibs or smart shirts.

Acknowledgement. The authors thank the Universidad Técnica de Ambato (UTA) and the Dirección de Investigación y Desarrollo (DIDE) for their support for the realization of this work through the research project "GENETIC PROFILE AS A DETERMINANT OF HEALTH AND METABOLIC RISK IN UNIVERSITY STUDENTS AFTER DOMICILE ISOLATION", code PFCHE17 and the research group PROMOTION OF QUALITY OF LIFE.

References

1. Linares-Vásquez, M.V.: How developers micro-optimize android apps. J. Syst. Softw. **130**, 1–23 (2017). https://doi.org/10.1016/j.jss.2017.04.018
2. Mbelwa, J.T.: Acceptability and use of mobile health applications in health information systems: a case of eIDSR and DHIS2 touch mobile applications in Tanzania. In: International Conference on Social Implications of Computers in Developing Countries. Springer, Cham, pp. 579–592 (2019). https://doi.org/10.1007/978-3-030-18400-1_48
3. Baños, R.: Novedosas herramientas digitales como recursos pedagógicos en la educación física. Granda: EmásF: Revista Digital de educación física **52**, 79–91 (2018)

4. Krauskopf, P.B.: World health organization academy: COVID-19 learning and who info mobile apps. J. Nurse Pract. **16**(9), 704–705 (2020). https://doi.org/10.1016/j.nurpra.2020.07.002
5. Birkmeyer, S.W.: Determinants of mHealth success: an empirical investigation of the user perspective. Int. J. Inf. Manag. **59**, 102351 (2021). https://doi.org/10.1016/j.ijinfomgt.2021.102351
6. Herren, O.M.: Emotion suppression, coping strategies, dietary patterns, and BMI. Eat. Behav. **41**, 101500 (2021). https://doi.org/10.1016/j.eatbeh.2021.101500
7. Briones-Villalba, R.: Effect of a physical activity program and nutritional education to prevent sugar sweetened beverages consumption and development of obesity among schoolchildren in Tijuana, Mexico. Revista Española de Nutrición Humana y Dietética, 22(3), 235–242 (2018). https://doi.org/10.14306/renhyd.22.3.519
8. Aznar Díaz, I.: Impacto de las apps móviles en la actividad física. Federación Española de Asociaciones de Docentes de Educación Física (FEADEF), 52. (2019)
9. Zhou, S.L.: Lessons on mobile apps for COVID-19 from China. J. Saf. Sci. Resilience (2021). In Press. https://doi.org/10.1016/j.jnlssr.2021.04.002
10. Torres, A.J., et al.: Análisis y utilidad de nuevas herramientas de aprendizaje. Del aula al ámbito de actuación profesional: apps y paquetes informáticos. Memorias del Programa de Redes-I3CE de calidad, innovación e investigación en docencia universitaria, pp. 2750–2760. Obtenido de (2017). https://dialnet.unirioja.es/servlet/articulo?codigo=6337448
11. Fernández García, C.V.: Fitbit, una app para medir la actividad física diaria. Rev. Cubana de Inv. Biomédicas **36**(4), 1–4 (2017)
12. Kumar, S.: Revisiting food delivery apps during COVID-19 pandemic? Investigating the role of emotions. J. Retail. Consum. Serv. **62**, 102595 (2021). https://doi.org/10.1016/j.jretconser.2021.102595
13. González, J.J.: Proyecto MÓVIL-ÍZATE: Fomento de la actividad física en escolares mediante las Apps móviles. Retos. Nuevas tendencias en Educación Física, Deporte y Recreación **30**, 3–8 (2016)
14. Leslie, M.: COVID-19 fight enlists digital technology: contact tracing apps. Eng. (Beijing, China) **6**(10), 1064–1066 (2020). https://doi.org/10.1016/j.eng.2020.09.001
15. Aznar Díaz, I.C.: Impacto de las apps móviles en la actividad física: un meta-análisis. Impact of mobile apps on physical activity: a meta-analysis. Retos nuevas tendencias en educación física, deporte y recreación **37**, 52–57 (2019)
16. Marques, G.D.: Impact of COVID-19 on the psychological health of university students in Spain and their attitudes toward mobile mental health solutions. Int. J. Med. Informatics **147**, 104369 (2021). https://doi.org/10.1016/j.ijmedinf.2020.104369
17. Pluymaekers, N.A.: On-demand app-based rate and rhythm monitoring to manage atrial fibrillation through teleconsultations during COVID-19. Int. J. Cardiol. Heart Vasculature, (28), 100533 (2020)
18. Aguilar, A.: Posibilidades de las aplicaciones móviles para el abordaje de la obesidad según los profesionales. Cross Mark, 1–5 (2015)
19. García, C.F.: Fitbit, una app para medir la actividad física diaria. Rev. Cubana de Inv. Biomédicas **36**, 1–4 (2019)
20. Chidambaram, S.E.: Observational study of UK mobile health apps for COVID-19. Lancet Digit. Health **2**(8), e388–e390 (2020). https://doi.org/10.1016/S2589-7500(20)30144-8
21. Mocha-Bonilla, J.A., Punina Chimborazo, B.F., Mocha Altamirano, K.I., Hidalgo Alava, D.J.: Development of an App for monitoring heart rate in people who practice regular physical activity. In: Botto-Tobar, M., Zambrano Vizuete, M., Díaz Cadena, A. (eds.) CI3 2020. AISC, vol. 1277, pp. 295–306. Springer, Cham (2021). https://doi.org/10.1007/978-3-030-60467-7_25

22. Valle-Leal, J.G.: Frecuencia cardiaca y actividad física en niños con obesidad del Sur de Sonora Enfermería Universitaria (15)4 (2018). https://doi.org/10.22201/eneo.23958421e.2018.4.546
23. Ernst, H.S.: Optimal color channel combination across skin tones for remote heart rate measurement in camera-based photoplethysmography. Biomed. Signal Process. Control **68**, 102644 (2021). https://doi.org/10.1016/j.bspc.2021.102644
24. Villa, J.G.-L.: Tests de salto vertical (I): Aspectos funcionales. Rev. Digit.: Rendimiento Deportivo **6**, 1–14 (2003)
25. Mocha-Bonilla, J.A., Guachimbosa, V.H., Santiago, C.G., Sánchez Guerrero, J.: Linear programming model applied to the optimization of nutritional diets for athletes. In: Auer, M.E., Tsiatsos, T. (eds.) IMCL 2019. AISC, vol. 1192, pp. 844–866. Springer, Cham (2021). https://doi.org/10.1007/978-3-030-49932-7_79
26. Ortega, J.P.: Comparación de dos métodos de detección para el registro de la frecuencia cardíaca: banda torácica vs camiseta técnica. Retos: nuevas tendencias en educación física, deporte y recreación **36**, 469–473 (2019)

Application of E-commerce in Ecuador's Economic Activities

Andres Palacio-Fierro[1,2](✉), Silvia Llamuca-Pérez[1], and Ximena Morales-Urrutia[3]

[1] Facultad de Ciencias Administrativas Y Económicas, Universidad Tecnológica Indoamérica, 180103 Ambato, Ecuador
andrespalacio@uti.edu.ec

[2] Universidad Camilo José Cela, Programa de Doctorado, 28692 Madrid, Spain

[3] Universidad Tecnológica Indoamérica, Centro de Investigación en Empresa, Sociedad Y Tecnología, 180103 Ambato, Ecuador

Abstract. Currently, e-commerce has become one of the most widely used forms of trade worldwide, favouring companies by increasing their sales; on the other hand, to customers or consumers by meeting their needs through online shopping in less time. The objective of the research focuses on analyzing the relationship between the use or application of e-commerce in companies according to their size. To meet the goal, the methodology used was built on the Pearson's correlation coefficient statistical technique, since it allows to obtain an association coefficient between variables that do not behave normally. The results point out that e-commerce is a form of exchange, which is being used by all types of companies analyzed, however, large companies are the ones that stand out, as they have the most resources to apply this type of trade.

Keywords: E-commerce · Enterprises · Ecuador

1 Introduction

In recent decades, new forms of exchange of products and/or online services have emerged [1, 2]; in this sense, e-commerce arises a set of activities to buy or sell, in which companies offer products through virtual platforms [3–5].

In this regard, Javid, Nazari, & Ghaeli [6] consider e-commerce as a transaction in which both the purchase and the sale is made through internet networks, functioning as intermediaries between consumers and producers. Thus, this type of trade has given retailers the opportunity to offer a new business concept through a new customer interface, interactive delivery systems [7], as well as a greater scope in terms of product selection, reducing infrastructure, expanding or reducing opening hours and a high degree of scalability [8].

In the same line, Turban [5] defined e-commerce as a process of transferring goods and/or information with computers, better known as internet and intranets. For its part, Clarke [9] mentioned that for e-commerce to work effectively, there need to be certain elements, such as: e-commerce support services that encompass interorganizational email,

M. Zambrano Vizuete et al. (Eds.): CI3 2021, LNNS 511, pp. 431–438, 2022.
https://doi.org/10.1007/978-3-031-11438-0_34

directories, commercial support systems for products, products and/or personalized services, management information, information systems, among others.

According to authors Daniel, & Wilson [10] the adoption of e-commerce involves a number of activities that could be implemented by companies, among which are to provide information about the company, about the goods and services offered, receive and place orders, receive payments, shipments of products and/or services, after-sales services, identification of suppliers, internal and external communication, exchange of documents, search for information, advertising, among others [11]. Therefore, the greater the scope of e-commerce in organizations, the greater the likelihood of gaining more commercial benefits [12].

In this regard, to explain the level of adoption of e-commerce, the development growth model designed by Richard Nolan throughout the 1970s is mentioned. In this model, the company faces six stages related to the use of the information system, among these are mentioned: initiation, contagion, control, integration, data administration, and maturity. In the same line, Rao, Metts, & Monge [13] mentioned the presence of four phases: existence, portal, addition of transactions, and integration of companies, to show the development of e-commerce in a corporation. Finally, Prananto, McKay, & Marshall [14] proposed a model linking the level of adoption of e-commerce with the required investment and potential benefit.

In this context, some of the aspects influencing the use of this type of trade by companies were identified: organizational preparation for the acceptance of e-commerce, external pressure, perceived advantages of e-commerce, easiness of use, perceived practicality, compatibility between e-commerce and organization culture, preferred labor values and practices [15, 16].

Thus, Turban [5] mentions that through the adoption of e-commerce, organizations gain certain advantages, such as: generating reduced costs, increasing sales, increasing productivity, reducing processing time, improving market reach, and increased customer loyalty.

Faced with this reality, the advantages proposed by e-commerce have made it a popular means of exchange, so, according to the IDC report in 2009 a total of $8 trillion for online purchases had been recorded, while, in 2013, this number doubled, amounting to a total of $16 trillion in purchases. However, this growth has not been replicated for all existing business types.

In this regard, Govindaraju, Wiratmadja, & Rivana [17] and Chiliya, Chikandiwa, & Afolabi [19] show that the use of e-commerce was superior in large companies in relation to small and medium-sized companies. In this regard, some academics have stated that there are inhibitory factors that hinder the use of this resource, for example, the lack of training of human talent, internal resistance, lack of support, insecurity in processes, unprepared business partners, internal restrictions [20]. Thus, Kotelnikov [21] considers that the lack of communication infrastructure, insufficient knowledge of ICTs, scarce financial resources, and poor legal structure, are factors that have not increased the use or adoption of ICTs by companies.

Similarly, Jones et al. [22] agree that factors such as security and privacy, lack of knowledge regarding the handling of e-commerce, and the high cost of initiation and maintenance have become the main barriers to the adoption of this type of trade in all kinds of organizations. Therefore, small and medium-sized companies have been relegated from this process of immersion in the different markets; they are also limited in their sales, job creation and contribution to the economic development of the sector in which they are developed [23–25].

2 Methodology

The data analyzed for research come from official statistical sources of the Ecuadorian Institute of Statistics and Census. The two selected variables, on the one hand, electronic business, corresponding to the independent variable (matches companies that have used ICT tools, within their business activities in the categories: supply chain management, logistics, inventory control, finance and budget management, human resources management, service and sales support, research and development, knowledge management); and size of companies, corresponding to the dependent variable, are related to the theoretical foundations set out in the study.

The sample selected for this study corresponds to the ICT Module in the 2015 Industrial Surveys. In addition, the timeframe for analysis covers the period 2012–2015. The selection of these variables responds to a double selection criterion: on the one hand,

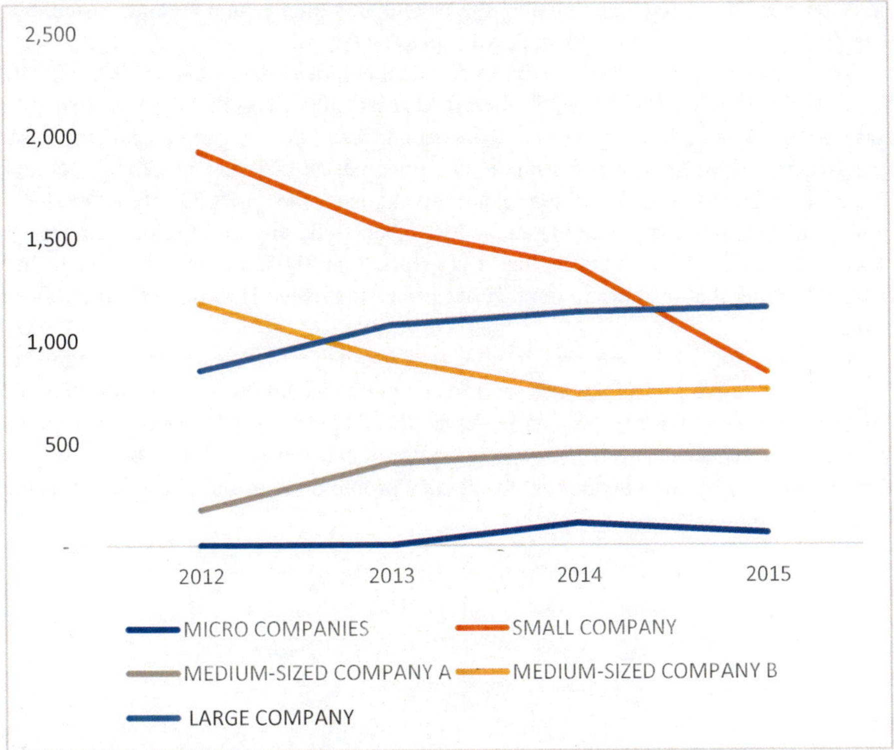

Fig. 1. Use of e-commerce in Ecuador companies. 2012–2015 period. Source: Ecuadorian Institute of Statistics and Census (2020)

the relationship presented by these and; on the other hand, the statistical availability of variables.

Pearson's correlation coefficient statistical technique was applied for data processing because it allows to obtain an association coefficient between variables that do not behave normally. It is calculated based on a series of assigned ranges. The values range from –1 to 1, with 0 being the non-correlation value, and the signs indicate direct and reverse correlation [26].

3 Results and Discussion

According to the results obtained, it is shown in Fig. 1 that e-commerce has become one of the most widely used means for the purchase and sale of products and/or services between companies (micro, small, medium and large companies).

In the period analyzed (2012–2015) the use of this type of trade has evolved differently for the different types of companies considered within the classification. The analysis highlights two important aspects: on the one hand, large companies have a sustained evolution over time, coinciding with what was stated by Govindaraju [17], Alam et al. [18] and Chiliya [19]. On the other hand, the types of companies that still struggle to implement e-commerce are micro and small companies, specifically highlighted are small businesses, which despite being the ones that started with a significant number, their sustainability over time did not evolve positively.

When analyzing the table, it is observed that the minimum correlation level is 0.50, i.e., it is slightly closer to 1 than 0. Therefore, the results show that there is a positive correlation, i.e., a high rating is given to the situation of knowledge management, R&D spending and sales service and support in companies in different productive sectors. Therefore, what has been proposed is met, by choosing the variables that allow determining the use of e-commerce in companies. When analyzing the significance, we see that it is 0.000 again, so it is true that it is greater than 0.05, and is even greater than 0.01, indicating that the correlation that has been established is most likely true (Table 1 and Table 2).

The execution of e-commerce in several of the economic sectors of Ecuador has been representative in sectors such as information and communication, financial and administrative activities and support. However, this fact is not appreciated in the remaining sectors, so other means of promoting the use of e-commerce should be considered through support by the government to identify problem areas and possible actions to take into account (Fig. 2).

Table 1. Variable correlations

		Company size	E-business - sales service and support	E-business - research and development	E-business - knowledge management
E-business - sales service and support	Pearson correlation	–	1	,573**	,571**
	Sig. (bilateral)	,000		,000	,000
	N	3134	3134	3134	3134
E-business - research and development	Pearson correlation	–	,573**	1	,654**
	Sig. (bilateral)	,000	,000		,000
	N	3134	3134	3134	3134
E-business - knowledge management	Pearson correlation	–	,571**	,654**	1
	Sig. (bilateral)	,000	,000	,000	,000
	N	3134	3134	3134	3134

**. Correlation is significant at level 0.01 (bilateral). Source: Ecuadorian Institute of Statistics and Census (2020)

Table 2. Variable correlations

		Company size	E-business - supply chain management	E-business - finance and budget management	E-business - human resources management
E-business - supply chain management	Pearson correlation	–	1	609**	,547**
	Sig. (bilateral)	,000		.000	,000
	N	3134	3134	3134	3134
E-business - finance and budget management	Pearson correlation	–	,609**	1	,648**
	Sig. (bilateral)	,000	,000		,000
	N	3134	3134	3134	3134
E-business - human resources management	Pearson correlation	–	,547**	,648**	1
	Sig. (bilateral)	,000	,000	,000	
	N	3134	3134	3134	3134

**. Correlation is significant at level 0.01 (bilateral). Source: Ecuadorian Institute of Statistics and Census (2020)

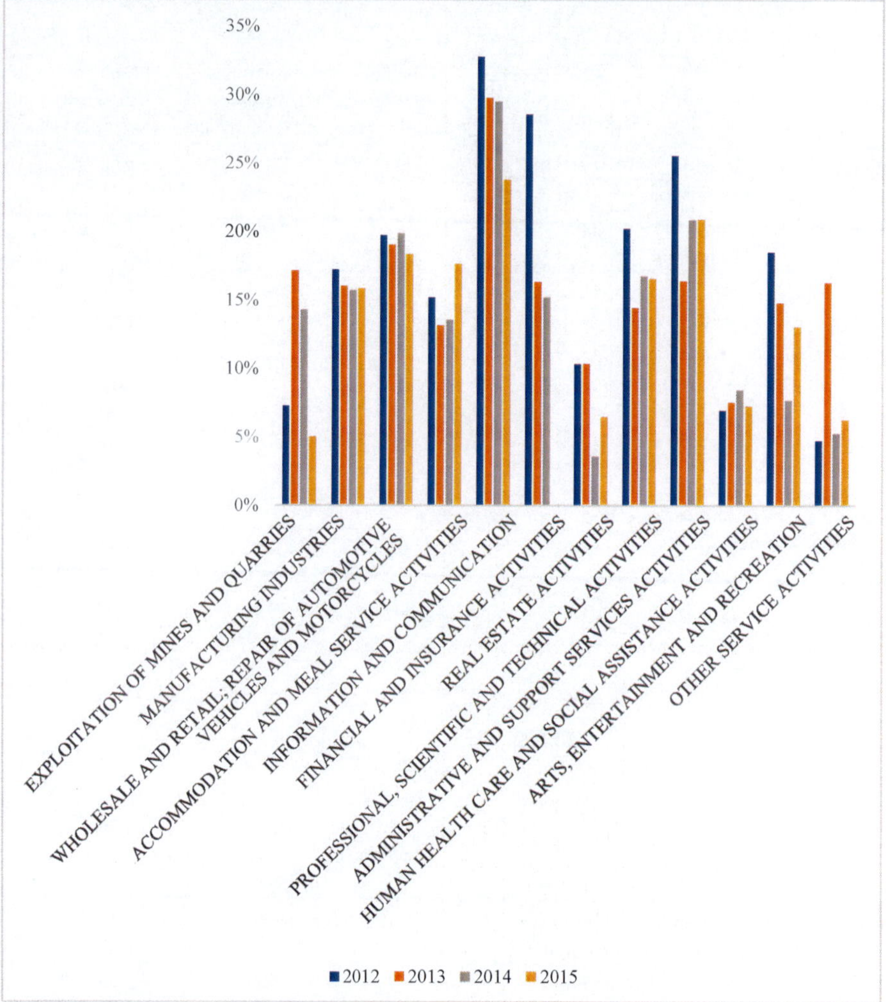

Fig. 2. Companies that carry out transactions over the internet. 2012–2014. Source: Ecuadorian Institute of Statistics and Census (2020)

4 Conclusions

In recent decades, e-commerce has undergone constant evolution worldwide and, particularly in Ecuador, becoming an engine of economic development for companies that implement it. The vertiginous changes that are present cause organizations to adopt instant forms of exchange, to be competitive against a market increasingly connected to technology. In addition, through e-commerce, specific tools are proposed to meet consumer needs more quickly and effectively.

In Ecuador, the adaptation of this type of trade has been progressive in recent years, therefore, the permanence and expansion of this will depend on the design and implementation of policies and guarantees granted by regulatory bodies to both entrepreneurs and users, with the aim of generating confidence in its use within society.

References

1. Bresciani, S., Giacosa, E., Culasso, F., Broccardo, L.: The family variable in the French and Italian wine sector. EuroMed J. Bus. **11**(1), 101–118 (2016)
2. Bresciani, S., Ferraris, A., Santoro, G., Nilsen, H.R.: Wine sector: companies performance and green economy as a means of societal marketing. J. Promot. Manag. **22**(2), 251–267 (2016)
3. Kim, H., Niehm, L.S.: The impact of website quality on information quality, value, and loyalty intentions in apparel retailing. J. Interact. Mark. **23**(3), 221–233 (2009)
4. Wang, W.T., Wang, Y.S., Liu, E.R.: The stickiness intention of group-buying websites: the integration of the commitment–trust theory and e-commerce success model. Inf. Manag. **53**(5), 625–642 (2016)
5. Turban, E.: Electronic Commerce 2010: A Managerial Perspective. Upper Saddle River: Pearson Education, New Jersey (2010)
6. Javid, E., Nazari, M., Ghaeli, M.: Social media and e-commerce: a scientometrics analysis. Int. J. Data Netw. Sci. **3**(3), 269–290 (2019)
7. Madan, K., Yadav, R.: Understanding and predicting antecedents of mobile shopping adoption: a developing country perspective. Asia Pac. J. Mark. Logist. **30**(1), 139–162 (2018)
8. Kalia, P., Kaur, N., Singh, T.: E-commerce in India: evolution and revolution of online retail. In: Mobile Commerce: Concepts, Methodologies, Tools, and Applications. IGI Global, 736–758 (2018)
9. Clarke, R.: The past, present and future of B2C eCommerce. In: Statement for a Panel Session, Sydney (2005)
10. Daniel, E., Wilson, H.: Adoption intentions and benefits realised: a study of e-commerce in UK SME. J. Small Bus. Enterp. Dev. **9**(4), 331–348 (2002)
11. Rahayu, R., Day, J.: E-commerce adoption by SMEs in developing countries: evidence from Indonesia. Eurasian Bus. Rev. **7**(1), 25–41 (2016). https://doi.org/10.1007/s40821-016-0044-6
12. Zhu, K., Kraemer, K.L.: Post-adoption variations in usage and value of e-business by organizations: cross-country evidence from the retail industry. Inf. Syst. Res. **16**(1), 61–84 (2005)
13. Rao, S.S., Metts, G., Monge, C.A.M.: Electronic commerce development in small and medium sized enterprises: a stage model and its implications. Bus. Process. Manag. J. **9**(1), 11–32 (2003)
14. Prananto, A., McKay, J., Marshall, P.: A study of the progression of e-business maturity in Australian SMEs: some evidence of the applicability of the stages of growth for e-business model. In: Proceedings of the PACIS, pp. 68–80. Adelaide (2003)
15. Wang, Y., Ahmed, P.K.: The moderating effect of the business strategic orientation on eCommerce adoption: evidence from UK family run SMEs. J. Strateg. Inf. Syst. **18**(1), 16–30 (2009)
16. Grandon, E.E., Pearson, J.M.: Electronic commerce adoption: an empirical study of small and medium US businesses. Inf. Manag. **42**(1), 197–216 (2004)
17. Govindaraju, R., Wiratmadja, I.I., Rivana, R.: Analysis of drivers for e-commerce adoption by SMEs in Indonesia. In Proceedings of the International Congress on Interdisciplinary Behaviour and Social Sciences 2014, (2015)

18. Alam, S.S., Ali, M.Y., Jani, M.F.M.: An empirical study of factors affecting electronic commerce adoption among SMEs in Malaysia. J. Bus. Econ. Manag. **12**(2), 375–399 (2011)
19. Chiliya, N., Chikandiwa, C., Afolabi, B.: Factors affecting small micro medium enterprises'(SMMES) adoption of e-commerce in the eastern cape province of South Africa. Int. J. Bus. Manag. **6**(10), 28–36 (2011)
20. Dubelaar, C., Sohal, A., Savic, V.: Benefits, impediments and critical success factors in B2C E-business adoption. Technovation **25**(11), 1251–1262 (2005)
21. Kotelnikov, V.: Small and medium enterprises and ICT: Asia-Pacific development Information Programme, e-Primers for the Information Economy, Society and Polity, APCICT (2007)
22. Jones, P., Packham, G., Beckinsale, M., Jahanshahi, A.A., Zhang, S.X., Brem, A.: E-commerce for SMEs: empirical insights from three countries. J. Small Bus. Enterp. Dev. (2013)
23. de Wit, G., de Kok, J.: Do small businesses create more jobs? New evidence for Europe. Small Bus. Econ. **42**(2), 283–295 (2013). https://doi.org/10.1007/s11187-013-9480-1
24. Ayyagari, M., Demirguc-Kunt, A., Maksimovic, V.: Small vs. young firms across the world, contribution to employment, job creation and growth, p. 41. The World Bank Development Research Group (2011)
25. Neumark, D., Wall, B., Zhang, J.: Do small businesses create more jobs?: New evidence for the United States from the national establishment time series. Rev. Econ. Stat. **93**(1), 16–29 (2011)
26. Martínez, R., Tuya, L., Martínez, M., Pérez, A., Cánovas, M.: El coeficiente de correlación de los rangos de Spearman caracterización. Habanera de Ciencias Médicas, 1–4 (2009)

Hearing Health Virtual Assessment Through Association Rules Mining Inside a College Community

Rubén Alejandro Ortiz(✉), María Isabel Subía, Eliana Acurio, and Hernán Barba

Escuela Politécnica Nacional, 170525 Quito, Ecuador
ruben.ortiz@epn.edu.ec

Abstract. This study quantified possible cases of hearing loss and determined possible causes through association rules mining. The data was collected through an online survey that evaluated the medical background of the participants, and a free online hearing test provided by the American company Phonak. The study sample consisted of 226 entries, among students, faculty, and staff members of the Escuela Politécnica Nacional, a public university from Quito, Ecuador. For the initial data treatment and conversion to binary tables of the audiometry results, a Python optical character recognition algorithm was used. Finally, the association rules were obtained using the Apriori algorithm, which was implemented through the arules package for R.

The results showed that 66.36% of the sample presents hearing loss in at least one ear. Furthermore, hearing loss causes in males are primarily related to prolonged use of headphones and loud noises, and these cases are mostly seen in younger individuals; while hearing loss causes in females are primarily related to family history of deafness.

Keywords: Hearing loss · Association rules · Optical character recognition · Audiometry · Ecuador

1 Introduction

Around the world, about 360 million people suffer from hearing loss [1]. Therefore, hearing loss is a health problem worldwide, to the extent that The World Health Organization (WHO) estimates that by 2050, 2.5 billion people will suffer from hearing loss to some degree, and at least 28% of them will require hearing rehabilitation [2].

A hearing threshold of 20 decibels (dB) or higher in both ears is considered healthy hearing. People unable to meet such threshold present hearing loss, which may have different degrees of impairment: acute, moderate, or severe. Furthermore, hearing loss can affect one or both ears, and affected people may respond differently to assistive devices, for example, hearing aids and cochlear implants [2].

Causes of hearing loss vary from preventable to non-preventable, and there are certain critical periods in life, like early childhood and old age, in which individuals are most

M. Zambrano Vizuete et al. (Eds.): CI3 2021, LNNS 511, pp. 439–449, 2022.
https://doi.org/10.1007/978-3-031-11438-0_35

susceptible to the effects of hearing loss [2]. Some hearing loss cases are avoidable by addressing the risk factors:

- Noise exposure in the workplace: Noise exposure makes up 22% of work-place related health issues [4], being the second most common risk factor in work environments, behind physical injuries [3].
- Recreational noise exposure: The WHO estimated that over a billion underage individuals are prone to develop hearing loss due to prolonged use of devices such as headphones at loud volume levels [5]. Smartphone use currently sits at 87% of the population in developed countries [6], which aggravates this risk factor. In addition, almost 50% of all portable music device users are listening to music at volume levels higher than the recommended, which is putting their hearing health at risk [5].
- Ear infections: It's a primary precursor for hearing loss cases, especially at a young age. Each year, 31 million patients with acute otitis media will most likely suffer from a developed chronic suppurative otitis media, including more than 7 million children. Additionally, more than 50% of patients affected by chronic suppurative otitis media may develop hearing loss [7].
- Ototoxic hearing loss: Medications such as aminoglycosides, used within infection treatments, and chemotherapeutic agents such as cisplatin [8] are common precipitants of ototoxic hearing loss. The fight against multi-drug resistant diseases turned antibiotics use into a growing concern in terms of the possible negative effects on hearing health [9].

Other risk factors include congenital infections like rubella and bacterial meningitis, which have had a decrease in infectivity rates in most parts of the world due to the development of newer and more effective vaccines, with higher uptake rates of immunization, and wider geographical coverage [1].

In Ecuador, there are more than 66 722 people that suffer from hearing loss [10]. However, due to the lack of accessible health examinations and lack of a culture of health issues prevention, many preventable and non-preventable hearing loss cases remain undiagnosed or untreated [3].

This study aimed to estimate possible cases and degrees of hearing loss in the college community of Escuela Politécnica Nacional from Quito, Ecuador, through an online survey that evaluated the medical background and exposure to loud noises, and a free online audiometry that evaluated the hearing function, provided by the American company Phonak. The sample consisted of 226 students, among students, faculty, and staff members.

The remainder of this paper is organized in the following way. Section two presents details of the implemented computational methodology for data processing and association rules algorithm. Section three illustrates and discusses experimental results. Finally, section four summarizes the main achievements of this paper.

2 Methodology

Due to the current global pandemic, considering that the college community is scattered throughout Ecuador, the data collection method consisted of an online Google Forms survey, developed under the guidance of a health specialist. The survey consisted of two sections. The first section included nine questions related to the following parameters: Age, gender, occupation, daily hours of headphones use, medication use, family history of deafness, medical history, previous hearing problems, and exposure to loud noises. On the other hand, the second section of the survey required the participants to upload the results (screenshot and manual input of data) of an online audiometry test, done through the open access webpage https://www.phonak.com/com/en/online-hearing-test.html.

The virtual hearing test does not replace a clinical audiogram performed by a certified medical doctor. However, this tool allows estimating hearing loss cases remotely and giving a first insight of hearing problems that can alert to the participants. Therefore, it is also an initial way to prevent more severe hearing loss. A visualization of the obtained results is shown in Fig. 1.

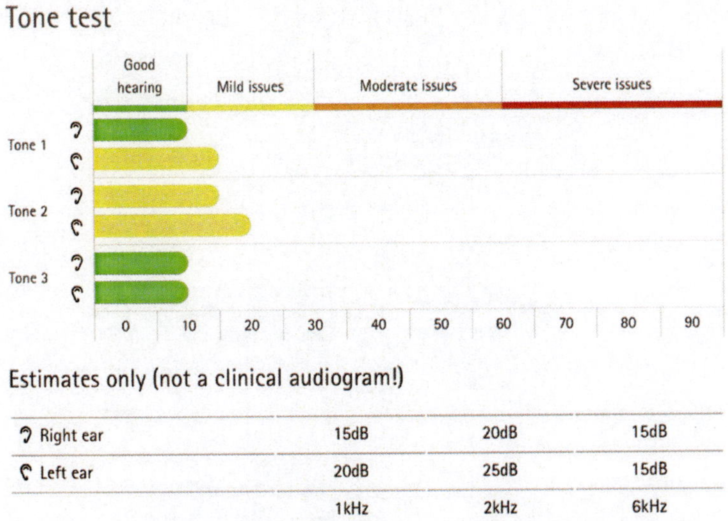

Fig. 1. Example of the results obtained from the audiometry online test.

The results include, in the upper section, a bar graph indicating the softest sounds heard (hearing level) by the individual at three different frequencies, for each ear: 1 kHz, 2 kHz, 6 kHz. The audiometry automatically identifies the scale for hearing issues as greater than 15 dB for mild issues, and greater than 35 dB for moderate issues. Since the online audiometry greatly depends on the quality of the audio output devices, the threshold considered for hearing loss cases in this study was 30 dB, providing a preliminary result on the hearing loss cases in the sample.

Due to the format of the audiometry results, the first step for the data analysis consisted of transforming the data from the screenshots uploaded to a.csv text file that could be used subsequently on the association rule analysis.

2.1 Audiometry Data Cleansing in Python

Optical Character Recognition (OCR) is a process that detects and classifies optical patterns contained in an image file, to identify and extract alphanumeric characters or others. The character recognition is accomplished through sequential steps of segmentation, feature extraction, and classification [11]. The programming language Python has an optical character recognition tool, called Python-tesseract, or Pytesseract, that recognizes and "reads" the text embedded in images [12].

The first step consisted of standardizing the screenshot formats into.png picture files that could be easily read by the OCR algorithm. This was done using the Python library PIL which provides the python interpreter with image editing capabilities, allowing to open each image and subsequently convert it to the desired format [13]. Then, a second algorithm reads each image and converts the text inside it to strings that can be easily segmented into the numerical data corresponding to the audiometry results. Finally, the required data is written in a.csv file. Flowcharts for both algorithms are shown below in Fig. 2 and Fig. 3.

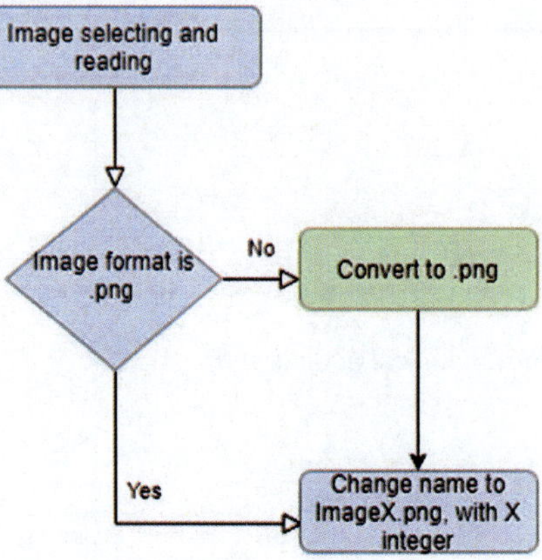

Fig. 2. Flowchart for image cleansing: Organizing data.

The purpose of asking for both a screenshot of the results and their manual input in the survey is to compensate for the percentage of inaccuracy in the OCR algorithm while optimizing the data analysis process through the export of data.

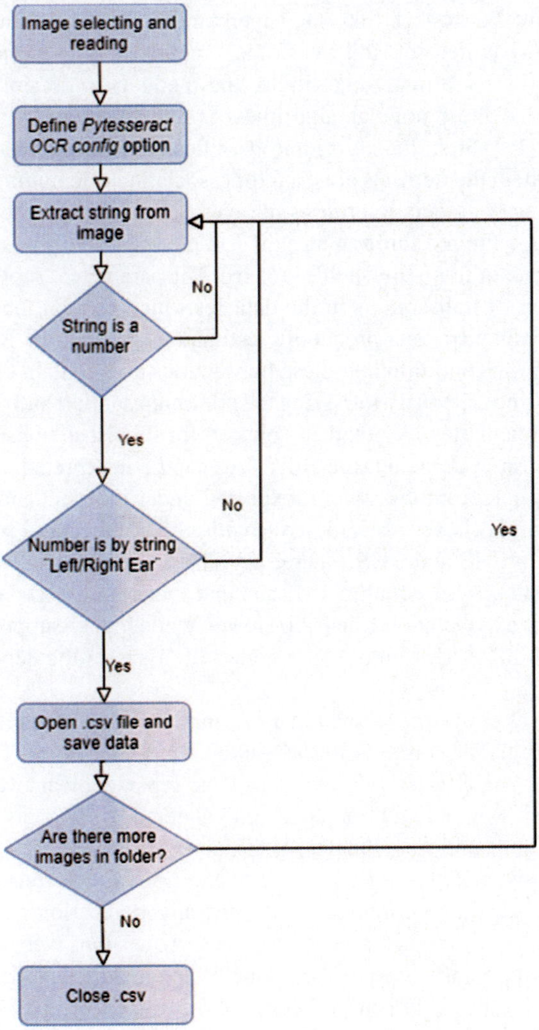

Fig. 3. Flowchart for data collection: Extracting audiometry values.

2.2 Association Rules Analysis in R

Data mining algorithms for association rules can find links between a set of items that occur simultaneously. Also, such algorithms are suitable to manage big data to yield significant association rules. In other words, association rules are statements that represent the co-occurrence of some items. Schematically, an association rule is represented by a Left Hand Side (LHS) of the association rule implying a Right Hand Side (RHS), as:

$$\{A, B\} \rightarrow \{C\} \tag{1}$$

where A, B and C are items in a dataset, and the variety of items is known as *itemset*. Thus, the event where a set of items occur, is known as a *transaction* [14]. To illustrate this scenario, in market basket analysis, supermarket products are items, an arbitrary group of products is an itemset, and a customer's purchase is a transaction.

Furthermore, there are many algorithms and methods to obtain association rules, however, one of the most popular algorithms is called Apriori [15], introduced by Agrawal in 1993 [17]. First, the algorithm identifies the itemsets that frequently occur simultaneously. Then, the itemsets are used for association rule mining. Proceeding this way ignores less likely association rules and saves computational resources. Afterward, association rules are filtered through a set of parameters called lift, support and confidence, that work as minimum thresholds [15, 16]. The parameter "support" of an itemset A is the proportion of transactions in the data set which contain the itemset A, while the parameter "confidence" is a probability estimate of finding the RHS of the rule in transactions under the condition that these transactions also contain the LHS [14, 18].

Nonetheless, simultaneously satisfying the minimum support and confidence thresholds can lead to quandaries. A practical solution for this issue consists of filtering or sorting through a third parameter, the "lift". Lift can be interpreted as the deviation of the whole rule support from the support expected under independence, given the LHS and the RHS support [14]. An association rule with a lift higher than 1 indicates a positive correlation between LHS and RHS. Therefore, a lift under 1 corresponds to a negative correlation. Naturally, a lift equal to 1 means that there is no correlation between the RHS and LHS of the association rule [16]. Hence, while higher support and confidence reflect the strength of an identified rule, a higher lift assesses the dependency between LHS and RHS [18].

The .csv file was converted into a binary format to ensure compatibility with association rules mining. The answers to the survey were adapted as TRUE or FALSE, according to each case. Gender was treated as three separate binary variables, allowing to analyze for men, women, and non-specified gendered participants. Occupation was interpreted as students and not students. Daily hours of headphones use were considered TRUE for answers equal or higher to 2 h, and FALSE otherwise. Medication use was treated as TRUE if the participants identified any medication taken regularly, and FALSE if the answer was left blank. The other survey questions were dichotomous, and directly adapted into binary format. Additionally, the audiometry data was converted through the criterion that a threshold volume of 30 dB is a symptom of hearing loss. Consequently, audiometry values equal or higher than 30 dB were interpreted as TRUE, for each hearing loss variable, while values lower than 30 dB were interpreted as FALSE.

In this study case, the Apriori algorithm was implemented through the arules extension package for R. This library provides all the needed infrastructure to manipulate datasets, use the mining algorithms and analyze the resulting itemsets and rules. A minimum support threshold of 10/[Data length] was taken in order to discard spurious association rules, considering a data length of 216 transactions. Consequently, association rules were sorted decreasingly as a function of confidence and lift. Besides, in order to obtain significant associations, we searched for maximal length rules. Thus, rules originated by fewer than three items were discarded.

Since our interest is rooted in the relationship between deafness variables and demographic-behavioral variables, association rules were further filtered to include only one deafness variable in the RHS. Hence, six separate analyses were run to cover both right (R) and left (L) ears, paired with either low frequency tone (1), medium frequency tone (2) and high frequency tone (3). Finally, association rules that were frequently found for different deafness variables were considered as part of the results.

3 Results and Discussion

The sample consisted of 226 responses to the online survey. From that data set, the audiometry data cleansing algorithm successfully recognized 204 results (90.3% of the total data), so manual collection of the missing data was required. The main reason for the unsuccessful recognition in the cleansing algorithm was that some audiometry images were too blurry, or too dark.

Once the data was collected and organized, a brief analysis of the main survey results was made, based on descriptive statistics and a further introduction to the main inferential results, which are later discussed and deepened by the association rules analysis. A percentage of 65.9% of the participants ages ranged between 20 and 25 years old, as it's shown in Fig. 4 as a histogram with 30 bins for better visualization. The trend suggests that most of the participants in the survey were students from the college community.

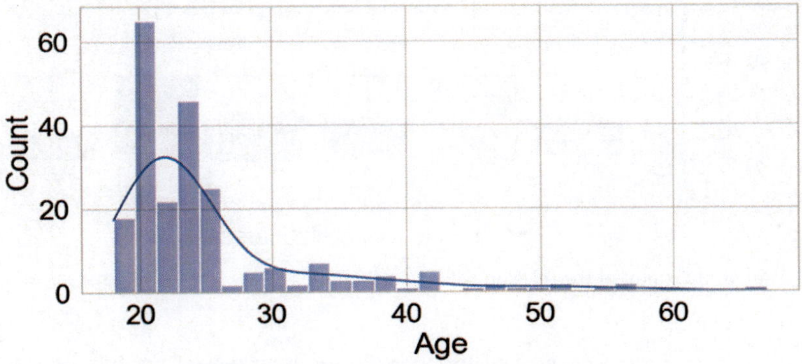

Fig. 4. Age distribution of participants.

The relationship between the daily use of headphones and age was explored. The results can be visualized in Fig. 5.

The median for an extended daily use of headphones of more than 5 h involves younger age individuals, which is consistent with the fact that young people, especially students, use devices like headphones for activities like virtual classes, hearing music, playing video games, etc.

Since the threshold considered for hearing loss cases was defined as 30 dB for each frequency (1 kHz, 2 kHz, 6 kHz), a scale of 1 indicates a possible case of hearing loss for one of the frequencies at one ear, and 6 indicates a possible case of hearing loss for all the frequencies at both ears. In Fig. 6, a box plot reflects the relationship between the presence of possible hearing loss cases (based on the relative scale stated before) and the ages of the participants.

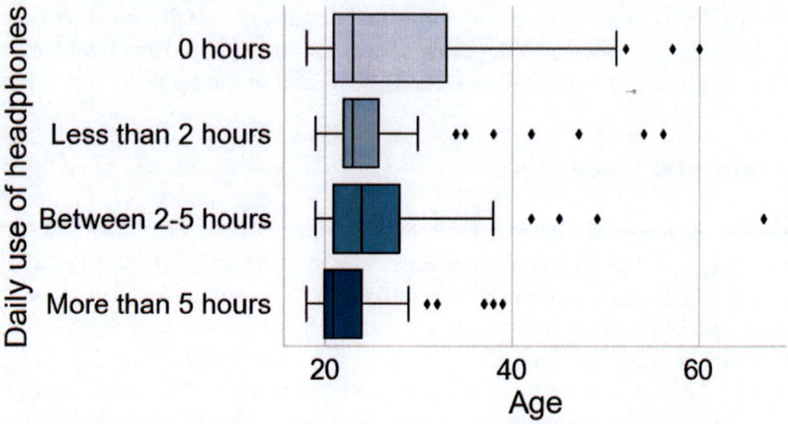

Fig. 5. Box plot of the relation between daily use of headphones and age.

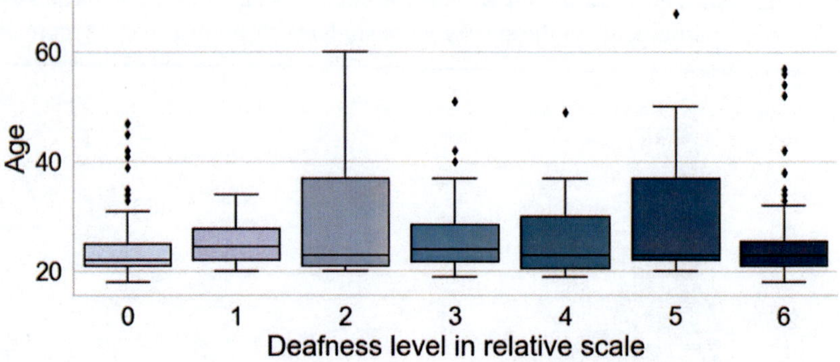

Fig. 6. Box plot of the relation between age and deafness level in relative scale.

In this case, the presence of values outside the whiskers of the box plot denotes numerous atypical values for deafness level 0 and 6, which is a consistent result considering the expected range of variability between audio devices used to take the audiometry test by the participants, and the fact that the sample size is extremely limited. Another interesting finding is that, for each of the deafness levels in the relative scale, the median age is lower than 30 years old.

Under the relative scale analysis, it was found that 66.36% of the sample presents a possible case of hearing loss in at least one ear for at least one of the frequencies tests with the audiometry. Table 1 shows that a total of three association rules were frequently found for all different deafness variables. Rules with a LHS equal to {Male, Student, Loud Noise} were found for each deafness variable, however, this result might be skewed, considering most participants were students. Rules with a LHS equal to {Male, Daily headphone usage, Loud Noise} were found for all 6 deafness variables, with an average lift of 1.2089 ± 0.0319. This result shows an existing correlation between deafness in males, frequent exposure to loud noises and high usage of headphones. Consequently, this association rule states that the origin of deafness in male students is related to frequent exposition to loud noises.

On the other hand, the most common rule for female participants had a LHS equal to {Female, Student, Family history of deafness}, appearing in 4 out of 6 deafness variables: LE2, LE3, RE2 and RE3. The rules had an average lift of 1.1088 ± 0.0346. This rule states that deafness in female students is commonly found when a family member has suffered from deafness. Additionally, this association rule could be read as a higher incidence of hereditary deafness in females, or a higher awareness of family health issues.

Table 1. Filtered association rules

LHS	RHS	Support	Confidence	Lift
Male, student, loud noise	LE1	0.1376	0.6122	1.2244
Male, daily headphone usage, loud noise	LE1	0.1009	0.5945	1.1891
Male, student, loud noise	LE2	0.1376	0.6122	1.2244
Male, daily headphone usage, loud noise	LE2	0.1055	0.6216	1.2432
Female, student, family history of deafness	LE2	0.0642	0.5384	1.0769
Male, student, loud noise	LE3	0.1330	0.5918	1.1946
Male, daily headphone usage, loud noise	LE3	0.1055	0.6216	1.2547
Female, student, family history of deafness	LE3	0.0642	0.5384	1.0868
Male, student, loud noise	RE1	0.1467	0.6530	1.2598
Male, daily headphone usage, loud noise	RE1	0.1055	0.6216	1.1992
Male, student, loud noise	RE2	0.1376	0.6122	1.2244
Male, daily headphone usage, loud noise	RE2	0.1009	0.5945	1.1891
Female, student, family history of deafness	RE2	0.0688	0.5769	1.1538
Male, student, loud noise	RE3	0.1284	0.5714	1.1863
Male, daily headphone usage, loud noise	RE3	0.0963	0.5675	1.1783
Female, student, family history of deafness	RE3	0.0642	0.5384	1.1179

4 Conclusions

In this work, we quantified possible hearing loss cases and their causes within the college community of Escuela Politécnica Nacional, through a survey with 226 participants, among students, faculty and staff members. The survey included an online audiometry, whose results were initially processed by an optical character recognition algorithm. Then, association rules were obtained using the Apriori algorithm and the results of both the survey answers and the audiometry.

There is a prevalence of hearing loss cases among young individuals of the sample. Furthermore, 149 people present a possible case of deafness for at least one frequency in at least one ear. Although the box plots for deafness and time of use show the presence of anomalous values, the trend shows that people aged between 20 and 25 are more susceptible to present hearing loss.

As a preliminary study, the obtained association rules have satisfactory lift parameters, and adequate confidence and support values. However, a more thorough study is suggested, in which hearing tests are performed under professional surveillance. In addition, collecting a greater sample is recommended to study a more representative part of the population of interest. On the other hand, regarding the optical character recognition for the audiometry data, the algorithm used could be further improved by better optical character recognition data sets and deep learning processes.

Acknowledgements. We thank Dr. Eduardo Carrera for providing a guide on the relative deafness scales and the relevant questions for the survey; We also thank Danilo Vásconez for his technical assistance on the elaboration of the figures for the descriptive statistics analysis.

References

1. WHO: Addressing the rising prevalence of hearing loss: a WHO report. world health organization, Geneva, Technical report (2018)
2. WHO: Deafness and hearing loss. https://www.who.int/news-room/fact-sheets. Accessed 09 June 2021
3. WHO: Global costs of unaddressed hearing loss and cost-effectiveness of interventions: a WHO report. world health organization, Geneva, Technical report (2017)
4. Key Work Health and Safety Statistics. https://www.safeworkaustralia.gov.au/doc/key-work-health-and-safety-statistics-australia-2020. Accessed 09 June 2021
5. WHO: Hearing loss due to recreational exposure to loud sounds a WHO review. world health organization, Geneva, Review (2015)
6. Poushter, J.: Smartphone ownership and internet usage continues to climb in emerging economies, Pew Research Center. Accessed 09 June 2021
7. Monasta, L.: Burden of disease caused by otitis media: systematic review and global estimates. PLoS One **7**(4), e36226 (2012). https://doi.org/10.1371/journal.pone.0036226
8. Mukherhjea, D.: The design and screening of drugs to prevent acquired sensorineural hearing loss. Expert Opin. Drug Discov. **6**(5), 491–505 (2011). https://doi.org/10.1517/17460441.2011.562887
9. Seddon, J.: Hearing loss in patients on treatment for drug-resistant tuberculosis. Eur. Respir. J. **40**, 1277–1286 (2012). https://doi.org/10.1183/09031936.00044812
10. Consejo Nacional para la Igualdad de Discapacidades Webpage. https://www.consejodiscapacidades.gob.ec/estadisticas-de-discapacidad/. Accessed 09 June 2021
11. Chaudhuri, A., Mandaviya, K.: Optical character recognition systems for different languages with soft computing. SFSC, vol. 352. Springer, Cham (2017). https://doi.org/10.1007/978-3-319-50252-6
12. Python Software Foundation: pytesseract0.3.7. https://pypi.org/project/pytesseract/. Accessed 09 June 2021
13. Python Software Foundation: Pillow 8.2.0. https://pypi.org/project/Pillow/. Accessed 09 June 2021

14. Hahsler, M., et al.: Introduction to arules – a computational environment for mining association rules and frequent item sets. J. Stat. Softw. (2005). https://doi.org/10.18637/jss.v014.i15
15. Reglas de asociación y algoritmo Apriori con R by Joaquín Amat Rodrigo, available under a Attribution 4.0 International (CC BY 4.0). https://www.cienciadedatos.net/documentos/43_reglas_de_asociacion
16. Fernández, C., et al.: Extracción de Reglas de Asociación en una Base de Datos Clínicos de Pacientes con Vih/Sida. In: "XXIV Congreso Anual de la Sociedad Española de Ingeniería Biomédica (CASEIB 2006)", 06 November 2006 - 08 November 2006, Pamplona, España, pp. 136–139 (2006)
17. Agrawal, R., Imielinski, T., Swami, A.: Mining association rules between sets of items in large databases. ACM SIGMOD Record. ACM. (1993). https://doi.org/10.1145/170035.170072
18. Guan, V.X., Neale, E.P., Tapsell, L.C., Probst, Y.C.: Identifying usual food choice combinations with walnuts: analysis of a 2005–2015 clinical trial cohort of overweight and obese adults. Front. Nutr. **7**, 149 (2020). https://doi.org/10.3389/fnut.2020.00149

Edufarmy: A Multisensory Educational Software System that Improves the Learning of Children with Dyslexia Using the Orton-Gillingham Approach

Shirley Mishell Pérez Quichimbo(✉), Erick David Barrera Quimbita, Milton Patricio Navas Moya, and Ximena López Chico

Universidad de las Fuerzas Armadas ESPE, Latacunga cl. Quijano y Ordónez y Hermanas Páez, Latacunga, Ecuador
smperez7@espe.edu.ec

Abstract. High rates of lack of motivation and lack of interest in learning and carrying out teaching and learning activities are some of the difficulties faced by children with dyslexia during their educational training. However, nowadays the use of technological tools has facilitated the process of reading and writing, providing didactic and interesting teaching for children with special educational difficulties. The objective of this article is to develop an interactive software using virtual reality tools, which implements the Orton Gillingham approach. This approach is mainly designed to work visual, auditory, and kinesthetic skills. By implementing the didactic game, it is possible to evaluate the technological acceptance and improvement of the condition of children with dyslexia in their reading and writing skills. The results of a study carried out in a psychological center prove that children with dyslexia have improved their condition favorably. It was proved that the stimulation of the senses by means of virtual reality allows them to improve their reading and writing skills, improving the areas affected by dyslexia such as background figure, visual-motor coordination, syllabication, laterality, and position in space.

Keywords: Technology · e-learning · Dyslexia · Children

1 Introduction

One of greatest social challenges facing society today is the integration of children with specific learning disabilities, with an emphasis on dyslexia. In essence, dyslexia are defined as an unexpected difficulty in reading fluency or accuracy that affects an individual's learning, educational level or chronological age. However, studies claim that a child with this disorder can be bright and talented despite a much lower reading level [1].

Deficiencies in word reading, spelling accuracy and fluency in children with dyslexia are accompanied or preceded by incorrect oral expression which tend to constantly cause

M. Zambrano Vizuete et al. (Eds.): CI3 2021, LNNS 511, pp. 450–463, 2022.
https://doi.org/10.1007/978-3-031-11438-0_36

feelings of stress and failure [2]. The British Dyslexia Association (BDA) estimates that 10% of the population is dyslexic and 4% are severely affected. The demand for children to be supported in their learning process is growing dramatically, because the traditional educational model is not able to meet the needs of all types of students, with or without disabilities [3].

In Ecuador, one of the main problems is the lack of a specific methodology that adapts to the individual differences of each student. Many parents need to look for specialists to provide methods and techniques to improve the condition of the child with dyslexia. But this requires time and constant perseverance in order to motivate and instill interest in children [4]. In recent years, technological advances have made possible the development of treatments focused on motivating children with dyslexia during their learning process. Among these treatments is the multisensory approach, which allows the practice of reading and writing using visual, auditory, kinesthetic and tactile resources. Using the multisensory approach provides greater interest when performing the activities and shows promising results with Students [5]. However, a characteristic challenge of the country is the lack of technological support tools that use virtual reality. This fact causes the community especially the parents or guardians of those affected to think that there are no treatment alternatives aimed at children with dyslexia [5]. The purpose of this research is to propose an analysis of computer games applied through virtual reality technology to support the psycho-pedagogical treatment of children with dyslexia [6]. The main characteristic of the use of this application is to contribute to the development of reading and writing skills and to reduce the incidence of dyslexia. It provides those affected from an early stage with interest and interaction with multisensory resources [7]. The research was based on the compilation and analysis of the bibliography of scientific publications, psych pedagogy books, and is complemented with a field research work in a psych pedagogical and psychological center where some important results of the development of the multisensory educational play system based on the Orton-Gillingham method were obtained [6].

2 State of the Art

Assistive technology drives a methodological change for children with special educational needs. The multisensory educational game system is composed of a web application that focuses on following the child's learning progress, and a virtual reality video game where the child can interact and play at any time. This video game is a perfect combination of Orton-Dillingham's multisensory and technological approach that avoids distractions or lack of interest of the children [8]. Developmental dyslexia (DDD) are a language disorder that primarily affects the ability to read and write. Children with dyslexia have problems in the rate and speed of learning written language, as well as in acquiring an adequate level of performance [9].

According to recent studies, children with dyslexia often use educational playful software to improve their condition. Over time, it was realized that technology can benefit students with special educational needs.

2.1 A Multisensory 3D Environment as an Intervention to Help Reading in Dyslexia

One of these investigations proposes the use of multisensory techniques with teaching methodologies based on phonics and phonemic awareness through the use of 3D environments implemented in interactive games to help improve reading in people with dyslexia [31].

2.2 Multiplatform Games for the Identification of Dyslexia in Preschool Children

Followed by another study that proposes as a basis the development of serious multiplatform games using a framework called PhoneGap, with the aim of identifying dyslexia in early ages in an interactive and entertaining way [32].

2.3 Multimedia Learning Based on Augmented Reality for Dyslexic Children

The research also points out a software that uses virtual reality technology in the teaching-learning process of people with dyslexia to help develop their cognitive learning and mental development, for which they created identifiers to generate 3D objects through a mobile device generating an interactive learning booklet [33].

2.4 Madrigale Application

Finally, we indicate a multimedia application whose objective is to help develop phonological skills and visuospatial attention in children between 7 and 9 years old through interactive educational games in a more attractive way, generating motivation in the learning process. Based on musical and ludic educational approaches [34].

3 Implementation

The development of educational software proposes an iterative approach and includes pedagogical computational aspects, which uses the methodology of Educational Software Engineering (MeISE) [30].

Figure 1 shows the structure of the application using the MeISE methodology. In the first stage, the definition of requirements, the preliminary analysis and the design of the different educational games are considered. In this process, the characteristics of the product to be made, the teaching, communication requirements and the architecture of the product to be built will serve as the basis for the solution of problems related to reading and writing and the pedagogical characteristics that support the slow development of vocabulary and delay in the development of speech with difficulties in articulating or pronouncing words are established. In the development stage, a detailed computational design for children with dyslexia is made involving motivational element such as music, sounds, animations and images. The software ends with an iteration plan, i.e., it completely covers some of the didactic objectives of the software, facilitating the process of both teaching and learning.

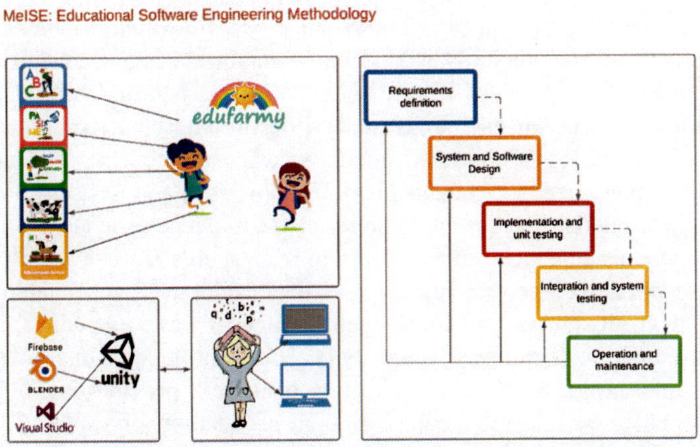

Fig. 1. MeISE educational software engineering methodology

Orton Gillingham's treatment of children with dyslexia emphasizes the individual introduction of each phonogram and all grammatical rules using visual elements, auditory and kinesthetic information where representations of print-sound correspondences are established and, the language units are introduced in a systematic sequence of increasing complexity of simple vowels and consonants through multi-syllable words [11].

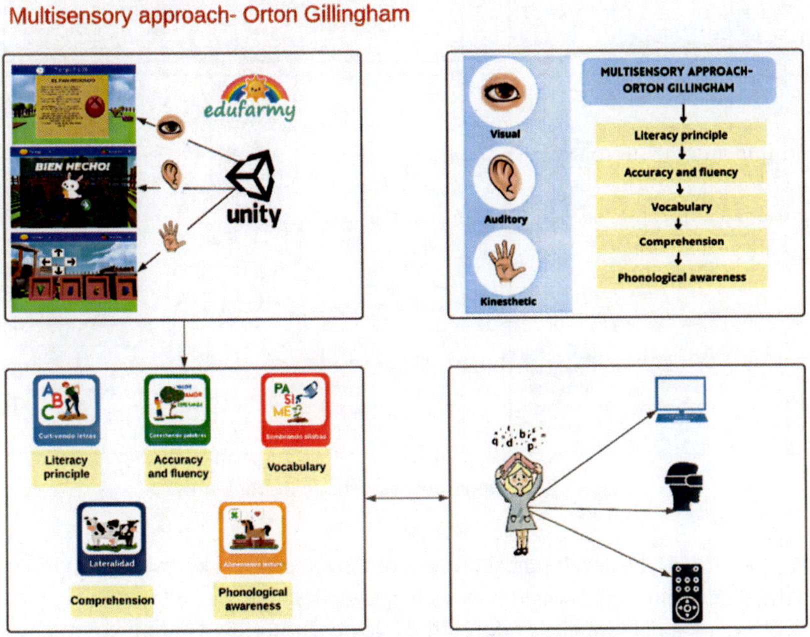

Fig. 2. Multisensory approach-Orton Gillingham

Figure 2 shows the first stage is the visual modality, which teaches the child to master the principles of literacy and to master information accurately and fluently. This is followed by the auditory modality where the child must learn the individual letter sounds by hearing the sounds, saying the sounds, and writing the letters that represent the sounds. The third stage consists of the kinesthetic modality, where the student experiences, relates and pronounces new concepts from his previous knowledge and focuses on writing and spelling. Included in this stage is the memory of the movements made by the muscles. Finally, through this approach, students learn to read, spell and write at their own pace.

In the execution of the project objective and the development of a reliable and robust application, the Unity tool is used. Subsequently, Blender was used mainly to model 3D objects and animation of different elements [12]. Beyond its dynamics, Visual studio was used for the programming control section because it provides a simple, modern and object-oriented language that combines high productivity and speed [14]. Finally, the hosting of real-time data uploaded to the cloud is saved in the in-dependent entity Firebase [15].

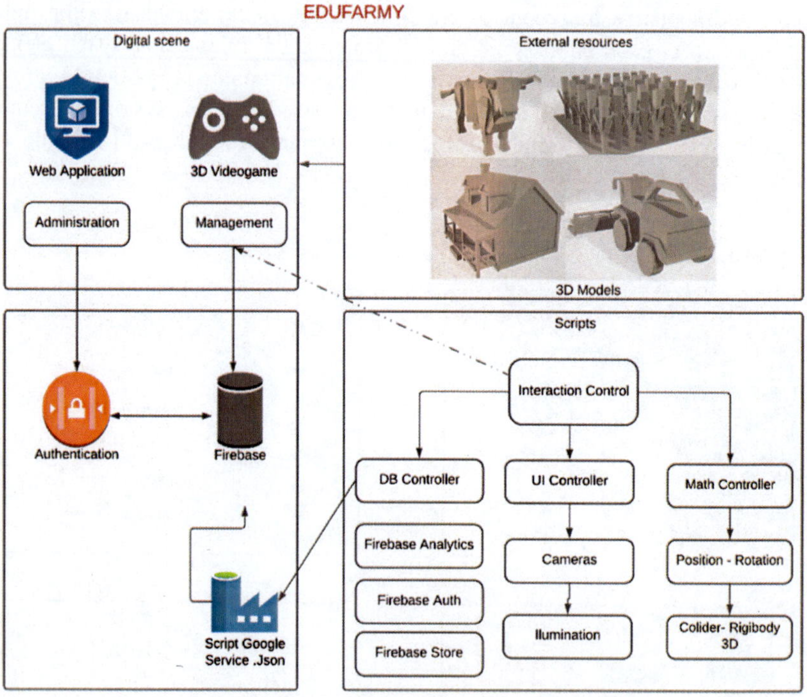

Fig. 3. Educational software diagram-EduFarmy

Figure 3 shows the development process of the educational software of a video game in a virtual environment, assisted by a web application that will be managed by a tutor, this will be a support for the child with dyslexia during his reading comprehension, grammar and reading fluency. The execution of this project will be done through 3D simulation graphic engines and character controller functionalities for a realistic effect within the video game. The project aims to be a virtual support and monitoring tool for children with dyslexia disorder, considering their special needs.

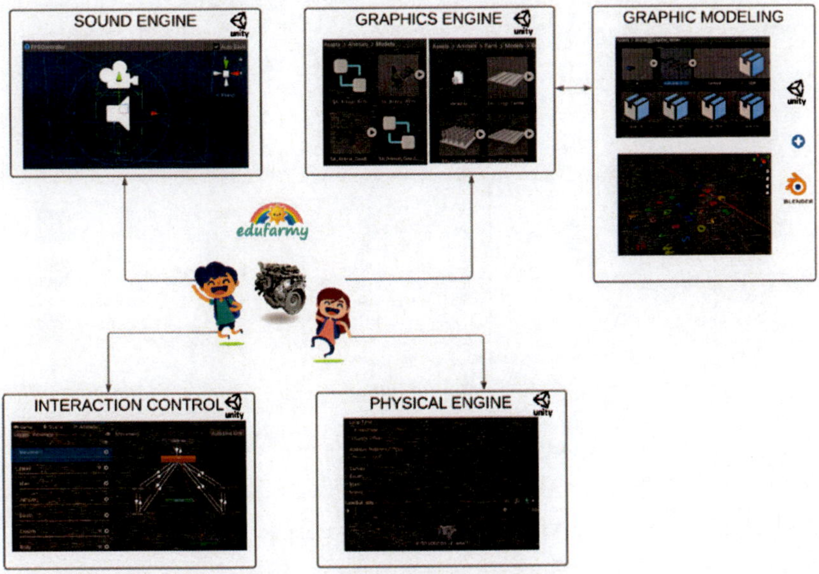

Fig. 4. EduFarmy game engine

According to the context defined in Fig. 4, the video game engine aims to address the teaching conditions identified in the study of learning disabilities of children with dyslexia, using multisensory tools to apply it within a virtual environment. Therefore, sounds and sound effects were augmented, which give credibility and verisimilitude to a farm environment, along with animal multimedia created by ChiquitinesTV, my Kindergarten, which involve the player emotionally. The graphic engine of the video-game was adapted to the characteristics of the child audience, with artistic and friendly designs. Blender was used to develop the 3D models and animate them. Next, the physics engine integrated in the videogame provides components that handle the physics simulation, such as collisions, forces and gravity. The Interaction Control is in charge of giving functionalities to the player to interact with buttons, cubes and physical elements that belong to the farm. Finally, the precise movement control of characters and their actions was defined to facilitate gameplay.

Figure 5 shows the structure of the software, which consists of a desktop application, virtual reality glasses that interact the virtual reality with the user of the application and a web application to manage the child's learning progress.

Requirements Capture

The functional and non-functional requirements identified in the design and gathering stage are presented below. These requirements serve to provide the application with functionalities that contribute to dyslexia in children, and thus, help in their special needs (Tables 1 and 2).

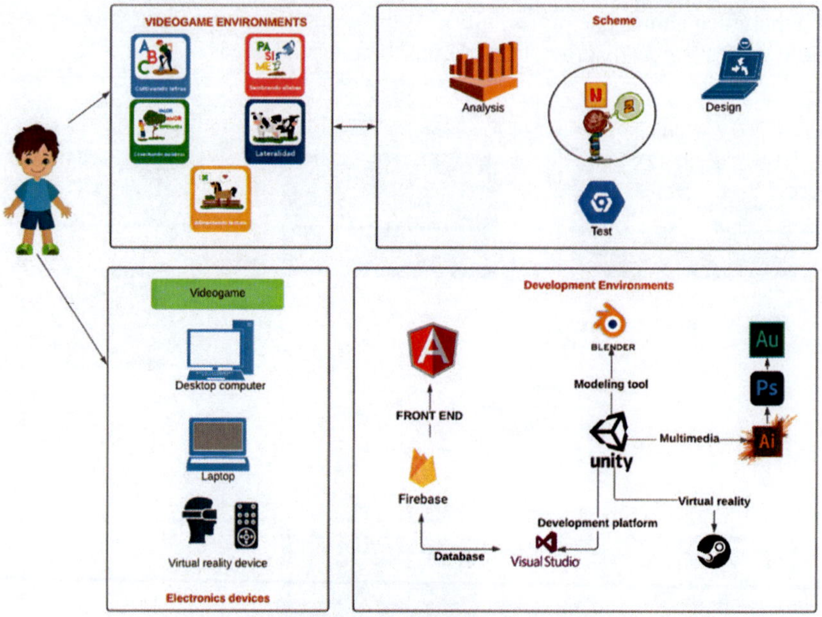

Fig. 5. Implementation scheme

Table 1. Requirements of the desktop application.

Functional	Non-functional
• It has a menu, in which the child chooses one of the mini games depending on the state in which their treatment is, and their level of learning • It shows levels focused on a farm where the child must illustrate their auditory, visual and kinesthetic skills • It has a three-dimensional environment that works with Virtual Reality	• It installs on any operating system since the video game is multiplatform • It contains motion sensors • Virtual Reality glasses provide entertainment and safety to the child

Table 2. Requirements of the web application.

Functional	Non-functional
• Allows the tutor to add the child's information • Allows the guardian to visualize the child's information • Allows the guardian to modify the child's information • Allows the guardian to delete the child's information	• Child information may only be modified by the guardian • The children's information will be stored in the Firebase database • The color of the application is in accordance with a child's environment

3.1 First Stage of the Videogame

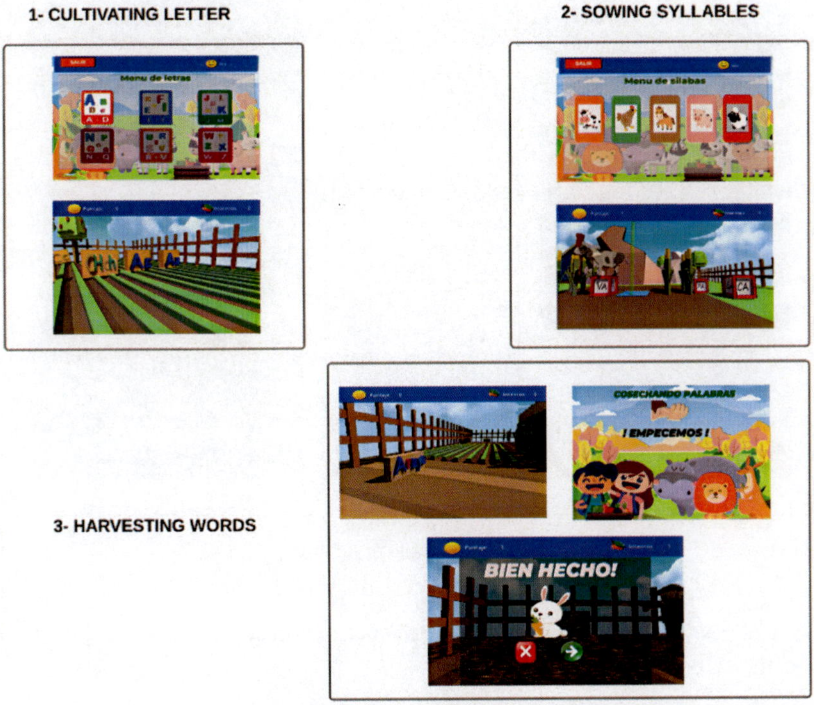

Fig. 6. Principal's game

Figure 6 proposes a space designed to work on dyslexia through virtual reality, implementing the set of letters of the alphabet, syllables, and words. At the beginning of each game, a multimedia sound is presented with its proper instructions. To keep track of a statistic, the child will be presented with their number of at-tempts and hits. The three main games represent the first stage of the Orton Gillingham Methodology, which helps the child develop classification skills and visual and auditory skills (Fig. 7).

3.2 Second Stage of the Video Game

The second stage of the virtual reality video game is aimed at working on visual, auditory, and kinesthetic skills. Therefore, the child is presented with the reading mini-game, and a laterality mini-game. At the beginning of each game, clear and summarized instructions are provided through multimedia audio.

In the reading game, the child will have a certain time to read a previously selected story, and then answer the questions in the questionnaire.

The game of laterality begins when the child correctly orients the word that is inside the cube. Then you must place the cube with its respective word in one of the two scenarios, to understand its meaning through visual skills.

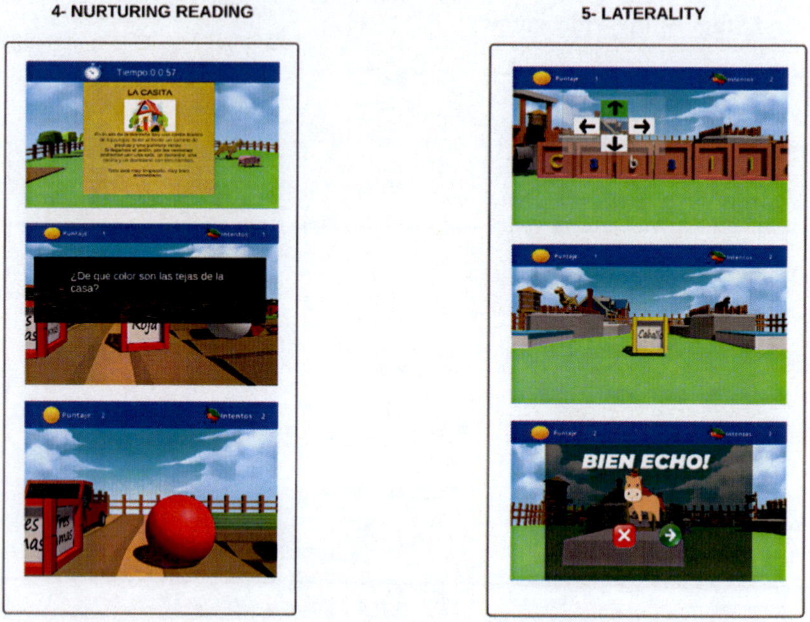

Fig. 7. Secondary game

Both games will count the number of attempts and successes that the child has made to solve the activities.

4 Evaluation of Results

To evaluate the results of the treatment progress, a 20-item test was applied, which uses a five-point scale (1 = very low, 2 = low, 3 = medium, 4 = high, 5 = very high) at the beginning and at the end of the experiment, each of these items evidenced the skills that the child should develop during the treatment. The children were trained on the functioning of the game, indicating the objectives of each one and how to carry them out.

At the end of the evaluation stage, the results obtained at the beginning and the end of the experiment were compared, showing a great difference in the results (Table 3).

The table above shows the averages of the scores obtained in the tests at the beginning and end of the treatment carried out on 12 children who were the ones who contributed in the research with the deadlines in the proposed therapy, for research purposes, as it can be observed the results of the application of the game in the treatment of the children during 6 months, there is an evident decrease in most of the child's special difficulties, only in 2 of the 20 items the same score was maintained, the same that are related to the difficulty of laterality, consequently it can be determined that the game had an effectiveness of 90% in the progress of the child's treatment, the same that can be increased with the use of more training time (Fig. 8).

Table 3. Results of the assessment of reading and writing skills

Items	Initial stage	Intermediate stage 1	Intermediate stage 2	Final stage
Items 1	4,083	3,851	3,538	3,25
Items 2	3,917	3,742	3,43	3,333
Items 3	4,083	3,45	3,313	3,167
Items 4	4,417	4,03	3,761	3,417
Items 5	3,917	3,901	3,909	3,911
Items 6	4	3,75	3,531	3,333
Items 7	4	3,81	3,529	3,333
Items 8	4,25	4,01	3.831	3,5
Items 9	4,583	4,12	3,881	3,5
Items 10	3,833	3,601	3,417	3,333
Items 11	3,333	3,143	2,993	2,917
Items 12	3,833	3,734	3,507	3,417
Items 13	4,167	3,85	3,709	3,667
Items 14	4,333	3,142	3,876	3,75
Items 15	4,667	4,306	4,007	3,833
Items 16	4,417	4,315	3,894	3,5
Items 17	4,583	4,263	3,926	3,75
Items 18	4,25	4,231	4,268	4,252
Items 19	4,167	3,806	3,607	3,5
Items 20	4,417	4,25	3.867	3,5

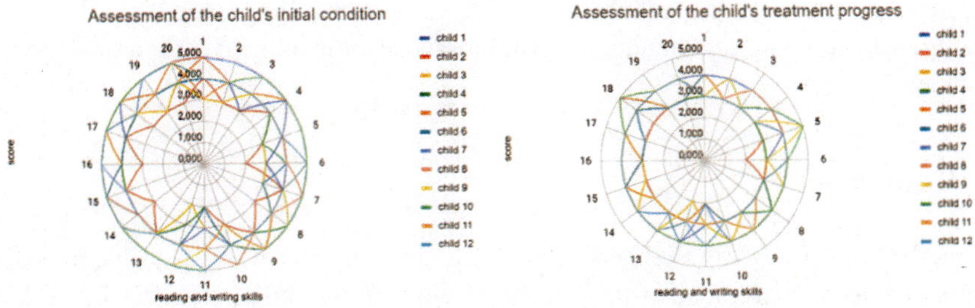

Fig. 8. Graphic representation of the play treatment.

The previous graph represents graphically the results of the progress of each child from the beginning to the end of the proposed treatment, as can be observed at the beginning, the values are within the pessimistic panorama, but as the treatment progresses it can be observed how these tend to be reduced presenting a more optimistic panorama of the progress of the treatment, It should be emphasized that dyslexia, being considered as a condition, cannot be eradicated but rather controlled and diminished, which is the purpose of this software, even though each child has a different way of learning, the playful game has managed to integrate each of these strategies allowing the progress of all of them to the same degree.

5 Discussion

The use of new technologies such as virtual reality in early stages of the teaching and learning process of reading and writing of children with special educational needs such as children with dyslexia, represents a tool for intervention and evaluation of their treatment, which shows substantial improvements in their condition. The proposed software was very well received by the children with dyslexia and the specialists in charge of their treatment. The unique approach of combining a virtual environment with a teaching methodology that fits the unique needs of children with dyslexia has obtained mostly positive results, evidencing progress in the treatment of children. Each one of the games allowed the training of the different literacy skills by the infants who are in the age range between 6 and 9 years old and suffer from dyslexia. The use of virtual reality in an interactive and simple game with friendly environments has allowed to capture to a greater degree their interest and commitment to the treatment, reinforcing their literacy skills, thus improving their confidence in this area of their education.

6 Future Works

For future work, it is proposed to improve the game by increasing the number of levels in each of its minigames, where it will be possible to determine precisely which skills the children need to improve in their treatment and practice them, allowing the specialists in charge of the treatment to better adapt the rehabilitation strategies for each of the children. In addition, the guidance within the game can be improved through artificial intelligence, allowing to simulate the continuous accompaniment of a tutor during the treatment.

7 Conclusions

Through the results of the study carried out with the children, it was possible to observe a remarkable improvement in their reading and writing skills, reducing by 13% the complications of dyslexia during the time determined for the study. In order to make a correct use of the desktop application, it was taken into account that the exposure in simulated environments should be progressive and during the time in which the child is treated, generating a greater interest and commitment in the therapy, which evidences the acceptance of the software by the child.

The play software allows the child to focus on different areas affected by dyslexia such as background figure perception, visual-motor coordination, syllabication, laterality and position in space. All these areas are mostly boosted by virtual reality which helps to involve all the child's senses during the treatment. In relation to the above, we can conclude that virtual reality contributes satisfactorily to the psycho-pedagogical field, specifically in the treatment of children with special educational needs. Technological tools create simulated environments that help to acquire motivation and interest in children at the school stage.

References

1. Kreutzer, J.S., DeLuca, J., Caplan, B. (eds.): Encyclopedia of Clinical Neuropsychology. Springer, Cham (2018). https://doi.org/10.1007/978-3-319-57111-9
2. Brimo, K., Dinkler, L., Gillberg, C., Lichtenstein, P., Lundström, S., Åsberg Johnels, J.: The co-occurrence of neurodevelopmental problems in Dyslexia. Dyslexia **27**(3), 1681 (2021). https://doi.org/10.1002/dys.1681
3. Pinto, F.S., Martins, V.F., Cibelle, A., Eliseo, M.A.: Requirements for the development of digital serious games for Dyslexics. In: 2019 14th Iberian Conference on Information Systems and Technologies (CISTI), Coimbra, Portugal, June 2019, pp. 1–6. https://doi.org/10.23919/CISTI.2019.8760627
4. Gooch, D., Vasalou, A., Benton, L.: Exploring the use of a gamification platform to support students with Dyslexia. In: 2015 6th International Conference on Information, Intelligence, Systems and Applications (IISA), Corfu, Greece, July 2015, pp. 1–6. https://doi.org/10.1109/IISA.2015.7388001
5. Broadhead, M., Daylamani-Zad, D., Mackinnon, L., Bacon, L.: A Multisensory 3D environment as intervention to aid reading in Dyslexia: a proposed framework. In: 2018 10th International Conference on Virtual Worlds and Games for Serious Applications (VS-Games), Wurzburg, September 2018, pp. 1–4. https://doi.org/10.1109/VS-Games.2018.8493407
6. McChesney, I., Bond, R.: Eye tracking analysis of computer program comprehension in programmers with dyslexia. Empir. Softw. Eng. **24**(3), 1109–1154 (2018). https://doi.org/10.1007/s10664-018-9649-y
7. Facoetti, A., et al.: Multiplatform games for Dyslexia identification in preschoolers. In: 2014 IEEE 11th Consumer Communications and Networking Conference (CCNC), ene. 2014, pp. 1152–1153. https://doi.org/10.1109/CCNC.2014.6940496
8. Jaffe-Dax, S., Raviv, O., Jacoby, N., Loewenstein, Y., Ahissar, M.: Towards a computational model of Dyslexia. BMC Neurosci. **16**(S1), O12 (2015). https://doi.org/10.1186/1471-2202-16-S1-O12
9. Ward, A., Bush, H., Braaten, E.B.: Reading Disorders/Dyslexia. In: Wilson, H.K., Braaten, E.B. (eds.) The Massachusetts General Hospital Guide to Learning Disabilities. CCP, pp. 21–37. Springer, Cham (2019). https://doi.org/10.1007/978-3-319-98643-2_2
10. Schlesinger, N.W., Gray, S.: The impact of multisensory instruction on learning letter names and sounds, word reading, and spelling. Ann. Dyslexia **67**(3), 219–258 (2017). https://doi.org/10.1007/s11881-017-0140-z
11. Ring, J.J., Avrit, K.J., Black, J.L.: Take flight: the evolution of an Orton Gillingham-based curriculum. Ann. Dyslexia **67**(3), 383–400 (2017). https://doi.org/10.1007/s11881-017-0151-9

12. Bruns, N.: Blender: Universelle 3D-Bearbeitungs-und Animationssoftware. Unfallchirurg **123**(9), 747–750 (2020). https://doi.org/10.1007/s00113-020-00836-0
13. Zawadniak, P.V.F., et al.: Micromouse 3D simulator with dynamics capability: a Unity environment approach. SN Appl. Sci. **3**(2), 1–11 (2021). https://doi.org/10.1007/s42452-021-04239-7
14. Wang, D.: Gamified learning through unity 3D in visualizing environments. Neural Comput. Appl. **29**(5), 1399–1404 (2017). https://doi.org/10.1007/s00521-017-2928-5
15. Hajian, M.: Deploying to Firebase as the Back End. In: Progressive Web Apps with Angular, pp. 9–27. Apress, Berkeley, CA (2019). https://doi.org/10.1007/978-1-4842-4448-7_2
16. Geiser, E., Kjelgaard, M., Christodoulou, J.A., Cyr, A., Gabrieli, J.D.E.: Auditory temporal structure processing in dyslexia: processing of prosodic phrase boundaries is not impaired in children with dyslexia. Ann. Dyslexia **64**(1), 77–90 (2013). https://doi.org/10.1007/s11881-013-0087-7
17. Galliussi, J., Perondi, L., Chia, G., Gerbino, W., Bernardis, P.: Inter-letter spacing, inter-word spacing, and font with dyslexia-friendly features: testing text readability in people with and without dyslexia. Ann. Dyslexia **70**(1), 141–152 (2020). https://doi.org/10.1007/s11881-020-00194-x
18. van Viersen, S., de Bree, E.H., Kroesbergen, E.H., Slot, E.M., de Jong, P.F.: Risk and protective factors in gifted children with Dyslexia. Ann. Dyslexia **65**(3), 178–198 (2015). https://doi.org/10.1007/s11881-015-0106-y
19. Gibby-Leversuch, R., Hartwell, B.K., Wright, S.: Dyslexia, literacy difficulties and the self-perceptions of children and young people: a systematic review. Curr. Psychol. **40**(11), 5595–5612 (2019). https://doi.org/10.1007/s12144-019-00444-1
20. Novita, S., Uyun, Q., Witruk, E., Siregar, J.R.: Children with dyslexia in different cultures: investigation of anxiety and coping strategies of children with Dyslexia in Indonesia and Germany. Ann. Dyslexia **69**(2), 204–218 (2019). https://doi.org/10.1007/s11881-019-00179-5
21. van der Kleij, S.W., Segers, E., Groen, M.A., Verhoeven, L.: Post-treatment reading development in children with Dyslexia: the challenge remains. Ann. Dyslexia **69**(3), 279–296 (2019). https://doi.org/10.1007/s11881-019-00186-6
22. Verpalen, J.M.P., van de Vijver, F.J.R.: Differences in neurocognitive aspects of Dyslexia in Dutch and immigrant 6-7- and 8-9-years old children. Springerplus **4**(1), 1–15 (2015). https://doi.org/10.1186/s40064-015-0874-1
23. Franceschini, S., Bertoni, S., Ronconi, L., Molteni, M., Gori, S., Facoetti, A.: Shall we play a game? improving reading through action video games in developmental Dyslexia. Curr. Dev. Disord. Rep. **2**(4), 318–329 (2015). https://doi.org/10.1007/s40474-015-0064-4
24. Fokides, E., Chronopoulou, M.-I., Kaimara, P.: Comparing videos and a 3D virtual environment for teaching school-related functional skills and behaviors to students with ADHD or developmental Dyslexia, displaying challenging behaviors: a case study. Res. Pract. Technol. Enhanc. Learn. **14**(1), 1–19 (2019). https://doi.org/10.1186/s41039-019-0117-0
25. Toffalini, E., Giofrè, D., Pastore, M., Carretti, B., Fraccadori, F., Szűcs, D.: Dyslexia treatment studies: a systematic review and suggestions on testing treatment efficacy with small effects and small samples. Behav. Res. Methods **53**(5), 1954–1972 (2021). https://doi.org/10.3758/s13428-021-01549-x
26. Ronconi, L., Melcher, D., Franchin, L.: Investigating the role of temporal processing in developmental dyslexia: Evidence for a specific deficit in rapid visual segmentation. Psychon. Bull. Rev. **27**(4), 724–734 (2020). https://doi.org/10.3758/s13423-020-01752-5
27. Eutsler, L., Mitchell, C., Stamm, B., Kogut, A.: The influence of mobile technologies on preschool and elementary children's literacy achievement: a systematic review spanning 2007–2019. Educ. Tech. Res. Dev. **68**(4), 1739–1768 (2020). https://doi.org/10.1007/s11423-020-09786-1

28. O'Brien, J., Ottoboni, G., Tessari, A., Setti, A.: Multisensory perception, verbal, visuo-spatial and motor working memory modulation after a single open-or closed-skill exercise session in children. J. Cogn. Enhancement **5**(2), 141–154 (2020). https://doi.org/10.1007/s41465-020-00189-x
29. Schouten, D.G.M., Smets, N.J.J.M., Driessen, M., Fuhri, K., Neerincx, M.A., Cremers, A.H.M.: Requirements for a virtual environment to support the social participation education of low-literates. Univ. Access Inf. Soc. **16**(3), 681–698 (2016). https://doi.org/10.1007/s10209-016-0502-z
30. Vicente y, P., Soledad, C.: Diseño de un juego serio para la mejora de la conciencia fonológica de los niños con dislexia, p. 10 (2016)
31. Broadhead, M., Daylamani-Zad, D., Mackinnon, L., Bacon, L.: A multisensory 3D environment as intervention to aid reading in Dyslexia: a proposed framework. In: 2018 10th International Conference on Virtual Worlds and Games for Serious Applications (VS-Games), Wurzburg, September 2018, pp. 1–4. https://doi.org/10.1109/VS-Games.2018.8493407
32. Facoetti, A., et al.: Multiplatform games for Dyslexia identification in preschoolers, p. 2
33. Bhatti, Z., Bibi, M., Shabbir, N.: Augmented Reality based Multimedia Learning for Dyslexic Children. In: 2020 3rd International Conference on Computing, Mathematics and Engineering Technologies (iCoMET), Sukkur, Pakistan, ene. 2020, pp. 1–7. https://doi.org/10.1109/iCoMET48670.2020.9073879
34. Di Tore, P.A., Di Tore, S., Ludovico, L.A., Mangione, G.R.: Madrigale: a multimedia application for Dyslexia and reading improvement gamifying learning experience. In: 2014 International Conference on Intelligent Networking and Collaborative Systems, Salerno, September 2014, pp. 486–491. https://doi.org/10.1109/INCoS.2014.48

Optimization in the Handling of Large Amounts of Data for Reading, Processing and Graphing EEG Data in Excel

Andrés E. Castillo R.[1], Yngrid J. Melo Q.[2](✉), Wilson G. Simbaña L.[2], Edgar A. Bravo D.[2], Wilmer R. Valles B.[2], and Luis M. Guallasamin P.[2]

[1] Universidad Politécnica Territorial de Aragua Federico Brito Figueroa, La Victoria, Venezuela
[2] Instituto Tecnológico Universitario Rumiñahui, Sangolquí, Ecuador
yngrid.melo@ister.edu.ec

Abstract. The article presented here refers to the research work done for reading, processing and plotting EEG signals, where physiological data of this type can be acquired from websites such as PhysioNet. Usually a lot of data is associated with EEG measurements in each patient. When you have large amounts of data in a spreadsheet and you want graphs that can be controlled interactively, or graphs that change from data to observation at the user's will using interactive controls, it is normal for the application to slow down and for the update processes to take several seconds, even minutes, which makes it difficult for the user to handle this type of application. Some techniques are shown to reduce the delays caused by transferring data from one site to another and their measurement is presented. The improvements are based on the use of VBA (Visual Basic for Application) and the advantageous use of certain structures that indirectly allow the processes to be carried out directly at the level of the available RAM memory, specifically using various VBA VARIANT.de VBA types. In this work we disclose ways to speed up the reading, transferring, processing and graphing of this data, so that the user experience is satisfactory from the point of view of speed of response. The main motivation is to develop them to be used in studies of specialties such as Software Development and Electronics in subjects such as Electromedicine.

Keywords: EXCEL · VBA · EEG signals · PhysioNet

1 Introduction

Currently, there are websites that manage databases of physiological signals that are available for access to the data stored there. One of these sites is PhysioNet [1] where such information is stored, this particular work is interested in the management of brain signals (EEG) which are measurements of electrical potentials made with various electrodes placed on the head.

The research refers to the use of spreadsheets such as EXCEL for reading, processing, transferring and graphing data. An application was developed that enables, through user interfaces, its use as a didactic resource for familiarization with EEG signals, as well as

M. Zambrano Vizuete et al. (Eds.): CI3 2021, LNNS 511, pp. 464–478, 2022.
https://doi.org/10.1007/978-3-031-11438-0_37

training to recognize them according to sex, pathology, age, in a flexible and low-cost way, within high quality teaching parameters. The main motivation is to develop them for use in the development of research projects in careers such as Electronics and/or Software Development, among others that need this type of signals.

Although there are specialized applications for measuring, processing and presenting signals related to human functioning, it is convenient to extend their use to spreadsheets because of the advantages of accessibility and because these tools, far from making data invisible, can put the student or professional in direct contact with them.

Current spreadsheets such as EXCEL, have capabilities for graphing, reading data from files, mathematical processing of data, and the power of programming supported by an advanced language such as Visual Basic Application [2]. In addition, they have graphical user controls, which allows the development of interactive and graphical applications within a spreadsheet. The particularity of EEG signals is that they can reach up to 21 measurement points and each measurement point can have a large amount of data, for example, the PhysioNet database presents a data per person of 21 points, where up to 91,000 samples per measurement point are stored. In addition, due to the need to obtain graphs where all the signals appear simultaneously in the same graph, these are vertically displaced so that the signals do not overlap, which causes the data to be duplicated, adding to each data a displacement factor that will vary for each signal. This means facing a spreadsheet with a relatively large amount of data.

In this order of ideas, when you have large amounts of data in a spreadsheet and you want to obtain graphs from these, which can be controlled interactively, or what is the same, graphs that change from data to observe at will of the user using interactive controls, at this point, it is normal that the application slows down and that the update processes last several tens of seconds, even minutes, which makes it heavy and annoying for the user to handle this type of applications. This was verified in works previously carried out at the Universidad Politécnica Territorial de Aragua (UPTA), to which the main author of this article belongs and in which both the main author and the secondary author participated as tutors. In these projects, two tools were developed to manage data from ECGs [3] and EEGs [4], in both cases, the data were extracted from the physionet.org site, and in both cases, it was possible to plot images, change the plotted data at the user's will, being able to choose at will the data of a particular electrode and other functionalities. However, the delays in data transfer ended up being annoying for the user of these tools.

The main problem, when handling EEG data, is that the data is large enough to slow down Excel and make the interactive experience unsatisfactory, because the response times to events are too long. This article presents better ways to handle the transfer, processing and plotting processes, using mainly RAM memory, not directly, but using a VBA variable type, the Variant type. Some techniques are presented to reduce the delays caused by transferring data from one place to another within Excel and the measurement tool is provided, so that the user can have a satisfactory interactive experience from the point of view of the speed of response.

1.1 EEG Signals

An EEG is the recording of the electrical activity produced by brain cells. It corresponds to a technique for studying this activity in the cerebral cortex, which is based on the excitatory or inhibitory activity of the pyramidal neurons located in that region [5]. It is a noninvasive technique, since the electrodes are placed on the scalp or in the cerebral or intracerebral cortex and record neuronal activity in real time.

Each electrode records the activity of a set of neurons and measures the potential difference with another reference electrode. There is currently an international system of electrode placement called the 10/20 system, which allows for comparison of recordings from different studies, as shown in the following figure (Fig. 1):

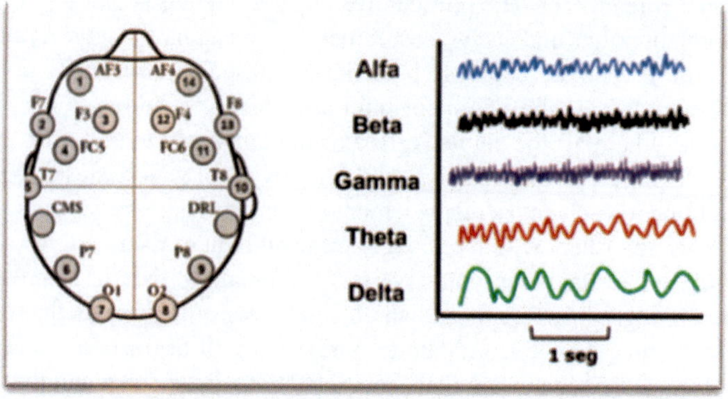

Fig. 1. Location of 14 electrodes according to the international 10–20 system. On the right schematic of different brainwave frequencies. Source [5]

According to [5] in reference to the recording of electrical activity, the EEG delivers five types of waves, which can be differentiated by:

a) Its frequency (number of times the wave repeats itself) which is measured in Hertz (Hz), therefore, 3 Hz means that the wave repeats itself 3 times in one second;
b) Its amplitude, which corresponds to the difference between the maximum and minimum voltage of the wave, which is measured in microvolts μV.

Nowadays there are different helmets that integrate electrodes that measure and send information, simplifying the processing of obtaining and placing the data. The use of these helmets has become very popular for research and experimental activities.

1.2 PhysioNet

PhysioNet [1], is a web-based resource that provides well characterized physiological signals and related open-source software to the biomedical research community. Established in 1999 under the auspices of the National Institutes of Health (NIH), this platform is managed by members of the MIT Computational Physiology Lab.

PhysioNet has three main and interdependent components:

1. A data archive (PhysioBank) of well characterized digital recordings of physiological signals, time series and related data for use by the biomedical research community.
2. Software library (PhysioToolkit) for the processing and analysis of physiological signals, detection of physiologically significant events using classical techniques and novel methods based on statistical physics and nonlinear dynamics, among many others.
3. A collection of Tutorials and educational material offering expert guidance on approaches to explore and analyze health data and physiological signals.

PhysioNet collects measurements from different institutes, a group of them are made up of measurements made on 290 patients, this was the one taken as a reference for this work. A user's data file consists of 10,000 records, each record has 16 fields.

An interesting fact is the integration of PhysioNet with ORCID, the unique digital identifier for researchers. Adding a researcher's ORCID ID to PhysioNet will help them connect professional work (datasets, software, etc.) to their public ORCID profile.

1.3 Spreadsheets and EEG

Today's spreadsheets far exceed the concept of grids with the ability to make calculations based on values placed in their cells. They have the capacity to store relatively large amounts of information since a sheet can have more than 15000 columns by 1 million rows. They can graph data, and have the ability to provide the user with interactive elements that allow the user to make choices dynamically.

Regarding the use of Excel as a platform to create didactic tools and as an application tool in the teaching of technical scientific subjects, authors have been involved in the elaboration of other articles as can be seen in [2, 6, 7]. Another important factor is the ability to create programs (macros) that can be activated at will by the user and that can perform highly complex processes on the data.

For example, in the case of an EEG signal extracted from PhysioNet, 22 meditions are obtained, with each measurement having 9,90000 samples. With this volume of data, the transfer, plotting and calculation times become long enough to be noticeable and unpleasant for the user.

One way to decrease these times is to make most of the time to be processed and transferred at the RAM level, and this can be achieved by using structures called Variant of the VBA language.

1.4 The EDF Format

The EDF format was conceived mainly to provide an open format for storing and exchanging long term EEG recordings. It was proposed at the 5th International Congress of Sleep Research in 1987. It was published in 1992 and since then it has become the standard for EEG and PSG recordings in commercial equipment and research projects.[8] The use of EDF format was to store EEG and PSG recordings in a standardized format.

The use of the EDF format was for EEG recordings and sleep recordings, the need to store data from other signals used in Clinical Neurophysiology laboratories was the reason for creating the EDF+ format. The EDF+ format was developed in 2002, and published in 2003 in Elsevier. It extends the capabilities of the EDF format, being able to support electrical and various other biological recordings obtained by transducers. The signals stored in EDF+ are generally: EEG, ECG, EOG, ERG, EMG, NC, MEG, MCG, EP, temperature, Light, Sound. EDF+ can also store events and annotations only without any signal.

Table 1. Information associated with the global header of an EDF file. Source [9]

Name	Function	Size (Bytes)	Format
Version	Format version of data	8	
Name	Name of the patient of EEG	80	
Record	Code name of the record	80	
Date	Show the date of the record	8	dd.mm.aa
Hour	Show the hour of the record start	8	hh.mm.ss
Header	Number of bytes in the register of header	8	
Reserved	Reserved space	44	
Samples	Number of data or samples	8	
Duration	Show the duration or scale of time	8	
No. channels	Number of channels (ns)	4	

The user has the possibility to record several events and annotations in a single EDF+ file, this is the fundamental difference between EDF and EDF+ files. An EDF file stores only uninterrupted recordings while an EDF+ file allows the storage of several non-sequential recordings in a single file. The similarity between formats allows reading, writing and editing software and can be developed relatively easily based on EDF software for both formats.

With respect to structure, the EDF file has the ability to contain a multichannel uninterrupted polygraph record. EDF files have two parts: a variable length header record that identifies the patient and the technical characteristics of the recorded signals and the records of the EEG signals captured by the medical equipment [9].

Regarding the header, it has two parts: global header of fixed size (256 bytes) and header of the signals of variable length.

The global header is in ASCII format and provides general information about the acquisition configuration and data recording. The information provided by this header is the patient identification, start date and time, the number and duration of the recorded data, the number of signals and the individual label of each channel, as shown in Table 1.

The signal header provides the information associated with each signal acquired with the polygraph equipment, see Table 2.

Table 2. Information associated with the signal header, ns: number of signals. Source [9]

Name	Function	Size (Bytes)
Channels	Identification (name) in all the channels	16*ns
Electrodes	Type of transducer used in each channel	80*ns
Scale amplitude	Scale of data in each signal	8*ns
Min. measure	Minimum physical measure possible	8*ns
Max. measure	Maximum physical measure possible	8*ns
Min. scale	Minimum digital amplitude on the Measure	8*ns
Max. scale	Maximum digital amplitude on the Measure	8*ns
Pre-filter	If filters are used	80*ns
Sample	Sample frequency	8*ns
Reserved	Reserved space	32*ns

$$scale\,factor = \frac{\text{Maximal Measurement} - Minimal\ Measurement}{Maximal\ Scale - Minimal\ Scale} \tag{1}$$

$$Offset = \text{Maximal Measurement} - scale\,factor * Maximal\ Scale \tag{2}$$

$$Real\ Sample = Sample\ stored * scale\,factor + offset \tag{3}$$

2 Methodology

For the development of the application, the phases of the software development life cycle were mainly followed, that is, analysis, design, construction and testing. The language used was VBA (Visual Basic for Applications). For space reasons, this article mainly shows the software construction stage and the acceptance tests performed.

As for the operation of the software, the following steps are taken:

1. Data readout using an open format for storing and exchanging long term EEG records.

2. Data processing through large blocks and the use of the variable type variable, to act first as a buffer where all the data is loaded and once loaded, assign it to a range of the Excel sheet.
3. Visualization of processed information.
4. Graphing of the information.

An important factor is loading data from files. Loading the data and placing them one by one on the sheet tends to result in excessive time in the order of minutes and tens of minutes. The best method is to load the data into memory and place them on the sheet in blocks, blocks in the order of 10000 data are recommended.

In this sense, the amount of data that can affect the machine will depend on the RAM memory installed, the speed of the processor and whether the machine operates at 32 or 64 bits. So here it is not a question of a specific number of data, but of an effective strategy to make the graphics update as fast as possible, without appreciable delays for the user of the sheet, even in conditions where the amount of data is such that it slows down the graphics update in the face of changes interactively imposed by the user. The following data taken from [10], give an idea of the capacity of an Excel workbook: number of rows: 1,048,576, number of columns 16,384, the number of sheets is indefinite and is limited by the RAM memory of the computer hosting EXCEL: So, it is possible to load Excel with considerable amounts of data.

Depending on the Operating System (Windows 7, Windows 10 for example), the number of bits it operates with (32 or 64 bits) and the specific capacity of the microprocessor used, a slowing down effect of the operations and therefore of the graphics update will start.

As the amount of data and the speed requirements to process it grow, the need to migrate to other solutions such as Power BI or migrate to Python and its power to handle Big Data begins to emerge.

There are strategies that work very well when the amount of data does not slow down normal Excel operations. One of them is to choose a range of cells to plot, consisting of two or more columns, where the first column corresponds to the X axis and transfer from sheets containing the data of interest to the plotting area, by combining indirect references, the use of lists or numerical controls, it is possible to trigger the transfer of data and display the graph of the data. If the data is very large, the transfer process can take so long that updating the graph would take so long as to be annoying to the user. And many programs with the same amount of data can perform an instantaneous update in less time for the user.

Macro language can also be used to transfer the data by loading it into an array and then placing the data in the plotting area. These can also become very costly processes in terms of wasted time.

One solution to this problem is to use the macro recorder and copy the data transfer, using the copy and paste tools and modifying the macro to suit your needs, this solution reduces the delay effects significantly. The disadvantage of this solution is that the data is transferred to the chart area without any changes.

Another solution if you want to modify the data or have more flexibility, is to work directly at the RAM level, make the necessary changes and download the data to EXCEL: The most expeditious way is to use Variant type variables to take ranges of Excel data, modify and return from the variable to the Excel cells that are convenient. Part of the code is shown below:

Another solution is to act on the ranges related to the data that the chart handles. By changing the ranges, we change the charts and this process tends to be the most efficient.

```
Dim aux as Variant
aux  = Range(EspecificaciónRangoA).value
'processes acting on the aux variable
Range(EspecificaciónRangoB).value = aux
```

3 Results and Discussion

According to what was expressed in the methodological section, the construction and management of the developed software is presented in this section.

In relation to data reading, part of the code that can extract the payload from the EDF files is presented. This is a necessary routine when one wishes to work with databases of EEG recordings [3] from research centers available on the Internet. The program takes into account the global header and the signal header, as presented in Tables 1 and 2, with the information of the signal header the scale factor and offset of each signal are calculated, so the program not only reads, but also converts the stored data into real sample data.

In addition, the program uses the variant type variables to store in memory the data read and after reading by blocks it places them directly, so that it reads in blocks at the memory level and transforms the data at this same level and then places them in the Excel sheet, which speeds up the processes, with respect to reading with Excel tools.

```
Sub LeerArchivoEDF(nombreArchivo As String)

  Dim datos()As Byte,l As Long, vectorMuestras() As Inte-
ger, _

            nSeñales As Integer, nRecords As Integer, du-
raciónRecord As Integer, _
            tamañoCabeceraBytes As Integer, idPaciente As
String, _
            etiquetas () As String, _
            tipoTransductor() As String, _
            dimensiones() As String, _
            minValorDimension() As Double, _
            maxValorDimension() As Double, _
            minValorDigital() As Long, _
            maxValorDigital() As Long, _
            datosPrefiltrado() As String, _
            muestrasXRecord() As Integer, _
            desplazamientoVertical() As Double, _
            separaciónEntreOndas As Double, _
            posRef As Long, _
            i As Integer, nSeñal As Integer, _
            nRecord As Integer, PosBytes As Long, _
            factorEscala As Double, offset As Double
  'file reading

      Open nombreArchivo For Binary Lock Read As 1
    l = FileLen(nombreArchivo)
    ReDim datos(l) As Byte
    Get #1, , datos
  Close #1

  separaciónEntreOndas=Worksheets("LecturaDa-
tos").Range("C2").Value
  'Llamando a subrutina para leer vector de datos globa-
les de la cabecera
  Call getParametrosCabeceraGlobal(datos, idPaciente,
nSeñales, _
                nRecords, duraciónRecord, tamañoCabece-
raBytes)

  MsgBox "Identificación paciente: " & idPaciente &
Chr(10) & _
          "nro Señales: " & nSeñales & Chr(10) & _
          "nro Records: " & nRecords & Chr(10) & _
          "duracion Records: " & duraciónRecord & Chr(10)
& _
          "Tamaño Cabecera en Bytes: " & tamañoCabeceraBy-
tes
```

In reference to the loading of data, VBA does not implement the concept of thread, as it has, for example, the Java language, and therefore it is not possible to activate independent processes that can at the same time that the data is loaded, interact with the graphical controls (GUI) that Excel provides.

Normally, it is not possible to interact until the process that is running is finished. However, if you load the data into memory, use arrays and match integer arrays with ranges that have the same size and structure of rows and columns, and also activate a function called DoEvents, which allows the workbook not to block, you have the same effect as if you launched a thread. The graphs would be loaded as the data is loaded and the execution times are greatly reduced. From the above, it means that performance and usability attributes of the program were improved.

As far as signal processing is concerned, this involves affecting a group or all of the data by some mathematical operation. Since the measurements involve millions of samples, in general processing directly using the spreadsheet is very time consuming.

In this case, the mathematical conversion of the data to be able to display them on the same graph, one on top of the other, is presented, this is called stack fitting, which involves for each signal, calculating its maximum and minimum value and adding a value to it, also adding a separation value so that the signals do not overlap on the graph (see Fig. 2).

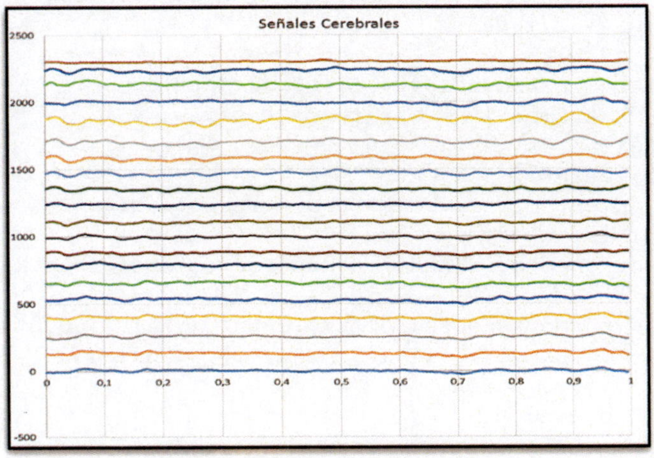

Fig. 2. Brain signals from software.

Below is part of the code for signal processing:

```
  Sub ponerSeñalesEnStack()
  Dim ns As Integer, i, j As Long, offset As Double
  'The aux variable declared as Variant is used to con-
tain data blocks directly in memory
  Dim aux As Variant, auxB As Variant, pos As String,
rangoTransferencia As String
  For ns = 0 To 21
     pos = rangoSimple(32, ns + 2)
     'offset=Worksheets("LecturaDa-
tos").Range("A1:V100000").Cells(32, ns + 2).Value
      offset = Worksheets("LecturaDatos").Range(pos)
      rangoTransferencia = rangoFilasColumnas(33, ns + 2,
33 + 910000, ns + 2)
  'Assigning a range to the aux variable directly loads
the signals from 'the sheet into memory in blocks.
      aux = Worksheets("LecturaDatos").Range(rangoTrans-
ferencia).Value 'With Worksheets("LecturaDa-
tos").Range("A1:V100000")
    '       aux = .Range(.Cells(33, ns + 2), .Cells(33 +
91000, ns + 2)).Value
    '  End With
      For i = LBound(aux) To UBound(aux)
          aux(i, 1) = aux(i, 1) + offset
      Next
      Calculate, DoEvents
  'Assignment of all data of a processed signal to a
range of the sheet
      Worksheets("STACK").Range(rangoTransferencia).Value
= aux
     ' With Worksheets("STACK").Range("A1:V100000")
      '     .Range(.Cells(33, ns + 2), .Cells(33 + 91000,
ns + 2)).Value = aux
     ' End With
  Next
  End Sub
```

Next, we present the results with respect to the visualization of the signals through the developed software. In Fig. 3 below you can see an example of changing the signal to display by manipulating a numerical control, in both figures can be seen in cell A1 the number of the signal, the numerical control is shown at the top of the image.

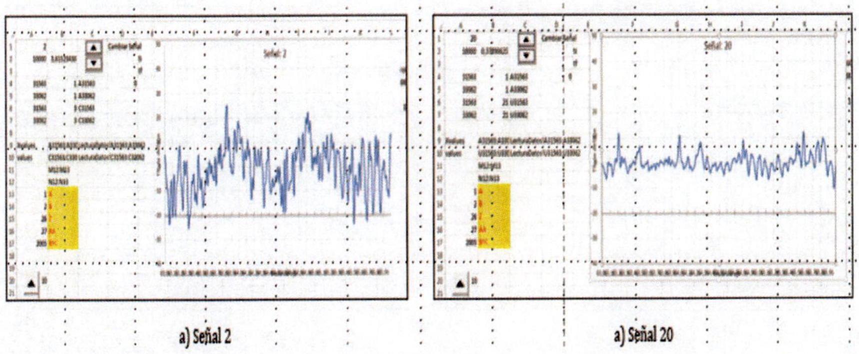

Fig. 3. Example 2 of changing the signal to be displayed by manipulating a numerical control.

Signal Shifting: using a numerical control the signal can be shifted to the left or right, as shown in Fig. 4.

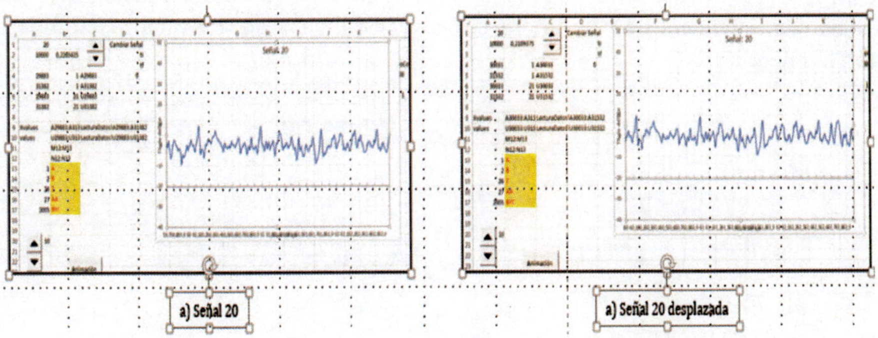

Fig. 4. Signal shifting.

Expand Contract: The same signal is presented in Fig. 5, but changing the amount of time it is presented in the graph, which gives an effect of contraction or expansion of the signal.

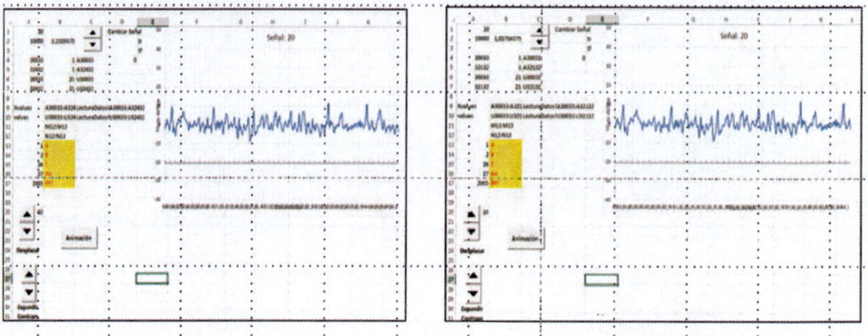

a) Señal 20 a) Señal 20 expandida

Fig. 5. Expand and contract signal.

Animation: The effect of seeing the signal in time as it would be seen in a real signal meter is achieved by pressing the animation button. See Fig. 6.

To measure the reading and processing times, the value of the time when the execution of the process begins is stored in a variable and at the end of the process the time is taken and the elapsed time will be the difference between this value and the value of the initially stored time variable.

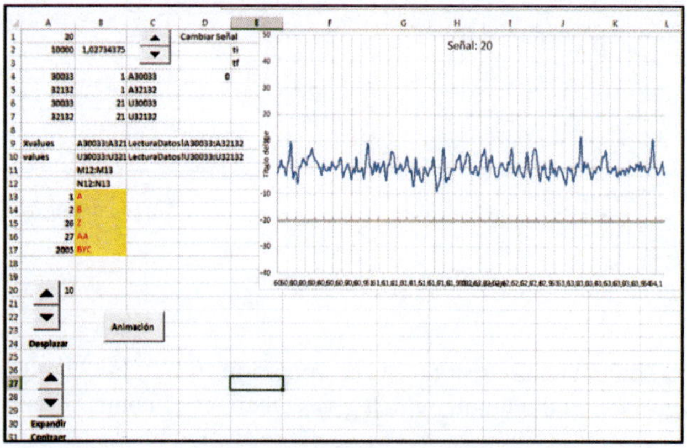

Fig. 6. Animation effect through the animation button

The following is the code for the reading process, which is very similar for the processes where the data is transformed.

```
  Private Sub CommandButton3_Click()
  Dim nombreArchivo As String, tiempo As Double, dir As
String
      'Capture of the name of the directory where the
calculation workbook is located.
      dir = ThisWorkbook.Path
      'Capture of the file name used and creation of the
complete file name
      'including the directory where it is
      nombreArchivo = dir & "\" & Range("C1").Value
      MsgBox "nombre Archivo: " & nombreArchivo
      'capture of the precise moment when the process be-
gins.
      tiempo = Timer

      Calculate

      Call LeerArchivoEDF(nombreArchivo)
      Calculate
      'Capture of the time at the end of the process,
calculation of the time of the process
      'formatting and presentation

       MsgBox "Total: " & Format((Timer - tiempo),
"#.###" & " segundos")
   End Sub
```

4 Conclusions

The software was built taking into account standard quality attributes for this type of development, such as simplicity, robustness, flexibility, performance, scalability and security.

A significant improvement in the performance of the book was obtained, in terms of loading, processing, transferring and graphing the data and in the reduction of the delay in the actions of modifying the data origin, changing the source data of the graphs, altering the limits of the X and Y axes of the graph. These improvements mean an increase in the usability and performance of Excel charting applications.

There are strategies that work very well when the amount of data does not slow down normal Excel operations, but if the amount of data and the speed requirements to process it increase, you should migrate to other solutions such as Power BI or migrate to Python and its power to handle Big Data.

The results in the time reduction will depend on each computer where it is tested, since not only the version of Windows used, the available RAM memory, the processor speed, but also specific configurations of the operating system, other installed services, so what we conclude here is that the techniques used improve the performance on each computer with respect to what it would have if these techniques were not used. These same techniques can be used for other types of signals, such as ECG signals.

The ease of programming, plotting and presenting data for input and output makes Excel a platform for creating mockups to support further development on more powerful platforms such as Python, Java or Matlab.

Having brain signals that can be manipulated, by electrode measurement, by patient, by pathology, can be a tool for either the student to learn or for the teacher to create configurations for evaluation.

References

1. PhysioNet. PhysioNet The Research Resource for Complex Physiologic Signals S.F. https://physionet.org/. Accessed 20 May 2021
2. Castillo, A.: Las Hojas de Cálculo como herramientas Didácticas, Interactivas, Gráficas, Funcionales y Adaptativas, in Congreso Pedagogía 2015, Editor, La Habana (2015)
3. Romero, E.: Desarrollo de una herramienta para simular la actividad eléctrica del Corazón, in PNF Electrónica. Univeridad Politécnica Territorial de Aragua, La Victoria (2017)
4. Cayafa, A.: Desarrollo de una aplicación basada en hojas de cálculo para simular la actividad eléctrica del Cerebro, in PNF en Electrónica. Universidad Politécnica Territorial del Estado Aragua La Victoria (2019)
5. Flores, F.: Electroencefalografía (EEG) y diversas manifestaciones del movimiento: una revisión del 2000 al 2017, in EmásF, Revista Digital de Educación Física (2018)
6. Simbaña, L.W.G., Castillo, R.A.E., Bravo, D.E.A., Guallasamin, P.L.M., Feria G., G.R.M.: Disruptive Use of Spreadsheets in the Teaching-Learning Process of Technical Scientific Subjects. In: Botto-Tobar, M., Zambrano Vizuete, M., Díaz Cadena, A. (eds.) CI3 2020. AISC, vol. 1277, pp. 362–373. Springer, Cham (2021). https://doi.org/10.1007/978-3-030-60467-7_30
7. Castillo, A.E.: Sistema Integrado Hoja de Cálculo, SIG y CAD, para calcular la facti-bilidad de una nueva estación de FM, in Seminario euro latinoamericano de Sistemas de Ingeniería., SELASI, I. (Ed), La Victoria.Aragua,Venezuela (2013)
8. Gade, A.: A simple format for exchange of digitized polygraphic recordings. In: Electroencephalography and Clinical Neurophysiology, pp. 391-393. Elsevier (1992)
9. Alvarez-Estevez, D.: European Data Format. S/F
10. Excel, T.C., Limite-en-libros-y-hojas-de-calculo-excel-2013–2016. http://trucosycursos.es/limite-en-libros-y-hojas-de-calculo-excel-2013-2016

Mobile App as an Alternative in the Process of Speech Therapy in Children with Cerebral Palsy

Dany Orbes[1], Juan Guevara[1], Paúl Francisco Baldeón Egas[2], and Renato M. Toasa[2](✉)

[1] Ingenious Works Ec, Ibarra, Ecuador
dany_orbes@ingeniousworks-ec.com
[2] Universidad Tecnológica Israel, Quito, Ecuador
rtoasa@uisrael.edu.ec

Abstract. Infantile cerebral palsy, known as ICP, is a disability that hinders the development of learning in infants. However, thanks to the technology we can overcome these barriers to learning, and even more when we can make use of technology accompanied by medical or therapeutic sciences, which is why this work aims to develop a prototype which we called EYEHOPE. This prototype represents a mobile application (mob-app) that is developed to meet the needs of learning at a low cost and with current technologies, such as: Eye tracking together with artificial vision. Furthermore, the developed prototype was tested in a comprehensive therapeutic center called NEUROCRECER, on patients with cerebral palsy, having good results with specialists in the therapeutic area. Thus, the prototype is able to provides great autonomy and improvement in pedagogical issues once it encourages the child with cerebral palsy to have the power to communicate and learn at the same time with a technological resource available to anyone.

Keywords: Cerebral palsy · Mobile app · Prototype · Learning

1 Introduction

One of the most common physical disabilities in children worldwide is cerebral palsy (CP). In recent years, "discoveries have been made in early diagnosis, prevention and treatment, altering the incidence, prognosis and response to treatment" [1]. "Based on the results found through different studies, it is stated that the treatment of cerebral palsy has continued to expand worldwide, providing physicians and families with the possibility of newer, safer and more effective interventions" [2].

"One of the main challenges to be solved is to develop an effective alternative in the communication process in children with this disability. With the advancement of technology, different applications have been created to improve the activities of children with disabilities, such as mobile applications that use different techniques such as gamification" [3], web-based intelligent educational systems [4], applications for improving

M. Zambrano Vizuete et al. (Eds.): CI3 2021, LNNS 511, pp. 479–489, 2022.
https://doi.org/10.1007/978-3-031-11438-0_38

parental attitudes, based on videos [5]. On the other hand, "developing an application that meets the requirements of these children is somewhat more complex, factors such as design, devices, audiovisual elements, among others, play a very important role".

In Latin America there is no joint epidemiological surveillance program for the evaluation of CP and the studies carried out are scarce, but they allow us to estimate the reality of this pathology in the countries of the region. "In Ecuador, a retrospective study of 127 children diagnosed with CP found that the main cause of the disease was perinatal asphyxia (77.2%), followed by postnatal (13.4%), prenatal (6.3%) and genetic (malformations) factors (3.1%)33" [6].

In this context, the purpose of this work is to generate an alternative means to improve the communication process in children with CP through the use of technology, for that we propose the development of an application for mobile devices that allows better communication between children and their therapists, This application incorporates current techniques of software development and agile methodologies, which allowed to identify the main requirements and cover them completely, as a case study we work with NeuroCrecer, which is an Integral Therapeutic Center for Early Stimulation, Neurological Rehabilitation for Children and Adults located in the city of Ibarra in Ecuador.

The rest of the paper is distributed in, Sect. 2 shows the literature review with related works, Sect. 3 describes the study case and development of the application, Sect. 4 shows test and results, Sect. 5 the results and finally in Sect. 5.1 the conclusions and future works.

2 Literature Review

The development of research related to techniques and tools that improve the learning process in children with cerebral palsy has evolved constantly and this is demonstrated by the literature, for the purposes of this work we chose to use high impact databases such as Scopus, Science Direct, Web of Science, IEEE since they store relevant works of high academic rigor.

Initially the paper titled: "An Evaluation of The Mobile Apps for Children with Special Education Needs Based on The Utility Function Metrics" [7], presented "the basic functional requirements of mobile applications for children with special needs, they analyzed 27 mobile applications available in different stores such as Google play, App Store, and in their analysis they mentioned that there are very few applications that meet the needs of this type of users, so it is important to develop this type of applications. On the other hand, applications have been developed to perform community surveillance of neuromuscular hip dysplasia in children with cerebral palsy" [8], allowing physicians to identify and treat disorders in a short time, this work is being done in the United States and is well received.

In 2008, Lindsay Pennington, conducted a study of the [9], which describes "a set of communication difficulties associated with CP", "also mentions that CP can cause alterations in sensory and cognitive development, as well as motor disorders". Problems in these areas of functioning can affect "the speech, language and communication development of children with this disorder". In the development of computer applications, in the study [10], "the development of an iPad application as an alternative means of communication for a child with CP and autism is shown, the authors focused on promoting key elements associated with the long-term use of alternative and augmentative communication, namely, targeted training of the student, teacher, educational assistant and parents, over the course of the school year".

Virtual reality (VR) environments have also increased, creating solutions to support children with PCs. "This type of application requires special features in the equipment since the consumption of computational resources is high" [11], in paper [12] presents "a study where they identify VR applications that support brain reorganization/plasticity, motor capacity, visual-perceptual skills, social participation and personal factors. On the other hand, immersive virtual reality is used to improve walking ability in children and youth with CP".

Finally, in 2021, "some works have been published, i-TelePT is a platform for telephysiotherapy for patients with CP, describing the content and structure of the module in an inclusive educational environment, describing the advantages and disadvantages of using technological platforms. Regarding communication, a system called "I-Talk" is developed for Alternative and Augmentative Communication (AAC) and to know its influence on the communicative skills of the subject, after several tests using the tool remarkable changes were identified since the patient can transmit wishes when he/she wants to drink, snack, eat and others". In [14], "a systematic evaluation of the effects caused by the rehabilitation process in patients with CP is carried out, this work is based on virtual reality technology".

As mentioned in previous paragraphs, there are different works that develop technological solutions for patients with CP, all based on scientific and academic content. In this sense, the need to continue with this line of research and technological innovation in the development of computer support systems is identified. For this reason, it can be assured that this proposed work will be of help to patients with CP in Ecuador and the world.

3 Case Study and Application Development

This section presents the case study focused on the area of teaching children with spastic cerebral palsy [15]. We propose to make a prototype that allows to improve learning with a mobile application and eyetracking, using a didactic methodology of augmentative and alternative communication systems, this methodology offers a pragmatic approach in language therapy, since it is about intervention instruments whose objective is the teaching of a structured set of codes different from speech, which allow representation functions and serve to carry out acts of communication.

These systems are commonly used communication boards. They are a set of graphic codes. However, the system could have photographs, representative drawings, abstract drawings or pictograms and written words. These codes have meanings, and they can use with a tool to build different expressions that users need in the communication [16].

In order to develop a prototype, we have divided into two areas such as: the first is the therapy that refers to an augmentative and alternative communication system as an applied methodology, and the second area is technical in the field of software and hardware required in development of the prototype.

3.1 Therapeutic Area

In this area, we worked together with the NEUROCRECER therapeutic clinic, they have a professional's group in the field of therapy pedagogy. Where we obtained qualitative data according to the Augmentative and Alternative Communication Systems teaching method, which will help us in the implementation of the algorithm.

According to Giraldo Jiménez, in his work entitled "Augmentative and alternative communication systems in intensive care, he talked about the need of patients who cannot communicate orally or language alterations. These people need a augmentative and alternative communication systems (AACS) as an alternative to non-functional speech and determining as a communication strategy, since it facilitate the communicative processes of the patient orienting themselves to the field of speech therapy, which is the discipline that combines health sciences, psychology and linguistics to help us with communication difficulties; (AACS) will allow overcoming barriers in various cases and that care is not limited only to physiological-anatomical aspects in terms of procedures for physical health care. The most important within procedure is the relationship between alternative-augmentative communication and the family" [17].

This method described above must be implemented in the development phase, using the benefits of eye-tracking as a strategy to increase the input of augmentative and alternative assisted communication, improving expressive communication and understanding of users.

3.2 Software and Hardware Technical Area

For the technical development of software and hardware it has been divided into several modules as shown in Fig. 1. These modules constitute the architecture of the system, among them we have processing and presentation modules.

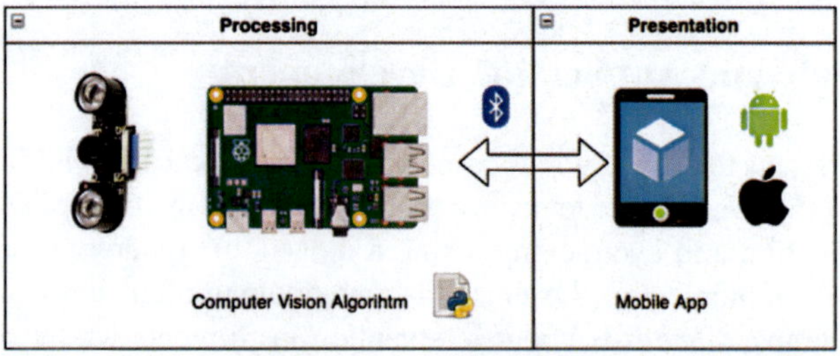

Fig. 1. System modules.

Processing Module: In this module, the images are processed using an artificial vision algorithm to detect the eye movement of the pupil in two coordinates. At this point we have a mesh according to the contour of the patient's eye, identifying the different directions such as: up, down, left, right and center, which in the case of the infant will help him to have a better visual dynamic handling much simpler to handle. The module also includes a self-calibration subroutine that is activated when detecting the patient's face, taking into account the 68 facial points.

Presentation Module: This module refers to the mobile application developed both in Android and IOs, where the user can view interactive games. The application has a focus on learning with grammatical content to build sentences using pictograms containing an order of pronouns, verbs and complements. The application use section contains more detail about the behavior of the app.

3.3 Application Operation

The application consists of a user-friendly interface for the therapist or user, in which it consists of few buttons and a much simpler handling. The following Fig. 2 shows a storyboard of the mobile application, it has a focus on practical functionality.

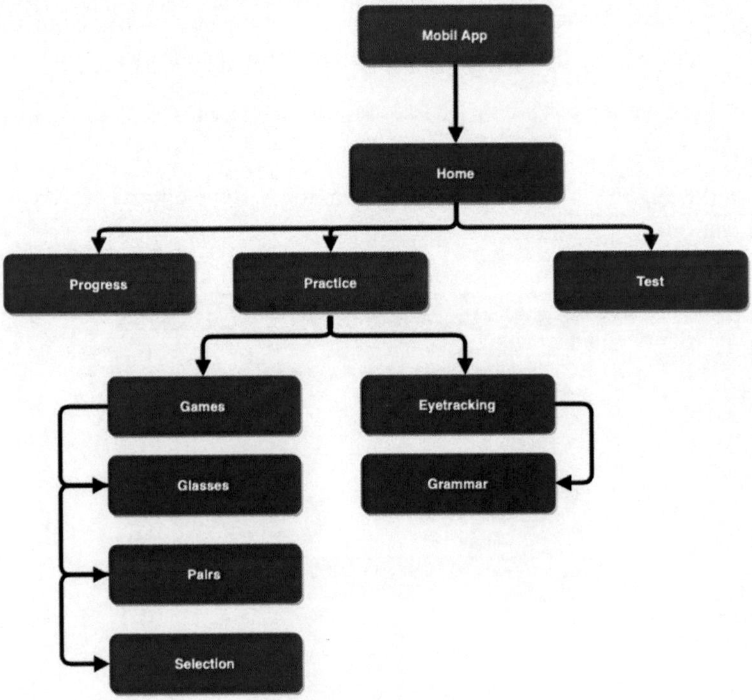

Fig. 2. Storyboard.

Home Screen: To manage the mobile application, we have a main menu with some options such as: progress, practice and test. The following Fig. 3 shows the home screen.

Progress: This option allows registration of users, data update and patient monitoring either general or particular.

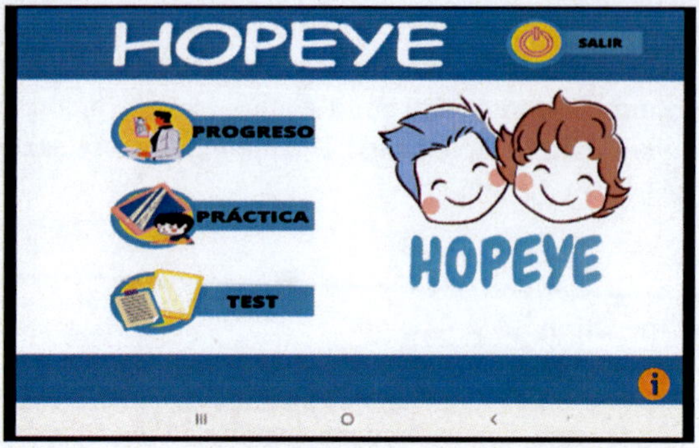

Fig. 3. Home screen.

Practice: This item offers activities to users, it can access games and the communication interface, which is carried out with a methodology set out in the therapeutic part.

Test: Finally, there is a test after each session, where we can analyze the process of the patient.

The following Fig. 4 shows progress screen, it has some options such as: monitor, registry and edit.

Fig. 4. Progress screen.

Monitor: This item shows the percentage of progress of the patient.

Registry: This item allows the entry of patient data.

Edit: This item allows modifying and updating the patient's data.

Practice Screen: This screen is divided into 2 segments of functionality, in which we have Games and eyetracking see Fig. 5, these two segments are under the functionality of eyetracking, where the child will develop their ocular dynamics, to be able to communicate by formulating sentences according to their needs.

Fig. 5. Practical screen.

Games: These games are developed under a didactic dynamic, where the child acquires different skills in eye tracking; there are three types of games (see Fig. 6):

a) *Find it*: this game has a simple mechanism, where there is a ball is hidden in one of the three glasses, then the glasses will move several times changing position, so the patient should never lose sight the ball and once the glasses stop the eyetracking will be activated to choose one, if the ball is there, the user will win the game, otherwise lost; This game has 3 levels of difficulty which will alternate colors until there is one and the speed of the game will progressively increase.

Fig. 6. Screenshots of the developed games.

b) *Pair:* This game allows the child to develop mental agility, which tries to improve the cognitive part of the child, also the game contains some sample and selection cards. The purpose of the game is memorizing a card indicated after selecting the correct card. It also has 3 levels of difficulty, having less time to memorize and hiding the cards in a better way.

c) *¿What is it?:* This game discriminates objects, which allows to improve the categorization of objects using color and shapes. During the game shows a card in the screen center and the user chooses, looking to the right or to the left, when the selection is okay, a check is displayed, otherwise one x.

Eyetracking: In this functionality, the application formulates sentences based on the methodology of augmentative and alternative communication systems, using pictograms as a way of giving graphic meaning to the sentences; this segment is made up of 3 types of tapes (see Fig. 7):

a) Pronouns: this contains the pronouns in which the sentence can begin, this contains 7 pictograms with each one of them.
b) Verbs: it contains the 10 most important actions that are used to link the pronouns and give the articulation with the necessary complement.
c) Complements: it consists of 25 pictograms where complements for a complete restructuring of the sentences are shown.

Fig. 7. Eyetracking screen.

4 Test and Results

The tests were developed in the NEUROCRECER therapy center under the company of the INGENIOUS WORKS technical team and highly trained therapists in speech therapy.

In the tests we had the voluntary support of Sebastian Cabrera (real patient), together with his family and his collaboration, the device can be calibrated until we get our expected result (see Fig. 8). In the field tests carried out in the NeuroCrecer clinic on the patient Sebastian, the specialists highlight a great advance at a cognitive level, improving his concentration and memory, before the prototype was implemented.

The results are qualitative because the tests were carried out with the requirements of the specialists and now, we are defining details for the delivery of the final prototype.

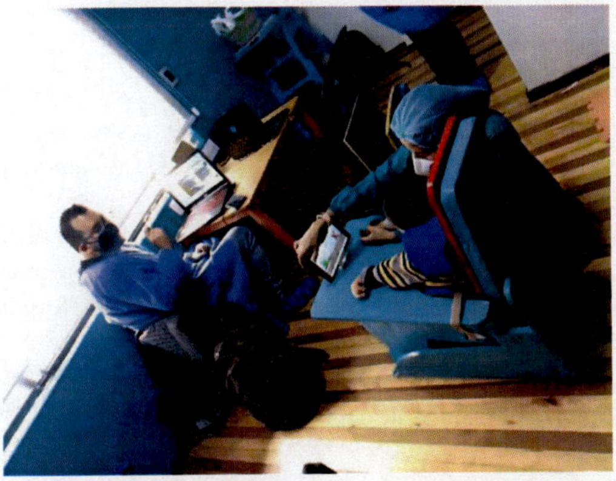

Fig. 8. Tests with real patient.

5 Conclusions and Future Work

The mobile application prototype in the learning area provides great autonomy and improvement in pedagogical issues since it encourages the child with cerebral palsy to have the power to communicate and learn at the same time with a technological resource available to anyone. This mobile application represents a tool that makes it easier for specialists to apply speech therapy using the method called SAAC, being a viable alternative in reference to the learning time, didactics, and dynamism that this application contains. family" [17].

This App has a Serial connection complement with an embedded system that allows for agility in the processing and allows its operation to be fluid because each device has its specific task and does not exist an oversaturation within the system. In the development of the application, the requirements of the patient and the specialist were considered, considering the field analysis for Neuropsychological and Neurodidactic development.

5.1 Future Work

Generate an online application where the pictograms are stored and is scalable to the patient's requirements, considering that each child with cerebral palsy has a different cognitive level. Use another data processing card, for example the Nvidia Jetson Nano card where it provides dedicated cores for graphics processing, which currently the raspberry pi does not count. In this way, performance when executing the computer vision algorithm could be improved. Create an ecosystem in the cloud where image processing could be performed and avoid the consumption of performance at the mobile application level and also avoid having additional hardware for image processing. It is intended to make the application for smart tv in order to have a larger size or resolution of images presented within the application, in this way the patient could improve the capture of the pictograms.

References

1. Novak, I., et al.: State of the evidence traffic lights 2019: systematic review of interventions for preventing and treating children with cerebral palsy. Curr. Neurol. Neurosci. Rep. **20**(2), 1–21 (2020). https://doi.org/10.1007/s11910-020-1022-z
2. Nelson, K., Ellenberg, J.: Children who 'outgrew' cerebral palsy. Pediatrics **69**(5), 1–5 (1982)
3. Toasa, R., et al.: A custom and dynamic game using gamification techniques to children from 4 to 5 years old. In: 2019 14th Iberian Conference on Information Systems and Technologies, pp. 1–5, June 2019. https://doi.org/10.23919/CISTI.2019.8760593
4. Peredo, R., Canales, A., Menchaca, A., Peredo, I.: Intelligent web-based education system for adaptive learning. Expert Syst. Appl. **38**(12), 14690–14702 (2011). https://doi.org/10.1016/j.eswa.2011.05.013
5. Romantika, I., Lusmilasari, L., Prabandari, Y., Syahrul, S.: Application of video-based health education in improving mother's knowledge and attitudes about behavioral problems among preschool children. Enferm. Clin. **30**, 172–176 (2020). https://doi.org/10.1016/j.enfcli.2019.07.071
6. Diaz, C., et al.: Prevalencia, factores de riesgo y características clínicas de la parálisis cerebral infantil. Arch. Venez. Farmacol. y Ter. **38**(6), 778–789 (2019)
7. Kraleva, R., Kralev, V.: An evaluation of the mobile apps for children with special education needs based on the utility function metrics. Int. J. Adv. Sci. Eng. Inf. Technol. **8**(6), 2269–2277 (2018)
8. Kulkarni, V., Davids, J., Bagley, A.: Smartphone app to enable community-based surveillance for neuromuscular hip dysplasia in children with cerebral palsy: development and application. Pediatrics **141**(1), 3–5 (2018)
9. Pennington, L.: Cerebral palsy and communication. Paediatr. Child Health (Oxford) **18**(9), 405–409 (2018). https://doi.org/10.1016/j.paed.2008.05.013
10. Desai, T., Chow, K., Mumford, L., Hotze, F., Chau, T.: Implementing an iPad-based alternative communication device for a student with cerebral palsy and autism in the classroom via an access technology delivery protocol. Comput. Educ. **79**, 148–158 (2014)
11. Toasa, R.M., et al.: Performance Evaluation of WebGL and WebVR Apps in VR Environments. In: Bebis, G., et al. (eds.) ISVC 2019. LNCS, vol. 11845, pp. 564–575. Springer, Cham (2019). https://doi.org/10.1007/978-3-030-33723-0_46
12. Snider, L., Majnemer, A., Darsaklis, V.: Virtual reality as a therapeutic modality for children with cerebral palsy. Dev. Neurorehabilitation **13**(2), 120–128 (2010)

13. Pahwa, P., Sharma, S., Mani, S.: The content and structure of tele-physiotherapy module (i-TelePT) for the management of children with cerebral palsy in an inclusive educational settings: a qualitative study. https://ejmcm.com/article_8327.html. Accessed 27 Apr 2021
14. Gao, C., et al.: Systematic evaluation of the effect of rehabilitation of lower limb function in children with cerebral palsy based on virtual reality technology, J. Healthc. Eng. (2021)
15. Lerma Castaño, R., Chanaga Gelves, M., Perdomo Urazan, D.: Neurodevelopmental approach to a case of spastic cerebral palsy level V using the Bobath approach. Fisioterapia, **41**(4), 242–246 (2019).
16. Villalobos, J.I.-L., Ruiz-Allec, L.D., Arrieta-Díaz, H., Leos-Ostoa, Y.: Terapia de lenguaje oral y comunicación aumentativa y alternativa en pacientes con parálisis cerebral espástica. Rev. Mex. Comun. Audiol. Otoneurología y Foniatría **5**(2), 47–52 (2016)
17. Jimenez, L.M.G.: Sistemas y estrategias de comunicación aumentativa y alternativa en cuidados intensivos: artículo de revisión. Areté **20**(2), 83–96 (2020)

Application for Registration of Candidates, Votes, and Statistical Diagrams

Aguas Luis[1(✉)], Camacho Diego[2], Quintana Alain[3], and Recalde Henry[1]

[1] Universidad Tecnológica Israel, Quito, Ecuador
laguas@uisrael.edu.ec
[2] Instituto Superior Tecnológico Vida Nueva, Quito, Ecuador
[3] Instituto Tecnológico Rumiñahui, Sangolquí, Ecuador

Abstract. The project presented below, was developed with one of the most widely used programming languages, we are talking about Csharp. Csharp is a language that adopts several features from other languages such as C + + or Java. Its use has become one of the main tendencies on the part of the developers since it has one of the most efficient and optimal compilers of the programming world, besides working with one of the most used Frameworks speaking of.NET Framework. The program has a database connection that we work with the MySQL engine, the system was made to work in the desktop version and its development was conceived in the Visual Studio programming IDE through the application of the.NET Framework with Windows Forms. In addition, it is important to emphasize that this project was made through the MVC architecture, MODEL-VIEW-CONTROLLER to exploit its features such as code separation, code reuse and that the application is scalable and easy to maintain in the future.

Keywords: MySQL · Csharp · NET framework windows forms · Desktop application · MVC · Visual studio

1 Introduction

To understand how our software works, it is necessary to establish some concepts involved in the development of our desktop application. First, we will talk about the architecture that is MVC: MODELS-VIEW-CONTROLLERS.

1.1 MVC Pattern

The MVC architecture proposes, regardless of the technologies or environments on which the system to be developed is based, the separation of the components of an application into three main groups (or layers): the model, the view, and the controller, and describes how they will relate to each other in order to maintain an organised, clean structure with minimal coupling between the different layers [1].

M. Zambrano Vizuete et al. (Eds.): CI3 2021, LNNS 511, pp. 490–502, 2022.
https://doi.org/10.1007/978-3-031-11438-0_39

1.2 The Model

In the Model layer we will always find a representation of the domain data, that is, those entities that will be used to store the information of the system we are developing. For example, if we are developing an invoicing application, the model will contain the classes Invoice, Customer or Supplier, among others. We will also find the business logic of the application, i.e., the implementation of the rules, actions and restrictions that allow us to manage the entities of the domain. It will therefore be responsible for ensuring that the system is always in a consistent and complete state [2].

1.3 The View

The views in the MVC design pattern, are in charge of establishing the interaction that the user will have with our application, all this is composed in our case of the forms that we have created for each window where different types of data will be entered by the end user and will allow the management by means of lists, buttons, text fields, menus, etc.

1.4 The Controller

The controller is perhaps the most important part of our system, as it holds the logic of our program. The controllers are the ones who process all the actions between the user and our project.

1.5 Csharp

Csharp is one of the most used languages in the world, it is one of the so called "high level languages" and belongs to the before mentioned.NET Framework, which has other languages such as Visual Basic, C+ + and C, with this type of computer language we can be able to create applications for both web and desktop.

1.6 NET Framework

To understand what C# is, it is essential to say what Microsoft.NET Framework is. It is a multi-language development environment designed by Microsoft to simplify the construction, distribution, and execution of applications for the Internet. It has essentially three components: a virtual machine (CLR: Common Language Runtime) that processes code written in an intermediate language (MSIL: Microsoft Intermediate Language), a class library (.NET library), and ASP.NET that provides the services needed to create web applications [3].

1.7 MySQL

MySQL is a Database Management System (DBMS) for relational databases. Thus, MySQL is simply an application for managing database files. It is also very remarkable that MySQL is open source, which means that it is free to use and can even be freely modified by downloading its source code. It has positively contributed to its development and continuous updates to make MySQL one of the most widely used tools for Internet-oriented programmers [4].

1.8 Microsoft Visual Studio

Visual Studio is a comprehensive development environment that allows us to work with several programming languages, allows us to create applications in very short times due to its simplicity and handling. This programming IDE allows us to develop in a more intelligent way because of the code suggestions provided and other benefits that can help us to develop in C# such as (Fig. 1).

1.9 Desktop Application

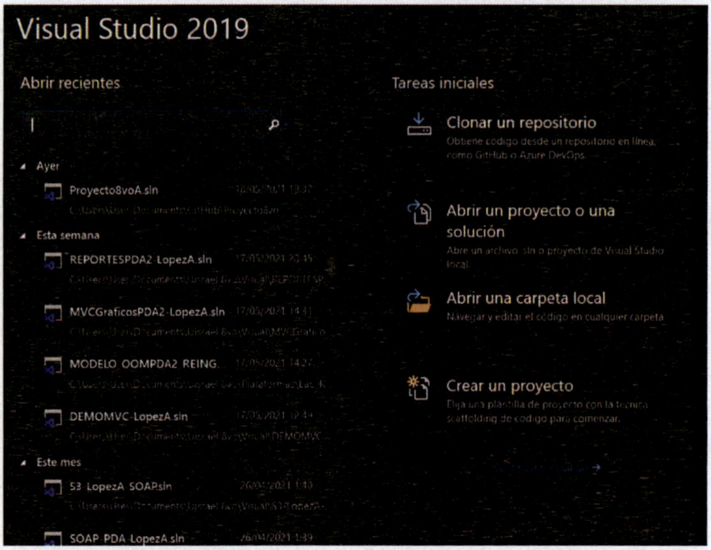

Fig. 1. Create visual studio project window

For the project that will be presented below, the Windows Forms desktop development version was used. This type of application only runs on the computer where the system was developed, or an executable of the programme can be made to install it on another machine. They work with databases that are also created on the same computer where the application is dev eloped (Fig. 2).

First, we proceed to create a project in Visual Studio, we create a new project by clicking on create. As we see in the figure below. After this, we choose the template for our project, which will be Windows Forms Application, and click next:

Finally, we choose a name for our project and click on create. In this way we will have our Windows Forms project.

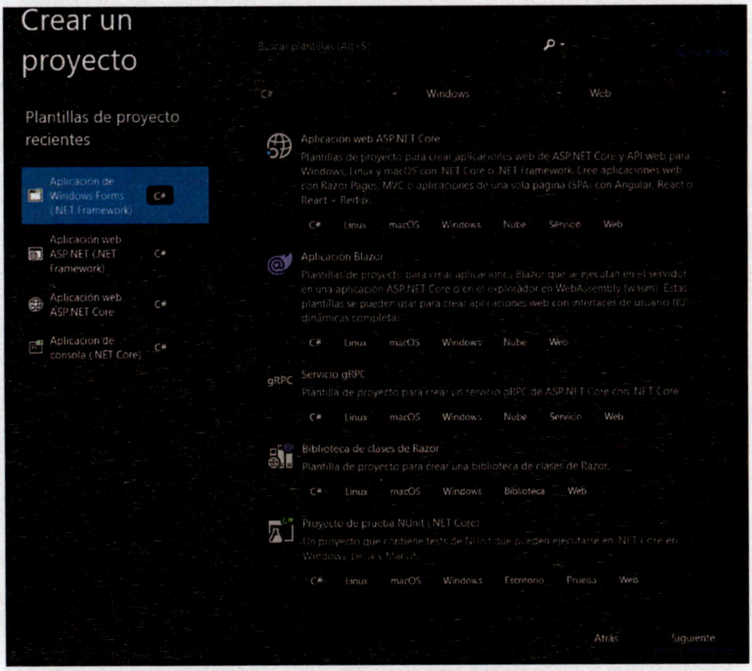

Fig. 2. Windows forms application window

2 Methodology

The development of this programme was carried out based on an experimental study. The purpose of experimentation is to identify the causes that produce certain results. Experimentation through software engineering allows developers to understand and comprehend all the variables that make up our system, based on theory and scientific evidence for software development. In this way, it is possible to understand how systems interact and to construct experimental research based on our own criteria that endorses our final product [5].

2.1 XP Methodology

For our project we have applied the agile programming methodology XP, this framework is of great help when we want to implement medium-sized systems where the requirements do not vary much during the development of the programme and every so often, versions of the system are obtained to check whether the user stories are being fulfilled or not.

Extreme Programming (XP) is an agile software development process, focused on good coding practices, clear communication, and teamwork. It is designed for small and medium-sized projects where requirements are changing. Therefore, it has a series of rules and recommendations that can be divided into planning and management, design, coding, and testing to produce software [4].

In the table below we see the user story oriented to add a new candidate, we have used the scrum methodology (Table 1).

Table 1. User Story 1-Add Candidate.

User Story	
Number: 001	User: Educational institution where voting will take place
Story name: Add candidate	
Priority: High level of compliance	Development risk: Medium complexity for the development team
Estimated points: Two weeks	Assigned Iteration: 2
Programmer in charge: Aguas Luis, Camacho Diego, Quintana Alain, Recalde Henry	
Description: The system should allow for the entry of basic data on a candidate for election	

In the next iteration we see the structure to create a new political party (Table 2).

Table 2. User Story 2: Add Political Party.

User Story	
Number: 002	User: Educational institution where voting will take place
Story Name: Add Political Party	
Priority: High level of compliance	Development risk: Medium complexity for the development team
Estimated points: Two weeks	Assigned iteration: 2
Programmer in charge: Aguas Luis, Camacho Diego, Quintana Alain, Recalde Henry	
Description: The system must allow the entry of the candidate's political party data and attach a photograph of the candidate	

Voting is the act by which an individual expresses his support for a candidate secretly, to guarantee the free decision of the voter and guaranteeing his privacy, so that his vote is not known by anyone else bearing in mind that the voting process can be carried out in a systematized way, in the following table we see the user story related to the voters (Table 3).

Table 3. User Story 3: Add Voters.

User Story	
Number: 003	User: Educational institution where voting will take place
Story name: Add voters	
Priority: High level of compliance	Development risk: Medium complexity for the development team
Estimated points: Two weeks	Assigned iteration: 3
Programmer in charge: Aguas Luis, Camacho Diego, Quintana Alain, Recalde Henry	
Description: The system must allow the user to enter the voters corresponding to the elections with their respective personal data	

The proposed system is more current and will be designed with agile methodologies, which takes an approach to decision making in the software project for that the project is iterative and incremental in the face of the requirements and solutions that evolve over time from the need for the project. SQL Server is used because the structure is more efficient (Table 4).

Table 4. User Story 4: Database Connection.

User Story	
Number: 004	User: System administrator
Story name: Database connection	
Priority: High level of compliance	Development risk: Medium complexity for the development team
Estimated points: Two weeks	Assigned iteration: 4
Programmer in charge: Aguas Luis, Camacho Diego, Quintana Alain, Recalde Henry	
Description: All data entered in the previous stories, such as adding candidates, political parties, and voters must be stored in a SQL database	

In this user story, the registration of the vote in the system has been considered, considering a level 5 interaction. For this it was made a methodology based on the dynamic key and the measures to be implemented for the different users who use the system to maintain an integrated application and that it can run in a way safe, which was also executed in the elaboration phase. With this it was concluded that this security model is decisive in the success or failure of the implementation of the project (Table 5).

A vote cannot be related to the voter who cast it. This is a requirement that appears in almost all possible scenarios. This is a requirement that is difficult to meet with current voting systems with ballots and auditors is the hypothetical anonymity in relation to

Table 5. User Story 5: Voting Record

User Story	
Number: 005	User: Educational institution where voting will take place
Story Name: Voting record	
Priority: High level of compliance	Development risk: Medium complexity for the development team
Estimated points: Two weeks	Assigned iteration: 5
Programmer in charge: Aguas Luis, Camacho Diego, Quintana Alain, Recalde Henry	
Description: The user will be able to vote through the system, which will be recorded in the database	

abstention. If required, it will mean that you can know how many and who vote, but not who participates (Table 6).

Table 6. User Story 6: Vote Validation.

User Story	
Number: 007	User: Educational institution where voting will take place
Story name: Statistical diagrams	
Priority: High level of compliance	Development risk: Medium complexity for the development team
Estimated points: Two weeks	Assigned Iteration: 7
Programmer in charge: Aguas Luis, Camacho Diego, Quintana Alain, Recalde Henry	
Description: The system must have statistical diagrams to show the voting results, which will be automatically modified with each vote	

The user interface should be friendly, it will be taken into account that the management of the program should be via keyboard and mouse for computer. Efficient algorithms were used, such as data encryption, report development, among others, a practical and easy-to-use solution was developed, taking as a reference the compliance of the requirements raised (Table 7).

Table 7. User Story 7: Statistical Diagrams.

User Story	
Number: 006	User: System Administrator
Story name: Vote validation	
Priority: High level of compliance	Development risk: Medium complexity for the development team
Estimated points: Two weeks	Assigned Iteration: 6
Programmer in charge: Aguas Luis, Camacho Diego, Quintana Alain, Recalde Henry	
Description: The system must validate that the user can enter only one vote for the election. The validation shall be done by means of the voter's ID card	

3 Results and Discussion

The first step is to verify that our programme has two types of access, the first is by means of a master identity card that will allow us to enter voters, candidates, political parties, and to visualise the voting results (Fig. 3).

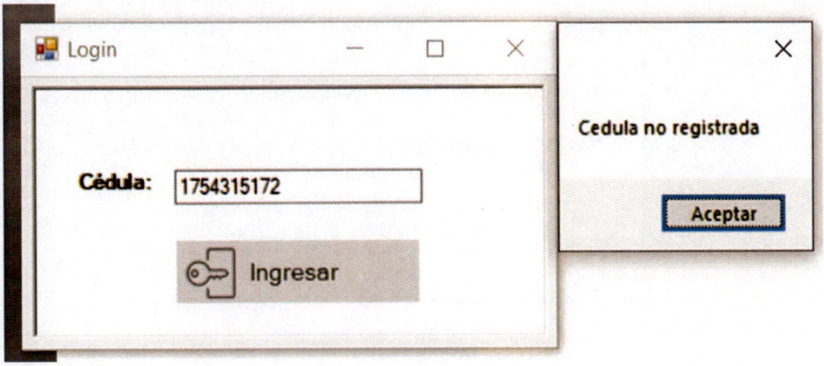

Fig. 3. Master identity card validation

As we can see in the previous image, we entered an ID that does not correspond to the Master user and is not registered; therefore, it does not allow us to enter the system to carry out a vote or to perform administrative actions.

To control this entry, a master user must be created. What we do is to enter wittmaster ID number, which in our case is 99999999999, and then we can access the system to type a new administrator user.

For this step, it is necessary to create the master user by making the following fields (Fig. 4):

We run the programme and enter the Master ID card. Once this card has been entered, the main window for system administrators is displayed. Next, click on "Añadir votantes" in order to enter a new master user with the following data: Name: Master, last name:

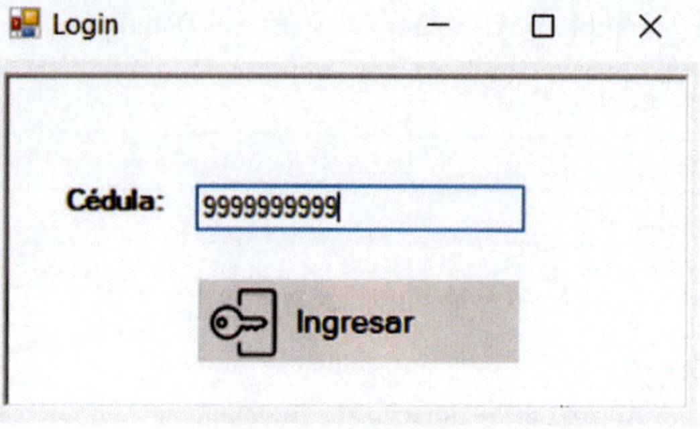

Fig. 4. Master ID entry

Root, ID: Master ID. The following window will open in order to enter the user, here we have an important validation field, due to the fact that an Ecuadorian ID validation algorithm was implemented, if the ID meets these conditions, the record is created and saved in the database.

With this step, a new Master user will be created and will be able to log in to the system. Enter again the ID created in Login (Fig. 5):

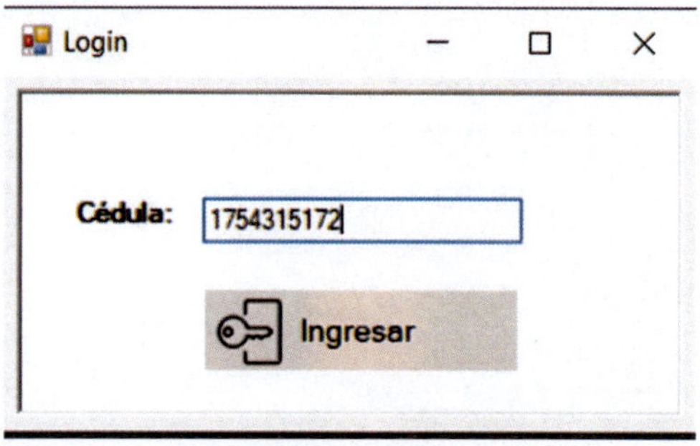

Fig. 5. Admin user created

In addition, as we will see below, we can log in to the system without any problem. The tests were carried out with data from fictitious lists, candidates, and voters. We started adding voters for the elections. These can be only voters or also candidates of political parties. For this reason, the voters were added first, as it is one of the foreign keys of the table Candidates. In order to add them we click on the option "Añadir Votantes" (Fig. 6).

The system will save a voter login (user) record including the IP address of the device to avoid double voting. We add the data (Fig. 7):

Next, we add a political party. In the main administration window, we click on "Agregar Partido Político" First the political parties must be created and then their candidates because the table Candidates has the foreign key of the table Political Parties in the database.

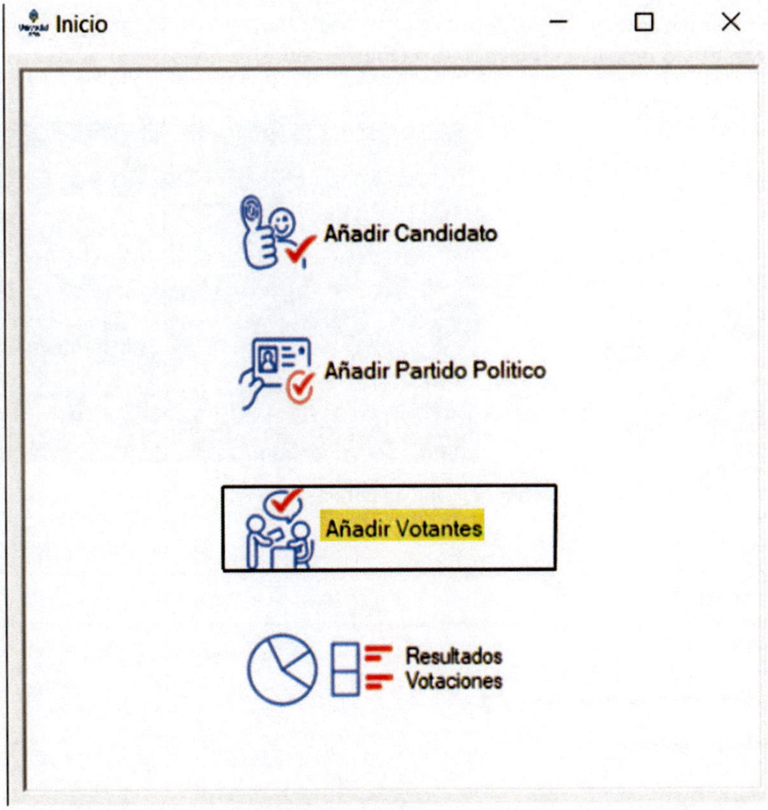

Fig. 6. Add voters

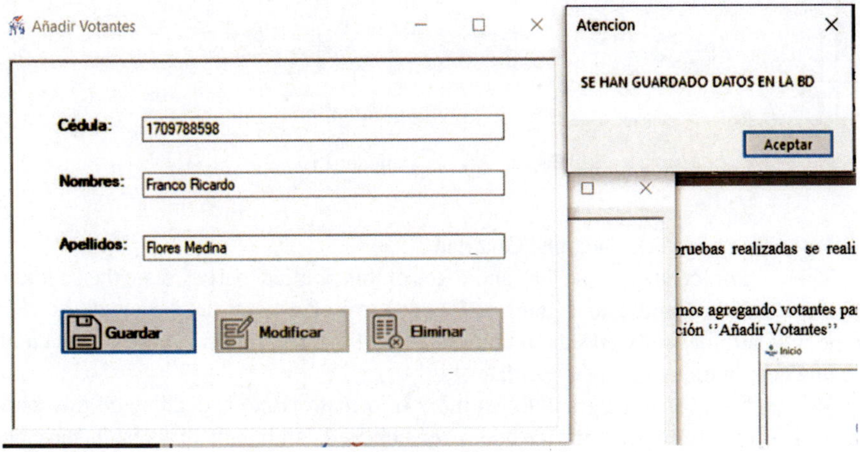

Fig. 7. Voter data.

We enter the data of the political party: We type the name of the party, and we have the possibility to upload an image of this (Figs. 8 and 9):

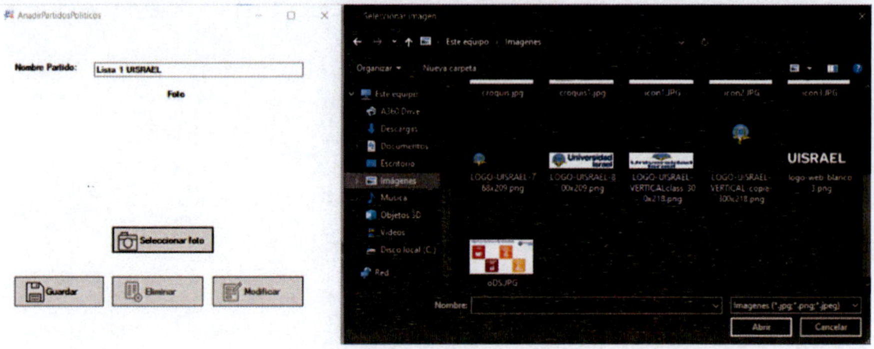

Fig. 8. Upload political party

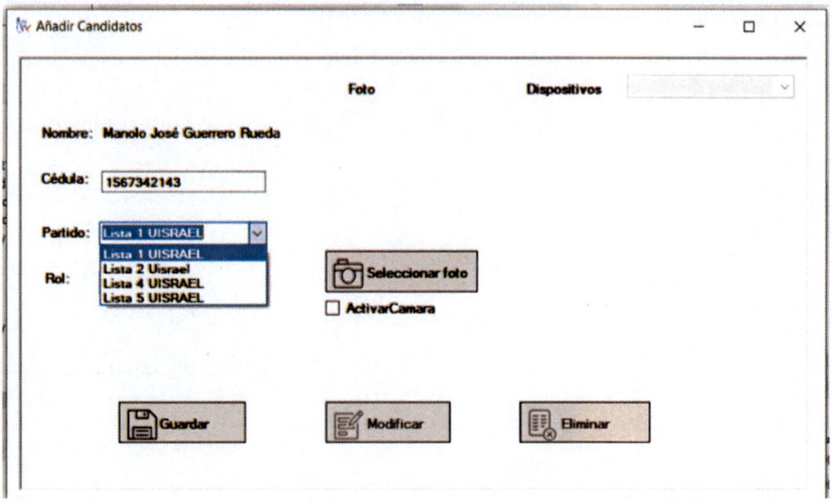

Fig. 9. Add a candidate-Lists

Open the image and click on "Guardar".

Then, we proceed to create the candidates of this political party.Create the candidate, click on "Añadir Candidato". Enter an ID card from the voter database and the user's name will automatically appear, as shown in Fig. 9. In the option "Partido", we choose the one corresponding to the candidate (Fig. 10).

We can hover over the graph to get more information that is detailed on the results. All aspects of system performance have been checked and satisfactory results have been obtained in relation to the programme's objective. The experimental part has been useful to correct errors within the versions executed.

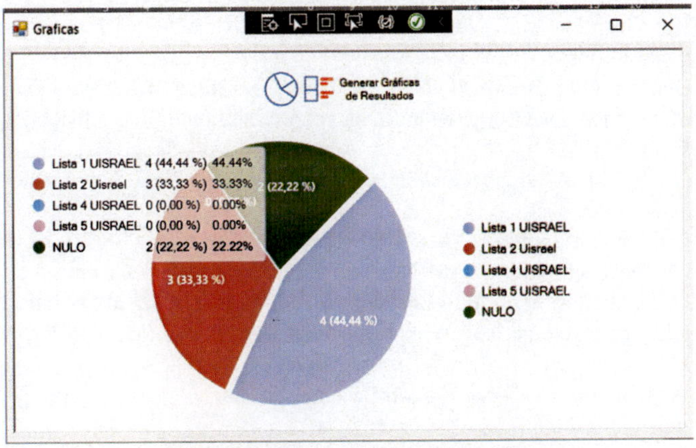

Fig. 10. Election results information.

The system will make the reports for the counting of the elections at the end of the election day. Working with forms is quite efficient when it comes to providing a responsive design that maintains a clear standard and current development trends. In programming nothing is set in stone, and it is recommended to learn Visual Studio with the Csharp programming language, as it is a very comprehensive environment that is constantly being updated and there is a lot of information on how to develop with this IDE, which helps a lot in learning.

Statistical graphs help us to better interpret and understand data. Learning to draw them using C# is quite good as we have more knowledge to develop more efficient programs. The methods of a class or object are implemented by means of member functions or methods. A method is a set of instructions that performs a certain task. The attributes of a class are the individual characteristics that differentiate one object from another and determine its appearance, state, or other qualities.

Model View Controller View (MVC) is a software architecture style that separates an application's data, user interface, and control logic into three distinct components. MVC has multiple views from the same model, allowing for much better reuse of developments and ensuring consistency between them. MVC is a software architecture approach used to separate code by its different responsibilities, maintaining different layers that are responsible for doing a very specific task.

The programme has a connection to a SQL database, this database must maintain a minimum of 3rd normal form normalisation to avoid inconsistencies in the data. Systems that are developed with robust databases are much more efficient in terms of data integrity, less redundancy, consistency of results and balance in obtaining data.

References

1. Madrid, U.P.d.: https://www.fi.upm.es/ 18 June 2018. https://www.fi.upm.es/docs/estudios/postgrado-admision/537_Seminario%20ESE.pdf
2. Sierra, F.J.C.: Microsoft C#: lenguaje y aplicaciones (2nd. ed.), RA-MA Editorial (2015)

3. Norte, U.T.d.: http://www.utn.edu.ec/. http://www.utn.edu.ec/reduca/programacion/poo/diagrama_de_clase.html
4. Ramírez-Bedoya, D.L., Branch-Bedoya, J.W., Jiménez-Builes, J.A.: Metodología de desarrollo de software para plataformas educativas robóticas usando ROS-XP. Revista politécnica **15**(30), 55–69 (2019)
5. Kasiak Tamara, G.D.A.: http://sedici.unlp.edu.ar/. 1 August 2012. http://sedici.unlp.edu.ar/handle/10915/18976
6. Deitel, P.J.: Cómo programar en C# (2a. ed.), Pearson Educación (2007)
7. Arias, Á.: Aprende a Programar ASP.NET Y C#, Publicado por IT Campus Academy (2014)
8. Sierra, F.J.C., Ceballos Sierra, F.J.: Enciclopedia de Microsoft Visual C#: interfaces gráficas y aplicaciones para Internet con Windows Forms y ASP.NET (4a. ed.), RA-MA Editorial (2015)
9. Merchán Manzano O.J.C.N.K.: http://dspace.uazuay.edu.ec/. (2017) http://dspace.uazuay.edu.ec/handle/datos/7036
10. kinsta.: https://kinsta.com/, 27th May 2019. https://kinsta.com/es/base-de-conocimiento/que-es-mysql/

Technology Trends

Information and Communication Technology in the Application of Strategies for Supply Chain Management in Business: A Systematic Review of the Literature

Aleixandre Brian Duche-Pérez(✉), Ana Olinda Marallano-Povis, Pilar Victoria Gálvez-Galarza, Brizaida Guadalupe Andia-Gonzales, Milena Ketty Jaime-Zavala, Marcela Candelaria Montesinos-Torres, and Ygnacio Salvador Tomaylla-Quispe

Universidad Católica de Santa María, 04001 Arequipa, Perú
aduche@ucsm.eu.pe

Abstract. The purpose of this review article is to carry out the balance of existing knowledge on the strategies at the upper, intermediate and lower levels, focused on ICTs for production processes, within the supply chain in the business field. For this reason, a review and analysis of the bibliography was carried out that allowed defining the ICTs in an operational aspect as a generator of a more efficient and collaborative coordination between the members of the organization and its suppliers and in an aspect of implementation that if not carefully evaluated It will produce high costs to the company. Regarding the management of the supply chain, it is considered as the central component in the operations of any company. From the analysis carried out, it is concluded that ICTs as tools will create advantages in the management of the supply chain from an adequate decision-making strategy.

Keywords: ICT · Management strategies · Supply chain · Business environment

1 Introduction

The "efficient organizations are integrated entities in which the different units, functions and levels support the strategy of the company and support each other" [1], consequently, the hierarchical structure of how companies work for decision making at a strategic, tactical and operational level, it is of vital importance since it allows making value judgments with anticipation and responsibility, which represents sustainable advantages for the success of the organization.

Decisions at a higher or strategic level are those that guide the direction of a company regardless of the field or sector to which it is dedicated. "The world of strategic decisions encompasses a multiplicity of approaches, interpretive analyzes and a myriad of theoretical and practical approaches…" [2]. So when companies implement ICT tools in their managements, areas or divisions, they become strategic forces that improve the performance of the supply chain.

M. Zambrano Vizuete et al. (Eds.): CI3 2021, LNNS 511, pp. 505–515, 2022.
https://doi.org/10.1007/978-3-031-11438-0_40

It should be noted that to implement ICTs there must be an integration process that allows connecting decisions and actions within the supply chain system. The integration strategy improves the exchange of information and knowledge, minimizes interruptions, reduces impacts to a minimum and achieves information to flow quickly leading to benefits such as: automation of services and processes, agile e-commerce, diversity of information [3, 4]; however others researchs demonstrated that the benefits are not materialized if there are inaccuracies in the implementation of ICTs [5].

On the other hand, at the intermediate or tactical level, strategies and decisions are formulated based on processes and performances that lead to obtaining short-term results, being monitored and evaluated by the managements that measure the degree of agility and progress of operational activities. It is at this point where tactical actions make the supply chain have the capacity to evolve in the face of the changes and uncertainties it faces on a day-to-day basis. Others specified that increasing agility increases response capacity, obtaining positive effects on performance and competitiveness [6].

Others authors are agree that there is articulation between ICTs and the supply chain, but these coincidences are raised within the framework of strategies, considering that ICTs are dynamic means that automate processes to achieve results and generate a qualitative impact [7, 8]; However, it must be taken into account that the technologies will not be enough, if the planning route is not defined in changing scenarios.

However, strategies framed at the lower or operational level are developed towards the quality of the information and how it is managed in the different actions of a company, especially those that have to do with: production, logistics, inventory, shipments, accounting, among others. Therefore, flexibility assumes an important role since it allows obtaining real and sufficient information from the environment for decision-making in versatile media. In that sense, when referring that inhibitors are obstacles that prevent a better performance of the company towards consumers, suppliers and customers due to a low capacity to build trust and share strategic information relevant and important for the development of new products or services [7, 9].

The problem arises when, when making decisions at a higher level, an adequate search for strategies is not carried out that, when executed, do not generate the expected results and that during the process financial resources of the company are committed; on the other hand, at the tactical level, the problems are located on the actions and measures when implementing ICTs, which, since they are not the ideal ones, expose the supply chain to inertia and the tendency of administrative and operational errors. A key factor for successful operational strategies is the quality of information, as it is a determining factor in flexibility in the face of the demand and need for a virtual market with an increase in new online commerce applications.

From what is indicated in previous paragraphs, the central problem of this review of the literature focuses on the following question: What result does the balance of existing knowledge reveal about strategies at a higher, intermediate and lower level, focused on ICT for processes? productive, within the supply chain?

2 Materials and Methods

This document is a systematic review of the literature carried out during September to December 2020. The main objective that guided the present one was: to take stock of the existing knowledge about the strategies at a higher, intermediate and lower level, focused on ICT for production processes, within the supply chain in the business environment, for which the following databases were used: Vlex, Science Direct, Base, Redib, Web of science, Scopus, Taylor & Francis online, Redalyc, Springer and Ebsco. The descriptors were "Information and Communication Technology", "Digital Technology", "ICT", "Information Technology" and "Supply chain management", with the boolean operator "AND" among the terms.

The following inclusion criteria were used to identify a document as valid: scientific articles in journals indexed in English reviewed under the double-blind peer-review system, published between 2009 and 2020, which are available for full-text review. Three reviewers, at different times, searched for documents in the identified databases using the proposed descriptors in order to triangulate, verify and validate the results obtained. Meta-analysis, review studies, books, book chapters, dissertations, theses, reports, journalistic articles, proceeding papers and non-scientific texts were not considered, as well as any other document where the inclusion descriptors selected within the keywords, objectives, results and conclusions.

Once the search and verification procedure of the results obtained according to the established methodology had been carried out, a sample of 8 documents was obtained.

The final sample of documents identified was categorized according to the following criteria (Table 1): author, main objective and population or study sample.

Table 1. Synthesis of articles published in databases from 2009 to 2020.

Author	Objective	Method/approach/paradigm	Population/sample
Aguilera, Colín, and Hernández (2013)	Analyze influence of Information Technologies in the current production processes of Small and Medium Enterprises-Manufacturing SMEs	Quantitative exploratory-correlational	247 SMEs in the manufacturing sector
Colin, Galindo, and Hernández (2015)	Analyze the relationship between ICT and supply chain strategies and management (SCM)	Quantitative descriptive - correlational	288 SMEs manufacture in Aguas Calientes
Kim (2017)	Analyze the correlation between integration technology information (IT), supply chain integration (SCI), and performance	Quantitative - Correlational	2,000 chain managers or production managers

(continued)

Table 1. (*continued*)

Author	Objective	Method/approach/paradigm	Population/sample
Scuotto, Caputo, Villasalero, and Del Giudice (2017)	Establish the Multiple Buyer-Supplier Relationship in the context of SME digital supply chain management	Quantitative - Relational	682 SMEs in supplier system category
Hugh, Wang, and Naim (2017)	Evaluate to what measure the types of IT flexibility in business performance	Exploratory	162 supply chain practitioner questionnaires
Díaz and Blanco (2018)	Analyze the adoption and use of information technologies in Cuban organizations	Quantitative comparative correlation-	140 workers from twelve companies and eleven budgeted entities
Soojung, Ryu, and Yang (2018)	Evaluate the interaction effects between the capabilities of the supply chain and information technology on the performance of the company	Quantitative - relational	142 managers of manufacturing companies
Sheko and Braimllari (2018)	Explore the relationships between SMC-IT inhibitors, IT enablers, information sharing, and information quality in supply chain management	Quantitative exploratory-correlational	183 business units, with 2 or more employees

3 Results

Eight articles met the inclusion-exclusion criteria established according to the proposed methodology. A greater number of publications were identified in the databases in Science Direct (02), Redib (02), Velex (01), Scopus (01), Taylor & Francis and Springer (01) one in each case. The writing publication language of the articles was English and Spanish (08). The investigations were developed in South Korea (03), Mexico (02), Albania (01), Italy (01), United Kingdom (01), most of the studies were carried out in 2018 (03), in the years 2017 were carried out (03) and 2013 and 2015 each with a publication.

From the results obtained, it has been possible to identify three types of objectives according to the central theme that makes up their nature:

Objectives aimed at analyzing information technologies in production processes: One author analyzed the influence of Information Technologies in current production processes of Small and Medium Manufacturing Company-SME On the other hand [10], others authors analyzed the adoption and use of information technologies in Cuban organizations (production processes) [11].

Objectives aimed at evaluating Information Technologies in the performance of the company: Two authors evaluated the effects of interaction between the capabilities of the supply chain and technology of the information on the performance of the company [7]. And other authors evaluated to what extent the types of IT flexibility in the performance of the company [12].

Objectives aimed at exploring the relationships between SMC-IT inhibitors, IT enablers: One research explored the relationships between SMC-IT inhibitors, IT enablers, information exchange, and the quality of information in supply chain management [9]. For its part, another study analyzed the relationships and correlations between information technology, strategies and supply chain management [8, 13]. And finally, other study established the Multiple buyer-supplier relationship in the context of SME digital supply chain management [14].

About the theoretical approaches used by the authors with respect to defining what ICTs are, we can find two studies who define ICTs in the framework of SMEs as a dynamic integration tool that allows the company to have immediate benefits with the business objective of improving operations such as production processes [15, 16].

On the other hand, one author theorize ICTs as a technique that facilitates the collection of vital information about key business processes and the exchange of before mentioned information about the functional area that allows internal and external integration with all the organization [13].

Likewise, on the use of ICT in production processes state that ICT integrates production processes in organizations which allow them to obtain important benefits in the control and delivery of the goods they produce and acquire [10]. They also mention that adequate communication with staff as well as with suppliers and with a successful application of ICT will lead to having a reliable handling of materials which will be in accordance with the policies and needs of the company.

About the ICT strategies applied to the management of the supply chain in the business environment, two types were identified. (a) Operational: One author stands out results that refer that an effective and collaborative coordination between the members of the organization and suppliers, in addition to the adequate use of ICTs, favors the establishment of strategies so that the handling of materials is reliable in accordance with the needs and policies of the companies, based on a design that will allow obtaining high productivity indices in operational actions and also have the support of the master production plan [10]. Therefore, the administration of operational activities must be planned so that they unfold without setbacks, being necessary to have work teams that function without interruptions or delays, so when applying strategies in production processes they should be controlled through statistical instruments, whose purpose will be to obtain a product produced with quality and that the adoption of technologies have been part of this process. (b) Implementation: One author specify that investments in ICTs are usually expensive, top management must propose a strategic vision for the effective use of technologies, avoiding unnecessary economic losses, the identification and analysis of actions must be support elements to influence, ¿ facilitate the adoption and use of ICTs, improving decision-making) [17]. Small and medium-sized companies require the implementation of strategies that ensure the identification of needs, depending on the context and demand of the company. This will allow the presence of business strategies that will determine the success or failure of a company. Thus, there are reactive companies where managers do not respond adequately to the changes offered by the market, nor do they respond to the demands of the environment, settling for an inferior performance, therefore they are companies without a specific strategy. While exploration companies have a greater probability of success as they explore market niches, innovation and the study of new trends are important for these companies; For this reason, when implementing strategies, some researchers refer that SMEs face risks that are related to human and financial resources, which show obstacles to selection, execution and application of ICT infrastructures related to the development of business strategies [18]. And other authors identified two types of interventions related to the implementation of ICT; these are: Pre-implementation interventions: Prepares the initial user on the characteristics of the system and the proper handling for each job. E, post-implementation interventions: They form a set of organizational and support activities, executed after the system is deployed, in order to strengthen the degree of acceptance and understanding of it by its potential users. For the implementation, coordination and cooperation in both types of interventions is necessary, between the IT provider company and the client organization [19, 20].

And, about the advantages of ICT in this field, five strategic advantages were identified: (a) Two studies specify that ICT are sources of competitive advantage in several business and management contexts, in the same way [21, 22], other author manifest that companies invest to establish infrastructure with ICTs with the in order to achieve and maintain competitive advantage. This shows that the implementation of technologies generates unique positioning with results and sustainabilit [23]. (b) Other author specify that a company cannot focus on all competitive advantages in price, quality, customer service, service time and flexibility, since each competitive advantage requires a type particular organizational structure and infrastructure [24]. (c) One study present an example of a business-to-business (B2B) model where buyers can choose an open electronic contracting system, stating that the efficiency created by a supply channel electronic (e-procurement) is more important than the supplier side creating a competitive advantage. On the other hand, an electronic contracting system based on the closed extranet, facilitates and is used to share strategic information of the suppliers through the system. These models depend on what the company determines and wants to achieve with the use of technology [25]. (d) The ICTs are a factor of competitive advantage and that advances have allowed many industry practices to flow within organizations which are essential in supply chain management and such operations are not possible without information technology management [26, 27]. (e) The supply chain of a company is a central component in the operations and strategies that in the long term determine positive growth [7].

Meanwhile, regarding supply chain management (SCM) includes product management, financial flow, supply source information, product assembly manufacturing as well as the management of the post-sale service, which generates an increase in productivity [8]. In conclusion, the researchers argue that supply chain management is a central component, having to integrate management and planning processes in order to satisfy customer demand.

And, regarding competitive strategies in this field, four types have been identified: (a) Provider network determine that the network of providers, consumers as well as other units favor the realization of transactions as well as the transformation of resources obtaining products or services [29]. This statement shows the integration of the parts that seeks to synchronize the requirements of the clients, generating a balance between the customer service, the investment in inventories and the unit cost of the product. (b) Focused on information management refer to the fact that the management of the service supply chain should focus mainly on information management, in the processes, in the capacity, having to be managed from the beginning of the chain to the end of the link, whose objective is the organizational success implemented through the strategy [30]. On the other hand, other author specify that the basic objective of information management is to organize and put into use the information resources of the organization (both external and internal) to allow it to operate, learn and adapt to changes in the environment [31]. The main actors in information management are the information professionals themselves, in close union with their users. Therefore, the benefits that an adequate information management generates for the organization will be: saving time in search of information; allows better use of information for decision making; increased productivity; improvement of knowledge; and improvement of internal and external communication of the organization). (c) Strategic collaboration and alliances affirm that strategic collaboration integrates all partners of the supply chain both intra-organizational and inter-organizational [32, 33]. In other words, partners, suppliers and clients must form a synergy to obtain better results, on the other hand the merger of suppliers also represents a valuable collaboration [34], which together with a strategic alliance of business with the organization allows the reduction of costs and the distribution of benefits [35]. And, (d) Digital supply platforms: they are spaces for interaction that respond to a need to save information a digital supply platform represents an area where to offer and sell personalized products [36]. Therefore, the use of digital platforms allows the supply chain greater visibility, closer into, real-time approach to the customer, optimizing timely deliveries, response to claims, returns, as well as better organization and planning of demand, the advancement of the different digital platforms allows shorter times and distances, positioning itself as a concrete competitive advantage in the market. Consequently, digital platforms are communication strategies that seek a multidirectional relationship between suppliers and consumers of products and services using different electronic business models and building emotional ties with end users. By offering and selling personalized products through digital platforms, clear advantages are evident (Power of decision of the user at the end; access to product information through catalogs; connection interface between users of a digital platform; use of the platform from different parts of the world; and fast interactive communication with the service provider) and disadvantages (There is no direct interaction with the end user; and depending on the location, the existence of a slow data transmission speed).

Finally, on the methodological approaches of the identified texts, it was found (03) of a quantitative-exploratory-correlational character, (04) of a quantitative-descriptive-correlational character, (01) of a quantitative-comparative-correlational character. And, regarding the size of the study sample, it was identified that (03) used ranges of samples between 247 to 682 workers, having as a common factor being SMEs in the manufacturing sector; (01) the comparative study considered 140 workers in total belonging to twelve companies and eleven budgeted entities; (03) cases that worked with samples of 142, 162 and 183, in each case. It should be noted that (01) takes as a very significant sample 2,000 production managers.

Regarding the use of research techniques and instruments, it was identified that the comparative quantitative study used an online survey which brings benefits in terms of time, costs, quality and scope. It is important to highlight regarding the studies in SMEs, despite having a common field of study, the instruments used differ from each other: (01) uses 03 surveys, one referring to Information Technologies of its own elaboration, the second survey on Information Technology Processes. Two articles uses a survey of own elaboration directed only to managers or owners of SMEs [37, 38]. Other authors used instruments adapted from other authors, such as the case of the study of inhibitors and facilitators of the supply chain, the SMC-TI questionnaire being adapted for the aspect of inhibitors from other authors who were an adaptation reference for the facilitators' aspect [28, 38]. For the case of the comparative study, a survey adapted was used, taking into account the peculiarities of Cuban organizations [19].

Finally, for the population of 142 managers of manufacturing companies, an adapted survey was applied from several authors used this instrument to evaluate the effects of interaction between the capacities of the supply chain and the characteristics of the electronic purchasing system, for which they took flexibility, agility, performance and electronic contracting as dimensions and also considered the types of system being open, closed, information sharing, information exchange, high cost of implementation, strong and weak access of potential partners through the use of the system [39–43].

4 Discussion

According to the results found, both theoretical and methodological, it has been identified that a model identified the variables of strategies, ICT and supply chain management (SCM) it was adequate because it demonstrated the integration of these and how they strengthen and improve the management of materials operation in manufacturing SMEs [7]. Although integration is conceptualized as a fusion that generates changes, the interaction generates action between the elements [6].

Carrying out a study of a relational quantitative approach concludes that the strategies when applied in the supply chain in relation to business results must be concrete, different and implemented separately and must be based on their environment so that an interaction between the actors leads to a competitive articulation.

On the other hand, it is necessary to indicate that ICTs and the performance of companies use financial performance as an indicator of result, however, this is not decisive since performance is also linked to customer satisfaction and marketing.

Regarding ICT, an author used a theoretical approach [9] which was compared to a structural equation model [45] that gave him the opportunity to demonstrate the interrelation of dependence between the variables of information technology, production processes and competitiveness. The data showed a high impact and relationship between the variables, which helped to determine that technological adoptions must be in accordance with the requirements of the company in order to generate benefits and ensure the product manufactured with quality and without defects.

5 Conclusions

From the review of the literature it is concluded that, for an analysis and description of the variables on information and communication technology, strategy and management of the supply chain, the quantitative and exploratory approach is the most appropriate.

Likewise, regarding the research design, both a cross-sectional and longitudinal line and an effect of data triangulation and information analysis can be used.

In the first case, it is determined that ICTs have a high and indisputable position in the company; that together with the capacities of the collaborators, they must generate and achieve productivity and sustainability effects that strengthen the company.

In case two, a supply chain that uses ICTs as a tool is identified, becoming digital supply chains whose network of actors and processes are interconnected and interdependent, working cooperatively from the supplier to the end user.

In the third case, strategies are defined as the horizon that seeks to achieve that the purposes are fulfilled. The strategies are oriented to dynamic, continuous, evolutionary processes and with times that allow adjustments to be made to achieve advantages in achieving results.

Finally, the most widely used research instruments have been the questionnaires and in terms of technique, it was the survey. In addition to this, it is advisable to complement the obtaining of results with case studies that will strengthen knowledge for business decision making.

References

1. Lingle, J.H., Schiemann, W.A.: From balanced scorecard to IS management. Manage. Rev. **5**(3), 56–61 (1996)
2. Kourdi, J.: Key strategies for making business decisions. Interdisc. Stud. Soc. Sci. **11**(2), 266–268 (2009)
3. Gualandris, J., Kalchschmidt, M.: A model to evaluate upstream vulnerability. Int. J. Logistics Res. Appl. **17**(3), 249–268 (2014)
4. Gualandris, J., Kalchschmidt, M.: Supply risk management and competitive advantage: a misfit model. Int. J. Log. Res. Appl. **26**(3), 459–478 (2015)
5. Cragg, P., King, M.: Small firm computing: motivators and inhibitors. MIS Q. **17**(1), 47–60 (1993)
6. Chakravarty, A., Grewal, R., Sambamurthy, V.: Information technology competencies, organizational, agility, and firm performance: enablig and facilitating roles. Inf. Syst. Res. **24**(4), 976–997 (2013)
7. Oh Soojung, O., Young, R., Yang, H.: Evaluate the interaction effects between the capabilities of the supply chain and information technology on the performance of the company. Technol. Manag. **20**(2), 91–106 (2019)
8. Colin, M., Galindo, R., Hernández, O.: Information and communication technology as a key strategy for efficient supply chain managament in manufacturing SMES. Procedia Comput. Sci. **55**, 833–842 (2015)

9. Sheko, A., Braimllari, A.: Inhibitors of information technology and the quality of information in supply chain management: a PLS-SEM analysis. Acad. J. Interdisc. Stud. **7**(3), 125–138 (2018)
10. Aguilera, L., Colín, M., Hernández, O.: The influence of information technologies in productive processes for greater competitiveness of the agua calientes SME: an empirical study. Manag. Develop. **5**(1), 40–68 (2013)
11. Díaz-Batista, J., Blanco-Fernández, Y.: Adoption and use of information technologies in Cuban organizations. Ind. Eng. **34**(3), 273–282 (2018)
12. Hugh Han, J., Wang Yingli, N.: Reconceptualization of information technology flexibility for supply chain management: an empirical study. Int. J. Prod. Econ. **187**, 196–215 (2017)
13. Kim, H.J.: Information technology and firm performance: the role of supply chain integration. Oper. Manag. Res. **10**(1–2), 1–9 (2017). https://doi.org/10.1007/s12063-016-0122-z
14. Scuotto, V., Caputo, F., Villasalero, M., Del Giudice, M.: A multiple buyer-supply relationsship in the context of digital supply chain management of SMEs. Prod. Plan. Control **28**(16), 1378–1388 (2017)
15. Devaraj, S., Krajewski, L., Wei, J.: Impact of electronic commerce technologies on operational performance: role of production information integration in the supply chain and operations management. J. Oper. Manag. **25**(6), 1119–1216 (2007)
16. Bardhan, I., Whitaker, J., Mithas, S.: Information technology production process outsourcing and manufacturing plant performance: the role of production process outsourcing. J. Manag. Inf. Syst. **23**(2), 13–40 (2006)
17. Jasperson, J.S., Carter, P.E., Zmud, R.W.: A comprehensive conceptualization of the post-adoptive behaviors associated with IT-enabled work systems. MIS Q. Manag. Inf. Syst. **29**(3), 525–557 (2005)
18. Devos, J., Van Landeghem, H., Deschoolmeester, D.: Rethinking IT governance for SMEs. Ind. Manag. Data Syst. **112**(2), 206–223 (2012)
19. Ventakatesh, V., Bala, H.: Technology acceptance model 3 and a research Agenda on interventions. Decis. Sci. **39**(2), 273–315 (2008)
20. Saga, V.L., Zmud, R.: The nature and determinants of IT acceptance, routinization, and infusión. Diffus. Transf. Implementation Inf. Technol. **45**, 67–86 (1994)
21. Bharadwaj, A.: A resource- based perspective on information technology capability and firm performance: an empirical investigation. MIS Q. Manag. Inf. Syst. **24**(1), 169–196 (2000)
22. Schlosser, F., Beimborn, D., Weitzel, T., Wagner, H.T.: Achieving social alignment between business and IT—an empirical evaluation of the efficacy of IT governance mechanisms. J. Inf. Technol. **30**(2), 119–135 (2015)
23. Monczka, R., Morgan, J.P.: Competitive supply strategies for the 21 century. Purchasing, 48–59 (2000)
24. Sander, N.R., Premus, R.: IT applications in supply chain organization: a link between competitive. J. Bus. Logist. **23**(1), 65–83 (2002)
25. Dai, Q., Kaufman, R.J.: Business models for Internet-based B2B electronic markets. Int. J. Electron. Commer. **6**(4), 41–72 (2002)
26. Nair, P.R., Raju, V., Anbuudayashankar, S.P.: Overview of information technology tools for supply chain management. CSI Commun. Comput. Soc. India **33**(9), 20–27 (2009)
27. Fu, Q., Zhu, K.: Endogenous information acquisition in supply chain management. Eur. J. Oper. Res. **20**(2), 454–462 (2010)
28. Tatoglu, E., Bayraktar, E., Golgeci, I., Koh, S.L., Demirbag, M., Zaim, S.: How do supply chain management and information systems practices influence operational performance? evidence from emerging country SMEs. Int. J. Log Res. Appl. **19**(3), 181–199 (2015)
29. Baltacioglu, T., Ada, E., Kaplan, M.D., Yurt, O., Kaplan, Y.C.: A new framework for service supply chains. Serv. Ind. J. **27**(2), 105–124 (2007)

30. Ellram, L.M., Tate, W.L., Billington, C.: Understanding and managing the services supply chain. J. Supply Chain Manag. **40**(3), 17–32 (2004)
31. Mesa, Y.R.: From information management to knowledge management. Information Organization: a determining factor in business management. Nat. Cent. Med. Sci. Inf. **14**(1), 37–61 (2006)
32. Flynn, B., Huo, B., Zhao, X.: The impact of supply chain integration on performance: a contingency and configuration approach. J. Oper. Manag. **28**, 58–71 (2010)
33. Zhao, X., Huo, B., Selen, W., Yeung, J.: The impact of internal integration and relationship commitment on external integration. J. Oper. Manag. **29**(1), 17–32 (2011)
34. Lai, F., Zhao, X., Wang, Q.: Taxonomy of information technology strategy and its communications for cocreating relational value: evidence from the logistics industry. MIS Q. Manag. Inf. Syst. **36**(1), 233–262 (2007)
35. Torkzadeh, G., Koufteros, X., Doll, W.J.: Confirmatory factor analysis and factorial invariance of the impact of information technology instrument. Omega Int. J. Manag. Sci. **33**, 107–118 (2005)
36. Dedrick, J., Xu, S.X., Zhu, K.X.: How does information technology shape supply-chain structure? evidence on the number of suppliers. J. Manag. Inf. Syst. **25**(2), 41–72 (2008)
37. Garcia, P.L.D., Martinez, S.M.C.: Innovation and Business Culture of MSMEs. Autonomous University of Aguas Calientes, México (2008)
38. Li, S., Lin, B.: Accessing information sharing and information quality in supply chain management. Decis. Support Syst. **42**(3), 1641–1656 (2006)
39. Swaford, P.M., Ghosh, S., Murthy, N.: Achieving supply chain agility through IT integration and fexibility. Int. J. Prod. Econ. **116**(2), 288–297 (2008)
40. Vickery, S., Calantone, R., Droge, C.: Supply chain fexibility:an empirical study. J. Supply Chain Manag. **35**(2), 16–24 (1999)
41. McGinnis, M.A.: Logistics strategy, organizational environment, and time competitiveness. J. Bus. Logistic **14**(2), 1–23 (1993)
42. Bruce, H.C.: Marketing performance measures: history and interrelationships. J. Mark. Manag. **15**(8), 711–732 (1999)
43. Wu, F., Yeniyurt, S., Kim, D., Cavusgil, S.T.: The impact of information technology on supply chain capabilities and business results: a resource-based view. Ind. Mark. Manage. **35**(4), 493–504 (2006)
44. Li, G., Yang, H., Sun, L., Sohal, A.S.: The impact of IT implementation on supply chain integration and performance. Int. J. Prod. Econ. **125**(1), 125–138 (2009)
45. Anderson, J.: Structural equation modeling in practice: a review and recommended two-step approach. Psychol. Bull. **103**(3), 411–423 (1988)

Customer Service Quality in Financial Entities a Factor for Generating Competitiveness in Times of Covid 19 Pandemic

Verónica Arévalo Bonilla(✉), Paúl Rodríguez Muñoz, and Franklin Daniel Aguilar E.

Instituto Tecnológico Universitario Rumiñahui, Sangolquí, Ecuador
veronica.arevalo@ister.edu.ec

Abstract. In a globalized world, companies are under increasing pressure to generate a competitive advantage over others. Satisfying the needs or expectations of customers, becomes a predominant factor in attracting consumers for financial entities, so it is necessary to know the value of customer perception and satisfaction in such services. This study is based on analyzing the predominant factors, to know the perception that customers have on the quality of service of the financial companies of the segment 3 (assets: greater than 5′000.000,00 up to 20′000.000,00), according to the classification of the Standard for the segmentation of the entities of the popular and solidary financial sector [1]. These factors are immersed in variables of tangible aspects that reflect the perception of the service, as well as in aspects of an intangible nature that are perceived and valued by the customer. Those ones allow generating a competitive advantage in the financial services sector. The research was based on a model of variables to identify the tangible and intangible factors in customers satisfaction and their perception of quality and satisfied needs. The reliability analysis tested the hypotheses proposed in the research, while the factor analysis was used to identify the variables, their correlation and the validity of the study with its factors, the application of multiple linear regression analysis, allowed observing the effect of the independent variables on the dependent variables of perceived quality and satisfied needs.

Keywords: Quality · Competitiveness · Customer service · Competitive advantage · Satisfaction

1 Introduction

Ecuador's financial industry has a variety of entities in the savings and credit union sector, providing similar services and products, to grant financial resources to their customers, to start a business or to expand their economic or personal activities.

According to the financial bulletin as of December 2020, published by the Superintendencia de Economía Popular y Solidaria (SEPS), there are 83 financial entities belonging to segment 3 of this sector, that's why the competition factor for attracting new customers and maintaining the loyalty of others, becomes an important element of

M. Zambrano Vizuete et al. (Eds.): CI3 2021, LNNS 511, pp. 516–527, 2022.
https://doi.org/10.1007/978-3-031-11438-0_41

analysis for savings and credit unions. At this point there are discussion factors that allow us to know, the perception that customers have of the services and products acquired, customers satisfaction becomes an objective of any organization, since it will depend on how satisfied the customers are, to achieve loyalty through the relationship maintained with them [2].

Due Covid 19 pandemic, the relationship with the customer and the service provided by financial entities has become more and more demanding, due people are working at home or lockdown, which hinders personalized service to the customer, therefore, improving the quality of service offered, is becoming a daily demand, understanding that customer service is anything what increases their level of satisfaction.

Satisfaction, or lack thereof, is the difference between how the customers expect to be treated and how they perceive how are being treated [3].

In this research we propose to analyze under a modeling, the correlation of the variables and factors that the customer perceives as satisfaction value, through the quality of service and satisfied needs, which will serve as evidence for decision making within the strategies to generate competitive advantages.

1.1 Quality of Service

There are studies carried out to measure perceived service quality based on scales and measures such as the SERVQUAL scale [4], where perceived quality is the result of the comparison between what the client considers that the service offers(expectation) and his perception of the performance of the service delivered [5].

In Latin America there is limited research in this regard, and in Ecuador a study was conducted based on this scale, that measures five dimensions: Reliability, responsibility, security, empathy, material and tangible goods, in a savings and credit union in the province of Manabí [6]. In this research, perceived quality was considered as the result of the consumer experience where the customer's expectations were also considered [7, 8].

1.2 Satisfaction

Customer satisfaction depends on the perceived performance of the product in relation to the buyer's expectations [9], and satisfaction will depend on whether or not, these expectations were met [10].

Another concept from by [11], as "the psychological outcome of a consumer experience, where the measure of satisfaction arises from the consistency in the answers to a several questions related to the degree of well-being that a person feels". The present research considers satisfaction, as the outcome of the service consumption experience.

1.3 Factors Determining the Quality of Service

The factors that are evaluated to measure service quality in the financial industry are derived from tangible and intangible variables, which the client perceives at the moment of experiencing the service [12].

"Intellectual capital is the combination of intangible assets, including the knowledge of staff, the capacity to learn and adapt, relationships with customers and suppliers, brands, product names, internal processes and R&D capacity, etc., of an organization, which, although not reflected in traditional accounting reports, it generates or will generate, future value for a sustained competitive advantage" [13].

The Following Factors are Evaluated in the Intangible Area
Employee service: refers to the attentiveness and skills provided by the financial entity's staff [13, 14]. Organizational capacity: intangibles such as: Culture, organizational structure, processes. Technological capacity: intangibles related to the development of activities and processes.

Tangible Factors
Physical infrastructure: Physical equipment and tangibles that the customer perceives when is being attended. The factors described above are part of the hypotheses to define the customer's perception of the quality of service, under the modeling proposed by [15].

1.4 Hypothesis

The proposed conceptual model is represented in Fig. 1, where employee attentiveness and organizational capability, are described as influencing factors on perceived capability to complying customer needs.

H1: Employee attentiveness and organizational capability and their direct influence on perceived quality.
H2: Employee attentiveness and organizational capability have a direct influence to complying customer needs.
H3: Technological capability directly impacts on perceived quality.
H4: Technological capability directly influences to complying customer needs.
H5: Physical infrastructure directly influences on perceived quality.
H6: Physical infrastructure directly influences to complying customer needs.

Perceived quality and complying customer needs, will be part of the outcome about satisfaction with the service received.

H7: Perceived quality has a positive influence on satisfaction.
H8: Complying customer needs have a positive influence on satisfaction.

The objective of this research is to determine the most significant variables perceived as quality in customer service at financial entities, in order to generate competitive advantage in the Covid 19 pandemic.

2 Methodology

The research was based on the confirmatory explanatory method, which consists of approaching the data to validate a hypothesis.

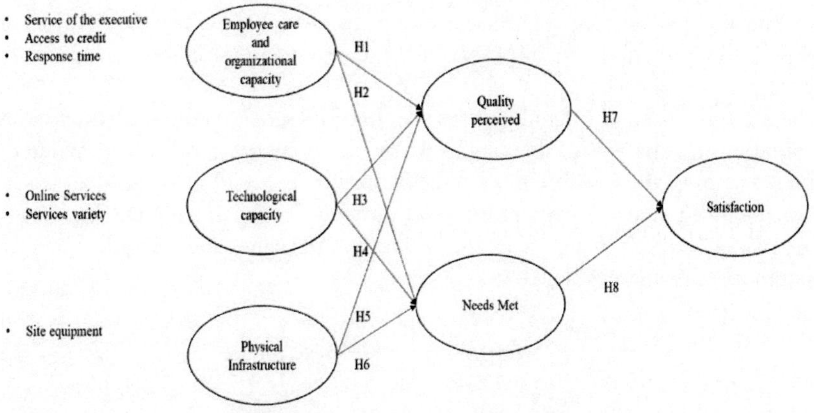

Fig. 1. Proposed conceptual model

The data used were obtained based on a survey done in the Cooperativa de Ahorro y Crédito Puéllaro Cía. Ltda., applied to a sample of 375 people, in different branches nationwide, with 9 samples considered invalid, applying a test based on the Likert scale with the following measurement scale: 5: very good; 4: good; 3: fair; 2: bad; 1: very bad.

By applying the sample formula, a 95% confidence level is considered, with a maximum accepted margin of error of 5%, obtaining a total of 375 samples (Table 1), for which the

$$n = \frac{N * Z^2 * p * q}{e^2(N-1) + z^2 * p * q} \tag{1}$$

Table 1. Shows the values considered for the Eq. (1).

Parameter	Confidence level values
N	15000
Z	1,96
P	50%
Q	50%
e	5%
n	375

The factorial confirmatory analysis was used for the analysis and reliability study [16], his makes it possible to contrast a model constructed in advance, in which the researcher establishes the total set of relationships between the elements that make it up, it is assumed that the researcher is able to predict as a priority, the data structure - preferably based on a well-established theory - and only needs to confirm that this structure can also be obtained empirically.

Technically, both strategies pursue a single objective: "to explain the covariances or correlations between data sets of observed or measured variables, through reduced data sets of latent variables or factors" [17].

The software used for data analysis was the SPSS IBM program, which allowed the tabulation and analysis of the data from the questionnaire applied nationwide in the several agencies of the Credit Union in the case study.

The reliability analysis was carried out through Cronbach's alpha, (Eq. 2) giving a value of .737, which is an acceptable value that measures the reliability of the test measurement scale (see Table 2).

$$\infty = \frac{k}{k-1}\left[1 - \frac{\sum Si^2}{St^2}\right] \quad (2)$$

Table 2. Feasibility statistics of the equation survey (2)

Cronbach's Alpha	No of elements
0,737	7

A summary of the confirmatory factor analysis obtained in the SPSS program is shown below.

Table 3 describes the KMO measure, Keiser-Meyer-Olkin, which measures the adequacy of the sample, and indicates that it is appropriate to apply the factorial analysis, which for the study was .836. This result indicates that when is closer to 1, then more appropriate it is, due to the high relationship that exists between the variables, in the same way it can be seen with Bartlett's test of sphericity, which indicates the correlation of the variables with a value close to zero, making the factorial model pertinent.

Table 3. Bartlett KMO test

Kaiser-Meyer-Olkin measure of sampling adequacy		0,823
Bartlett's test for sphericity	Approx. chi-squared	531,466
	Gl	21
	Sig.	<,001

In relation to the Chi-square index that serves to test the referred hypotheses, it was determined that the alternative hypothesis H3 was rejected, leaving the other variables as valid. (H1, H2, H4, H5, H6, H7, H8).

Following with the analysis, by means of the anti-image correlation matrix, it was detected that all the items are feasible, since they affect in a positive way, as their value is close to 1, which is counteracted by the results of the communalities by extraction, in this case it is suggested that item 6 be removed for detailing a value not so close to 1.

However, it is an item that after its analysis, it was decided to keep it since it does not affect the correlation matrix.

3 Results

In the result of the total variance explained described in Table 4, it was noted that, it details 3 components with the respective items and reflected in the rotated component matrix, whose values are close to 1, and detailing the three constructs observed in the model proposed.

Table 4. Total variance explained

Component	Initial eigenvalues			Sums of loads squared by extraction			Sums of loads squared by rotation		
	Total	% variance	Accumulated %	Total	% variance	Accumulated %	Total	% variance	Accumulated %
1	2,985	42,649	42,649	2,985	42,649	42,649	2,434	34,768	34,768
2	1,044	14,913	57,562	1,044	14,913	57,562	1,231	17,584	52,353
3	,806	11,514	69,076	,806	11,514	69,076	1,171	16,723	69,076
4	,688	9,824	78,900						
5	,549	7,841	86,741						
6	,505	7,221	93,962						
7	,423	6,038	100,000						

According to the results obtained through the confirmatory factor analysis, it was found that the validity and reliability of the constructs estimated are in accordance with the principles proposed.

Variables Related to the Survey Questions

P2: The training of the person in customer service.
P4: Service rating.
P4: Service rating.
P5: Service in the credit area.
P6: Service in the cashier area.
P10: Preference for online transactions.
P11: Usefulness of the website.
P12: Comfortable physical space.
P14: Service comparison (Fig. 2).

The results of the model show that with high levels of statistical significance (p-value < 05), the hypotheses H1, H2, H4, H5, H6, H7, H8 are fulfilled. The results indicate that the constructs: employee service and organizational capacity; technological

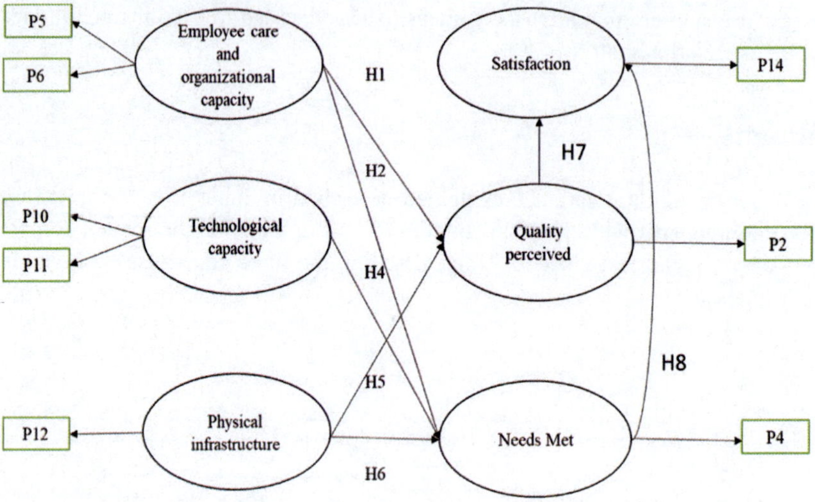

Fig. 2. Below shows the conceptual model with the variables observed in each of the factors.

capacity and physical infrastructure, affect the perceived capacity and the needs covered of the customer, except for hypothesis 3 where technological capacity does not affect the perceived quality, a phenomenon that is explained by the market segment to which the Credit Union is, since it is a population that is mostly located in rural areas with low access to Internet.

The results in Table 5 present a summary of the values of the Chi-square analysis, for the hypotheses (H1-H2, H4, H4, H5, H6, H7, H8), according to the values presented in the Chi-square test, yield values greater than the limit according to the degree of freedom and the estimated error of 5%. Hypothesis H3 was eliminated for not having correlation variables.

Table 5. Chi-square tests

	Value	gl	Bilateral asymptotic significance
Pearson's Chi-square H1	143,003	6	<,001
Pearson's Chi-square H2	95,609	6	<,001
Pearson's Chi-square H4	8,174	2	0,017
Pearson's Chi-square H5	136,606	6	<,001
Pearson's Chi-square H6	177,669	6	<,001

The following is the rotated component matrix, where the factor analysis is reduced to three components, corresponding to questions P4, P2, P6, P11, P12 for component one; P14, P5, for component two and finally P10 for component three, all of which are correlated with each other (see Table 6).

Table 6. Rotated component matrix

	Component		
	1	2	3
14. If you compare this service with another institution, you believe that the Credit Union's service is	,198	,790	-,077
4. How would you rate the overall service of the Credit Union?	,669	,403	,105
2. The person who attended you showed that is	,806	,069	,021
5. How would you rate the service you receive in the credit area?	,124	,766	,219
6. When you have gone to the cashier's office, how would you rate the service you received?	,611	,286	,149
10. Would you like to perform your transactions by Internet?	,043	,054	,950
11. What do you think about the usefulness of the Credit Union's web page	,615	,261	,444
12. The place where you waited to be attended to, is	,738	,040	-,046

Once the results of the components were obtained, the multiple linear regression analysis was applied, which allowed us to observe the relationship between the variables

Table 7. Regression statistics and analysis of variance (3)

Regression statistics	
Multiple correlation coefficient	0,788
Determination coefficient R^2	0,621
R^2 adjusted	0,617
Standard error	0,427
Remarks	368,00
Analysis of variance	

	Degrees of freedom	Sum of squares	Mean square average	F	Critical value of F
Regression	4,00	107,81	26,95	148,03	0,00
Residual	361,00	65,73	0,18		
Total	365,00	173,53			

	Coefficients	Standard error	Statistic t	Probability	Lower 95%	Upper 95%	Lower 95,0%	Upper 95,0%
Intercepción	0,608	0,128	4,757	0,000	0,357	0,860	0,357	0,860
P5 Variable X 1	0,089	0,021	4,266	0,000	0,048	0,131	0,048	0,131
P6 Variable X 2	0,290	0,033	8,831	0,000	0,225	0,354	0,225	0,354
P11 Variable X 3	0,119	0,026	4,629	0,000	0,068	0,169	0,068	0,169
P12 Variable X 4	0,285	0,036	7,948	0,000	0,215	0,356	0,215	0,356

of the components. The results of Table 7 and the Eq. (3) used are detailed below.

$$P4 = f(P5,\ P6,\ P11,\ P12) \tag{3}$$

The variables analyzed in Table 7 are: Service in the credit area, service in the cashier area, usefulness of the web page and comfortable physical space, which have an impact on the needs met of the customer, and at the same time are statistically significant in explaining the behavior of this variable.

It should be noted that in this model, the variable P10 (Internet transaction preferences) is not statistically significant, which is why it was not considered.

In the same way, perceived quality was analyzed in terms of the following variables described in Table 8 and the Eq. (4) used are detailed below.

$$P2 = f(P5, P6, P12) \tag{4}$$

The results in the multiple linear regression model, it's noted that the perceived quality is dependent on the employees' service component and the organizational capacity, reflected in the variables of service in the credit area, service in the cashier's area and physical infrastructure (see Table 8).

Therefore, the customers perceive quality through the way they are served by the representatives of the areas of the Credit Union, whether in cashier's area, investments or credit, as well as whether their needs were met through the skills of the executives.

Table 8. Regression statistics and analysis of variance

Regression statistics								
Multiple correlation coefficient				0,70				
Coefficient of determination R^2				0,49				
R^2 adjusted				0,48				
Standard error				0,48				
Remarks				368,00				
Analysis of variance								
	Degrees of freedom		Sum of squares		Mean square average		F	P-Value F
Regression	3,000		80,014		26,671		114,112	0,000
Residual	362,000		84,609		0,234			
Total	365,000		164,623					
	Coefficients	Standard error	Statistic t	Probability	Lower 95%	Upper 95%	Lower 95,0%	Upper 95,0%
Interception	1,057	0,144	7,333	0,000	0,774	1,341	0,774	1,341
P5 Variable X 1	0,063	0,023	2,688	0,008	0,017	0,109	0,017	0,109
P6 Variable X 2	0,271	0,036	7,459	0,000	0,200	0,343	0,200	0,343
P12 Variable X 3	0,333	0,039	8,467	0,000	0,256	0,411	0,256	0,411

The technological capacity by using the web page, does not have an impact on the perception of quality of the customers, but it does seek to satisfy their needs.

Finally, it was noted that the organization's equipment and the comfort felt in the facilities, impacts in the quality and needs of the customers.

Additionally, Table 9 demonstrates hypotheses H7 and H8, through the Measure of Sampling Adequacy (KMO), so the perceived quality is reflected in the training of the financial executive, which impacts customer satisfaction, as well as the needs covered, as reflected in the rating of the service received and directly influencing consumer satisfaction.

The above confirms the results obtained in the survey, where it is evident that 73.06% say that financial executive of the Credit Union is very skilled (P2), influencing in the quality received, while the qualification for the service (P4) is very good with a result of 72.70%, impacting on the customer requirements fulfilled. Finally, a comparison of the service with other financial entities (P14) was carried out and it was found that 62.15% said that the service is better than that of the competition, which has an impact on customer satisfaction.

Table 9. KMO and Bartlett's test

Kaiser-Meyer-Olkin measure of sampling adequacy		0,586
Bartlett's test for sphericity	Approx. chi-squared	122,699
	Gl	3
	Sig.	0

4 Conclusions

Customer satisfaction is an important factor to be considered in the quality management of any organization. Globalization and business competitiveness seek to maintain customer loyalty and attract new ones, so the companies constantly focus their efforts on service or product innovation, but without analyzing whether their customers' needs are being satisfied through perceived quality; therefore, when evaluating the quality of service, the customer's perceptions should be considered, which may vary depending on the segment and its needs.

For the case studied, according to the descriptive data, the perceived quality is defined in 62.15%, according to the skills of the executive who attended the customer, and at the same time it is shown that their needs were met, achieving an acceptance of very good in the service received, which is why it can be concluded that customer satisfaction is at a high level.

Regarding the conceptual model, the confirmatory factor analysis showed that the constructs, the measurement scale and the correlation of variables are well defined, as detailed in the results of each of its components, showing that the perceived quality and the needs covered are affected by: personal service and organizational capacity, technological capacity, and physical infrastructure, which have a direct impact on customer satisfaction.

It is concluded that financial companies should focus their efforts on knowing which variables are important to maintain customer satisfaction, which differ from one organization to another, since they are influenced by market segmentation.

In the present study it was observed that the technological capacity through the use of the web page is not very relevant, due to the fact that a large part of the sample is located in rural areas with low access to internet, a contradictory situation regarding to the changes in the needs and preferences in times of the Covid19 pandemic.

On the other hand, intellectual capital, as an intangible component, is a very relevant factor since it relates the training or qualification of staff to serve a customer, the kindness in the service, and the time of assistance will generate a positive impact on the perceived quality.

This research will be a contribution to the financial market, since through a model approach that allows analyzing the study variables and their behavior, it will allow making decisions based on evidence, focusing on the customer, the processes and the correct allocation of resources to generate competitiveness through quality.

References

1. Página de inicio de SEPS. https://www.seps.gob.ec/estadisticas?productos-estadisticos. consultado por última vez el 22 Sep 2021
2. Denton, D.K.: Calidad en el servicio a los clientes. Ediciones Díaz de Santos (1991)
3. Pérez Rodríguez, M.D.: Calidad de servicio y Atención al Cliente (2a ed.). Editorial ICB (2012)
4. Spiros, G.: Antecedents to perceived service quality: an exploratory study in the banking industry. Int. J. Bank Mark. **21**(4), 168–190 (2003)
5. Botero, M.M., Peña, P.: Calidad en el servicio: El cliente incógnito. Elsevier. es/sumapsicol, vol. 13 no.(2), pp. 217–228 (2008)
6. Loor-Moreira, A., Calderón-Salazar, J., Arias-Arias, A.E.: Factores que determinan la satisfacción del socio con respecto al servicio en una Cooperativa de Ahorro y Crédito de Manabí. Factors that determine the satisfaction with respect of service in the credit union of Manabí. CIENCIA UNEMI **10**(25), 36–44 (2017). https://doi.org/10.29076/issn.2528-7737vol10iss25
7. Macías, M.E.U.: Modelo de gestión de la calidad en el servicio al cliente: propuesta para las grandes superficies. Revista Lebret **5**, 333–354 (2013)
8. Berry, L.L., Bennett, D.R., Brown, C.W.: Calidad de servicio. Ediciones Díaz de Santos (1989)
9. Philip, K., Gary, A.: Marketing. Décimo cuarta edición. PEARSON EDUCACIÓN, México (2012)
10. Mateos de Pablo Blanco, M.A.: Atención al cliente y calidad en el servicio. COMM002PO. IC Editorial (2019). https://elibro.net/es/lc/ister/titulos/124251
11. Varela, J., Lévy, J.P.: Modelización con estructuras de covarianzas en ciencias sociales: temas esenciales, avanzados y aportaciones especiales. Netbiblo (2006). https://dialnet.unirioja.es/servlet/libro?codigo=402086
12. De la Fuente Mella, H., Díaz Bravo, I.: Análisis de los factores determinantes de la calidad percibida del servicio prestado por una cooperativa de ahorro y crédito: una aplicación basada en modelos de ecuaciones estructurales. Ingeniare. Revista chilena de ingeniería **21**(2), 232–247 (2013). https://doi.org/10.4067/S0718-33052013000200007

13. Medina, S., González, M., Pérez, H.: El concepto de capital intelectual y sus dimensiones. A. J. **13**, 16 (2007)
14. Godoy, J.N.: El capital humano en la atención al cliente y la calidad de servicio. Observatorio laboral revista venezolana **4**(8), 23–35 (2011)
15. Carvajal, S.A., Leguina, A., Paulina, E.: Factores determinantes en la percepción de la imagen y calidad de servicio y sus efectos en la satisfacción del cliente un caso aplicado a la banca chilena. **19**(2), 2013 (2007)
16. Batista-Foguet, J.M., Coenders, G., Alonso, J.: Análisis factorial confirmatorio. Su utilidad en la validación de cuestionarios relacionados con la salud. Med. Clínica **122**(1), 21–27 (2004). https://doi.org/10.1157/13057542
17. Bollen, K.: Structural Equation with Latent Variables. Wiley, Hoboken (1989)

Contexts of Digital Tools in Marketing Management in MSMEs for Their Permanence in the Market

Paulina Renata Arellano(✉), Franklin Daniel Aguilar, Alain Quintana Bornot, and Luis Alberto Carrera

Instituto Tecnológico Universitario Rumiñahui, Sangolquí, Ecuador
paulina.arellano@ister.edu.ec

Abstract. Globalization has led that competitiveness among Ecuadorian companies is becoming difficult, as abroad companies with a higher level of development are now in the country, that is forcing to local companies to develop digital tools that allow them to make an adequate marketing campaign, to ensure their permanence in the market by finding new customers. In order to know, how is the performance of MSMEs in Ecuador in relation to management of digital tools that allow their permanence in the market, a survey was conducted to 166 MSMEs, duly registered in the Superintendence of Companies, Securities and Insurance of Ecuador, these 166 companies are in services, commercial sectors, and industry. The results show that 70.50% [1] of the companies have a website for digital marketing, however, the frequency of use of digital tools is limited to a weekly and biweekly frequency. This reduces the possibility of attracting new customers, with the disadvantage that many do it empirically, without having the appropriate software, and with partially trained staff for this purpose. Therefore, it is about training the human talent of MSMEs, for the proper management of digital tools, with a greater frequency of use them to increase a presence of more than 90% of the companies formally registered in the country.

Keywords: e-commerce · Internet · Social networks · Marketing

1 Introduction

Currently, the increasing level of competitiveness of different business sectors, in addition to changing market trends and consumer demands, are factors that have contributed to the fact that e-commerce is becoming increasingly relevant and is widely considered as an alternative [1] used in the exchange of goods, as it stimulates commercial activity, which provides consumers with more opportunities to develop negotiations and business with greater market coverage [2].

The Internet offers to the world a new market, whose assets are mainly knowledge and information [3]. Manufacturers or suppliers of goods or services and users have access, to transfer information and entertainment in a simple and economical way, for commercial or social purposes. In practice, thousands of companies around the world

M. Zambrano Vizuete et al. (Eds.): CI3 2021, LNNS 511, pp. 528–541, 2022.
https://doi.org/10.1007/978-3-031-11438-0_42

have many years of experience in using the Internet as a channel for sales, promotion and distribution through e-commerce [4].

In Ecuador, according to the Ecuadorian Institute of Statistics and Census (INEC), in 2012 there were 542,000 companies, of which only 232 had Internet presence, and in the sectors where large companies are located, highlighting the financial sector which communicates with customers through electronic channels such as ATMs, online banking, mobile banking, through which offers its customers access to their accounts for online transaction requests, purchases, payments, transfers, etc., as well as at the institutional level, to carry out online negotiations in the national and international market [5].

E-commerce in Ecuador has become more dynamic in recent years due to the increased access of citizens to the Internet, a significant increase in the use of mobile devices, as several product stores have implemented their shopping and sales processes by technological platforms that allow customers to shop from their homes or offices [1, 2].

On the other hand, it should be emphasized that there is currently a marked preference for up to 60% of the economically active population to use new communication technologies as a tool for searching for goods and services, not being the trade of products and services indifferent to this trend, being able to find numerous web pages and sites specialized in the diffusion and marketing of all kinds of products [6].

However, according to an international research developed by the Latin American Institute of Electronic Commerce (2018), although the use of technological tools for the development of commercial and advertising activities has increased in recent years, only 10% of companies in Latin America sell online through websites specialized in selling or through their web sites.

Readiness in Latin America in 2016, highlighted the increase in online commerce opportunities due to a systematic development of monetary and online connectivity, taking into account the facilities provided by digital marketing, as well as the growing trend in the massification of its use, greatly increased the possibilities of the customer when choosing products and services on online, being able to compare prices, models, quality, as well as other alternatives from the comfort of the customer's home or office, that in many cases may receive more quickly and conveniently the requested product [7].

Although the use of digital tools for marketing and e-commerce are relatively recent, the means for its realization are diverse due to the exponential growth that have digital tools, following there will be a brief review of the historical evolution of digital marketing [8], as well as the description of each of these alternatives, which can be adapted according to customer characteristics and user requirements, pointing out that a tool that is successful for one company may be unsuitable for another, however it is necessary to manage all the alternatives and use them according to the market, characteristics of the audience to be addressed and the economic possibilities of the company for the implementation of the digital marketing campaign [9].

1.1 Marketing in the Globalization Age

Modern commerce is characterized by a marked tendency towards economic globalization, hence a higher level of competitiveness in the supply of goods and services in regional and international markets, a phenomenon that is largely due to the emerging of digital marketing since the mid-1990s [10].

This phenomenon led to the trade democratization, which years ago was in the hands of a few transnational companies, allowing small and medium-sized traders to buy products and services without intermediaries through the new communication technologies; this phenomenon has developed systematically up to the present and is considered a global trend [11].

The supply of goods and services is currently enhanced by the daily use of digital marketing, in such way that the customer can be informed almost instantly about new products, sales, easy payments, and also ensuring better shipping management from manufacturers and sellers to customers [12].

As Porter states, despite the fact that companies develop their activities in the same competitive environment, their success and profitability will be conditioned to the marketing strategies of the products and services offered, these strategies will allow greater competitiveness in the market [13].

It is evident that those companies that are using digital marketing, will have better opportunities to gain the market by placing their goods or services almost immediately to the consumer, who will look characteristics, prices and details of the product offered, through the systematic use of new communication technologies.

As stated by [14], only those companies that generate new knowledge, disseminate it within themselves and incorporate it, into their technologies, products and services will be able to achieve success, i.e., regardless of the quality and prices of the product offered, it will be the use of new marketing technologies that guarantee the permanence of goods and services in a highly competitive market.

The above shows the importance of digital marketing strategies in the age of digitization, as a stage of the human progress, which has created a dependence on new communication technologies for almost all human activities.

1.2 Digital Relationship Marketing

Digital relationship marketing is the implementation of strategies aimed to anticipating, knowing, foreseeing and satisfying the needs of current and potential customers, improving and developing the products and services of the company [15]. It is noteworthy that digital relationship marketing aims to create and strengthen long-term business relationships using information and communication technologies as the main platform.

The trend to use digital marketing makes it a space for interaction between marketers, clothing manufacturers and customers, opening an opportunity for more information about tastes, interests and needs of customers, so new products can be improved or created to achieve a higher level of customer satisfaction.

1.3 Integration of Digital Media with the Marketing Strategy

The organization of ICTs through an associated marketing strategy, supported by digital tools, makes it possible to have a clear picture of the market you want to enter. There is a bi-directional relationship that allows to gather the appropriate characteristics, for example, to determine the purpose: priorities, opinion interests, behavior, impact of purchases, etc. and also to know the competitors [16]. The strategy used in relationship marketing is to find the best customers: those who cover more purchases, more often, etc.

This information will be used to establish direct marketing in terms of advertising, using digital tools helps to manage materials, create knowledge, and manage interactive activities that drastically improve budget spending. Among the digital marketing tools are: social media, audiovisual advertising, telemarketing, Web mobile (applications), email, specialized platforms, online and offline media.

1.4 Digital Tools and MSMEs

The trend to use digital media has attracted the attention of medium and large companies, being currently a trend, digital catalogs are developed seasonally to ensure customer loyalty, providing easy payments, sales and free deliveries as alternatives to maintain a highly competitive market [17].

MSMEs are currently trying to relive the national economy, it is essential to remain at the forefront in the use of technological tools to perform marketing activities; unfortunately, most MSMEs keep traditional management models resulting in stalling of business development, which does not adapt the changes in the market.

Despite the great advantages of e-commerce, there are several aspects in which MSMEs do not exploit this technology; aspects such as lack of knowledge about the possibility of expanding the supply of their products, low level of knowledge of technology, lack of training process, lack of decision of the owners and resistance to the use of technology to sell their products, as well as other elements that restrict the management of e-commerce as a tool for business growth [2].

Due Ecuadorian companies must be more competitive, because the challenges of new competitors, it is necessary to develop digital tools that allow the best marketing to attract new customers, therefore the purpose of this research was to describe the performance of Ecuadorian MSMEs, in relation to the management of digital tools, highlighting its advantages and disadvantages, to design strategies to improve their use.

2 Methodology

A descriptive research was applied, due to the characteristics of the study, a mixed qualitative and quantitative approach was applied, the characterization was performed by using open questions for gathering the field information, the survey technique was applied to the main actors of the managerial and administrative process of a sample of 166 MSMEs duly registered with the Superintendence of Companies, Securities and Insurance of Ecuador. For the data gathering, a survey was used as a tool for compiling quantitative data to reflect present conditions of the impact of the management of digital tools and strategies for marketing of medium and small Ecuadorian companies, prior to that, a survey to the workers of the companies, allowed a diagnosis about the digital tools that they were using.

The scope of the research was correlational, seeking to determine the dependence between management strategies and the management of digital tools, to improve the marketing process and its impact on the economic and financial performance of small and medium Ecuadorian companies.

2.1 Survey Sample (Population)

The research was conducted on MSMEs duly registered with the Superintendence of Companies, Securities and Insurance of Ecuador. The survey sample was the 166 companies registered as MSMEs, without considering their size or their business line.

The survey had open questions, focused to the digital tools known and usually applied for the marketing of MSMEs, taking into consideration the technical capabilities of the staff, as well as the effects noted due the application of these practices, in terms of improvements in marketing processes and economic and financial performance of the business.

2.2 Procedure for Information Processing and Analysis

Statistical analysis was performed to identify trends, using descriptive statistical parameters by pie charts, and Excel for data processing and Sigmaplot software for graphics.

3 Results and Discussion

Prior to the analysis of the technological skills of employees and digital equipment capacity of MSMEs, first of all, a stratification of small and medium-sized industries in Ecuador was carried out, highlighting that according to what was observed in Fig. 1, the activities are carried out in the service and commercial sectors, with 41.57% and 32.53% respectively, which coincides with what is observed in other Latin American economies, where economic development corresponds to the service and trade sector, unlike the first world economies where the industrial sector is privileged.

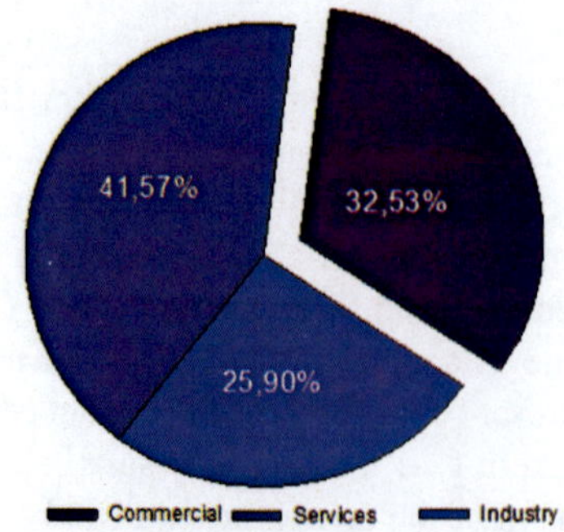

Fig. 1. Classification of MSMEs in Ecuador according to the activity performed

As shown in Fig. 1, according to the economic activity, most of the companies are engaged in supplying services, followed by selling goods activities, which means into a greater focus on marketing activities rather than supply chain activities, which highlights the importance of actions to disseminate information to ensure a market niche that allows effective products selling, highlighting some authors [18], that digital tools help to attract new customers.

To carry out marketing activities in today's world, a company must have digital tools. This situation was increased in the period 2020–2021 by the pandemic of COVID-19 [6], in this way the results expressed in Fig. 2, indicate that 92.77% of the companies have internet service, where a high percentage of employees have access to personal computers, it's evident that 84, 34% of the companies have more than one computer with internet connection, giving a high average level to the use of digital media, from the 70.48% of the companies that have internet access, the 73.48% of them have a formal website, however only 45.78% of them have online orders and only 33.87% have online sales catalogs and only 39.46% have links to other websites.

The results show that despite the technological capacity of the companies that have computers and websites, they are underutilized, so they do not formally carry out e-commerce for selling the products through their websites. It has been shown that digital marketing campaigns improve the position of the companies [19], based on the exploration of new market niches [20].

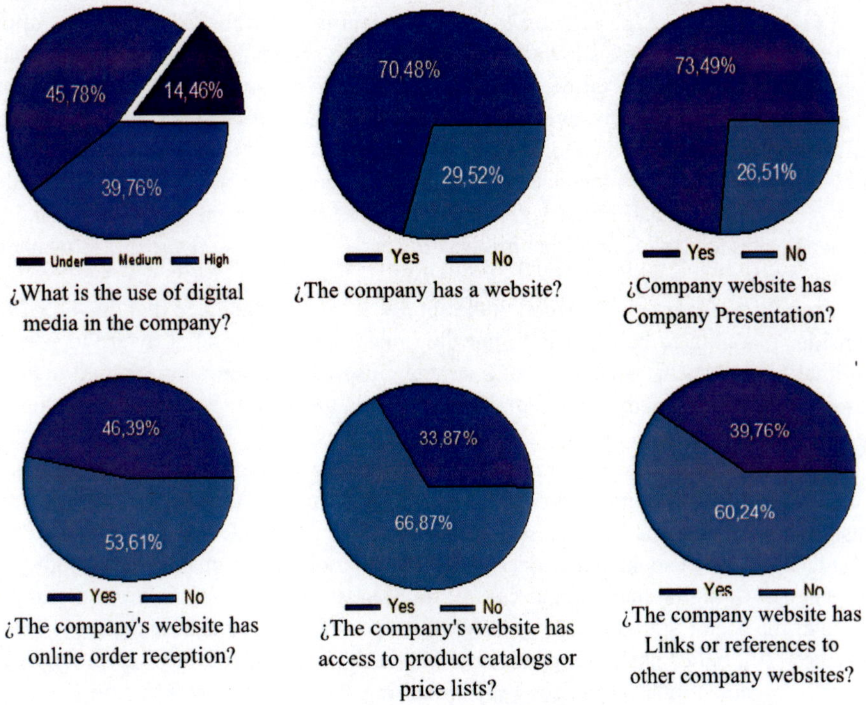

Fig. 2. Technological capacity for the use of digital tools of Ecuadorian MSMEs.

One of the weaknesses for marketing plans using digital tools of Ecuadorian MSMEs is the frequency of internet use, it is known that digital marketing processes are more dynamic and changing [21], however most companies only connect weekly, biweekly and even monthly which creates a great disadvantage when it comes to capturing new markets (Fig. 3).

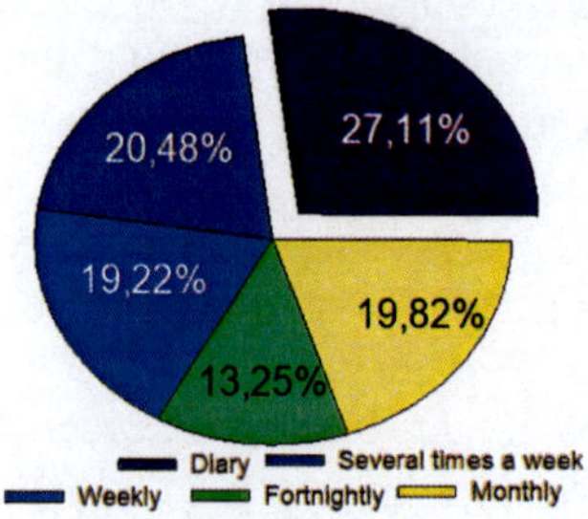

Fig. 3. Frequency of use of digital media by Ecuadorian MSMEs.

The information presented shows the frequency that MSMEs update digital media, highlighting that within the strategies, it is necessary to increase to a daily frequency [22], since there is a need to establish direct contact with consumers [23], so that they know the features, prices and sales offered.

On the other hand, in addition to the infrequent use of websites, their applications for marketing are scarce as shown in Fig. 4, where it is shown that although 71.69% have company advertisements in the web, only 39.76% have a suggestions and complaints mailbox, only 54.22% have advertisements for job applications by web, only 45.78% use software to manage the website, 78% use software to manage the supply chain, 50.60% use software for logistics management, 75.30% use computer programs for accounting management, only 58.43% use software for process management and control, 60.28% use web pages to provide customer information and 62.05% use software for information management.

From the information provided in the survey, it can be concluded that most MSMEs consider planning to be important, but they make limited use of software to achieve optimal management and thus ensure strategic use of economic, human and material resources of the company, which implies that during the planning process it is necessary to generate strategies to increase the use of these tools.

Most MSMEs manage information and marketing through their websites [24], as well as to attracting human talent [25] and provide timely and reliable information to customers [26], for this it is necessary that the company has a Community Manager for the management of social networks and any digital tool that is considered relevant [27].

The use of digital media requires staff training, several authors highlight the importance of the use of digital media by observing a partial improvement in the provision of services [30], better positioning and visualize of the business [31], as well as improvement in organizational processes [32] increasing the customer satisfaction [33], also increasing the ability to adapt to market changes and greater participation in them by capturing new niches [20] (Fig. 5).

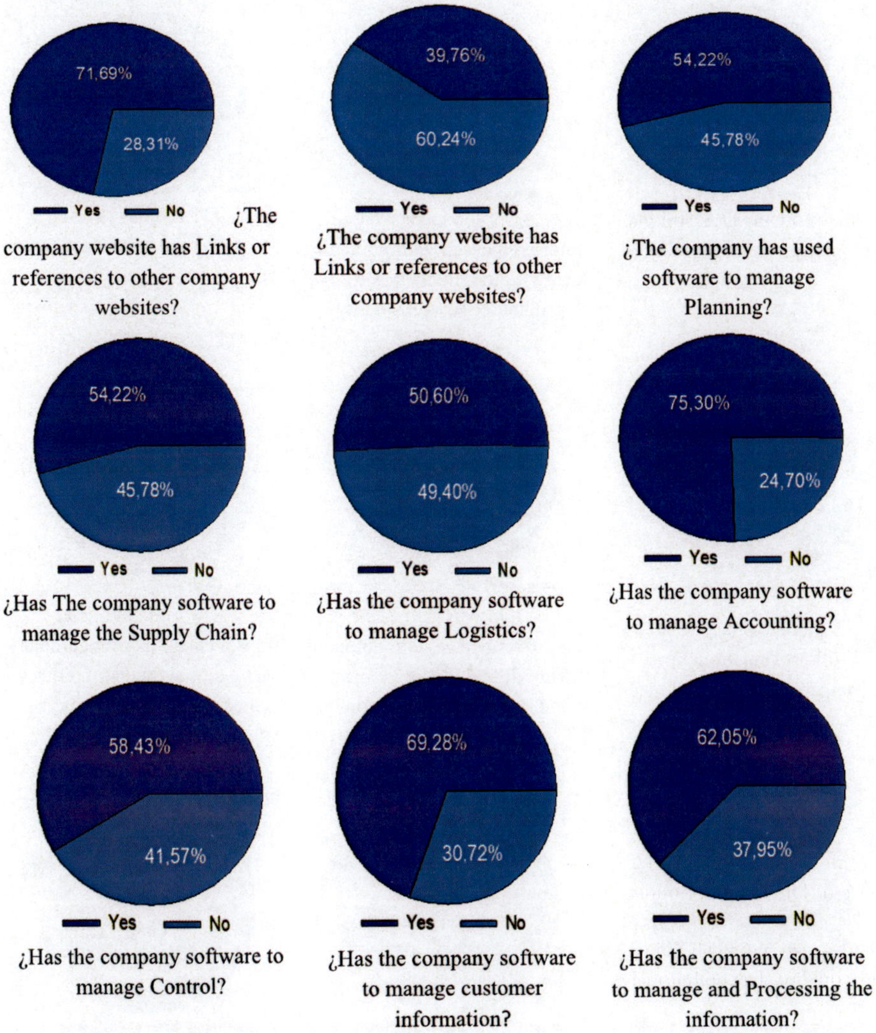

Fig. 4. Tools used for business management and marketing of Ecuadorian MSMEs.

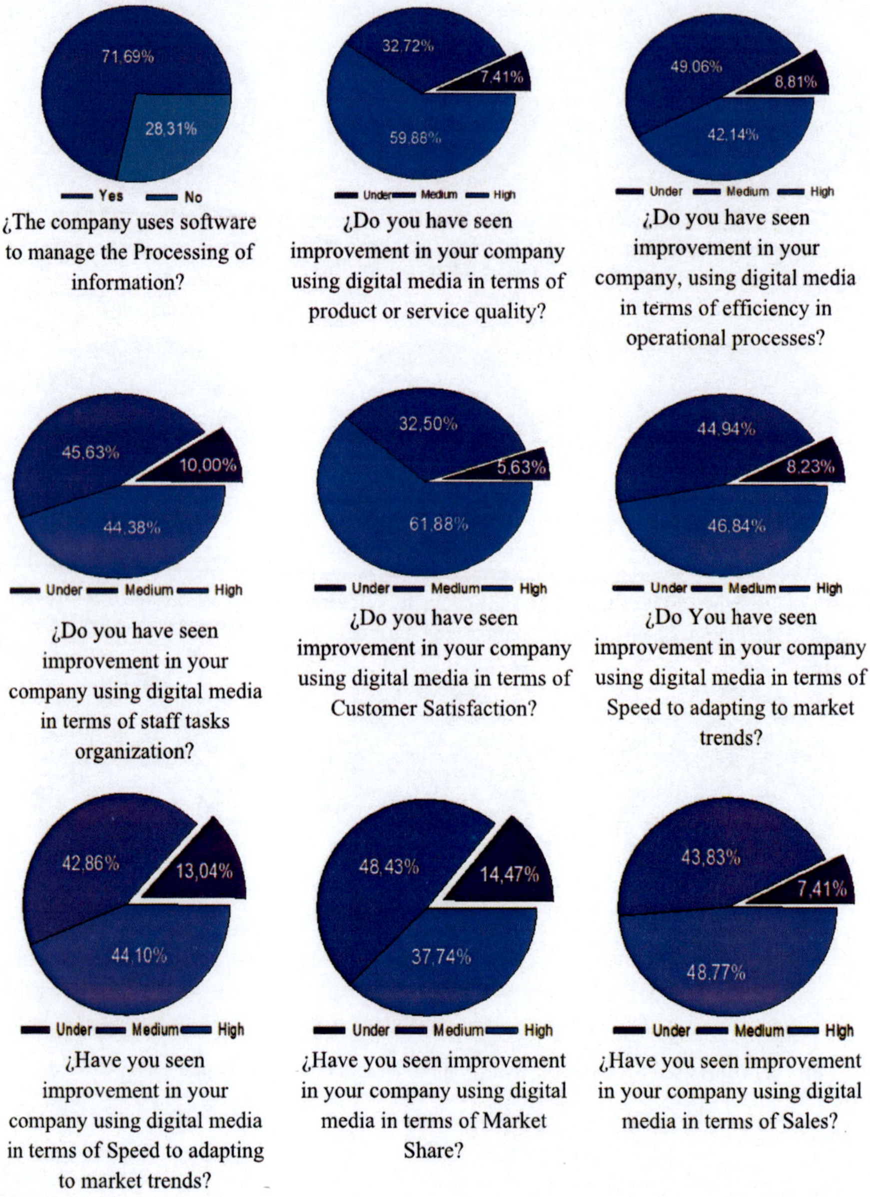

Fig. 5. Corporate improvements obtained by the management of digital networks in Ecuadorian MSMEs

From the digital tools, the most used are social networks such as Facebook, Instagram, LinkedIn, and Youtube, with 46%, however within the companies the use of a website, where the users feedback is left, it's poorly positioned, so it's necessary to relaunch them to improve the services provided by the companies, digital interaction with customers allows a better understanding of the customer, and therefore to design better marketing strategies [35] (Fig. 6).

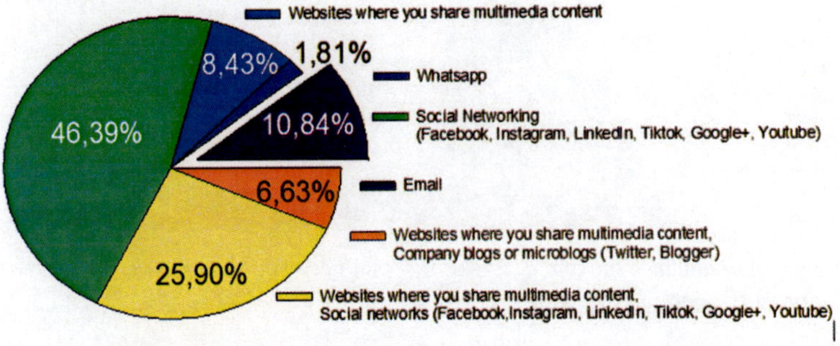

Fig. 6. Digital tools most used by Ecuadorian MSMEs

Finally, from the point of view of performance, digital tools have improved the profitability of Ecuadorian MSMEs by 46.20%, productivity by 40.51% and staff motivation and satisfaction by 44.03%. Several authors agree that this results in a better organizational climate [36, 37], which lead to better functioning of companies and the achievement of objectives and goals of the organization (Fig. 7).

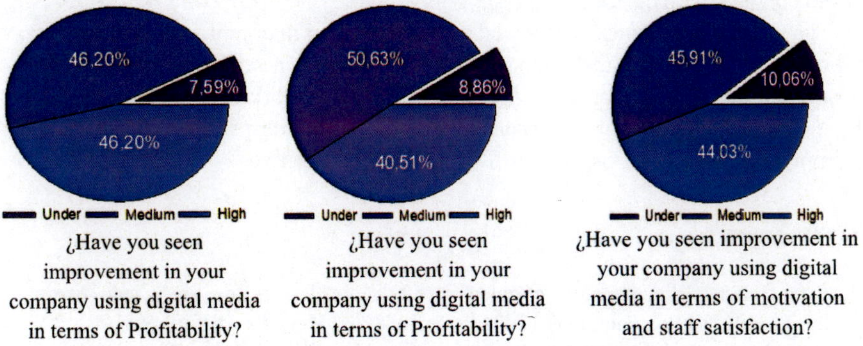

Fig. 7. Digital tools most used by Ecuadorian MSMEs

According to the results obtained, a proposal for the development of digital tools should be made with the objective of improving the performance and competitiveness of the companies, in order to achieve a better positioning of them, to this effect the proposal is shown in Fig. 8.

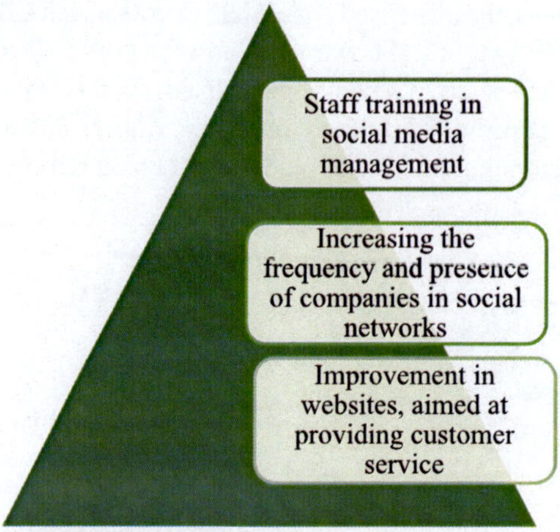

Fig. 8. Proposal to enhance the management of social networks to improve the competitiveness and operation of Ecuadorian MSMEs.

4 Conclusions

From the technological point of view, Ecuadorian MSMEs have the necessary IT infrastructure for the consolidation of an efficient digital platform; however, they lack the necessary applications and frequency of use for digital marketing to enable the positioning of companies through the capture of new market niches.

MSMEs have partially trained personnel in the management of digital tools, however, a continuous training program should be developed to increase staff capacities especially in the management of digital marketing and management of social networks, and to set online contacts with customers and suppliers.

The use of digital tools has allowed an improvement in organizational processes, visibility and positioning of the companies, as well as in economic performance, evaluated through profitability and productivity, as well as in the working environment; however, the standards are below the performance of other companies, so these indicators should be improved to make Ecuadorian MSMEs more competitive.

References

1. Cruz, N.K.E.: El comercio electrónico en el Ecuador. J. Sci. Res. Rev. Cienc. Invest. **2**(6), 29–32 (2017). https://doi.org/10.26910/issn.2528-8083vol2iss6.2017pp29-32
2. Pesántez-Calva, A.E., Romero-Correa, J.A., González-Illescas, M.L.: Comercio electrónico B2B como estrategia competitiva en el comercio internacional: desafíos para Ecuador. INNOVA Res. J. **5**(1), 72–93 (2020). https://doi.org/10.33890/innova.v5.n1.2020.1166
3. Fransi, E.C., Salla, Y.M., Ramón, N.D.: Cooperativismo 2.0: presencia en Internet y desarrollo del comercio electrónico en las cooperativas oleícolas de Cataluña. REVESCO. Rev. de Estud. Cooperativos, (124), 47–73 (2017). https://doi.org/10.5209/REVE.54926
4. Fernández-Portillo, A., Sánchez-Escobedo, M.C., Jiménez-Naranjo, H.V., Hernández-Mogollón, R.L.: importancia de la Innovación en el Comercio Electrónico. Universia Bus. Rev. **47**, 106–125 (2015)

5. Alvarado-Gastiaburo, Á.C., Vergara-Díaz, N.S.: El desafío del comercio electrónico en la economía del Ecuador. Polo del conocimiento **3**(1), 67–86 (2018). https://doi.org/10.23857/pc.v3i1%20%20Mon.681
6. Rodríguez, K.G., Ortiz, O.J., Quiroz, A.I., Parrales, M.L.: El e-commerce y las Mipymes en tiempos de Covid-19. Espacios, **41**(42), 100–118 (2020). http://w.revistaespacios.com/a20v41n42/a20v41n42p09.pdf
7. Suominen, K.: El comercio digital en América latina: ¿qué desafíos enfrentan las empresas y cómo superarlos?, serie comercio internacional, comisión económica para América latina y el Caribe (CEPAL), Santiago (2019). https://www.cepal.org/sites/default/files/publication/files/44976/S1900842_es.pdf
8. Figueroa González, E.G.: Rentabilidad y uso de comercio electrónico en las micro, pequeñas y medianas empresas del sector comercial. RIDE Revista Iberoamericana para La Investigación y el desarrollo educativo, **6**(11), 285–308 (2015). http://www.ride.org.mx/index.php/RIDE/article/view/177
9. Bricio Samaniego, K., Calle Mejía, J., Zambrano Paladines, M.: El marketing digital como herramienta en el desempeño laboral en el entorno ecuatoriano: estudio de caso de los egresados de la Universidad de Guayaquil. Revista Universidad y Sociedad, **10**(4), 103–109 (2018). http://scielo.sld.cu/scielo.php?pid=S2218-36202018000400103&script=sci_arttext&tlng=en
10. Horta, D.M.: Impactos socio-políticos de la globalización en América Latina. Revista Austral de Ciencias Sociales, (6), 19–34 (2017). http://revistas.uach.cl/index.php/racs/article/view/1126
11. Mendoza, P.M.P.: El Comercio Internacional y las Nuevas Tecnologías. Revista de Investigación Formativa: Innovación y Aplicaciones Técnico-Tecnológicas, **1**(1), 28–40 (2019). http://ojs.formacion.edu.ec/index.php/rei/article/view/145
12. Moro Vallina, M., Rodés Bach, A.: Marketing digital: Comercio y marketing. Ediciones Paraninfo, SA. Madrid, España (2014)
13. Rengel, R., Eyerkaufer, M.L., Schnorrenberger, D.: Framework estratégico para sustentação da vantagem competitiva em empresas de serviços contábeis: análise da complementaridade entre as perspectivas de Porter e das capacidades dinâmicas. Revista Catarinense da Ciência Contábil **18**, 1–19 (2019). https://doi.org/10.16930/2237-766220192839
14. Ordoñez, E., Barrera, H., Abril, J.: La Sociedad del conocimiento enfocado al capital intelectual empresarial. 593 Digital Publisher CEIT, **4**(6), 74–94 (2019). https://dialnet.unirioja.es/servlet/articulo?codigo=7144058
15. Álvarez, A.T.V., Vasco, G.H.T.: El marketing relacional y la rentabilidad en los servicios prestados por tres concesionarios automotrices en la provincia de Tungurahua del Ecuador. Revista UNIANDES Episteme, **4**(2), 135–149 (2017). https://dialnet.unirioja.es/servlet/articulo?codigo=6756364
16. Corrales, A.M.S., Coque, L.M.P., Brazales, Y.P.B.: El marketing digital y su influencia en la administración empresarial. Dominio de las Ciencias, **3**(4), 1161–1171 (2017). https://dialnet.unirioja.es/servlet/articulo?codigo=6325485
17. Gutiérrez-Leefmans, C., Nava-Rogel, R.M.: Mercadotecnia digital y las pequeñas y medianas empresas: revisión de la literatura. Enl@ ce: Revista Venezolana de Información, Tecnología y Conocimiento, **13**(1), 45–61 (2016). https://www.redalyc.org/pdf/823/82346016004.pdf
18. Barón, S., Fermín, S., Molina, E.: Estrategias de mercadeo basadas en el marketing digital orientadas a la captación de nuevos clientes de las PYMES. Caso de estudio: Grupo Inter Game, CA. Revista Enfoques, **2**(7), 164–174 (2018). https://revistaenfoques.org/index.php/revistaenfoques/article/view/39
19. Corrales, A.M.S., Coque, L.M.P., Brazales, Y.P.B.: El marketing digital y su influencia en la administración empresarial. Dominio de las Ciencias **3**(4), 1161–1171 (2017). https://doi.org/10.23857/dom.cien.pocaip.2017.3.4.jul.%2037-53

20. Chica, C.A.P., Álvarez, J.C.E., Zurita, I.N., Clerque, S.E.M.: El Marketing como estrategia de posicionamiento en empresas de servicios. Dominio de las Ciencias **5**(3), 784–802 (2019). https://doi.org/10.23857/dc.v5i3.988
21. Patiño, E.: Planeación estratégica de mercadeo y relaciones de transferencia en el ecosistema digital. Revista ESPACIOS, **39**(50) (2018). http://www.revistaespacios.com/a18v39n50/18395013.html
22. Arellano, J.M.V., Rubio, S.A.R., Lares, G.G.: Marketing digital y las finanzas de las pymes. Revista de Investigación en Tecnologías de la Información: RITI, **6**(12), 100–106 (2018). https://dialnet.unirioja.es/servlet/articulo?codigo=7107413
23. Pollán, M.E.M., Fernández, N.P.: Herramientas de Marketing Digital y competencia: una aproximación al estado de la cuestión. Atlantic Review of Economics: Revista Atlántica de Economía, **3**(3), 3 (2019). https://dialnet.unirioja.es/servlet/articulo?codigo=7407247
24. Pérez, M.I.M.I.R., Carreras, A.B.L., Bustamante, M.J.A.H.: Uso e impacto de las redes sociales en las estrategias de marketing de las PyME´ s. Revista de Investigación Académica Sin Frontera: División de Ciencias Económicas y Sociales, (19) (2018). https://revistainvestigacionacademicasinfrontera.unison.mx/index.php/RDIASF/article/view/47
25. Castro, P.E.G., et al.: Procesos de reclutamiento y las redes sociales/Recruitment and social networks. RIDE Revista Iberoamericana para la Investigación y el Desarrollo Educativo, **6**(12), 60–76 (2016). http://ride.org.mx/index.php/RIDE/article/view/226
26. Canossa Montes de Oca, H.: Sistemas de información de marketing en la planificación estratégica de la era digital. Redmarka. Revista de Marketing Aplicado, **23**(2), 43–60 (2019). https://doi.org/10.17979/redma.2019.23.2.5708
27. Elorriaga-Illera, A., Usin-Enales, S., García-Azpuru, A.: Evolución del community manager en las pymes del País Vasco y Navarra. El profesional de la información (EPI), **27**(3), 651–658 (2018). http://www.elprofesionaldelainformacion.com/contenidos/2018/may/18.html
28. Aparicio-Gómez, W.O., Aparicio-Gómez, C.A., Niño, J.F.H.: El aprendizaje móvil (m-learning) como herramienta formativa para la empresa. Revista Internacional De Pedagogía E Innovación Educativa, **1**(1), 69–102 (2021). https://editic.net/ripie/index.php/ripie/article/view/27
29. Armijos Delgado, N.M.: Marketing digital: una herramienta para potenciar y promocionar las Mypimes en el ámbito internacional. INNOVA Res. J. **4**(1), 1–8 (2019). https://doi.org/10.33890/innova.v4.n1.2019.764
30. Prieto, M.F.: La prestación de servicios a través de plataformas digitales. El caso Deliveroo. Temas laborales: Revista andaluza de trabajo y bienestar social, (151), 193–217 (2020). https://dialnet.unirioja.es/servlet/articulo?codigo=7464150
31. Torres, R., Rivera, J., Cabarcas, R., Castro, Y.: La efectividad del uso del marketing digital como estrategia para el posicionamiento de las pymes para el sector comercio en la ciudad de Barranquilla. Investigación y Desarrollo en TIC, **8**(2), 3–5 (2017). http://revistas.unisimon.edu.co/index.php/identic/article/view/2947
32. Perdigón Llanes, R., Viltres Sala, H., Madrigal Leiva, I.R.: Estrategias de comercio electrónico y marketing digital para pequeñas y medianas empresas. Revista Cubana de Ciencias Informáticas, **12**(3), 192–208 (2018). http://scielo.sld.cu/scielo.php?pid=S2227-18992018000300014&script=sci_arttext&tlng=pt
33. Laroussi, A.A.: Análisis de la relación entre la comunicación integrada de marketing y la satisfacción del cliente. El caso Mcdonald's. Revista de Estudios Empresariales. Segunda Época, (2), 32–52 (2020). https://150.214.170.182/index.php/REE/article/view/5839
34. Sosa, R.M.S., Torres, M.D.C.N., Aparicio, C.G.M.: Marketing digital para los consumidores del siglo XXI. Hitos deficiencias económico administrativas, **65**, 24–37 (2017). https://core.ac.uk/download/pdf/236397755.pdf

35. Quiñonez, C.J.M., Rodríguez, T.C.C.: Redes sociales factor predominante en la fidelización del cliente. Dominio de las Ciencias **3**(4), 519–532 (2017). https://doi.org/10.23857/dom.cien.pocaip.2017.3.n%C3%BAm.%204.oct
36. Henao, A.E.P.: La comunicación organizacional en la gestión empresarial: retos y oportunidades en el escenario digital. Revista GEON: Gestión-Organización-Negocios. **7**(1), 9–25 (2020). https://dialnet.unirioja.es/servlet/articulo?codigo=7828548
37. Muñoz, E.M., Gracia, T.J.H., Torres-Flórez, D., Velázquez, M.D.R.G., Arango, O.M.: La realidad comunicacional, liderazgo y clima laboral de la pequeña empresa industrial textil en el Estado de Hidalgo. Revista GEON (Gestión, Organizaciones Y Negocios) **5**(1), 76–90 (2018). https://doi.org/10.22579/23463910.15

Strategic Plan Deployment Model as a Strategy Management System

Amanda Martinez[1](✉) and Kevin Alvarez[2]

[1] Instituto Tecnológico Superior Universitario España, ISTE, 180109 Ambato, Ecuador
amanda.martinez@iste.edu.ec

[2] Ecuador Universidad Tecnica de Ambato, UTA, 180103 Ambato, Ecuador

Abstract. The work proposes a business case analysis, under the application of the BSC and Kaizen model, relating a model of deployment of strategic plans based on relationships between strategies and projects. The model begins with the formulation of strategies, converts them and assigns them into individual work plans by area that are finally implemented through levels of management. Three types of levels are defined: managerial projects, internal projects, and kaizen projects, which are related both hierarchically and hierarchically; that is to say, the results of each one have a direct influence on all levels until the achievement of the strategies is achieve. A case study is presented in a manufacturing company that begins with the balanced scorecard and the relationships between strategies, a strategy map that expands the strategic objectives in the projects. The strategy implementation map is mentioned, which displays the strategies at managerial levels and finally the performance evaluation. The results allow demonstrating the successful application of the model since it is possible to relate strategies, work plans, and projects for effective implementation.

Keywords: Strategic plan · Strategic management · Operational efficiency · Work plans · Strategic performance

1 Introduction

Conceptualizations of strategic management focus on integrating strategic planning and performance measurement. In other words, while strategic planning helps identify strategies for change it does not necessarily induce the implementation of these strategies. Strategic management 'is concerned with ensuring that strategy is implemented effectively' [12].

However, leaders my fail to link their strategic planning efforts to other critical decision making processes. It is observed that, having a strategic planning system in place is not a guarantee that an organization will achieve its strategic goals and objectives as planned [11].

Indeed, only 18 percent of organizations report being highly effective with strategic initiatives. There are two fundamental reasons for this poor effectiveness of strategic initiatives: 1) the strategy itself may be flawed, and 2) there is not a clear bridge from strategic planning to implementation [2].

M. Zambrano Vizuete et al. (Eds.): CI3 2021, LNNS 511, pp. 542–554, 2022.
https://doi.org/10.1007/978-3-031-11438-0_43

Leaders need to balance between the structural dimension (hierarchy, budgets, plans, control, and procedures) and personal dimension (human resources). These dimensions must not be in conflict and should be balanced in order to achieve an improvement in overall performance [7, 15].

The performance of any organization (for-profit, non-for-profit, micro, macro, etc.) is becoming an annoyed headache for stakeholders [21]. Different disciplines, management approaches, and studies explored varied reasons and justifications for success or/and failure of organizations not being able to achieve their ultimate strategic objectives, or have not fulfilled, at least, their required performance [1].

Strategic planning and management increase operational efficiency by providing a clear roadmap that facilitates alignment of activities and resources to realize well set and realistic goals leading to high efficiency in business operations [11]. A company achieves a superior profitability in its industry when achieving higher prices or lower costs than its competitors; this is achieved through the operative effectiveness or the strategic positioning [6].

Strategy formulation is a crucial dimension of organizational agility building. The definition of the strategy allows the organization to undertake a course of action based on a clear understanding of 'why' acting (rationale) and 'what' to do, along with the assessment of the organizational preparedness to act [24]. Strategies need to reflect current conditions within the organization and its environment while the development of objectives and annual work plans require both management and staff input [11].

An efficient administrative model in an organization requires an operation with a systemic approach where people or work groups respond to a structure and planning designed for their reality [18]. Efficiency in process implementation to achieve goals plays a pivotal role for a significant number of companies [26]. In this context, it is important to investigate the performativity of strategy by asking how strategy 'works in practice', focusing on the ways in which strategy has implications for organizational reality [14].

Acknowledging the difficulty in sharing strategic plans with staff, the strategy management system helps to show how each staff member's job contributes to the overall objectives of the organization. The strategies developed in a functional or operating unit should be aligned with that of the organization [25].

Organizational goal systems comprise a number of interrelated organizational goals, which are pursued in the long term or for a certain period [3]. Besides, strategic performance measures (SPM) are considered to be important for translating strategy into measurable objectives and, if appropriately designed and communicated, can facilitate strategy implementation, align management deci sions and actions with strategic goals, and enhance organizational performance. plans". SPM in such systems can help organizations set and achieve strategic objectives, align individual behaviours and attitudes with strategic objectives, and, ultimately, enhance organizational performance [16].

Performance management is a process of two-way communication between the team staff and its managers on the issue of performance. In this process, it helps set the team staff' performance goals through the communication foundation, then improve the team staff's performance ability by counseling the defects and deficiencies toward a higher level of goal [4, 22].

The goal of this research is to proposes a Strategic Plan Deployment Model based on relationships between strategies and projects. The model integrates the strategic plan with the execution of projects according to hierarchical and heterarchical levels for the effective implementation of the strategy. The model that is presented applies to manufacturing industries, however it can be applied in service companies with their respective adjustments.

The rest of the paper is organized as follows: Sect. 1 begins with a brief synopsis of the principles applied in this study followed by a description of the methodology used. Section 2 describes the strategy mapping, strategy implementation and performance evaluation applied to a case study. Section 3 shows the discussion of the results. Finally, Sect. 4 rounds up the paper with the conclusions.

2 Material and Methods

2.1 Strategic Plan

The strategic plan is an integrated document in the business plan that reflects the economic-financial, strategic and organizational planning that a company has to address its objectives and achieve its future mission.

Throughout its development, the strategic plan points out the bases for the operation of the company in line with the achievement of future objectives, even though it does not know what the future will be. Besides, it will define what actions will have to be carried out at the business level to be able to face the challenges that are emerging and finally achieve the objectives previously set. Strategic thinking is a way to achieve a position of greater advantage as it allows creating favorable scenarios for better performance [19].

Strategic planning is the process of determining the optimal future for an organization and the changes required to achieve the future. Organizations that engage in a through and transparent strategic planning process can expect the following outcomes: a shared vision of the future among the board and the staff, a common understanding of the mission, agreement on major goals for the next three to five years and how to achieve them [13].

2.2 Strategic Management

Strategic management is the extension of strategic planning, which incorporates work plams, budgeting, information and control systems [20].

The strategic management system is corporate in nature and can express the comprehensive and holistic nature of strategic planning, in order to raise awareness of the need for plans from all units and from the various functional areas (marketing plans, human resources plans, new product development plans) are incorporated into a general business plan [20].

A strategy allows an organization to configure its resources and capabilities to meet the needs of the environment to achieve competitive advantage. The process of undertaking a strategy is strategic management. All organizations set goals they want to achieve. Strategic management is about analysing the situation facing the firm.

This analysis will allow managers to formulate strategies for dealing with the situation or challenge facing the company. A point worth noting is that these elements are co-dependent; that is, in formulating a strategy and organization must also consider how that strategy will be implemented [10]. Table 1 shows the three interdependent elements.

Table 1. Strategy Management Elements adapted from [10]

Elements	Description
Strategy analysis	It allows managers to evaluate how well the company is positioned
Strategy formulation	Is a process of continuously asking questions for select the best strategy
Strategy implementation	Strategies need to be communicated, understood, and properly coordinated with stakeholders inside and outside the organization

2.3 Deployment Model

The research proposes a strategic plan deployment model as a system in which strategic objectives are integrated with projects whose activities are directly related. From the definition of the strategic plan with a two-year planning horizon, management meetings are held in which a committee is formed to review the strategies and their implementation.

When the strategic plan is understood by the managers of each area in the organization, they are delegated to carry out work plans that must be linked to the operational plan and the strategic objectives; i.e., the projects proposed must necessarily contribute to a strategy.

For the budgeting, the work plans and operations plan must be completed, since the necessary expenditure and investment needs will be defined from them.

The proposed model is based on the interdependence between strategy for-mulation and strategy implementation based on projects defined in the work plans to be deployed in a binding manner at all hierarchical levels, these are disseminated in all functional areas and internal projects and kaizen projects are proposed that align to the company's strategies and support the work plans. The model shown in Fig. 1 presents the relationships given from strategic planning to performance evaluation that is formed from the internal projects and kaizen projects that support the work plans, and these in turn the strategic objectives.

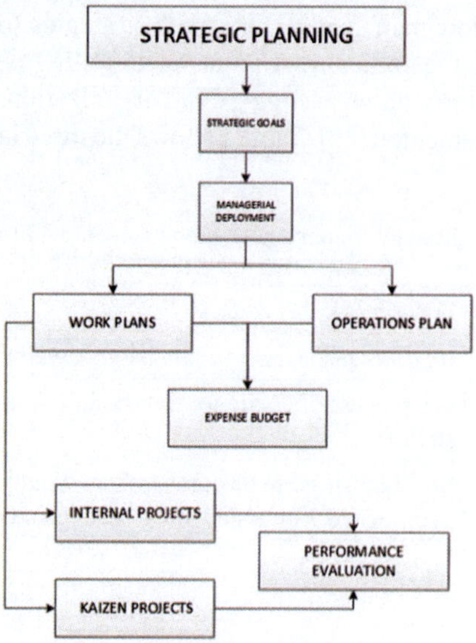

Fig. 1. Strategic deployment model

2.4 Strategic Deployment Levels

The implementation of the strategic objectives is achieved from the assignment of projects by levels within the strategy management system. Each level corresponds to a degree of responsibility (hierarchical) that will be linked to its later and previous level; this allows the relationship to be given both vertically and horizontally.

In Fig. 2 it is observed that the implementation of the strategy begins with the objectives, continues to the managers, follows the headquarters and super visors and reaches the workforce. The flow of project allocation and the flow of information is generated in both directions; that is, a strategy will be achieved only if there is the interaction of all levels in the company. The definition of each level proposed in this research is described as follows:

1. Level 1. Management Projects: These are projects that are entrusted to the manager of the area and that appear in their work plan, these require analysis and financial evaluation to determine the added value that will impact the assigned objectives.
2. Level 2. Internal Projects: These are projects run by plant managers or supervisors and are part of your individual work plan.
3. Level 3. Kaizen Projects: These are projects with multidisciplinary work teams in which the workforce takes a leading role.

To ensure that internal projects and activities are carried out using the same method, within the strategic plan deployment model are defined eight continuous steps that must be respected to achieve the stated objectives and ensure that the result of each is aligned to the strategy. All the levels shown in Fig. 2 follow a project flow that is indicated in Table 2.

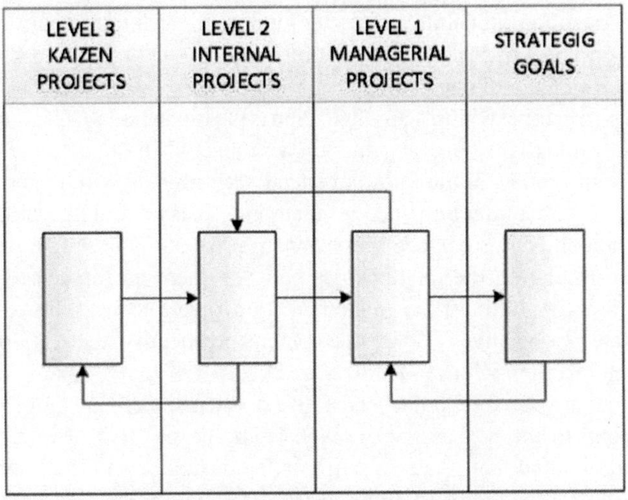

Fig. 2. Levels of strategy implementation

Table 2. Standardization to implement strategic projects

Activity	Description
Opportunity selection	Identify the opportunity for improvement and describe it
Clarfy/Subdivide	Obtain data that identifies the problem
Root Causes	Problem analysis
Plan	Raise the activities, responsible and compliance dates
Do	Carry out the proposed activities
Check	Verify the results achieved
Act	Standardize and correct
Improvement	Propose new opportunities for improvement

2.5 Strategic Mapping

This investigation presents the application of the proposed model in a company in the footwear sector in Ecuador. This sector must have strategies that are correctly implemented and allow it to achieve competitive advantages in the face of the massive inflow of foreign products that reduce global income. In this context, the research uses the principle of defining strategy from strategic maps, specifically the Balance Scorecard tool [5], whose approach is a short and complete method of developing a strategic plan. It is a system that allows to manage the strategy in objectives and to evaluate the business results.

This information is inserted in the Strategic Plan where the SWOT analysis is previously carried out, pointing out, for example, the internal aspects such as business, commercial, operational strengths and weaknesses, and in external aspects threats such as changes in consumption habits, consumer buying, the economic political context of the country, economies of scale and in opportunities: the incursion into new markets or new manufacturing technologies.

The initial perspective within a Strategic Plan is the Financial; that is, to have financial benefits, to add value to the company and to maintain control of costs and expenses. As a result, the driving forces of the future financial performance will be generated within the perspectives of Customer Satisfaction, Internal Processes and Learning Growth.

In Fig. 3 the company's balance scorecard from the study is shown, in which from the strategies formulation the relationships between them are described. This makes it possible to identify initially the implementation routes to be followed in order to achieve the strategic objectives [17]. In the financial perspective, there is 1 main strategic objective aligned to the increase of profits and growth of the company, from which 3 related objectives are derived. In the perspective of clients, there are 5 objectives related to each other and to the next financial level. In the perspective of internal processes 3 objectives are defined that relate as much as the customer and the financial level. It is important that this relationship exists because the internal processes must seek the operational and financial efficiency that allows an adequate control production cost.

Finally, in the perspective of learning and growth, 3 objectives are defined related to improving the training of workers. Defining the strategies in each perspective is necessary the deployment of them both at the corporate and functional level in which it is important teamwork and participation of all hierarchical level.

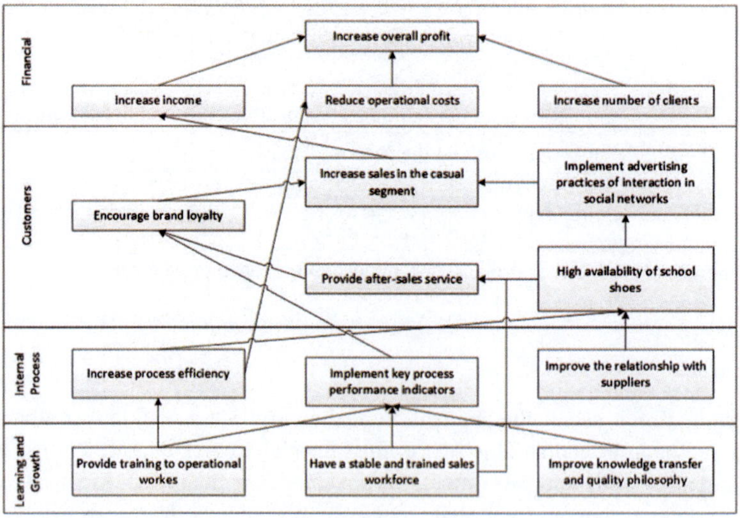

Fig. 3. Balance scorecard

2.6 Deployment Levels and Relationship

Once the strategic objectives are defined, and from the existing relationships between perspectives [8], the strategic deployment map is drawn up in Fig. 4, whose head will be

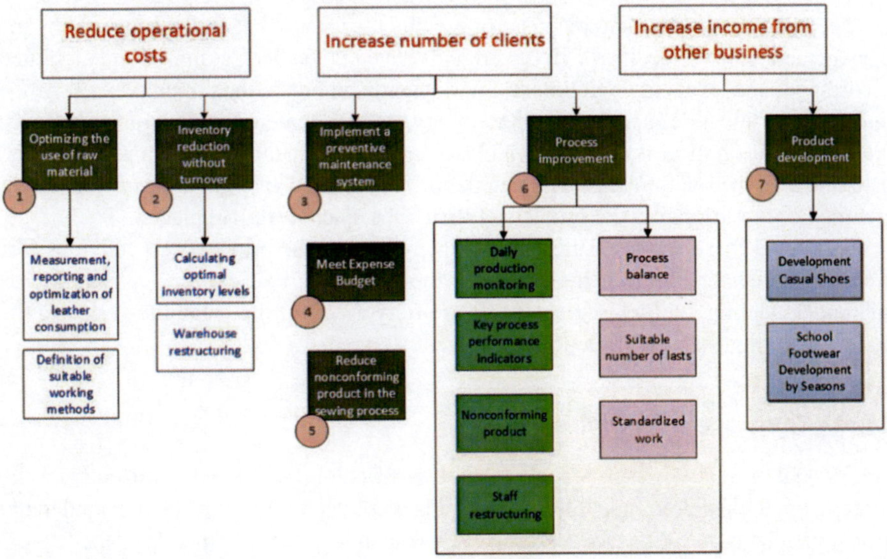

Fig. 4. Strategic map

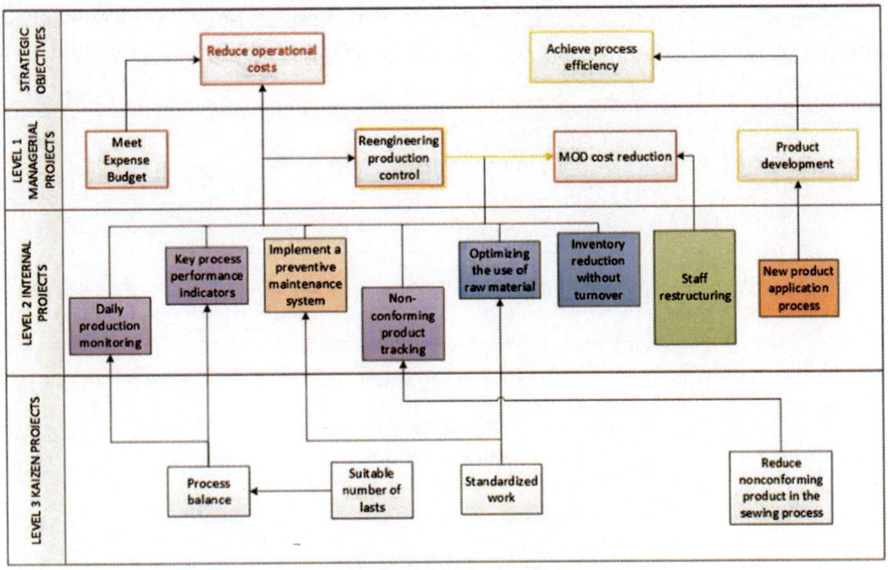

Fig. 5. Strategy implementation map

the financial objectives and projects aligned to the achievement of these objectives are proposed downwards. This research shows the strategic deployment in the production area because of its importance in measuring the product cost and the effectiveness in achieving the strategies.

The strategic deployment map is proposed by the area manager and approved by the management committee, in which the projects have not yet been assigned and classified according to the levels of deployment. Subsequently, in Fig. 5, the projects are classified into the three levels defined in the methodology of this research. For each of the projects, the assignment analysis is carried out and they are located on the strategy implementation map in which the relationships between them are identified and only those whose result is directly proportional to the projects of their subsequent level are placed.

Following these steps, assures the company that all the projects and activities that are carried out aim at the achievement of the strategies, being a tool of easy access and understanding, all hierarchical levels are fully aware of the value they bring to the company during the execution of its activities.

2.7 Performance Evaluation

The evaluation matrix describes in each row a different project and is raised from the strategy implementation map, the relationships presented in this must be reflected in the matrix and the monitoring of each project corresponds to the responsible of the level.

The performance evaluation of each project or strategic initiative is carried out by means of a matrix in Table 3 in which the perspective [12], strategic objective and type of project or level to which it belongs are identified.

The matrix performs the quantitative analysis of each project by defining a monitoring indicator, its formula, unit of measure and trend to control (see Fig. 6). In this case, all

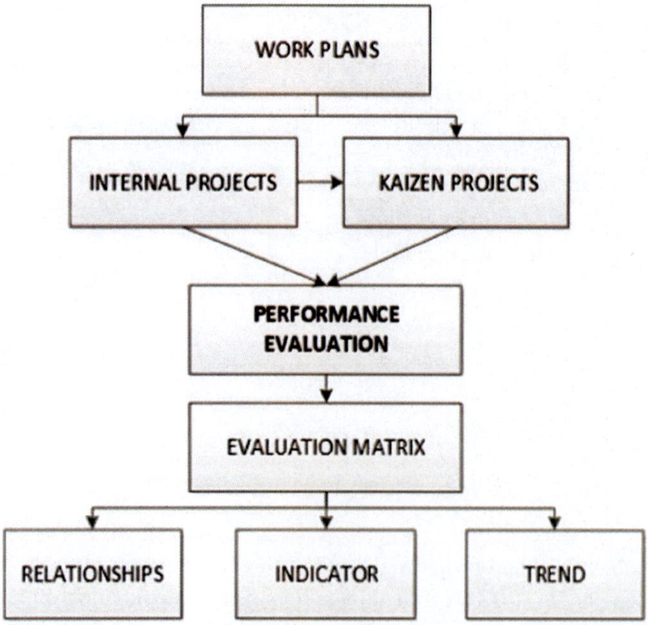

Fig. 6. Evaluation map

levels follow the same method of executing activities and the results are periodically evaluated in a meeting with participation from the managers to the workforce.

The participation of all levels in the evaluation matrix allows both the vertical (hierarchical) and horizontal (heterarchical) implementation of the strategy. This ensures that the execution of the activities is understood and supported by all, provides clarity on the impact of the decisions that are made and increases work empowerment.

Table 3 shows how the correct implementation of the strategies is ensured by keeping under constant analysis the relationships between perspectives, objectives, initiatives and levels; that is, the model proposed in this research is based on the continuous search for relationships between strategies from the scorecard balance, strategic map, strategy implementation map and evaluation matrix.

Table 3. Evaluation matrix

Perspective	Strategic objective	Strategic initiative	Type	Unit	Indicator	Formula	Achieved value	Trend
Financial and process	Reduce operational costs	Optimizing the use of raw material	Level 2	%	Leather consumption	Current/previous	93%	Decreasing
Financial and process	Reduce operational costs	Inventory reduction without turnover	Level 2	$	Dollars of inventory	Dollar amount	$5238.12	Decreasing
Financial and process	Reduce operational costs	Implement a preventive maintenance system	level 2	Min	Unplanned stoppages	Sum of time	30 min	Decreasing
Financial and process	Reduce operational costs	Meet Expense Budget	Level 1	%	Meet expense budget	total spent/total budgeted	98%	Decreasing
Financial and process	Reduce operational costs	Reduce nonconforming product in the sewing process	Level 3	%	Quality index	Product OK/total production	3,2%	Decreasing
Financial and process	Reduce operational costs an achieve process efficiency	Reengineering production control	Level 1	$	Product cost	Cost	$8,12	Decreasing
Financial and process	Achieve process efficiency	Product development	Level 1	#	Products developed	Quantity	1	Increasing

3 Discussion

The implementation of the strategic plan is vital to make the strategies effective for the company and to achieve the desired competitive advantage [9, 23]. The results of the research through the deployment model based on work plans by functional areas that are linked to internal projects and kaizen projects show that the implementation can be given from identifying the existing relationships between strategies and projects.

Maintaining a system of distribution and assignment of projects by levels in which relationships and alignment to the strategy are verified, allowed the execution of activities to be done with the prior analysis of the impact it will have on the company's strategic plan. This system allowed the linking and understanding of the work teams towards the strategy and work empowerment. The strategic plan deployment model allows to relate all the strategies since the scorecard balance sheet survey with a first analysis of the implementation routes, followed by the strategy map adapted to each functional area and the strategy implementation map that assigns strategic initiatives or projects to the three levels defined: managerial projects, internal projects and kaizen projects.

It is observed in Fig. 7 that the model allowed to achieve with 2 strategic objectives, 4 managerial projects, 8 internal projects and 4 kaizen projects, various relationships between levels, as follows: 5 relations between strategic objetives and managerial projects, 15 relations between managerial projects and internal projects, 7 relations between internal projects and kaizen projects, i.e., as the levels advance, the relationships are greater than the number of projects to be carried out. This system ensures that projects are aligned with the organization's strategies.

It is found that one of the limitations of the research is that the analysis of the model is carried out in manufacturing companies, so the evaluation of the model in service companies is proposed as future research. In addition, a deployment model can be proposed through horizontal levels and investigated on the relationship of the strategy implementation with the performance evaluation of the human resources department.

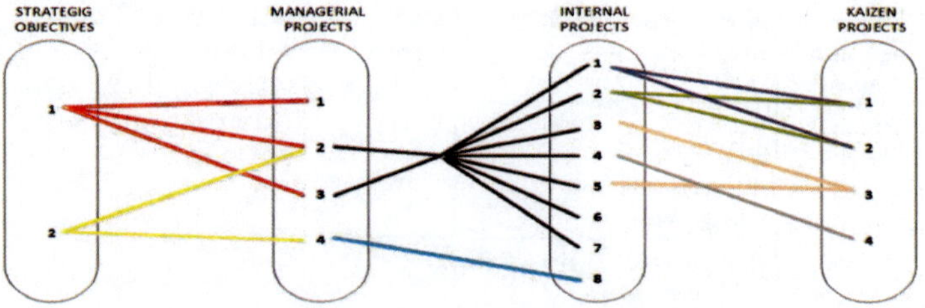

Fig. 7. Relations on strategy implementation

4 Conclusions

The strategic plan may be correctly defined, but if there is no deployment methodology with clear guidelines in the organization, the desired results are not achieved satisfactorily. It is important that strategic objectives are understood by staff and relate directly through organizational levels to individual work plans and objectives. Information should flow both ways and the necessary tools should be in place to document the results and impact of each project on the strategic objective.

The research is carried out in a manufacturing company and a case study is obtained with the application of the strategic plan deployment model by which it was possible to obtain different relationships between objectives, work plans and projects. The model allows to integrate strategies and projects from the definition of the BSC, the deployment of strategy implementation map and the performance evaluation with which the objective of this research is achieved. In this sense, traditional approaches focus on the analysis and formulation of strategies, while the focus of the model is towards the implementation of these strategies. The definition of levels allows the proper distribution and allocation of projects and the involvement of staff, this makes the strategic plan be understood, implemented and evaluated.

Acknowledgment. The authors would like to thanks to Instituto Tecnológico Superior Universitario España for the support, encouragement and facilities provided to undertake this research.

References

1. Ali, A.A.: Strategic planning–organizational performance relationship: Perspectives of previous studies and literature review. Int. J. Healthcare Manag. **11**, 8–24 (2018)
2. Aucoin, B.M.: Missing pieces in strategic planning and execution: the talent development perspective. IEEE Eng. Manage. Rev. **46**(4), 26–31 (2018). https://doi.org/10.1109/EMR.2018.2878691
3. Bergmann, A., Strecker, S.: Toward modeling strategic plans: requirements language design considerations. In: 2018 IEEE 20th Conference on Business Informatics (CBI), vol. 02, pp. 11–20 (2018). https://doi.org/10.1109/CBI.2018.10042
4. Bryson, J., Crosby, B.: Understanding strategic planning and the formulation and implementation of strategic plans as a way of knowing: the contributions of actornetwork theory. Int. Public Manag. J. **12**, 172–207 (2009)
5. Falcón, V.V., Jácome, D.L.: El balanced scorecard como herramienta de gestión organizacional. Revista Científica Ecociencia **6**, 1–24 (2019)
6. Fuertes, G., Alfaro, M., Vargas, M., Gutierrez, S., Ternero, R., Sabattin, J.: Conceptual framework for the strategic management: a literature review descriptive. J. Eng. 6253013 (2020)
7. Gorchakova, E.N., Sazonova, V.L.: Problems of corporate management systems integration. In: 2017 International Conference "Quality Management, Transport and Information Security, Information Technologies" (IT QM IS), pp. 418–421 (2017). https://doi.org/10.1109/ITMQIS.2017.8085850
8. Cortés, G., Martell, A.G.C., A.V.A.H.M.A.T.R.E.R.: Aplicación de la planeación estratégica y fundamento normativo en una empresa procesadora de carne. Revista Mexicana de Agronegocios (2016). https://www.redalyc.org/articulo.oa?id=14149188003
9. Guiliany, J.G., Durán, S.E., Portela, E.C., Pulido, R.A.P., Cali, E.G., Marcano, A.I.P.: Proceso de planificación estratégica: Etapas ejecutadas en pequeñas y medianas empresas para optimizar la competitividad (2017)
10. Henry, A.: Understanding Strategic Management. Oxford University Press, Oxford (2018)
11. Kabeyi, M.J.B.: Organizational strategic planning, implementation and evaluation with analysis of challenges and benefits for profit and nonprofit organizations. Int. J. Appl. Res. **5**, 27–32 (2019)

12. Kools, M., George, B.: Debate: the learning organization—a key construct linking strategic planning and strategic management. Public Money Manag. 40, 262–264 (2020)
13. Llivisaca, J., Jadan, D., Guamán, R., Arcentales-Carrión, R., Peña, M., Siguenza- Guzma, L.: Key performance indicators for the supply chain in small and mediumsized enterprises based on balance score card. TEST Eng. Manag. **83**, 25933–25945 (2020)
14. Lord, G., Markert, K.: The Manual of Strategic Planning for Cultural Organizations: A Guide for Museums, Performing Arts, Science Centers, Public Gardens, Heritage Sites, Libraries, Archives and Zoos. Rowman & Littlefield Publishers (2017)
15. Merkus, S., Willems, T., Veenswijk, M.: Strategy implementation as performative practice: Reshaping organization into alignment with strategy. Organ. Manag. J. **16**, 1–16 (2019). https://doi.org/10.1080/15416518.2019.1611403
16. Perrin, J.M.: Strategic planning from the bottom up: a unit strategic plan that pushes change. J. Libr. Adm. **57**, 712–722 (2017)
17. Pollanen, R., Abdel-Maksoud, A., Elbanna, S., Mahama, H.: Relationships between strategic performance measures, strategic decision-making, and organizational performance: empirical evidence from canadian public organizations. Public Manag. Rev. **19**, 725–746 (2017)
18. Barrizonte, J.L.Q., Bastida, E.J.L., Alonso, K.R.: Planeación estratégica con enfoque prospectivo para la editorial Universo Sur. Revista Universidad y Sociedad **7**, 160–167 (2015). http://scielo.sld.cu/scielo.php?script=sci_arttext&pid=S2218-36202015000300024&nrm=iso
19. Rodriguez, E.: Administrative model for the organization of agroindustrial microenterprises at universidad tecnica del norte. TEST Eng. Manag. **83** (2020)
20. Rodríguez, E., e libro, C.: Planificación estratégica: fundamentos y herramientas de actuación. Elibro Catedra, Editorial Brujas (2016)
21. Romero, A.: Dirección y planificación estratégicas en las empresas y organizaciones: un manual práctico para elaborar un plan estratégico. Díaz de Santos (2004)
22. Suarez, L., Recalde, M., Páez, J., Quispe, J., León, R.: Economic growth and capital market development in ecuador. TEST Engineering Management 83 (2020)
23. Ting, W.: Applying the balanced score card in the team strategic performance management. In: 2011 2nd International Conference on Artificial Intelligence, Management Science and Electronic Commerce (AIMSEC), pp. 990–993 (2011). https://doi.org/10.1109/AIMSEC.2011.6010534
24. Maldonado, G.V., Erazo, M.A.: El reto de la planificación estratégica en las pymes. Revista Publicando **3**(8), 335–344 (ago 2016). https://revistapublicando.org/revista/index.php/crv/article/view/245
25. Wahyono, W.: A conceptual framework of strategy, action and performance dimensions of organizational agility development. Ind. Commercial Training **50** (2018)
26. Wehde, M.: Practical application of a balanced scorecard as a strategic management system. In: 2018 IEEE Technology and Engineering Management Conference (TEMSCON), pp. 1–7 (2018). https://doi.org/10.1109/TEMSCON.2018.8488423

Supervised Machine Learning Algorithms for LOS/NLOS Classification in Ultra-Wide-Band Wireless Channel

Juan Minango[1(✉)], Wladimir Paredes-Parada[1], and Marcelo Zambrano[1,2]

[1] Instituto Superior Universitário Rumiñahui, Sangolqui, Ecuador
juancarlos.minango@ister.edu.ec
[2] Universidad Técnica del Norte, Ibarra, Ecuador

Abstract. Ultra-Wide-Band (UWB) channels have attracted attention due to its potential in constructing accurate indoor positioning system (IPS) such as residential environment. Although promising, UWB channels experience some problems inherent to indoor environments particularly when in both the transmitting and the receiving antenna there is no a direct line of sight which producing an estimation error on the positioning accuracy. Thus, in this paper, it is evaluated and compared the application of different machine learning algorithms (MLA) to UWB channel classification in order to detect if a UWB channel is a line of sight (LOS) or a not line of sight (NLOS) channel. The MLA are capable to work on different characteristics which improves channel classification. The attained results prove the validity of the different MLA in order to classify between LOS/NLOS channels where the best performing classifier was obtained by the Gradient Boosting on the set of simulated UWB channel realizations.

Keywords: Ultra-wide-band · Machine learning · Channel classification · LOS · NLOS

1 Introduction

Ultra-Wide-Band (UWB) channels have attained ample attention in wireless communication in order to develop Indoor Positioning System (IPS) with high accuracy [1] once in base-band, the pulse of UWB channels have a high time resolution which produces a better accurate distance measurement. Besides, since the signal spectrum of UWB channels consist of multiple sub-bands, it is possible to increase the data rate and wall-penetration abilities [2]. Nevertheless, indoor environments such as residential, are consist of various objects such as desks, computers, furniture, walls and so on. Thus, the direct LOS from the transmitting to receiving antenna for UWB signals might be obstructed and only reflected signals, that is None-Line-Of-Sight (NLOS), could achieve the receiving antenna.

M. Zambrano Vizuete et al. (Eds.): CI3 2021, LNNS 511, pp. 555–565, 2022.
https://doi.org/10.1007/978-3-031-11438-0_44

IPS achieves its best performance under LOS channels with an positioning accuracy can attain exact levels [3]. On the other hand, under NLOS propagation conditions will produce the measured distance between the transmitting and receiving antenna to be larger than the current distance, thus, a decreasing of position accuracy, once the reflected signals received will produce additional ranging bias in the position measurements. Thus, it results evident that the problem of NLOS channels condition is important for the impact of IPS resolution [4]. In this context, it is required to identify and classify between NLOS and LOS channels in order to avoid NLOS channels that influences IPS's performance [5,6].

There are plenty techniques to detect NLOS propagation conditions and deal with this trouble. Statistical evaluation of the estimated position measurements under LOS/NLOS propagation conditions is analysed in [7,8], while other approaches considered the channel impulse response (CIR) [9,10], where the principal idea behind these approaches is that the energy of the first multi-path component is significantly greater than the energy of the delayed multi-path components under LOS conditions. This distinction tends to shorten under NLOS propagation condition [11]. Moreover, NLOS detection can be implemented also at a higher logic level that uses the range estimation where map of the indoor environment is required.

Supervised machine learning algorithms (MLA) have found its application in wireless networks. These MLA have been considered for jamming detectors [2] and antenna selection [8]. On the other hand, there are several MLA available, such as Random Forest, Gradient Boosting, Support Vector Machine (SVM), Gaussian Naive Bayes, Decision Tree, KNearest Neighbor, and so on for classification tasks; where these MLA have their advantages and disadvantage. In [12], a SVM classifier was used to mitigate NLOS channels. However, this SVM classifier was not compare with the others MLA.

Motivated by the employ of MLA for channel classification, we compare the performance of MLA to classify UWB channels under LOS and NLOS propagation condition due to the ability of MLA to perform over different features and to determine patterns which will aid into the classifier process. This paper is organized as follows. Section 2 introduces the UWB channel model used into this paper during the experiment together with the channel realization simulator employed for the generation of LOS and NLOS propagation condition. Section 3 explains the LOS and NLOS channel classification methodology for MLA which includes the channel data preparation and training phase. Performance and comparison evaluations of the different MLA considered in this paper are shown in Sect. 4. Section 5 presents the conclusions.

2 Channel Model and Simulation Environment

The Ultra-Wide Band (UWB) channel model is considered once it presents an extremely wide transmission bandwidth which allows higher data transmissions, precise position location, higher multiple access capacities and covert communications. Despite these advantages, UWB channel can give rise to new effects at the receiving antenna due to its inherent large bandwidth in comparison to narrow-band channel models. Thus, a negligible number of multi-path components (MPC) superpose inside each resolvable delay interval, in this manner the central limit theorem (CLT) is not more appropriate, and the fading amplitudes are no-follow longer Rayleigh distribution.

Once the UWB channel model can achieve a waveform with a wide up of 7.5 GHz, its MPC are apart by around 133 pico seconds (ps), which is equivalent to a 4 cm path-length difference, its can be independently determined at the receiving antenna.

2.1 Power Delay Profile (PDP)

In the time-domain, the CIR of a UWB channel can be written as:

$$h(t) = f(t) * \sum_{l=1}^{L} \sum_{k=1}^{K} \alpha_{k,l} \delta(t - T_l - \tau_{k,l}), \tag{1}$$

with $f(t)$ being the path frequency dependence, L the total number of clusters and K the total numbers of rays, $\alpha_{k,l}$ denotes the coefficients of the k-th ray in the l-th cluster, and $\tau_{k,l}$ represents the arrival time belongs to the k-th ray of the l-th cluster at the arrival time T_l.

The Eq. (1) at the frequency-domain is given by:

$$H(\omega) = F(\omega) \sum_{l=1}^{L} \sum_{k=1}^{K} \alpha_{k,l} \exp(-j\omega(T_l + \tau_{k,l})), \tag{2}$$

with $F(\omega)$ being the frequency dependence of rays arrivals and is denoted by:

$$F(\omega) = C_0(\omega/\omega_0)^{-\kappa} \tag{3}$$

with C_0 being a constant value, κ denotes the frequency dependence of the path-loss, and ω_0 represents the frequency.

Moreover, by employing Taylor series, (3) can be expanded as follow:

$$F(\omega) \approx F(\omega_c) + F'(\omega_c)(\omega - \omega_c), \tag{4}$$

with ω_c being the center frequency.

The cluster arrival times are modelled by a Poisson distribution given by:

$$p(T_l|T_{l-1}) = \Lambda_l \exp[-\Lambda_l(T_l - T_{l-1})], \ \ l > 0, \tag{5}$$

where Λ_l denotes the cluster rate.

On the other hand, the arrival time $\tau_{k,l}$ is modelled with the mixture of two Poisson distribution given by:

$$p(\tau_{k,l}|\tau_{(k-1),l}) = \beta\lambda_1 \exp[-\lambda_1(\tau_{k,l} - \tau_{(k-1),l})] + (1-\beta)\lambda_2 \exp[-\lambda_2(\tau_{k,l} - \tau_{(k-1),l})], \tag{6}$$

where $k > 0$, β denotes the mixture probability and, λ_1 and λ_2 represent the ray arrival rates.

Cluster Powers: The PDP follows a exponential distribution in each cluster. Its equation is given by:

$$E[|\alpha_{k,l}|^2] = \frac{\Omega_l}{\gamma_l}\exp(-\tau_{k,l}/\gamma_l), \tag{7}$$

with Ω_l being the energy of the l-th cluster, and γ_l represents the intra-cluster decay time constant.

Finally, Ω_l follows an exponential decay given by:

$$10\log(\Omega_l) = 10\log(\exp(-T_l/\gamma_l)) + N_{cluster}, \tag{8}$$

with $N_{cluster}$ follows a normally distributed variable with standard deviation $\sigma_{cluster}$.

2.2 Small-Scale Fading

The Nakagami-m distribution with parameters (Ω, m) is considered for the fading amplitude $\alpha_{k,l}$, which is given by:

$$p(\alpha_{k,l}) = \frac{2}{\Gamma(m_{k,l})}(\frac{m_{k,l}}{\Omega_{k,l}})^{m_{k,l}}\alpha_{k,l}^{2m_{k,l}-1}\exp(-\frac{m_{k,l}}{\Omega_{k,l}}\alpha_{k,l}^2), \tag{9}$$

where $\Gamma(\cdot)$ is the Gamma function, $m \geq 1/2$ is the Nakagami m-factor and $\Omega_{k,l}$ is the mean-square value of the amplitude.

The m factor has a mean μ_m and standard deviation σ_m, where μ_m and σ_m have a delay dependence given by:

$$\mu_m(\tau) = m_0 - k_m\tau \tag{10}$$

$$\sigma_m(\tau) = \hat{m}_0 - \hat{k}_m\tau \tag{11}$$

Furthermore, note that for the first component of each cluster, the m factor assumes a deterministic behaviour given as follow:

$$m = \tilde{m}_0. \tag{12}$$

2.3 Simulation Environment

Based on the above description to generate the data set for UWB channels in LOS and NLOS conditions, we have employed Matlab software where we randomly generate 10000 UWB channel realizations using the parameters giving in Table 1.

The parameters of Table 1 correspond a residential environment that cover a distance from 8 to 25 m with a carrier frequency of $f_c = 6$ GHz. The justification of these parameters can be found in [4]. During the simulation processes, we first select either LOS and NLOS by choosing the appropriate parameter values from Table 1. Remarks that each of these UWB channel generated by our simulator includes multiple taps with its corresponding delay and magnitude.

Our simulator aims to generate a data set that is employed to perform UWB LOS and UWB NLOS channel classification by using different machine learning algorithms. It is considered that the receiving antenna knows perfectly the channel impulse response.

Table 1. Simulator parameters of UWB channel model generation.

Power delay profile	LOS	NLOS
Λ_l [1/ns]	0.047	0.12
γ_1 [1/ns]	1.54	1.77
γ_2 [1/ns]	0.15	0.15
β	0.095	0.045
Γ [ns]	22.61	26.27
k_γ	0	0
γ_0 [ns]	12.53	17.50
$\sigma_{cluster}$ [dB]	2.75	2.93
κ	1.12 ± 0.12	1.53 ± 0.32
Small-scale fading	LOS	NLOS
m_0 [dB]	0.67	0.69
k_m	0	0
$\hat{m}_0$ [dB]	0.28	0.32
$\hat{m}$	0	0
$\tilde{m}_0$	NA	0.69

Figure 1 shows the channel magnitude of three channel realization for UWB LOS and UWB NLOS, respectively versus the time. In Fig. 1, we can appreciate that LOS channels present a higher magnitude in comparison to NLOS channels once there is a line of sight between the transmitting and receiving antenna. Furthermore, it is observed that higher magnitude of the MCP for LOS channel realization is concentrated from 0 ns to 10 ns, whereas for NLOS channel realizations the higher magnitude of MCP is more spread in the considered time interval.

3 LOS and NLOS Channel Classification

We consider a supervised machine learning problem where a set of labeled channel realization is employed to train an ML model. Once the ML model is trained, it will be employed to classify a new input channel realization as LOS or NLOS which represents a binary classification problem.

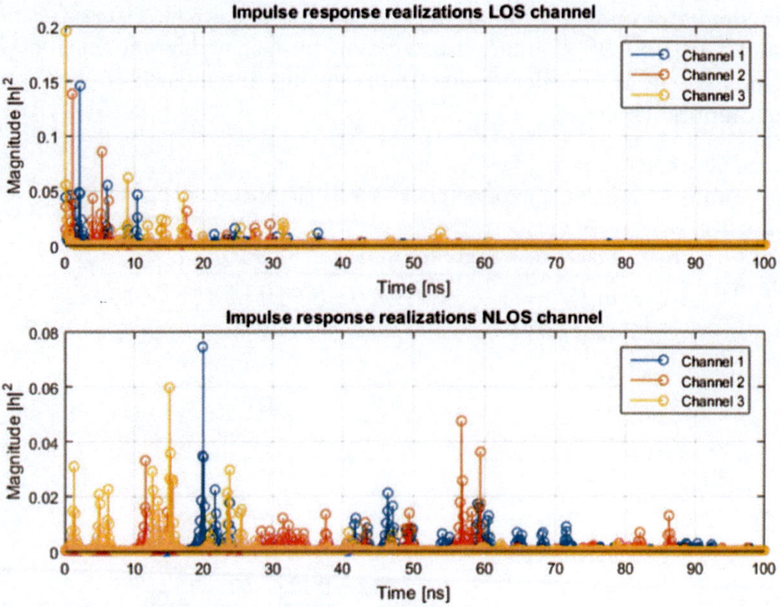

Fig. 1. Magnitude of three UWB channel realizations for LOS and NLOS.

3.1 Training Phase

Employing the labeled channel realization set generated through the simulator described in the section, we proceed to randomly select 80% of the total channel set for training and 20% for testing.

As follows we describe the measurements (features) considered during the training phase.

1. *Mean Excess Delay (MED):* The MED is calculated as follow

$$\tau_{MED} = \frac{\sum_{l=1}^{L}\sum_{k=1}^{K}(T_l + \tau_{k,l})\left|\alpha_{k,l}\right|^2}{\sum_{l=1}^{L}\sum_{k=1}^{K}\left|\alpha_{k,l}\right|^2}. \tag{13}$$

2. *Root Mean Square (RMS):* The RMS can be obtained by

$$\tau_{RMS} = \frac{\sum_{l=1}^{L}\sum_{k=1}^{K}((T_l + \tau_{k,l}) - \tau_{MED})^2\left|\alpha_{k,l}\right|^2}{\sum_{l=1}^{L}\sum_{k=1}^{K}\left|\alpha_{k,l}\right|^2}. \tag{14}$$

3. *K Factor:* It is given in dB and computed by

$$K_{factor} = 10\log\left(\frac{\left|\alpha_{1,1}\right|^2}{\sum_{l=2}^{L}\sum_{k=2}^{K}\left|\alpha_{k,l}\right|^2}\right). \tag{15}$$

4. *Channel magnitude:* It represents the distance between the maximum and minimum peak of a UWB waveform that represents the channel (see Fig. 1).

5. *Signal paths over 10* dB: Number of MPC that are within 10 dB of the peak amplitude.
6. *Signal paths of 85%:* Quantity of MPC, whose carries 85% of the total energy.

Figure 2 shows the six measurements considering as features during the training phase for the UWB channel realizations of LOS and NLOS. Note that these figures includes both LOS and NLOS channel samples obtained by the simulator where we have considered the same number of samples for LOS and NLOS channel which represents a balanced class.

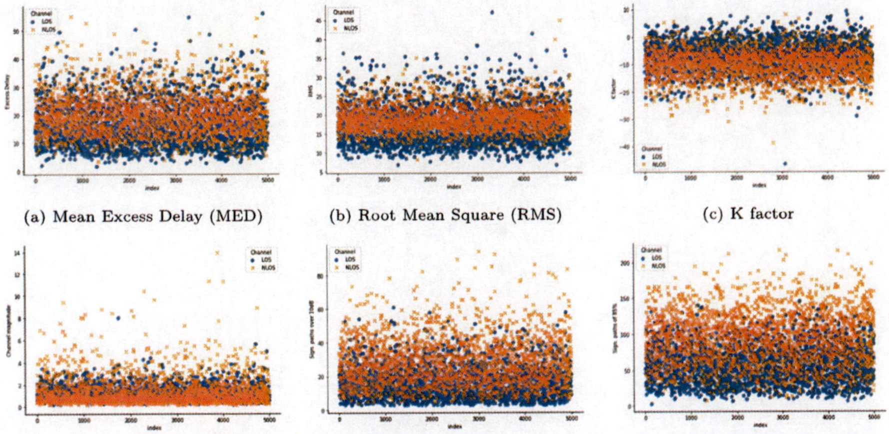

(a) Mean Excess Delay (MED) (b) Root Mean Square (RMS) (c) K factor

(d) Channel magnitude (e) Signal paths over 10 dB (f) Signal paths of 85%:

Fig. 2. Training phase measurements (features) for LOS and NLOS channel realizations

Once we define the percentage and the features employed during the training phase, we export the sets of channel realizations from Matlab .mat file to Python software to train the different machine learning algorithms (MLA) in order to classify the set of test channel realizations as LOS or NLOS. In this paper, we have trained 18 MLA.

4 Performance Evaluation of LOS and NLOS Channel Classification

Once the MLA were trained, we proceed to compare their performance during the testing phase which corresponds to 20% of the total set of channel realizations.

4.1 Testing Phase

In order to compare the different MLA, we considered the test accuracy as the comparative parameter. Thus, Fig. 3 shows the accuracy comparison among the

MLA considered during the test phase. From this figure, it is evident that the Random Forest Classifier achieves the best accuracy whereas Passive Aggressive Classifier is the it is the worst classifier for LOS and NLOS classification.

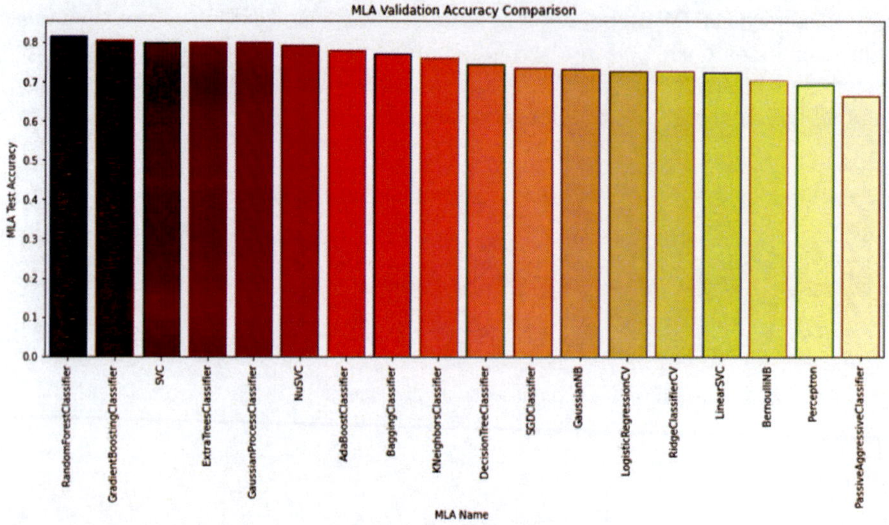

Fig. 3. Accuracy comparison among MLA classifier during the testing phase.

A better comparison parameter to check any MLA classifier is the AUC (Area Under The Curve) of a ROC (Receiver Operating Characteristics) curve. This curve is a performance measurement for classification problem with different thresholds, where AUC is the degree of separability and ROC denotes the probability curve.

In our case, AUC informs how much an MLA classifier is capable of distinguishing between NLOS and LOS channels, where if the AUC is high, the MLA is better. Table 2 compares the accuracy and AUC of the MLA classifiers.

Based on the results shown in Table 2, we determine that the Gradient Boosting Classifier (GBC) achieves the best performance. Thus, henceforth we consider the GBC. Figure 4 represents the confusion matrix of the GBC where it is observed that for the test channel realizations we have 195 samples that were wrong classified which represents false positive (FP).

To improve the performance GBC, we need to reduce the number of FP. Thus, we employ different learning rates for the GBC. The learning rate represents an important hyper-parameter of the GBC.

Table 3 shows the accuracy comparison of the GBC with different learning rates. From Table 3, it is evident that a learning rate of 0.2 produces a better accuracy. Thus, we select this learning rate. Then, we compute a confusion matrix with this new configuration of the GBC which is presented in Fig. 5.

Table 2. Accuracy and AUC of the MLA classifiers.

MLA	Accuracy	ACU
Random Forest Classifier	0.806	0.803
Gradient Boosting Classifier	0.805	0.825
Extra Trees Classifier	0.803	0.803
SVC	0.801	0.800
Gaussian Process Classifier	0.800	0.800
NuSVC	0.791	0.791
Bagging Classifier	0.790	0.792
Ada Boost Classifier	0.778	0.776
K Neighbors Classifier	0.760	0.761
Decision Tree Classifier	0.746	0.747
Gaussian NB	0.731	0.734
Logistic Regression CV	0.726	0.726
Ridge Classifier CV	0.725	0.725
Linear SVC	0.724	0.724
SGD Classifier	0.724	0.728
Bernoulli NB	0.703	0.703
Perceptron	0.691	0.685
Passive Aggressive Classifier	0.630	0.628

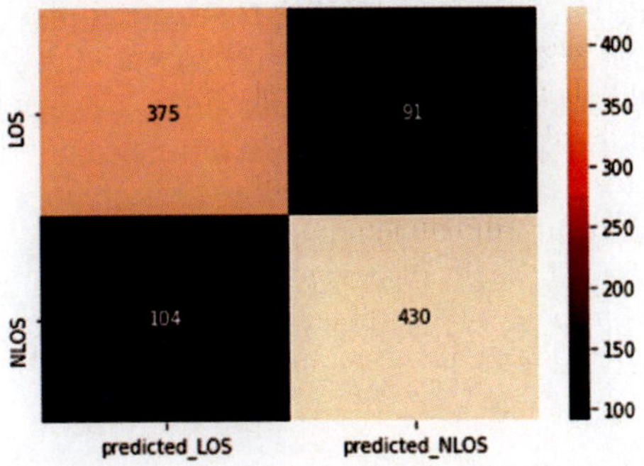

Fig. 4. Confusion matrix of the GBC.

Table 3. Accuracy of the GBC with various learning rates.

Learning rate	0.08	0.15	0.18	0.25	0.28	0.3
Accuracy	0.806	0.805	0.806	0.824	0.808	0.810

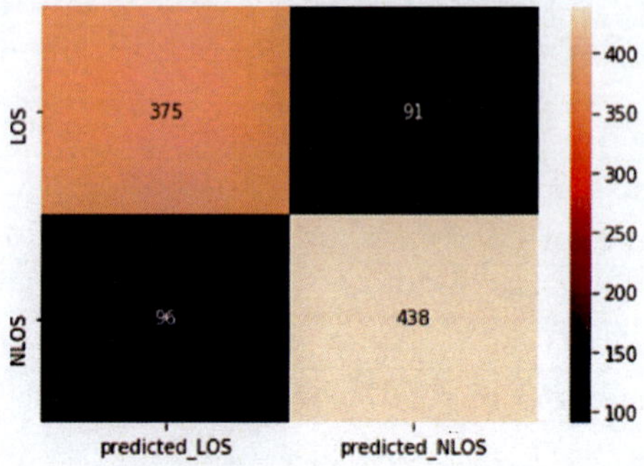

Fig. 5. Confusion matrix of the GBC with learning rate of 0.25

From Fig. 5, we note that the number of FP is reduced from 195 to 187 samples of channel realizations which were wrong classified. Note that with this new learning rate we obtain a small improvement in the performance of GBC.

5 Conclusion

In this paper, we have compared different machine learning algorithm (MLA) classifiers in order to perform UWB channel model classification between LOS and NLOS. Firstly, we have developed a simulator for UWB channel realizations for both LOS and NLOS in a residential environment with the aim of build a channel realizations data set that is then used to train and test the MLA. During the training phase we have considered different measurements for the classification process. Furthermore, during the testing phase we have evaluated the performance of MLA by mean of test accuracy and AUC-RCO curve where we have determined that the Gradient Boosting Classifier (GBC) achieves the best performance. To further improve the performance of GBC, we have trained it for various learning rate achieving a better performance in the classification process between LOS and NLOS channels.

The simulated channel dataset and MLA codes, for reproducibility purpose, can be found in https://github.com/jminango20/NLOSandLOSChannel_MachineLearning.

References

1. Puñal, O., Aktaş, I., Schnelke, C., Abidin, G., Wehrle, K., Gross, J.: Machine learning-based jamming detection for IEEE 802.11: design and experimental evaluation. In: 2014 Proceeding of IEEE International Symposium on a World of Wireless, Mobile and Multimedia Networks, pp. 1–10 (2014)
2. Molisch, A.F.: Ultra-wideband communications: an overview. URSI Radio Sci. Bull. **2009**(329), 31–42 (2009)
3. Lee, J.-Y., Scholtz, R.: Ranging in a dense multipath environment using an UWB radio link. IEEE J. Sel. Areas Commun. **20**(9), 1677–1683 (2002)

4. Alsindi, N., Duan, C., Zhang, J., Tsuboi, T.: NLOS channel identification and mitigation in ultra wideband ToA-based wireless sensor networks. In: 2009 6th Workshop on Positioning, Navigation and Communication, pp. 59–66 (2009)
5. Gururaj, K., Rajendra, A.K., Song, Y., Law, C.L., Cai, G.: Real-time identification of NLOS range measurements for enhanced UWB localization. In: 2017 International Conference on Indoor Positioning and Indoor Navigation (IPIN), pp. 1–7 (2017)
6. Sang, C.L., Steinhagen, B., Homburg, J., Adams, M., Hesse, M.: Identification of NLOS and multi-path conditions in UWB localization using machine learning methods. Appl. Sci. **10**, 3980 (2020)
7. Khodjaev, J., Park, Y., Malik, A.: Survey of NLOS identification and error mitigation problems in UWB-based positioning algorithms for dense environments. Ann. Télécommun. **65**, 301–311 (2010)
8. Zhang, J., Salmi, J., Lohan, E.-S.: Analysis of kurtosis-based LOS/NLOS identification using indoor MIMO channel measurement. IEEE Trans. Veh. Technol. **62**(6), 2871–2874 (2013)
9. Kram, S., Stahlke, M., Feigl, T., Seitz, J., Thielecke, J.: UWB channel impulse responses for positioning in complex environments: a detailed feature analysis. Sensors **19**, 5547 (2019)
10. Maranó, S., Gifford, W.M., Wymeersch, H., Win, M.Z.: NLOS identification and mitigation for localization based on UWB experimental data. IEEE J. Sel. Areas Commun. **28**(7), 1026–1035 (2010)
11. Maali, A., Mimoun, H., Baudoin, G., Ouldali, A.: A new low complexity NLOS identification approach based on UWB energy detection. In: 2009 IEEE Radio and Wireless Symposium, pp. 675–678 (2009)
12. Li, W., Zhang, T., Zhang, Q.: Experimental researches on an UWB NLOS identification method based on machine learning. In: 2013 15th IEEE International Conference on Communication Technology, pp. 473–477 (2013)

Social Entrepreneurship as a Detonator of Quality of Life and Sustainable Development Because of the Mayan Train Project in Rural Communities of the Municipality of Escarcega

José Alberto Sánchez López(✉), Dulce María de Jesús Delgado Cih, Sagrario María Quijano Gutierrez, and Geidy de los Ángeles Gómez Xul

Tecnológico Nacional de México Campus Instituto Tecnológico Superior de Escárcega, 24350 Escárcega, Campeche, México

alberto_sanchez@itsescarcega.edu.mx

Abstract. Social entrepreneurship is a recently created term that has managed to become popular in an accelerated way, as it aims to be one of the strategies to attack poverty and marginalization of a population where the market has failed and the action of the State is non-existent or ineffective. Under this great premise, the objective of this project is to develop a proposal for social entrepreneurship congruent with the Mayan Train project, which takes advantage of the region's resources and, in order to, supports the population, progressively reduces existing poverty levels as well as improve the quality of life of certain population. This project is approached from the qualitative approach and proposes to carry out a descriptive and interpretive analysis on the cases of social entrepreneurship existing in the region and concludes with specific proposals for integration strategies to the Mayan Train project, or else, with the creation of a model of social entrepreneurship congruent with the Mayan Train project. It was found that 63% of entrepreneurs seek to create jobs, and 7% promote ethical practices in the markets, however, despite the fact that 70% of those surveyed know the term, only 19% of them carry out or have carried out social enterprises. Similarly, it was obtained that 67% of those surveyed contribute to solving a social problem, of which 30% generate jobs, 11% provide training to society, another 11% promote ethical practices in markets and only 4% contribute by supporting other companies.

Keywords: Social entrepreneurship · Quality of life · Train Maya

1 Introduction

Mexico is a country that is territorially divided into states or federative entities with 3 levels of government: federal, estatal and municipal, since 2018 the country has a government headed by Andrés Manuel López Obrador, who is promoting the "Mayan Train" project in the Yucatan peninsula, which is part of one of the study elements.

The place of interest for this research is the town of Escárcega located in the homonymous municipality of the state of Campeche. Currently Escárcega is a town whose main

M. Zambrano Vizuete et al. (Eds.): CI3 2021, LNNS 511, pp. 566–577, 2022.
https://doi.org/10.1007/978-3-031-11438-0_45

source of development is being an important communications node, since federal highway 186 from Villahermosa, Tabasco joins it, and continues east to Chetumal, Quintana Roo; also, Federal Highway 261 that continues north, connecting it with Champotón, Campeche and Mérida in the state of Yucatán.

According to the poverty and evaluation report published by CONEVAL [1] Escarcega, however being an important communications node, is one of the 5 municipalities with the most people living in poverty in the state of Campeche with a percentage of 59.1% and 11.3% in extreme poverty.

The government of Mexico seeks to reactivate the economy by maximizing tourism, therefore, the population of Escárcega must be prepared to face the changes and take advantage of the business opportunities that will arise with the implementation of the Mayan Train project. By the way, Gerardo Islas in an article for [2] affirms: "the project will detonate a high social utility, by virtue of the fact that urban and commercial growth around the stations will be managed through sustainable communities, renovation and rearrangement territorial projects that will promote the growth of the region for the main benefit of its inhabitants".

In this way, it is clear that Escárcega will undergo a great economic transformation when such a public investment project is generated. The proposed research project includes the social component because Escárcega is located in an area of high marginalization and it is expected to produce benefits for social entrepreneurs, but also due to the influence of the sustainable dimension, it is desirable that the business proposals that are generated in Escárcega are friendly to the environment, which is one of the premises of social entrepreneurship.

1.1 Background

Much has been said about the term social entrepreneurship, however, it should be noted that this is not a term of our century and, contrary to what one might think, it has its origins in the nineteenth century with the emergence of the social economy. In an article published on the [3] they mention that "Bill Drayton is known as the father of social entrepreneurship, since it was he who coined the term social entrepreneurship, to refer to people who, in an innovative way, they solve important problems in society". Similarly, various investigations have shown that "social entrepreneurship is presented as a form of employment that in practice becomes one of the strategies to attack poverty and marginalization in developing and emerging economies" [4] and that "the social enterprise contributes to the sustainable rural development of its community and improves the quality of life of its inhabitants" [5].

In the state of Campeche there are 5 municipalities with the highest poverty values. The town of Escárcega is one of them, in which, the poverty levels are 59.1% and 11.3% of extreme poverty, according to data from [1] the specific problem is that in the state of Campeche, the population has remained in economic and social stagnation, a situation that has not allowed taking advantage of the opportunities for development that public or private entities offer to start a business and/or to make it grow effectively [6]. This means that there is a broader and deeper problem, the consequence of which is manifested in the ravages of poverty, which migration and other serious social effects for the population are derived from. Another interesting fact about the problem of poverty in Campeche is that, in most households, the income obtained is spent on consumer goods (daily expenses to satisfy basic needs), not allowing savings or investment.

The same has also happened in other countries, since governments have failed in the way of encouraging the economy of the people. Such is the situation of the Choco Andino Biosphere Reserve, which in a study found that there is "the precarious capacity to govern natural resources" [7].

Faced with this situation, the government has promoted strategies to combat poverty, although efforts have been insufficient to present encouraging results, such is the case of programs aimed at eradicating poverty in a comprehensive manner, meeting the most urgent needs of the population. A clear example is represented in the "Programa de canasta basica" implemented in the state of Campeche as a precedent to the "Progresa" program, which was later replaced by the "prospera" strategy [8].

The government of Mexico seeks to reactivate the economy by maximizing tourism, so the population of Escárcega must be prepared to face the changes and take advantage of the business opportunities that will arise with the implementation of the Mayan Train project.

In a population with high poverty rates, the low income of families does not allow income or investment. This is the case of Escárcega, a municipality in which poverty reaches values of 59.1% and 11.3% according to data from the 2020 Poverty Report published by [1] and in which the population has remained in economic and social stagnation [6]. The foregoing explains that the very diverse forms of economic activities that occur in the town of Escárcega have not been sufficient or adequate to generate development at the family and business level.

Currently, it seeks to reactivate the state's economy by maximizing tourism, so the population of Escárcega must be prepared to face the changes and take advantage of the business opportunities that will arise with the implementation of the Mayan Train project.

One way to achieve the development of both families and at the company and society level is through social entrepreneurship, however it is not identified that in Escárcega exists evidence of this form of entrepreneurship, which represents an opportunity for the development of this research already that various authors have shown that "social entrepreneurship is presented as a form of employment that in practice becomes one of the strategies to attack poverty and marginalization in developing and emerging economies" [4] and that "the social enterprise contributes to the sustainable rural development of its community and improves the quality of life of its inhabitants" [5].

In this way, the proposed research project includes the social component because Escárcega is located in an area of high marginalization and it is expected to produce benefits for social entrepreneurs, but also due to the influence of the sustainable dimension, it is desirable that the proposals business generated in Escárcega are friendly to the environment, which is one of the premises of social entrepreneurship. In this context, it is necessary to find cases of social entrepreneurship or existing social enterprises in the region and produce a "model" form or general procedure with recommendations and strategies to carry out social entrepreneurship according to the Mayan Train project in the municipality of Escárcega.

1.2 Theories and Entrepreneurship

Entrepreneurship is a term that has became popular in the 80s and 90s. According to [9] when looking for the relationship of entrepreneurship from the methodological point of view with the different sciences, it is possible to find a relationship between some basic and applied sciences and entrepreneurship. For example: there is a direct relationship with the Economy and Management; and an indirect relationship with Mathematics, Psychology, Sociology and Biology.

There are different theories or approaches to studying entrepreneurship. According to [10] the most common are:

Inverted triangle theory
Schumpeter theory
Austrian School Theory
Howard Stevenson theory
Allan Gibb's theory

[11] affirm that the Social Cognitive Theory (Bandura, 1986) is useful in the analysis of education for entrepreneurship because it includes a greater diversity of elements and typology of entrepreneurship.

1.3 Entrepreneurship

Analyzing the concept of entrepreneurship involves various economic and social aspects, since by nature this term has its origin in those areas. In opinion of [12] "understanding the phenomenon of entrepreneurship requires, first of all, to analyze it from its evolution as an economic and social aspect, to then study it from various perspectives, and ultimately propose its analysis from the cultural and psychoanalytic point of view". For this reason, in order to complete analysis of the concept of entrepreneurship it is necessary to emphasize the social and economic dimensions.

Regarding the concept of entrepreneurship, this is a term that has been widely used as a global research object. This word can have different interpretations depending on the field in which it is applied, and its origin goes back to the meaning of terms used to refer to people capable of taking the initiative and discovering new horizons. In this regard, [12] points out that the word "entrepreneurship" is derived from the French term entrepreneur, which means being ready to make decisions or to start something.

There are several ways to start a business, one of them is social entrepreneurship (object of this study). "The distinction between social and commercial enterprises is not dichotomous, but rather more accurately conceptualized as a continuum that varies from the purely social to the purely economic" [13].

Finally, from a broader approach, [14] affirms that entrepreneurship can be understood "as the development of a project that pursues a certain economic, political or social purpose, among others, and that has certain characteristics, mainly that it has a share of uncertainty and innovation."

1.4 Entrepreneur

According to [15] the word entrepreneur comes from the French entrepreneur (pioneer), initially being used to refer to these adventurers like Columbus who came to the New World without knowing with certainty what to expect.

There is some confusion between the words entrepreneur and businessman, when in fact they are terms that express a different meaning. Both terms are generally taken as synonyms, when in reality the differences between one and the other concept are duly marked [19].

Social entrepreneurship, various investigations have shown that "social entrepreneurship is presented as a form of employment that in practice becomes one of the strategies to attack poverty and marginalization in developing and emerging economies" [4] and that "the social enterprise contributes to the sustainable rural development of its community and improves the quality of life of its inhabitants" [5].

Social entrepreneurship is a type of entrepreneurship which is characterized by the generation of social projects "a common element in different definitions of social entrepreneurship is the search for solutions to social problems".

[13] present a very complete definition of social entrepreneurship "refers to a specific type of entrepreneurship that seeks solutions to social problems through the construction, evaluation and pursuit of opportunities that allow the generation of sustainable social value, reaching new and stable balances in relation to social conditions, through direct action carried out by non-profit organizations, companies or government agencies".

1.5 Strategies to Combat Poverty

The government has promoted strategies to combat poverty, although efforts have been insufficient to present encouraging results, such is the case of programs aimed at eradicating poverty in a comprehensive manner, meeting the most urgent needs of the population. A clear example is represented in the "Programa de Canasta basica" implemented in the state of Campeche as a precedent to the "Progresa" program, which was later replaced by the "prospera" strategy [8].

The government of Mexico seeks to reactivate the economy by maximizing tourism, so the population of Escárcega must be prepared to face the changes and take advantage of the business opportunities that will arise with the implementation of the Mayan Train project. However, in Campeche there are only 13 social tourism companies that bring together a total of 581 workers. "These social enterprises offer various activities, among which are boat tours through estuaries and islets, sport fishing, tours of interpretive trails, swimming in cenotes, sighting of flora and fauna and a great variety of artisan workshops" [16].

2 Methodology

The methodology used in the project consisted of surveying the cases of social entrepreneurship existing in the municipality of Escárcega, with the aim of knowing them and evaluating which ones are the most appropriate;

Subsequently, 6 of the best companies to work on the project were selected, the entrepreneurs were interviewed to learn more about their entrepreneurial experience, the impact and the benefits to create new forms of social entrepreneurship appropriate to the Mayan Train project.

Research approach and type
This project was approached from a qualitative paradigm. "It does not correspond to a quantitative study since a causal or correlational analysis was not carried out" [13], on the contrary, in this research a descriptive and interpretive analysis was carried out on the evidence of social enterprises in the local context.

In the second stage, it is intended to end with specific proposals for integration strategies for the Mayan Train project.

The research carried out is exploratory, descriptive, analytical and in the field.

2.1 Population and Sample

The study was carried out in the municipality of Escárcega, in the state of Campeche, Mexico and its unit of analysis are economically active men and women who are developing one of the forms of social entrepreneurship.

By consulting the official INAES database, the directory of social enterprises that have received financing from the federal government of the Mexican Republic was located. Data from 46 social entrepreneurship projects were observed for the period of years between 2015 and 2019 financed in the municipality of Escárcega. By reviewing the specific data of each project, it was found that the projects correspond to 27 organizations or forms of entrepreneurship that for the purposes of this research are considered "social enterprises" themselves that meet the definition of "social entrepreneurship" established by [13] who define it as a "specific type of entrepreneurship that seeks solutions to social problems through the construction, evaluation and pursuit of opportunities that allow the generation of sustainable social value, reaching new and stable balances in relation to social conditions., through direct action carried out by non-profit organizations, companies or government agencies" [17].

In conclusion, the finite population is 27 social enterprises distributed in rural communities of the Escárcega municipality.

2.2 Data Collection Instrument

In the first stage of the project, the instrument used was surveys, these were applied in order to know the cases of social entrepreneurship existing in the Escárcega region. The second stage consists of selecting 6 of the best social enterprises and interviewing those responsible, with the intention of knowing more about the entrepreneurial cases and being able to develop their own proposal congruent with the Mayan Train project.

2.3 Data Analysis Procedure

The survey was applied to 27 companies, located in the center of the city of Escárcega, Campeche. The application was carried out in person, in printed surveys (as a sample), without However, the responses were captured on a digital platform (google form), where the responses received were concentrated.

The first step that was carried out was to concentrate the responses in the google form, which generated an Excel file for analysis.

As a second step, the questions and answers were separated, one for each sheet where a number was assigned to each answer, remaining as follows;

a) For the proportional questions, the following numbers were assigned;
   Yes = 1 No = 2
b) For the linear questions, they were grouped according to the coinciding data, so the numbering remained from 1 to 5.

As a third step, we proceeded to obtain the positioning measures mean, median, mode, variance, and standard deviation.

3 Results

3.1 Social Entrepreneurship

Social entrepreneurship can be considered as a strategy to attack the marginality of peoples; Because of this, it is important to know how well employers know the term.

A total of 27 entrepreneurs in the Escárcega region were surveyed, seeking to know how positioned the term social entrepreneurship is and how many of them know its meaning. It was found that 70% of the respondents do know the meaning of social entrepreneurship, however, of the total of respondents, only 19% of them have or have had social entrepreneurship projects.

They were questioned about the general benefit of their entrepreneurship, regardless of whether they have or have had social entrepreneurship projects and it was found that, of the total number of respondents, 67% of them contribute to solving some social problem. Of the 67% of those who answered that they do contribute to the improvement of social problems, 30% contribute by granting jobs, 11% by providing training, another 11% by providing low-cost work materials, and a final 11% by providing digital and advertising services. While 4% contribute helping other companies.

Of the total respondents, 63% provide work or employment to society, 22% provide care for the environment, 7% provide products at low prices, 4% provide training and finally 4% do not provide any type profit, which means that 89% have a commitment to the social environment while 7% have a commitment to the natural environment.

Corporate social responsibility is the result of the commitment by companies to respect the environment and promote sustainable development practices, hence the importance of knowing the contribution to the environment that Escárcega ventures have. It was found that, of the total number of respondents, only 10% contribute to the improvement or preservation of the environment. Of the 10% of the companies that contribute to the improvement or preservation of the environment, 37% do so by not generating pollution and making proper use of waste, while the other 63% did not specify how they contribute.

3.2 Quality of Life

Quality of life represents a multidimensional term of social policies that means having good "objective" living conditions and a high degree of "subjective" well-being, and also includes the collective satisfaction of needs through social policies in addition to satisfaction individual needs.

One of the important aspects that help to measure the level of quality of life in the population is health, which is why it was questioned about the access of workers to health services and it was found that 78% of workers did not have insurance, 15% have IMSS and 7% have private insurance.

Another quality of life indicator is access to education, surprisingly it was found that all of the respondents do not contribute to education through the granting of scholarships to workers or their children to continue with their studies. Regarding housing, it was found that no undertaking grants or facilitates housing loans for workers.

Regarding employment, it was found that 30% of the enterprises do not address the problem of unemployment in the population in any way, 26% do so with the opening of new establishments, 22% with an increase in production and finally 22% do so through training programs.

4 Discussion

Campeche is one of the states of the Mexican Republic with alarming poverty rates, it is currently awaiting for a project that will positively impact the state and specifically the municipality of Escárcega, which is why it is intended to take advantage of this project to propose new forms of economic development through social entrepreneurship.

Proposals will seek a way to develop productive projects for rural communities in the municipality of Escárcega. To develop it, it is necessary to identify and know the cases of social entrepreneurship existing in the region, to know the social impact they represent and in which areas they manage to improve the quality of life of society, that is why the study groups were the 27 cases of entrepreneurship in the Escárcega region.

A set of existing social enterprises in the Escárcega region was located in the database of the National Institute of Social Economy (INAES), giving a total of 47 projects, which were in charge of 27 companies.

These companies were our population to study, since the objective of the study is to know the existing cases of social entrepreneurship to develop a proposal that can be adapted to the Mayan Train project with the intention of triggering economic development and achieving a better quality of life for rural communities in the municipality of Escárcega.

"The field of social entrepreneurship is a young field; However, its promise to help alleviate social problems is so desired by governments, academics and society in general that it has become popular, but without having established a great umbrella theory" [18], however, it is a Research conducted in 2017 concludes that there are three schools of thought for social entrepreneurship.

The first one is the School of Innovation, whose definition of social entrepreneurship revolves around "individuals who work on the innovative solution of social problems; In the same way, there is the western school of social business, which includes non-profit businesses that generate their own sources of income through the sale of products or services, while fulfilling a social mission, finally there is the school of Asia's social business, which shares the same definition as the western social business school" [18]. In the same way, that does not mean that entrepreneurship theories are alien to him.

There are numerous theories of entrepreneurship that may be applicable to social entrepreneurship, among them, and one of the most applicable is the one created by Andy Freire who exposes the inverted triangle theory where "the fulcrum is the entrepreneur, who needs two components additional to arrive: the business idea with market viability and capital" [10].

Two exceptions of great importance within the quality of life variable stand out from the results obtained in the research. The first of them is related to education, since all the respondents affirmed that they do not provide any type of support to the workers or their children as extra help for them to continue with their studies; Another alarming result is access to housing, since, in the same way as in the previous case, the total number of respondents affirmed that they did not grant any housing credit to their workers.

Finally, of the total of those surveyed, 78% of them do not provide any type of insurance to their workers. These events have a negative impact on the quality of life of workers since in order to enjoy a satisfactory quality of life, one must have access to the 5 factors. These data confirm the poverty statistics presented by CONEVAL in 2017.

[20] Conducted an investigation where he concluded that "social entrepreneurship is a concept that designates a phenomenon whose goal is to solve a social need directly related to the productive sector", in the same way he mentions that "social entrepreneurship has become a factor of change for disadvantaged actors due to the following characteristics:

- Generates employment and development for the vulnerable population.
- Improves the quality of social services.
- Promotes ethical practices in the markets (fair trade)
- Sensitizes new entrepreneurs who want to enter the productive sector"

It is clearly observed that the characteristics mentioned by the author are relevant and applied in the Escárcega community since when questioning how social enterprises impact the environment, 63% of them seek to generate jobs, and 7% promote ethical practices in markets, however, despite the fact that 70% of those surveyed know the term, only 19% of them carry out or have carried out social enterprises. Similarly, it was obtained that 67% of those surveyed contribute to solving a social problem, of which 30% generate jobs, 11% provide training to society, another 11% promote ethical practices in markets and only 4% contribute by supporting other companies. For his part, [21] mentions that "social entrepreneurship tends to occur successfully, in sectors of the economy where the market has failed and State action is non-existent or ineffective."

Therefore, "the concept of social entrepreneurship [...] must always be located within a precise sociocultural and institutional context, taking into account its space-time dimension" [20]. Another important point and according to [22] is that "social entrepreneurship is vital to unlock economic inclusion and the growth of developing economies" that is why the importance of perform social entrepreneurship projects in the Escárcega region before the imminent arrival of the Mayan Train project.

Very linked to these terms and, above all, to the thesis proposed by Mballa, the quality of life enters whose concept must be recognized as "multidimensional since said concept is influenced by factors such as: lifestyle, housing, satisfaction in school and in employment, as well as economic situation" [23] related to these factors, it was obtained that in the enterprises of the Escárcega region, 78% of the workers do not have access to basic health services 15% have IMSS and 7% private insurance; On the other hand, in terms of education, no enterprise grants educational support to its workers or their children, in the same way, no enterprise grants housing loans. In terms of unemployment, 30% said they did not address the issue of unemployment in any way, 26% did so by opening establishments, 22% through training programs for young people and a last 22% increasing their production to generate more opportunities labor.

Due to this, it will be opportune to investigate in more depth the reason why the enterprises and companies in the town of Escárcega do not provide support to society in terms of quality of life. It is necessary to delve into what are the causes of such a low level of quality of life in the locality, in order to design strategies that help to detonate the economic development of companies, that brings an improvement in quality of life factors.

Social entrepreneurship as a trigger for quality of life and sustainable development is a subject very little studied, it could even be said that it is a pioneering research, due to this, the existing information is very limited; Although the variables studied are of common interest and one has a direct relationship with the other, they had never before been related in the same project.

5 Conclusions

The Mayan Train project is a great platform that will benefit the population of Escarcega, it intends to be an economic detonator for the population, but in addition to that, it is a latent opportunity to develop social entrepreneurship projects to increase the quality of life of the population and manage to get Escárcega out of the backlog.

Very interesting data were obtained regarding social entrepreneurship and quality of life; One of the most surprising was the scare, almost nil, contribution of the enterprises to the quality of life of the population, as well as the little development of social enterprises in the region.

These results open a new line of research to propose models that seek to improve these areas and thereby manage to get the population of Escárcega out of backwardness and poverty.

It can be concluded that in the town of Escárcega there are limited cases of social entrepreneurship, and that these contribute to the solution of a social problem in a scarce way through the generation of jobs, training for society, promotion of ethical practices in the markets and the support to other companies.

Currently working on the design of the proposal that allows to generate a model of social entrepreneurship adaptable to the Mayan Train project. It is observed that this project can generate environmentally friendly social enterprises, since Escárcega is located in a nature protection zone, such social enterprises could be supported with other strategies of the federal government such as "sowing life" and "young people building the future".

References

1. CONEVAL: Poverty report and evaluation 2020. Campeche. CONEVAL, Mexico City (2020)
2. Islas, G.: The Mayan Train and its multiplier effect. Forbes Mexico (2020). https://www.forbes.com.mx/el-tren-maya-y-su-efecto-multiplicador/
3. Wiese Foundation: How did social entrepreneurship start in the world?, 31 July 2018. https://www.fundacionwiese.org/blog/es/inicio-del-empredamientos-socialen-el-mundo/
4. Valdés, J.A.: Socual Entrepreneurship: a necessity to alleviate poverty. XV Research Congress on the third sector. Mexico (2015)
5. Vázquez-Maguirre, M., Portales, L.: The social enterprise as a detonator of quality of life and sustainable development in rural communities, pp. 255–284 (2014)
6. Santos, V.R., Barroso, F.G., Ávila, O.J.: Result of the measurement and analysis of entrepreneurial skills for the creation of agribusinesses. Experience of the Mayan area of Campeche. Mexican agribusiness magazine, 35 (2014)
7. Heredia-R, M., Falconí, V., H-Silva, J., Amores, K., Endara, C.A., F-Ausay, K.: Technological innovation for the sustainability of knowledge and natural resources: case of the Choco Andino Biosphere reserve. In: Botto-Tobar, M., Zambrano Vizuete, M., Díaz Cadena, A. (eds.) CI3 2020. AISC, vol. 1277, pp. 464–476. Springer, Cham (2021). https://doi.org/10.1007/978-3-030-60467-7_38
8. Levy, S., Rodríguez, E.: No inheritance of poverty. The progress-opportunities program of Mexico. Inter-American Development Bank (2005)
9. Hernán, G.C., Gálvez, A.E.: Entrepreneurship model in Red-Mer. Application of entrepreneurship theories to business networks. Academy. Latin Am. J. Adm. (2008)
10. Gómez, L.A.: Entrepreneurship theories (2012). http://uniajc.pbworks.com/w/file/fetch/54060386/THEORIES_OF_EMPLOYMENT.pdf. Total entrepreneurial capacity and its factors
11. Osorio, T.F., Pereira, L.F.: Towards a model of education for entrepreneurship: a view from cognitive social theory. Adm. Notebooks **24**(43) (2011). http://www.redalyc.org/articulo.oa?id=20521435001
12. Rodríguez, A.R.: New perspectives to understand business entrepreneurship. Thought and Management, 26. Obtained from Thought and Management (2009). https://rcientificas.uninorte.edu.co/index.php/pensamiento/article/view/864/508
13. Guzmán Vásquez, A., Trujillo Dávila, M.A.: Social entrepreneurship - literature review. Manag. Stud. (2008)
14. Formichella, M.M.: The concept of entrepreneurship and its relationship with education, employment and local development. Buenos Aires (2002)
15. Castillo, A.: State of the Art in Teaching Entrepreneurship. Chili (1999)
16. González, G.E.: Ecotourism and social participation in conservation: perception of community actors and perspectives associated with an ecotourism enterprise in Campeche, Mexico, Mexico (2018)
17. Hernández, R., Fernández, C., Baptista, M.: Investigation methodology. McGraw-Hill/Interamericada Editores, México D.F. (2010)
18. Bravo Monge, C.: The schools of thought of social entrepreneurship. Entrepreneurial spirit, 26 (2017)
19. Buscardo, C.A., Saavedra, G., Camarena, A.: Towards an understanding of the concepts of entrepreneurs and entrepreneurs. Sum Bus. (2015). https://doi.org/10.1016/j.sumneg.2015.08.009
20. Mballa, L.V.: Training of Social Entrepreneurs for disadvantaged citizens: an option to fight poverty and social marginalization in San Luis Potosí. Successful practices in the implementation of innovation and local competitiveness policies, 347 (2015)

21. Ramírez, Á.: What is social entrepreneurship? Central America (2008). http://www.schwabfoundseoy.org/es/competitions/competition/75
22. Hung Kee, D.M., Ali Kazmi, S.S., Hashim, M.: Social entrepreneurship and its impact on economy: in perspective of Pakistan. Int. J. Acad. Res. Account. Financ. Manag. Sci. (2016)
23. Velarde Jurado, E., Ávila Figueroa, C.: Assessment of quality of life. Public Health 350 (2002)

Marketing 2.0 as a Tool for Microenterprise Growth

Jessica Quispe(✉), Estefani Segura, David Chacón, and William Ortega

Instituto Tecnológico Vida Nueva, Quito 170126, Ecuador
jessica.quispe@istvidanueva.edu.ec

Abstract. Nowadays, due to the different economic changes in companies, marketing 2.0 has become one of the most viable strategies for organisational development within the markets, as this economic model tends to show a competitive growth in the industries. Lack of knowledge is a key factor that prevents organisations from growing and gaining the competitive advantages that companies should have when they enter the market. Its object of study allowed to consider the relevant theories of several authors, which ratified the results obtained, where it is indicated that organisations depend on innovation, interaction and the development of new tools for micro-enterprise growth that allow them to survive in a changing market, however, they must also contribute to the proper management of Marketing 2.0 and its recognition by stakeholders. This research determined the characteristics and the relationship that marketing has had on the growth of microenterprises obtaining optimal results that allowed the identification of the key factors of application to the study group of 243 microenterprises in the South of Quito.

Keywords: Microenterprise growth · Strategies · Innovation · Marketing 2.0

1 Introduction

Organisations have grown significantly in the markets due to the different strategies that have to be applied. According to the new needs, most of the organisations in the environment are considered highly competitive, but what is it that prevents many of them from growing? This is one of the questions that people and entrepreneurs often ask themselves.

The new needs of the markets and the different economic, political and social difficulties that the world is going through generate a search for new strategies that allow overwhelming growth. This is the reason why marketing 2.0 has become one of the most important organisational tools in the industry worldwide, allowing companies to make their products and services known to a larger number of consumers in the market, through the use of technology. Nowadays, digital marketing has a dynamic added value because through its broad approach and methodological tools it makes companies more competitive According to the authors [1] digital marketing opens up new possibilities for interacting with customers because each user is no longer a mere receiver of information

M. Zambrano Vizuete et al. (Eds.): CI3 2021, LNNS 511, pp. 578–586, 2022.
https://doi.org/10.1007/978-3-031-11438-0_46

but a co-creator, which is given by the personalisation and immediacy of content. Therefore, by relating the articles to the study variables, it is possible to determine which are the tools for micro-enterprise growth that small companies manage and are unaware of, limiting them to sustained development due to inadequate application and the restriction of their organisational capacities, and with this information it is also possible to identify what companies think about the applicability of Marketing 2.0 in their businesses.

1.1 Marketing

Marketing is known as the set of operations and resources aimed at distributing products and services that aim to satisfy the needs of customers in a profitable way in the medium, short or long term, through the use of various media, looking for important alternatives for SMEs, where access and satisfaction in the economic environment of Ecuador is facilitated [2].

Digital Marketing

Digital marketing is the use of new technologies, through the use of applications that help the development of activities, which can achieve the purchase and retention of consumers in the market and at the same time considering the importance of strategically planning the use of digital media [3].

[4], classic author of the Marketing discipline defines it as a business and social issue, which allows several people to obtain what they need and long for in the creation and exchange of values with others, which can be determined as a cause comprising a set of interconnected steps that allow the achievement of results from the point of view of the company, and this recognises captivate customers and seek the satisfaction of their desires and needs.

Web 2.0

Web 2.0, evolving from Web 1.0 which allows to distinguish the different contents handled by websites, which can be controlled by companies such as the media.

It is also known as the social web that has given users access, allowing them to connect, build relationships and converse. According to the authors [1], digital marketing opens up new possibilities for interaction with customers, because each customer no longer only receives information, but many create content, which is given by the personalisation and immediacy of the content. Among the best known are: blogs which are known as tools where several writers leave messages and at the same time readers can interpret and leave their opinions about a topic in general, also considered are wikis which are methods in which people can interact with information and create a community and finally social networks in which they are formed by people, who can communicate among other users and handle different topics such as personal and work experiences.

According to [5] there are some myths and realities in which six scenarios related to communication are established, from which it can be synthesised that there is impoverishment of language, democratisation of management, low level of users, ignorance of digital media, scarce content and technological rubbish.

Social Media Marketing

This tool seeks interactive participation between user and company, where content creation is allowed compared to traditional marketing, which has made people pay more attention to seek alternatives to capture attention using social networks, which allows them to create advertising campaigns with the aim of communicating a brand, service, or product [6].

Relationship Marketing

[7] states that due to the extinction of service marketing, relationship marketing is born, understanding this type of marketing as the management of the buyer, this case usually occurs due to the diagnosis and analysis of knowledge, i.e. that allows to know the needs and different behaviours in their external environment, considering that these characteristics vary rapidly, due to the different points of view of the companies and the satisfactory variations of customers, therefore, the personalisation that can be externalised in social media are important factors for relationship marketing to be successfully applied in companies.

Currently the lack of effectiveness and efficiency, determines that companies cannot survive in the market, this is like going to market barefoot, without being able to compete economically with other companies, it is also necessary to be based on key indicators that help to improve the efficiency and effectiveness in relation to the customer and the management of the company with the buyer.

Marketing 2.0

Due to the rise and development of marketing 2.0, companies gather tools and indicators such as: market research and digital media, which enable the analysis of the level of customers and what is their behaviour in relation to the creation and consultation of information, which favours the interaction on the web of the customer and company, which through the development of content in different digital segments promotes a sustainable work and that meets the needs of consumers.

1.2 Microenterprise Growth

The micro-enterprise growth of companies is oriented towards market management, which is an essential element for the organisation that wants to feel committed and above all that provides information in the development of a value offer, as well as satisfying the needs and preferences of the market, as a source of reasonable competitive advantages for the organisation. [9] states that relationship management is required as a dose of knowledge between customers and companies. Therefore, both approaches focus their attention on effectively allocating resources in the organisation.

Nowadays, knowing the business environment of organisations has become one of the most important pillars that allows for disruptive growth in the market, applying micro-enterprise growth tools that focus on developing strategies for positive change in industries.

At present, the growth of microenterprises is due to the need for entrepreneurship due to high unemployment rates, causing many people to start up a microenterprise, which is a determining factor for the development of the country due to its representativeness. As [10] states, the greatest generator of economic income of the countries are destined in the microenterprises since they promote the social and monetary stability of all of them, thus promoting the increase of the wealth of the countries in Latin America in a 95% of its economic resources comes from the microbusinesses, but the great challenge that the companies have in the market is to maintain, to be coupled and to develop in the sector that they develop since they are in the way with some social, economic and cultural problems, that is to say the great industries do not complete their market [11].

Growth Factors

Social Responsibility

[12], states that a socially responsible organisation is one that apart from offering products and services at low prices and of good quality, generates economic income, sources of employment, and pays taxes to the state in the right way, challenges its own creativity and innovation, allowing the identification of the problems that afflict the community and the sector that surrounds it.

Stakeholders

Stakeholders (customers, workers, employees, financial institutions, society, government, suppliers, competitors, and intermediaries) have a positive or negative impact within companies, and this minimises business growth to some degree. The application of stakeholder groups helps to determine the degree of social responsibility of each stakeholder group, which helps to identify one of the strategic resources for the survival of organisations.

Growth Prospects

Innovation

Technological, social, and cultural innovation is one of the strategies that companies must manage, as the growth of any organisation depends on its capacity for techno-logical adaptation. Being in a globalised world where all potential customers are at the click of a button and this therefore facilitates the processes of buying, selling and distribution of products to all companies determines that organisations that do not have the capacity for innovation and change will not be able to have a competitive advantage in the markets. This will lead to lower revenues and a loss of stability with-in the company, which is why innovation is not only a strategy, but a necessity for many companies.

Economy

The economy within organisations has been constantly changing due to the various technological and digital changes that they have undergone, as this brings with it consequences that lead to companies having difficulties in increasing their capital. The economy will depend a lot on the way in which companies apply disruptive growth strategies in the markets, as nowadays these factors are becoming increasingly competitive.

Product Diversification
Product and service diversification has become a management strategy to maximise the actual and potential market, in order to have and achieve greater growth and profitability within the industry to which it belongs. This allows the identification of new market niches where companies can enter and achieve the expected success. This can also happen when companies, regardless of their size, create new products or services to satisfy all customer needs and thus increase the competitive capacity of the organisations within the markets.

Growth Strategies
Micro-enterprise growth strategies, as the integration of fundamental aspects, should enable the creation and identification of resources and investment in research and development (R&D), marketing and acquisition of new techno-logical equipment, which help the survival of enterprises, whereby growth factors are also considered to be internal such as age, motivation and knowledge management and external such as competitors and stakeholders [13].
Knowledge Management
Knowledge management is necessary to create an environment of socialisation, where individuals can amplify the relationship between individuals and abstract innovation, such as a new administrative system or innovative organisational structure, and this allows the creation of interest groups inside or outside the organisation. All of the above according to [14] is managed as a repetitive, spiralling loop, which provides the facilitating conditions for knowledge creation as mentioned by Nonaka.
Dynamic Knowledge Creation
As [15] states it is necessary to be able to form intelligent organisations, dynamic in knowledge, these must be structured in a dynamic and innovative way and be in constant learning, where they can be based on two basic and significant frameworks such as the techno-cognitive framework of support and the strategic-tactical frame-work of sustenance; both interacting synergistically and in an articulated manner.

1.3 Marketing 2.0 in Business Growth

Applying this tool [16] states that it is the impetus for microenterprises to have competitive advantages over the traditional and ambiguous way of applying advertising and selling products, all processes can be fully measurable, when applying and executing a digital marketing campaign where the expected impact of the strategy can be measured, the repercussion it has had and even the return on investment, which allows adjusting the implemented strategy as necessary, thus allowing to improve the results obtained, also improves the growth of 88% of SMEs.

2 Method

2.1 Design

This research has an exploratory qualitative approach and bibliographic documentary analysis in databases of classic and contemporary authors. A record of 80 indexed articles was found, of which those directly related to the proposed topic were analysed.

In addition, a survey was carried out with companies in the commercial sector located in the south of Quito, where the search for potential solutions to the proposed problem was implemented, identifying factors and changes that may occur in the business environment.

2.2 Participants

A total of 243 companies from the commercial sector of South Quito, Guamaní sector, participated in a survey consisting of 14 questions that identified the reasons and characteristics of the implementation of Marketing 2.0 in business growth.

2.3 Procedure

The research variables marketing 2.0 and business growth were determined, and through the operationalisation of variables the applicable survey questions were identified.

The purpose of the questions was to find out the variables and factors of the use of the Marketing 2.0 tool in relation to business growth.

2.4 Statistical Analysis

The analysis of the surveys was carried out in which Cronbach's Alpha was considered with a minimum value of 0.80 for the application of the survey to be acceptable. In addition, multiple linear regression was applied, establishing a dependent variable and three independent variables for the study.

2.5 Ethical Considerations

All companies received the survey information as well as the objective of the re-search and the questions. The companies gave their written and verbal approval. Once their consent was received, the surveys were implemented (Table 1).

3 Results

Table 1. Cronbach's alpha calculation

Cronbach's alpha	No. of elements
0,813	14

Note: Results expressed as a percentage

The survey was established by 14 items constituting a value of 0.813 which is in the good range. With this value check, the questions are applied to the 243 companies (Table 2).

When applying the multiple linear regression, it can be verified that the independent variables have a 56% and 66% linear relationship between the dependent variable which is marketing 2.0, so it can be determined that the proportion is acceptable in dependence on business growth.

Table 2. Cronbach's alpha calculation

Model	Non-standardised coefficients		Standardized coefficients	t	Sig.	95.0% confidence interval for B	
	B	Standard error	Beta			Lower limit	Upper limit
(Constant)	2,276	0,177		12,840	0,000	1,927	2,626
How often do you use 2.0 marketing strategies to promote your product?	0,233	0,054	0,290	4,272	0,000	0,125	0,340
How much has competition influenced your business growth?	−0,030	0,051	−0,036	−0,580	0,562	−0,130	0,071
How often have you used social media to promote your products?	0,024	0,053	0,030	0,444	0,657	−0,081	0,129

a. Dependent variable: Do you think the application of marketing 2.0 is important for business growth?

Note: Results expressed as a percentage

4 Discussion

According to the results, it can be considered that microenterprises generate con-tent in a decentralised way, as [18] states so the use of web 2.0 is necessary to enable them to generate targeted advertising, based on the knowledge of the personal tastes of their consumers, with the use of social networks like (Twitter, Facebook, LinkedIn, etc.) because each member of the microenterprises publishes on their own initiative their satisfactions, preferences, hobbies, phobias and illusions and externalises them through these media.

The marketing 2.0 variable has a relationship of more than 50% with the business growth variable. By using it as a tool, microenterprises can create content to publicise their products and the company. This causes customers to receive, interpret and ex-pose through social networks their positions regarding the product they want, or desire and the micro-entrepreneur must use the tools and strategies according to current technology, with options on the market that are intuitive, easy to handle and with free functionality.

Michael Porter's theory mentions three strategies that companies can apply to gain a competitive advantage in the market: cost leadership, differentiation and market focus or segmentation, also known as Porter's generic strategies, which can be applied according to the company's line of business and the focus that investors wish to give the organisation in the future.

Cost leadership: Michael Porter mentions that "cost leadership strategy is based on a company aiming to be the lowest cost producer in its industry sector", It can therefore be deduced that in order to develop the cost leadership strategy, companies de-ploy scale production to reduce the final production cost of the product, selecting quality raw materials and technology according to the high production level of the company [19].

Differentiation: refers to a company being unique in the market or sector, whose characteristics are valued by the customer, identifying the similar or common tastes and preferences of a segment of consumers to satisfy the identified needs, creating products with exclusive features and benefits for which the price of the product is higher than those of the competition. In order to obtain the differentiation strategy, it is possible to work on innovating the product itself, improving sales channels, distribution and the marketing approach [20].

Targeting: This strategy emphasises looking for a different picture among companies in order to compete. The company selects a specific market segment to create products or provide services by adjusting all strategies to the specific target group [20].

Taking into account the definitions of Michael Porter's theories on generic strategies, it is established that companies can select one of the three established strategies for their continuity due to their practical application, because it is not possible to unify the strategy of cost leadership with differentiation, since differentiation requires a strong insertion of capital for the innovation and development needed in the products, marketing application or distribution channels to obtain a competitive ad-vantage of differentiation, by inserting a large amount of capital, the principle of cost leadership cannot be met as the costs of products and processes will be higher than those of other companies.

Nor can the relationship of the market segmentation strategy or cost leadership approach be taken, because by taking a specific segment to which needs are sought to be satisfied, a high economic investment in studies and specific developments for a specific population will be used. This means that, as they are not exponential products, it will not be possible to produce at scale and therefore the principle of cost leadership will not be fulfilled.

5 Conclusions

Taking into account the information analysed in the research, it is determined that over time the paradigms of implementation and operation of digital marketing are disappearing, which is why the implementation of marketing 2.0 which allows companies to differentiate themselves in the market.

Digital marketing in companies has been reduced due to low economic investment, technological changes and changing consumer behaviour, taking into account the aforementioned aspects, organisations do not apply or develop digital tools that allow them to establish a customer-company relationship.

56% and 66% of the questions on business growth showed a linear relationship on the marketing 2.0 variable, which allows us to determine that companies do show a high importance to remain in the competitive market, so it is necessary to apply it as a necessary tool within the company.

References

1. Maqueira, S., Bruque, J.: El nuevo marketing en la Web de las Redes Sociales., Alfaomega. Méxic (2012)
2. Moschini, S.: Las claves del marketing digital. colección: periodismo de vanguardia (2012)
3. Martinez Beltran, M., Sanchez Mosquera, A.: Repositorio Institucional UCC: Herramientas del marketing digital como factores de éxito de la cadena wal-mart méxico en correlación con grupo exito en colombia. Repos. Inst. UCC (2019). https://repository.ucc.edu.co/handle/20.500.12494/15640
4. Kotler, P., Keller, K.: Dirección de Marketing (2012)
5. Celaya, J.: Web 2.0: Marketing o realidad (2007)
6. San Millán Fernández, E., Medrano García, M.L., Blanco Jiménez, F.J.: Social media marketing, redes sociales y metaversos. Universidad, Soc. y Mercados Glob. - I Congr. Nac. Mundos Virtuales -Metaversos Web 3D y Redes Soc., pp. 353–366 (2008)
7. Chica, J.: Del marketing de servicios al marketing relacional. Rev. Colomb. Mark. **4**(6), 60–67 (2005). https://www.redalyc.org/articulo.oa?id=109/10900606. Accessed 14 Apr 2020
8. Martínez, C., Monserrat, A., Alhama, J.: El paradigma relacional en las estrategias de las organizaciones de marketing de destinos. Interactividad y función social en las webs turísticas oficiales/The relational paradigm in the strategies used by destination marketing organizations interactivity. Rev. Lat. Comun. Soc. **72**, 374–396 (2017). https://doi.org/10.4185/RLCS-2017-1170
9. Álvarez, G.: Diseño web y posicionamiento SEO (2016). https://www.webbbli.com/tag/gonzalo-alvarez/
10. Agyapong, D.: Micro, small and medium enterprises' activities, income level and poverty reduction in Ghana–a synthesis of related literature. Int. J. Bus. Manag. **5**, 196 (2010)
11. Alcaide, C., Bernués, J.C., Díaz-Aroca, S., Espinosa, E., Muñiz, R., Smith, R.: Marketing y Pymes. Las principales claves de marketing en la pequeña y mediana empresa (2013)
12. Dyllick, K., Hockerts, T.: Más allá del caso de negocio para la sostenibilidad corporativa Estrategia empresarial y medio ambiente. Bus. Strateg. Environ. **1** (2002)
13. Aguilera Castro, A., Puerto Becerra, D.P.: Crecimiento empresarial basado en la Responsabilidad Social Adriana Aguilera Castro. Pensam. Gestión **32**, 1–26 (2012). http://www.scielo.org.co/scielo.php?pid=S1657-62762012000100002&script=sci_arttext&tlng=en
14. Seung, W.Y., Ji, H.S.D.H.L.: Beyond-the-learning-process.pdf. Perform. Improv. Q. **22**(3), 49–69 (2009). http://www.performancexpress.org/wp-content/uploads/2011/11/Beyond-the-Learning-Process.pdf
15. Francisco, J.: Gestión dinámica de conocimiento organizacional. Enl@ce Rev. Venez. Inf. Tecnol. y Conoc. **12**(2), 55–78 (2015)
16. Torres, R., Rivera, J., Cabarcas, R., Castro, Y.: La efectividad del uso del marketing digital como estrategia para el posicionamiento de las pymes para el sector comercio en la ciudad de Barranquilla. Rev. UNISIMON-Investigación y Desarro. en TICs **8**(2), 3–5 (2017). http://revistas.unisimon.edu.co/index.php/identic/article/view/2947
17. Parise, P., Guiñan, S.: Marketing using web 2.0. (2008)
18. Caro, J.L., Luque, A., Zayas, B.: Nuevas tecnologías para la interpretación y promoción de los recursos turísticos culturales. **13**, 931–945 (2015)
19. Porter, M., Kramer, E.: Estrategia y sociedad (2006)
20. Porter, M.E.: Ventaja Competitiva - Creación y sostenibilidad de un rendimineto superior, p. 573 (2010)

Strategic Management of Human Talent and Job Performance: Policies and Human Talent Subsystems for the Growth and Development of MIPYMES

Emanuel Bohórquez(✉), Verónica Benavides, William Caiche, and Arturo Benavides

University Península of Santa Elena – UPSE, Av. Principal La Libertad, Santa Elena, Ecuador
ebohorquez@upse.edu.ec

Abstract. At present, organizations and companies face new challenges that involve human capital and their work performance, being essential the strategic management of human talent, which allows capturing, retaining and valuing human capital, being the most important asset for their invaluable contribution by providing development and productivity.

The main problem of this research is to diagnose the current situation regarding the strategic management of human talent and job performance in the company "Stone Distributor".

For the study, a qualitative and quantitative approach was used with a descriptive research scope, supported by a bibliographic research method, which involved both the manager, administrative staff and workers. Taking as the main results obtained in terms of human talent policies that 48% and 44% of the aforementioned personnel disagree with the discipline and open-door policies respectively. Regarding the human talent subsystems, 44% and 32% of those surveyed disagree with the performance evaluation and recruitment and selection subsystem; while 36% state that they are undecided with the training and development sub-system.

Finally, a recommendation is established to prepare and implement a strategic plan for human talent (human talent policies and subsystems), which allows the improvement of the perception of the degree of conformity regarding the application of processes for the management of human talent, as well as the skills, knowledge, effectiveness and job performance of all employees.

Keywords: Strategic management · Human talent · Job performance

1 Introduction

In today's world, human talent is a important component in companies, because it is the basis of their productivity and boom. As stated [1], organizations over time are promoting changes in their structures, investments, technologies, markets, among other important aspects to be more competitive in globalized environments, however, human resources as the center of business dynamics has been little considered and valued in the face of

M. Zambrano Vizuete et al. (Eds.): CI3 2021, LNNS 511, pp. 587–599, 2022.
https://doi.org/10.1007/978-3-031-11438-0_47

these changes, despite the fact that the extensive literature of administrative sciences throughout more than a century shows that human resources are the most valuable asset.

For this, the strategic management of human talent is important, [2] he highlights it as the importance of employee valuation, thinking in terms of relationships means taking the next step and valuing those relationships. This focus on workforce performance or human capital is a relatively new concept for both line managers and HR professionals, but it is also increasingly a feature of the increased effort to assess intangible asset performance.

Studies carried out in Latin America specifically, detail that a common problem is identified in many Latin American countries, and it is the loss of talents to which companies are exposed, for which from the management of human talent there must be a strategic management that seek to minimize the occurrence of this problem, and to allow assertively disseminating the organizational objectives, seeking that the entire team of collaborators actively participate, generating a sense of belonging [3].

Currently, in the Santa Elena peninsula, although success does not only depend on job performance, this has become a important element to achieve the effectiveness, efficiency and above all the objectives of a company, but many times it is interrupt-ed by the lack of elaboration and implementation of a strategic plan for human talent.

Thus, we have the "Stone Distributor" where it was possible to verify relevant characteristics that negatively affect the performance of the personnel, which has been obtained through preliminary information. Generally, it is possible to mention not having a strategic plan for human talent, within which factors such as an inadequate work environment and a visionary culture are derived; as well as the lack of training both in the operative and administrative areas, the policies established empirically in the company are not complied with; In addition, the load of hours of work of the personnel is directly affecting the productivity of this company. Thus, productivity and therefore sales decrease or maintain a standard level with the company where they provide their services and empowerment of their activities.

This research is structured as follows: the second section presents the principles and theories regarding the strategic management of human talent and job performance; the third section details the methodology used; the fourth section presents the results and discussion. Finally, the fifth section shows the conclusions and recommendations.

2 Strategic Management of Human Talent and Job Performance

2.1 Strategic Management of Human Talent

Highlighting the contribution of [4] the planning of human resources is a key and integrating movement of human resources management, becoming a necessary compass for the entire operation, because it contains subsystems and factors that if not are managed properly have a direct impact on staff and their performance.

Also, evaluating the contribution of [5] the strategic management of Human Talent is the formulation, implementation of policies and human resources skills, producing skills and behaviors of the workers that are necessary for the company reaching its strategic goals.

In conclusion, human talent management is an indispensable tool for the management of human talent in any type of organization or company regardless of its size, in order to achieve the objectives, set through the application of policies and practices focused on human resources.

Human Talent Policies

In relation to the applied human resources policies [6] it contributes that they are significant, to a great extent, they are used to attract, retain and motivate employees, in general, it is established that when these objectives are achieved, it produces an effect positive in the results, because it influences the discipline, organization and compliance of the collaborators.

Likewise, [7] points out that human talent policies are guidelines that are established to manage functions and safeguard that these are carried out in accordance with the planned objectives. These form an administrative orientation to prevent employees from taking actions that are inappropriate for their area or that represent a danger to the success of the specific functions or tasks assigned according to the requirements of the job.

Considering the contribution of [5] on the organizational policies are those that the members of the organization must fulfill within are the Human talent policies. There are three types of policies focused on human talent:

Discipline policies: concerning the fulfillment of the processes set for the organization by the individuals that comprise them.

Open door policies: the collaborators themselves are respective, managing to access higher levels with ease, to draw up concerns and suggestions.

Policies on diversity: related to safeguard and work with people of different characteristics, respecting the social composition.

Human Talent Subsystems

We mention [8] stating that a subsystem is a continuous process that provides the information and the appropriate tools to the employees to manage their activities, becoming the fundamental axis for the management of an optimal human talent management system.

The human talent management subsystems are the links that interrelate human development with organizational development, it is also responsible for shaping the necessary conditions for people to progress comprehensively within the organization and contribute to its permanence [9].

There are different types of subsystems depending on the appreciation and contribution of the author, such as the following:

We mention the article in the scientific journal [10] citing Idalberto Chiavenato (1993) analyzes Human Resources Administration as a process that is constituted by integrated interdependent subsystems distributed as follows:

HR provision subsystem: This contains market research, recruitment and selection, and manpower.

HR application subsystem: This subsystem includes the analysis and description of positions, performance evaluation, integration or induction and the movement of personnel.

HR maintenance subsystem: Includes compensation, hygiene, social benefit plans and safety at work, investigations and surveillance of personnel.
HR development subsystem: Includes preparations and personnel progress plans.
HR control subsystem: Includes the data system, HR research system. and the HR audience

The Human Talent Management Systems, according to [11] are subdivided into: Recruitment and selection subsystem, training and training subsystem, evaluation and performance subsystem, Safety and hygiene subsystem.

Recruitment and selection subsystem: Recruitment can be defined as a set of instructions used to attract sufficient and suitable candidates for a specific position in an organization.

Training and training subsystems: it comprises the training, education and development that were part of a suitable and professional education.

Evaluation and performance subsystems: performance evaluation is used as an indicator of the quality and fulfillment of the professional profiles of human resources. The design of the techniques of the companies and their manuals, as part of the responsibility of the human talent area, is part of the success of the progress of all the personnel.

Health and safety subsystem: Information from job analysis is also excellent in identifying health and safety considerations.

[12] states that the dynamic element of the proposed model are the subsystems that make up the integrated system for the development of human talent, such as:

Integration Subsystem: it is very important to have the right people to develop the different functions and activities that the company needs, for this reason the selection of personnel must be demanding according to the criteria considered and necessary that, through excellent development, guarantee optimal development and operation of the organization.

Recruitment and selection subsystem: it is an important process within human talent management since it allows attracting potential candidates who may fill a vacancy within the organization. The purpose of the selection is to choose and classify among the recruited candidates, the most suitable that meet the needs of a position within the organization, with the aim of maintaining and increasing the effectiveness of staff performance.

Compensation subsystem: this subsystem seeks to incentivize employees so that their individual needs can be met, through fair remuneration and benefits that adhere to the law.

Development subsystem: it is used for the development of processes that enhance the skills, knowledge and skills of the employee for the benefit of the company. Training: Training means preparing the employee to better hold a position.

Human resource audit subsystem: brings together the different processes used to control and evaluate the activities carried out by each member of the organization, verifying the results obtained.

2.2 Job Performance

According to [13] it holds that job performance is subject to the skills and knowledge provided by the worker's competencies, in the processes of achieving the company's objectives. This can be verified through performance reviews, through which employers take into account elements such as organizational skills, productivity, leadership ability and time management to study each employee individually.

In addition, we value the contribution of [14] on job performance involves the interaction of human relationships, emotions, competences, feelings; very important elements in the direction to the fulfillment of goals and objectives. A high job performance will be one that achieves such powerful emotional ties and that they are capable of sacrificing some particular interests for the common welfare and is capable of resisting dangers and persevering until achieving the objectives, engaging in continuous improvement as it is linked to the initiative and the capacity for innovation.

Therefore, [15] contribution that the term work performance refers to what a worker actually does in a company and their performance in their work activities, however, the aspects that the individual has such as efficiency are necessary, quality and productivity with which it performs.

Job performance is the way in which an individual develops within the organization, the appropriate relationships, skills and competencies at the time of carrying out an assigned activity, which is why a large part of organizations evaluate performance to measure the degree of effectiveness and efficiency with which they carry out their activities.

Job Performance Factors

There are various factors with which the performance of human talent can be related, making it the most appropriate as stated by the author [16], such as:

Competences: declares that it is the knowledge of effectively developing a task, using the necessary capacities, experiences, skills and understanding, as well as the properties that facilitate the solution of contingent contexts and problems.

Knowledge: [17] that knowledge is an operation of reality, but of the reality that belongs to the human being, and such reality is fickle, therefore, knowledge is equivalently changing.

Skills: [18] tribute that it is the ability to carry out conceptual concepts, it is the ability to work together with the members of the organization, understand them intellectually to study them, diagnose and motivate them, both individually and in group situations.

Leadership: [19] leadership is the step where an individual influences a group of people and instills, motivates and directs their responsibilities so that they achieve the established objectives, the essence is to have followers.

Measurement of Job Performance

Work performance is a variable that can be measured with various indicators, among these we have: efficiency, effectiveness, effectiveness, productivity, quality, customer satisfaction, among others, these are essential to weigh human talent [20]:

Effectiveness: [21] states that this implies efficiency and effectiveness, that is, the achievement of programmed results in time and costs as optimized as possible. It means doing the right thing with great accuracy and not having any unnecessary use of time or money.

Quality: [22] the quality of a product or service lies in the disposition of the product for human consumption, in other words, reaching the customer's expectations in the specific use of the product.

Customer satisfaction: [23] satisfaction is the level of acceptance that a product provides to the consumer according to their needs, leaving as a result the optimal state of mind on the level of satisfaction generated by a good or service acquired.

3 Methodology

The "Stone Distributor" is a company that began its activities in 1981, located in the canton of the Libertad, starting as a small stand dedicated to the sale of slaughtered chickens and becoming today a company dedicated to the commercialization and distribution of meat products, expanding its product lines such as beef and pork; managing to attract more than the clients of family consumption, other types of market such as hotels, restaurants, boats among others.

The main objective of this research is to diagnose the current situation regarding the strategic management of human talent and job performance in the company "Stone Distributor", being a non-experimental study of a non-probabilistic nature carried out between 2018 and 2019.

Thus, for the present research a qualitative and quantitative approach was used with a descriptive research scope, relying on a bibliographic research method related to the strategic management of human talent and job performance, as well as field research that in This study was applied by collecting primary information on site provided by the manager, administrative staff and workers belonging to the company "Stone Distributor", applying information gathering tools such as interviews and surveys that allowed identifying the qualities and characteristics of the problem raised, as well as for its analysis, understanding and measurement of results.

For the development of this project, a person represented by the manager was considered as the population for the interview; while for the survey there was a population of 65 people made up of administrative personnel and workers; in this case, the sample corresponds to the total population as it is finite and achievable, which will contribute with their opinion about the strategic management of human talent and job performance in the company "Stone Distributor".

The main research instrument were the surveys, which were applied to the 65 collaborators; It consists of 26 closed questions, structured in three parts: the first with sociodemographic data, a second section with questions related to the strategic management of human talent and a third section with questions oriented to job performance, the last two sections with Likert scale options.

4 Results and Discussion

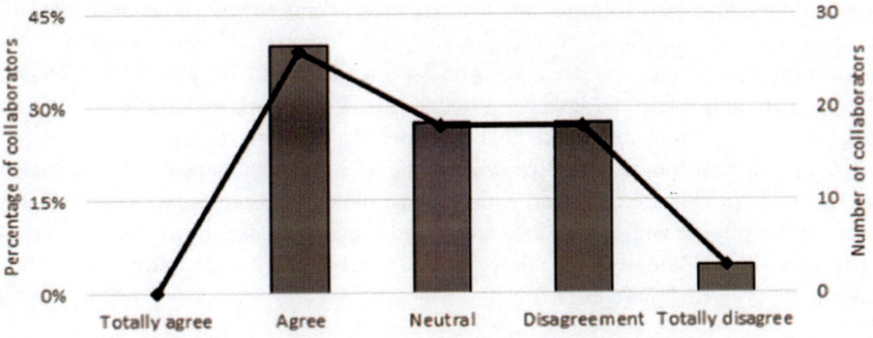

Fig. 1. Strategies for the management of human talent

In Fig. 1, referring to the strategies for the management of human talent, it can be highlighted that 40% of the staff of the "Stone Distributor" affirm that they agree with the strategies for the management of human talent, while 28% they are undecided and also in disagreement, being able to explain because these strategies are not formally established but rather empirically and are not applied constantly; culminating with 4% that indicate they totally disagree justified by the ignorance given by the little time that said personnel has working.

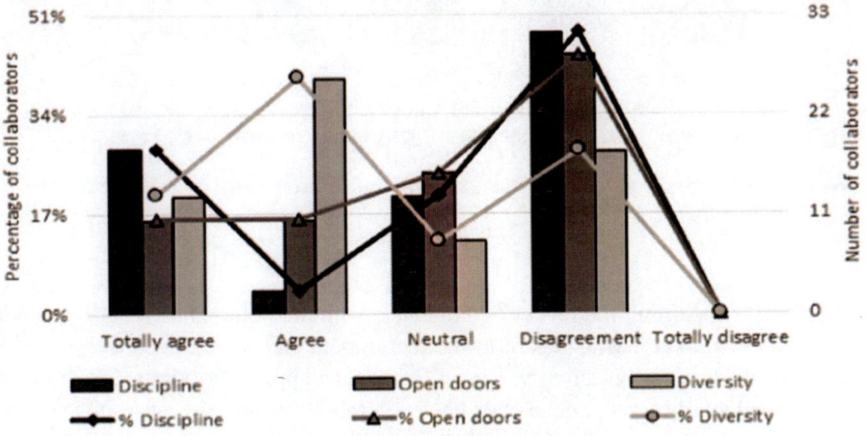

Fig. 2. Human talent policies

Figure 2, related to the degree of compliance with the human talent policies, in terms of discipline policies 48% of the staff mention disagreeing, because there are discipline human talent policies but they are empirically and they are not written; on the other hand, 28% declare that they totally agree, indicating that these types of policies are not complied with by the staff in general and 4% agree that at the time of their hiring, they verbally expressed them.

In relation to open door policies, 44% mentioned disagreeing because their daily tasks do not allow them a suitable moment in which concerns or problems can arise; while 24% are undecided because they consider that due to their degree of work experience they cannot provide their opinion to managers on a specific issue; finally, 16% for the category of agreement and totally in agreement define that if they have had the opportunity to speak to their superiors and that their suggestions are taken into account.

Finally, in correspondence to the diversity policies, it was obtained that 40% declare that, if these policies are complied with and that they work together regardless of sex, race, among others; while 28% indicate they disagree because they observe a scarce application of it and finally 12% are undecided because they are unaware of some type of formally proposed policies.

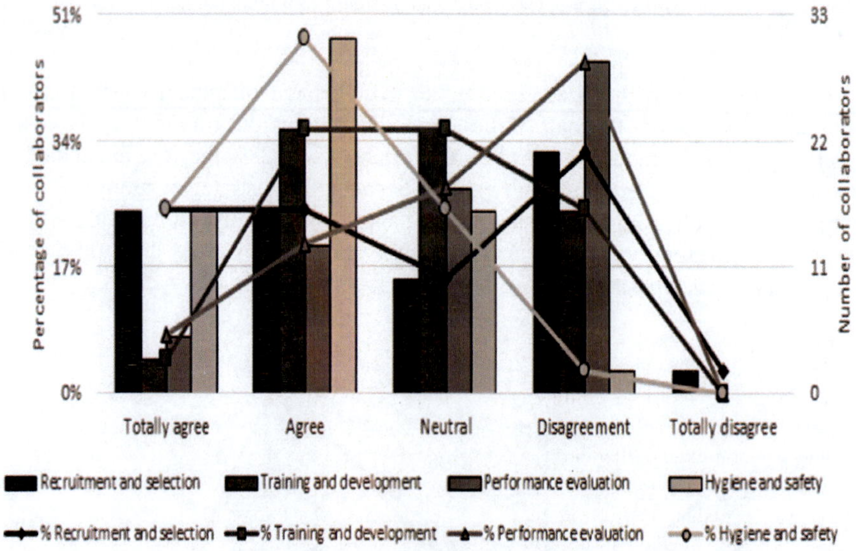

Fig. 3. Human talent subsystems

In Fig. 3, concerning the level of compliance with the human talent subsystems in the "Stone Distributor"; with regard to the recruitment and selection subsystem, 32% of respondents disagree because there is no protocol and job descriptions; 24% indicate they agree and totally agree; on the other hand, 4% totally disagree because at the time of being hired the company did not have a recruitment and selection process of personnel.

In relation to the training and development subsystem, 36% state that they are undecided because several members are new and have not yet received some type of training except for induction of the teams to be managed; while with the same 36% they agree, mentioning that there have been trainings although they have not been constant, mainly in customer service and merchandising; finally, with 4%, the respondents state that they totally agree because they have received training but that this process is carried out empirically.

Regarding the performance evaluation subsystem, 44% disagree as there is no performance evaluation; likewise, 28% indicate that they are undecided, commenting that through direct observation, managers make corrections to staff; and with 8% the respondents state that they fully agree, stating that they are informally evaluated by the heads of areas.

Finally, due to the hygiene and safety subsystem, 48% and 24% are shown to fully agree and agree respectively, stating that employees are provided with utensils and supplies necessary for the development of their daily activities in the operational area, the company complying with safeguarding the integrity of the staff and finally with 4% they disagree due to the lack of knowledge of the proper use of the equipment they have at their disposal.

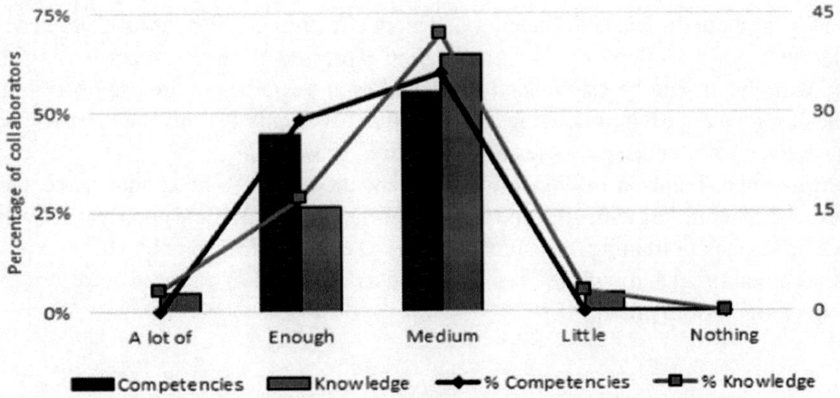

Fig. 4. Competencies and knowledge for job performance

Figure 4, referring to the degree of contribution made by skills and knowledge for job performance; regarding skills, 56% and 44% consider that they contribute normally and sufficiently respectively, being able to point out, on the one hand, that there would be an adequate management of skills for the use of machines and equipment assigned to them for the development of their activities, and on the other hand, it would be necessary to take actions that allow other collaborators to improve their skills for the proper development of their daily activities.

Regarding knowledge, 64%, 28% and 4% state that their knowledge contributes moderately, sufficiently and much respectively, with this the lack of training and / or seminars for the acquisition and / or strengthening can be considered in a majority of new knowledge.

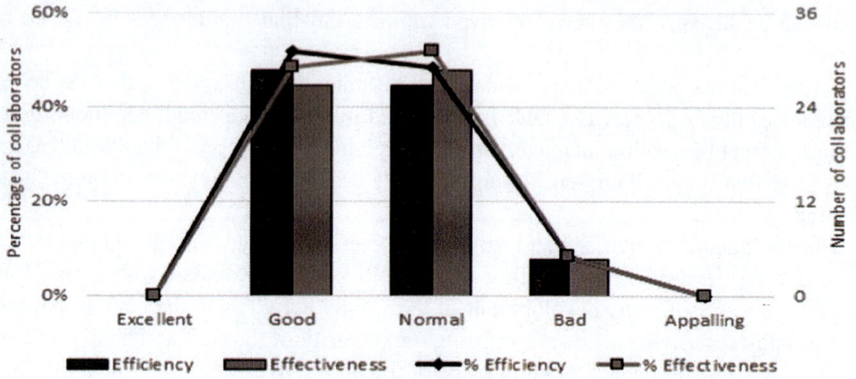

Fig. 5. Work efficiency and effectiveness

In Fig. 5, about the level of efficiency and work effectiveness, in relation to efficiency, similar weightings of 48% and 44% can be seen expressed as good and normal respectively, with this, it can be established that supplies and equipment are used necessary for the development of their activities; on the other hand, only 8% are classified as bad, because sometimes unnecessary resources are used or wasted.

On the other hand, in reference to efficiency, the existence of similar percentage values can be mentioned with 48% and 44% qualifying it as normal and good respectively, analyzing the lack of training to ensure that employees can fulfill their obligations within the established time; and with 8% it is weighted as bad because on some occasions the objectives are not fully met.

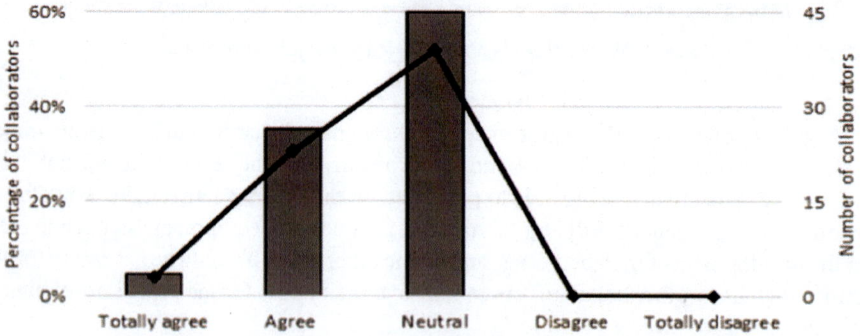

Fig. 6. Strategic management of human talent and job performance

In Fig. 6, about the degree of conformity about how the strategic management of human talent influences the work performance of the collaborators in the "Stone Distributor", 60% are shown to be undecided, explaining by the fact of not existing formally policies and application of human talent subsystems; on the other hand, 36% and 4% agree and totally agree, appreciating compliance with the current way of managing human talent by managers, despite being informal and empirical.

5 Conclusions and Recommendations

Today, in a globalized, changing and competitive world, it is imperative that managers recognize the importance of the strategic management of the human talent that works in their companies, regardless of the size of a business, it is a key factor for an adequate job performance of your employees.

In this research it has been possible to diagnose the current situation regarding the strategic management of human talent and job performance in the company "Stone Distributor".

To conclude, regarding the strategic management of human talent, we began by expressing about the degree of conformity with the human talent policies, it is thus, that for discipline policies 48% of the staff mention disagreeing, because they are empirically and are not written; In relation to open door policies, 44% mentioned disagreeing because their daily tasks do not allow them to present concerns or problems; while 24% are undecided because they consider that due to their degree of work experience they cannot provide their opinion to managers. In correspondence to the diversity policies, it was obtained that 40% declare that if these policies are complied with and that they work together regardless of sex, race, among others.

Continuing, with respect to the recruitment and selection subsystem, 32% of respondents disagree because there is no protocol and job descriptions. In relation to the training and development subsystem, 36% state that they are undecided because several members are new and have not yet received some type of training except for induction of the teams to be managed; while with the same 36% they agree mentioning that there have been trainings although they have not been constant. Regarding the performance evaluation subsystem, 44% disagree as there is no formal process and application of performance evaluation. Due to the hygiene and safety subsystem, 48% and 24% show that they fully agree and agree respectively, stating that employees are provided with utensils and supplies necessary for the development of their daily activities in the area. operative.

The application of strategies for the management of human talent in the "Stone Distributor" is rated favorably with only 40% indicating agreement, although they are not formally established but empirically established and are not applied constantly; leaving a difference of 60% who would be dissatisfied and which would lead to consider the implementation of actions to reduce such results.

As a result, it is concluded regarding the degree of contribution made by the skills and knowledge of employees for job performance, that 56% and 44% consider that their skills contribute in a normal and sufficient way respectively, while 64% and 28% state that their knowledge contributes moderately and sufficiently respectively.

In relation to efficiency, similar weightings of 48% and 44% can be observed, expressed as good and normal respectively. And in reference to efficiency, the existence of similar percentage values can be mentioned with 48% and 44% qualifying it as normal and good respectively.

Finally, it can be concluded that the degree of conformity about how the strategic management of human talent has influenced the work performance of the collaborators in the "Stone Distributor" reaches 60% of undecided, explaining by the fact of not existing formally policies and processes for human talent subsystems.

According to the findings obtained, it is recommended that the directors of the "Stone Distributor" develop a strategic plan for human talent (human talent policies and subsystems) according to the needs of the business, which, once approved, must be shared with all staff and executed formally and fairly; with which it is possible to improve the degree of conformity in terms of the application of processes for the management of human talent, as well as the skills, knowledge, effectiveness and work performance of all collaborators.

The present investigation was limited in some cases by the limited availability of time on the part of the collaborators due to the rotation of schedules and the various activities assigned to them. Future research could compare and contribute to the discussion of results in other contexts in terms of strategic planning of human talent and its influence on work performance, as well as a basis for correlational research. In addition, the recommendations are expected to make viable and improve the management of human talent for the benefit of all employees.

References

1. Briceño, M.: El Talento Humano: Un Capital Intangible que otorga valor en las organizaciones. Int. J. Good Consci. **7**(1), 57–67 (2013)
2. Torre, C., Delgadillo, H., Fregoso, J.: La gestión estratégica del talento humano como ventaja competitiva en las organizaciones del sector hotelero. Red Internacional de Investigadores en Competitividad **9**(1), 1071–1089 (2015)
3. Vargas, A., et al.: Importancia de la planeación estratégica en las áreas de gestión humana de las organizaciones. Fundación Universitaria Luis Amigó **3**(1), 116–122 (2016)
4. Córdova, M., Urresta, A.: Diseño de un plan Estratégico de recursos humanos para fomentar el trabajo en equipo de la empresa Luxus Blumen S.A con la aplicación a las PYMES. Universidad Politecnica Salesiana, Quito (2015)
5. Alles, M.: Dierección estratégica de recursos humanos: gestión por competencias. Granica, Buenos Aires (2015)
6. Garcìa, J., Durá, S., Prieto, R.: Polìticas de gestión de talento humano el desarrollo de competencias. FACE (Facultad de ciencias económicas y empresariales) **17**(2), 130–141 (2017)
7. Chiavenato, I.: Comportamiento organizacional. McGraw-Hill, México (2015)
8. Saavedra, R., Castro, L., Restrepo, O., Rojas, A.: Planificación del desarrollo. Universidad de Bogotá Jorge Tadeo Lozano, Bogotá (2017)
9. Caicedo, V., Acosta, A.: La gestión de talento humano y el ser humano como sujeto de desarrollo. Revista Politécnica **8**(14), 105–113 (2012)
10. Obando, E.: La administración de recursos. Universidad Latina de Costa Rica, Costa Rica (2015)
11. Amador, A.: Administración de recursos humanos. Su proceso organizacional. Universidad Autónoma de Nuevo León, México (2016)
12. Guamán, M., Guamán, P., Ruíz, E.: Sistema de gestión de talento humano para el sector avícola de la provincia de Tungurahua. Universidad Técnica de Ambato, Ambato (2017)
13. Alarcón, M., Espinoza, K.: Plan estratégico para la gestión del talento humano para la hacienda hostería Milliguayco localizada en el cantón Pallatanga provincia de Chimborazo. Universidad Laica Vicente Rocafuerte de Guayaquil, Guayaquil (2015)
14. Oscco, H.: Gestión del talento humano y su relación con el desempeño laboral del personal de la municipalidad distrital de Pacucha - Andahuaylas - Apurímac, 2014. Universidad nacional José María Arguedas, Perú (2015)

15. Mazariegos, M.: Motivación y desempeño laboral. Universidad Rafael Landívar, Quetzaltenango (2015)
16. Capuano, A.: Evaluación de desempeño: desempeño por competencias. Invenio **7**(13), 139–150 (2004)
17. Esparza, R., Rubio, J.: La pregunta por el conocimiento. SABER. Revista Multidisciplinaria del Consejo de Investigación de la Universidad de Oriente **8**(4), 1–9 (2016)
18. Stephen, R.: Comportamiento Organizacional. Pearson Educación, México (2013)
19. Méndez, J.: Teoría del liderazgo: material para el curso de administración de proyectos II. Instituto Tecnológico de Costa Rica, Costa Rica (2014)
20. Rojas, M., Jaimes, L., Valencia, M.: Efectividad, eficacia y eficiencia en equipos de trabajo. Revista Espacios **39**(6), 1–15 (2018)
21. Mejía, C.: Indicadores de la efectividad y eficacia. Planning S.A., Medellín (1998)
22. Lizarzaburu, E.: La gestión de la calidad en Perú: un estudio de la norma ISO 9001, sus beneficios y los principales cambios en la versión 2015. Revista Universidad & Empresa **18**(30), 33–54 (2015)
23. Vivas, G.: Satisfacción de los clientes. Universidad de Belgrano, Argentina (2013)

Prediction of Diseases in the Elderly in Manabí Through Big Data Technologies

Fernando Alfredo Reyes Reyes[1,2](✉), Marely Del Rosario Cruz Felipe[1,2], and Jorge Parraga-Alava[1,2]

[1] Instituto de Posgrado, Universidad Técnica de Manabí, Portoviejo, Ecuador
Freyes3632@utm.edu.ec

[2] Facultad de Ciencias Informáticas, Universidad Técnica de Manabí, Portoviejo, Ecuador

Abstract. In the last decade, data collection and analysis have increased considerably in many fields of society. Big Data analysis has begun to play a fundamental role in the advancement of practices and research, as well as in the area of health. In this sense, data mining is a tool that allows you to find useful behavior patterns for decision-making in medicine or clinics, as well as predicting diseases. CVDs are one of the main causes of death in older adults and in the province of Manabí there are no studies reported on the analysis of risk factors for this disease in the elderly population. This research compares the performance of 13 learning algorithms to predict the risk of cardiovascular diseases (CVD) in the population of Manabí, the dataset used is composed of 604 participants, from a study by the National Institute of Statistics and Census (INEC) on health, the well-being of the elderly. The performance of the model was measured by the area under the curve (AUC), obtaining that the model with the best performance is the Random Forest, from which the prediction was made and the value of the variables could be obtained through the Shapley index or more significant risk factors in CVD. In Manabi, these are: if you have shortness of breath, the value of stature_altura if you have taken medication and may or may not have persistent dizziness.

Keywords: Big Data · Data mining · Prediction algorithms · Random Forest

1 Introduction

Cardiovascular diseases are one of the major causes of risk in older adults in the world. In Ecuador, according to surveys by the National Institute of Statistics and Census [1], it is shown that the patients who were admitted to the hospital and died due to CVD is 14.3% in men and 15.0% in women in the Costa region. In the province of Manabí, significant data regarding this disease are presented in most adults between 65 and 85 years of age [1].

For the prevention and detection of causes of disease such as CVD (Cardio Vascular Disease) in the elderly, analysis of data from surveys has been developed, patient data in isolation and in many cases, they are not analyzed with the level of detail and with the haste that this requires.

M. Zambrano Vizuete et al. (Eds.): CI3 2021, LNNS 511, pp. 600–611, 2022.
https://doi.org/10.1007/978-3-031-11438-0_48

At present there are Big data technologies that are applied in different solutions, optimizing the data analysis time and performing a deep analysis of the same as prediction models. This applied to the analysis of cardiovascular diseases would allow to develop of prediction models for this disease and obtain which are the most significant risk factors in this disease, allowing measures to be taken to improve the quality of life of older adults.

In the last decade, data collection and analysis have increased tremendously in many fields of society. Big Data analysis has become a fundamental factor in the advancement of research and medical practices. It has provided tools to store, manage, examine and assimilate large volumes of data, these can be structured and unstructured produced by current hospital systems [2]. Recently, Big Data analysis has been applied to aid the disease surveillance and exploration process. With proper management, improvements in the quality, quantity, storage, and analysis of health data could lead to significant improvements in many health outcomes.

In treating a patient, his care is not only based on his own health data, his medical history, the results of medical examinations, etc., but also on data and reports on the health of thousands of people [2]. In this sense, artificial intelligence and data mining are tools that allow finding useful behavior patterns for decision-making in medicine. To do this, it uses classification and prediction methods.

There are several studies where Artificial Intelligence and data mining techniques are applied where it is applied to health. In [3] machine learning was used to characterize the risk of suffering from type 2 diabetes mellitus. In which machine learning is used to improve discrimination by analyzing complex interactions between massive variables in this way testing the capacity of algorithms that were used. The GBM model worked better. Found the 10 main variables in all the methods used, showing competence to predict the risk of DM2.

In [4] health at an international level, the "Comparative prediction analysis within cancer databases" stands out, which is responsible for comparing the precision and performance of supervised and unsupervised clustering algorithms, for that matter. of the unsupervised, the most relevant is the K-means. And the multilayer Perceptron. On supervised models, the SVM performed better. About breast cancer, use a Wisconsin Diagnostic Database (BCW).

In [5] a predictive study for the diagnosis of cardiovascular diseases is presented. In this study, Bayesian networks and decision trees are used to perform prediction tasks related to this disease. As a result of the study, an alternative is presented to determine whether a clinical procedure should be performed on a patient based on variables such as blood pressure, gout, hypothyroidism.

The study of pressure in cerebral aneurysms is carried out in [6] where the relationship of shear stress with pressure in cerebral aneurysms and the prediction of the risk of rupture using artificial intelligence tools was studied. What it does are two studies, the first of which is performed fluid dynamic simulation (CFD) for 60 human brain aneurysms which 30 previously ruptured and the other 30 not ruptured, to study the behavior of shear stress, the results showed a linear behavior. between the averaged time and the aspect ratio, while the different pressures increased and in the second study, it showed that the morphological and fluid-mechanical parameters are recognized as the main risk factors for aneurysm rupture. For this, 5 Machine Learning models were used to predict the risk of aneurysm rupture, where the model that had the best response was adaptive reinforcement (AdaBoost).

Machine learning is used to determine cardiovascular risk as in [7] it is used to characterize cardiovascular risk. With which he tested a machine learning technique, such as random survival forest modeling, and obtained that this technique works better than the established risk scores, better precision is obtained to predict cardiovascular events.

To create the machine learning model, the Bayesian Naive Bayes (NB) classification algorithm was used to diagnose CVD and Bayesian Networks as a tool to identify the cardiovascular risk factors with the highest incidence and thus establish priorities for primary prevention.. Once the risk estimate is obtained, it will serve as a useful key point in decision-making for the treatment of cardiovascular disease [8].

Some research carried out applies machine learning to healthcare solutions such as [9] where machine learning is a discipline of artificial intelligence that can create systems that learn automatically. Learning means that from millions of pieces of data you can identify complex patterns. The one that learns is an algorithm that reviews the data and is capable of predicting future behavior automatically, it also implies that these systems are optimized autonomously, without human intervention.

Another application of neural networks is in [10] where the influence of emotions in the science of learning is analyzed, for this they used the LSTM machine learning method, which interprets characteristics and extracts them from one or more images of an individual. The results indicated that negative mood states can influence a low percentage of accuracy of responses in an assessment.

There are also several types of machine learning, unsupervised learning, and supervised. Unsupervised learning is a method that adapts to observations to find groups of data that have similar characteristics, in this learning it does not have a priori knowledge, as indicated [11].

Supervised learning as proposed [11] in this learning model seeks to induce models capable of predicting the value of a certain dependent variable from independent variables, this algorithm creates a function that assigns the inputs to the desired outputs. A standard formulation of the supervised learning task is the classification problem.

Machine learning can learn non-linear interactions repeatedly from millions of data using algorithms [12], which have been applied in various fields, such as disease risk assessment and prediction in [7, 13] an investigation presents the performance of two unsupervised machine learning-based algorithms for the detection of premature ventricular contraction beats [14].

Another interesting finding in [15] focuses on predicting the likelihood of patients having heart disease using medical profiles such as age, gender, blood pressure, and blood sugar. Which uses the IHDPS system and compares three algorithms, such as DT, Naïve Bayes, and Neural Network. They use Cleveland this database contains information on heart disease. I perform a division of the records into two data sets. To avoid bias, the records from each set were randomly selected, the models were trained and validated against a test data set. And he obtained as a result that the most efficient model is Naive Bayes in predicting heart disease.

In the previous works, there is a limitation, it is that they allow evaluating some models and using 2 to 4 prediction algorithms, this work proposes the evaluation with 13 prediction algorithms for the detection of cardiovascular diseases. Another important element is the quality of the data source since they are a significant element when applying a model. In the Manabí region, despite the existence of data collected from population surveys, no study is reported in which this type of analysis is carried out for the prediction of CVD.

The objective of this work is to predict diseases, for which a series of machine learning algorithms are evaluated to predict the risk of cardiovascular disease (CVD) in elderly people in the Manabí region, identifying the important variables and evidencing the performance of the model used.

2 Methodology

A hybrid methodology is proposed between Crisp-DM (Cross - Industry Standard Process for Data Mining) [16], and KDD process (Knowledge Discovery Databases), with which it is better adapted to our data mining analysis, which will allow us to the characterization of the patients. It comprises four phases: context analysis, data collection, modeling, evaluation. As shown in the diagram in Fig. 1 These steps included in the methodology will allow the optimal development of a proposal that can provide a solution to the problem of analysis and prediction.

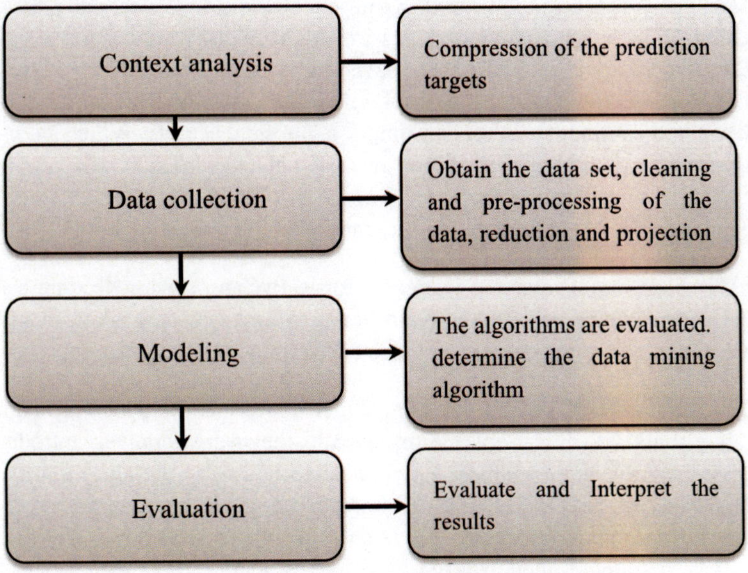

Fig. 1. Design of the applied methodology.

Phase I: Context analysis. The objective is precisely to contribute to the early detection of diseases in the elderly so that they can be prevented, so a predictive model is designed based on traditional risk factors and also information from tests carried out, from [1].

Phase II: Data collection the participants of this study came from a database that has 5236 participants and 841 variables, from which only the one that belonged to the Manabí province was extracted, from these 36 most significant variables were selected for CVD disease, based on the analysis of the specialist in the subject, an exploration of the data is carried out, which verifies that the data is not incomplete or missing. The selection of the variables to be used is made, this was done with the Power Bi program, which is a tool that has a large number of options to import data from very different sources such as Excel or Access sheets to data from a website that allows it as a Google Analytics or Facebook account as well as having connections to the most used Big data and relational sources.

With the Power BI tool, the information type is reviewed, so cleaning of the data was carried out in which records that contained null, incomplete values, values that exceeded the normal range, variables that they do not provide information, and if there is a need to recode variables, the column labels were changed to understand what values were involved.

Subsequently, we proceeded to select the attributes that provide the most significant or relevant information, on the type of data and the possible values. What was obtained, 3 different groups according to their nature? These are the objective characteristics: data such as age, biological sex, height, weight which are measured (days, binary, cm, kg). The second group characteristic of examination is values of physical and physiological examination. Of those who have blood pressure (high and low), cholesterol, and levels of glucose, triglyceride, vitamin B12, these have been classified as (normal, above normal, and well above). And the last group, they are the subjective ones, is information about lifestyle, such as smoking, alcohol consumption, physical activity. Which are measured in binary values. A dichotomy was carried out on variables such as sex, smoking, physical activity, alcohol consumption, race, level of education (if they have studied, such as kindergarten or other) if they live alone, so these variables are categorical.

Phase III: Modeling. The model is created for it, PyCaret statsmodels are used, here the environment is prepared for the entire learning process. This allows handling the pre-processing steps of the data and divides the data into test and training sets.

Phase IV: Evaluation. CVD risk assessment models were developed in the study, using 13 ML algorithms with all variables. Subsequently, they were iteratively introduced into a progressive number of variables, automatically chosen by the algorithm itself. All algorithms were trained and tested by 10 times cross-validation, this was repeated 100 times, the performance of all models was calculated with the test samples, the parameters of all models were determined by a 10 times cross-validation and a search of all training data.

Model performance refers to the ability to identify who is at risk of developing cardiovascular disease. Sensitivity, specificity, positive predictive value, and area under the curve (AUC) are used to assess discrimination. Sensitivity equals the recovery rate, true positives, and represents the proportion of positive samples that are correctly identified. Specificity indicates that the true negative sampling rate can be detected correctly.

3 Results and Discussion

To determine the risk factors in CVD, the following was done: First, the model is created for it, the statistical models of PyCaret are used, here I prepare the environment for the entire learning process. This allows handling the data preprocessing steps and dividing the data into tests and the training set can be observed in the Table 1. The data are divided as mentioned [17]. The machine learning models rely on different secondary data sets, one for training and the other for testing to evaluate the model. The training data sets are the basis for learning the models and the test set is used with the final fitted models to evaluate which model is better. To do this, the available data is divided into a training set and a test set. For our case, the division is 80% and 20%.

Table 1. Configuration and data division.

	Description	Value
0	session_id	592
1	Target	Diagnosed heart problems
2	Traget Type	Binary
3	Label Encoded	0.0: 0, 1.0: 1
4	Original Data	(604, 12)
5	Missing Values	True
6	Numeric Features	6
7	Categorical Features	5
8	Ordinal Features	False
9	High Cardinality Features	False
10	High Cardinality Method	None
11	Transformed Train Set	(421, 13)
12	Transformed Test Set	(182, 13)
13	Shuffle Train-Test	True
14	Stratify Train-Test	False
15	Fold Generator	StratifiedKfold

CVD risk assessment models were developed using 13 ML algorithms with all variables. They were entered interactively in a progressive number of classified variables, selected by the algorithm itself, all models were trained and tested through cross-validation of 10 times during each process, which was repeated 100 times, the performance of all models was calculated about the testing sample.

3.1 Model Performance Comparison

Once the setup runs, we can use compare_models to briefly evaluate the performance of all the models in PyCaret's model library. There are 13 classifiers available. Compare_models function trains all models and evaluates their performance against 6 common ranking metrics.

In the comparison of the algorithms, all the available variables are used, all the models to predict the risk of CVD, showing a predictive performance with ACCURACY that ranges between 0.862 and 0.8599, the RF model performed better, and I present better AUC (0.6927), ACCURACY (0.8626) positive predictive value, the data that the RF model maintained a strong predictive performance (Table 2).

Table 2. Comparison of algorithms.

	Model	Acu	AUC	Recall	Prec	F1	Kappa	MCC	TT (Sec)
rf	Random Forest Classifier	0.8626	0.6927	0.0167	0.1000	0.0286	0.0255	0.0383	0.526
et	Extra Trees Classifier	0.8554	0.6339	0.0333	0.2000	0.0571	0.0364	0.0586	0.476
Knn	K Neighbors Classifier	0.8530	0.4911	0.0167	0.1000	0.0286	0.0094	0.0164	0.123
ridge	Ridge Classifier	0.8483	0.0000	0.0333	0.1500	0.0536	0.0216	0.0308	0.023
Lr	Loght Regression	0.8461	0.6682	0.1167	0.2750	0.1617	0.1041	0.1135	0.605
lightgbm	Light Gradient Boosting Machine	0.8388	0.6575	0.1033	0.2583	0.1450	0.0812	0.0894	0.105
gbc	Gradient Boosting Classifier	0.8318	0.6642	0.0833	0.2250	0.1136	0.0483	0.0617	0.140
Ada	Ada Boost Classifier	0.8224	0.5865	0.1667	0.2700	0.1971	0.1090	0.1150	0.133
lda	Linear Discriminant Analysis	0.8224	0.6726	0.1833	0.2867	0.2176	0.1259	0.1316	0.028
svm	SVM –Linear Kemel	0.7792	0.0000	0.1167	0.1332	0.0695	0.0160	0.0211	0.026

(continued)

Table 2. (*continued*)

	Model	Acu	AUC	Recall	Prec	F1	Kappa	MCC	TT (Sec)
dt	Decision Tree Classifier	0.7393	0.4923	0.1500	0.1226	0.1340	−0.0168	−0.0171	0.024
nb	Naïve Bayes	0.2015	0.5963	0.8967	0.1382	0.2393	−0.0036	−0.0183	0.023
qda	Quadratic Discriminant Analsis	0.1945	0.4937	0.9100	0.1389	0.2408	−0.0020	−0.0090	0.027

Create Model. A model is created with the selected algorithm using the create_model (Create_model) function. In Table 3 they show the data with cross-validation results of 10 times and 6 different metrics. The mean is the value we get from the compare_models (compare_models) function.

Table 3. Model creation.

	Accuracy	AUC	Recall	Prec.	F1	Kappa	MCC
0	0.8605	0.8333	0.0000	0.0	0.0000	0.0000	0.0000
1	0.86.05	0.7027	0.0000	0.0	0.0000	0.0000	0.0000
2	0.8810	0.5486	0.0000	0.0	0.0000	0.0000	0.0000
3	0.8571	0.8194	0.0000	0.0	0.0000	0.0000	0.0000
4	0.8571	0.6389	0.0000	0.0	0.0000	0.0000	0.0000
5	0.8571	0.6806	0.0000	0.0	0.0000	0.0000	0.0000
6	0.8571	0.7407	0.0000	0.0	0.0000	0.0000	0.0000
7	0.8810	0.5671	0.1667	1.0	0.2857	0.2553	0.3825
8	0.8571	0.6921	0.0000	0.0	0.0000	0.0000	0.0000
9	0.8571	0.7037	0.0000	0.0	0.0000	0.0000	0.0000
Mean	0.8626	0.6927	0.0167	0.1	0.0286	0.0255	0.0383
SD	0.0093	0.0883	0.0500	0.3	0.0857	0.0766	0.1148

The table shows the training and evaluation of the model which uses cross-validation. The output shows fold scores such as Accuracy, AUC, Recall, F1, Kappa. It is observed that the mean score of all the models coincides with the score printed in compare_model (), this is because the metrics in the score tables are mean scores of all the iterations.

We proceed to analyze the relationship between the variables. Python's Seaborn package allows you to easily build a correlation map, which is shown in Fig. 2. According to the data shown in Fig. 2, which shows the correlation between the different variables, it is observed that the attributes that are most correlated with each other are the two systolic and diastolic blood pressures, with the highest correlation value 0.98, there are others Attributes that show correlation are, persistent headache and persistent dizziness so it has a special value, and they are those that imply the objective value. This vision gives us an idea of what could be the factors that are related to the presence of the disease. And they provide us with the influence that one variable may have concerning another, something that would allow a more in-depth analysis of the variables considered as risk factors for this analyzed disease.

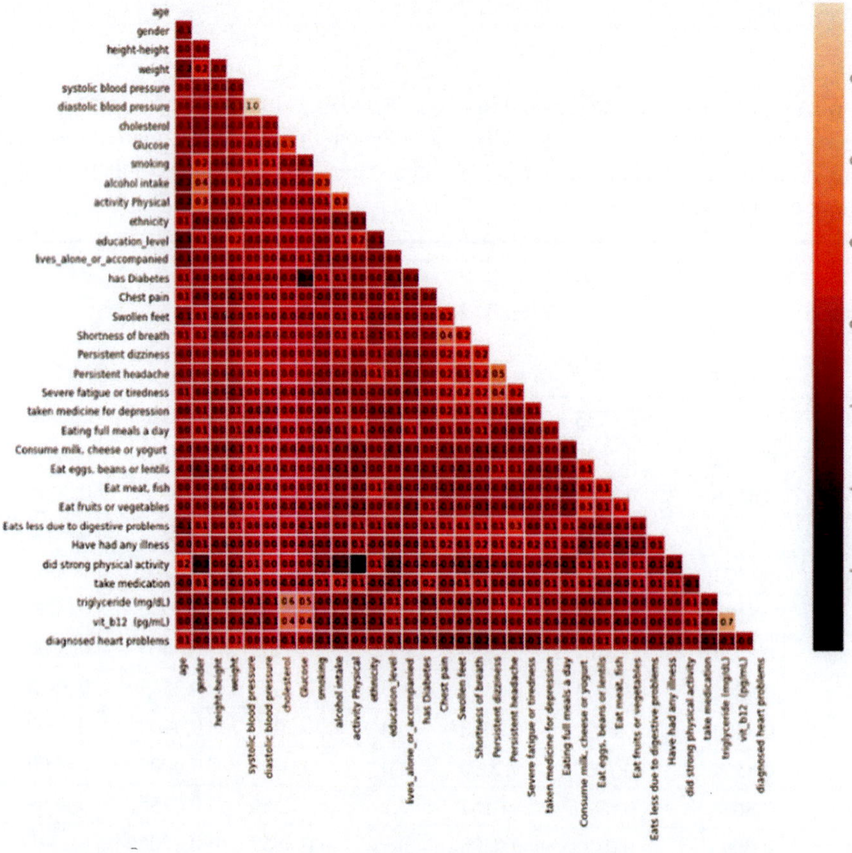

Fig. 2. Correlation between variables.

These correlations are not sufficient to determine risk factors, it should be applied according to other studies such as in [2] the analysis of complex interactions between massive variables, which studies the importance of the variables, where it was estimated using an additive explanation approach of Shapley. This is a unified approach to explain the output of any machine learning model. Which takes into account not only the feasibility but the value of the variables.

Applying the Shapley index to explain the risk factors that most influence CVD, the results shown in the graph in Fig. 3 are obtained.

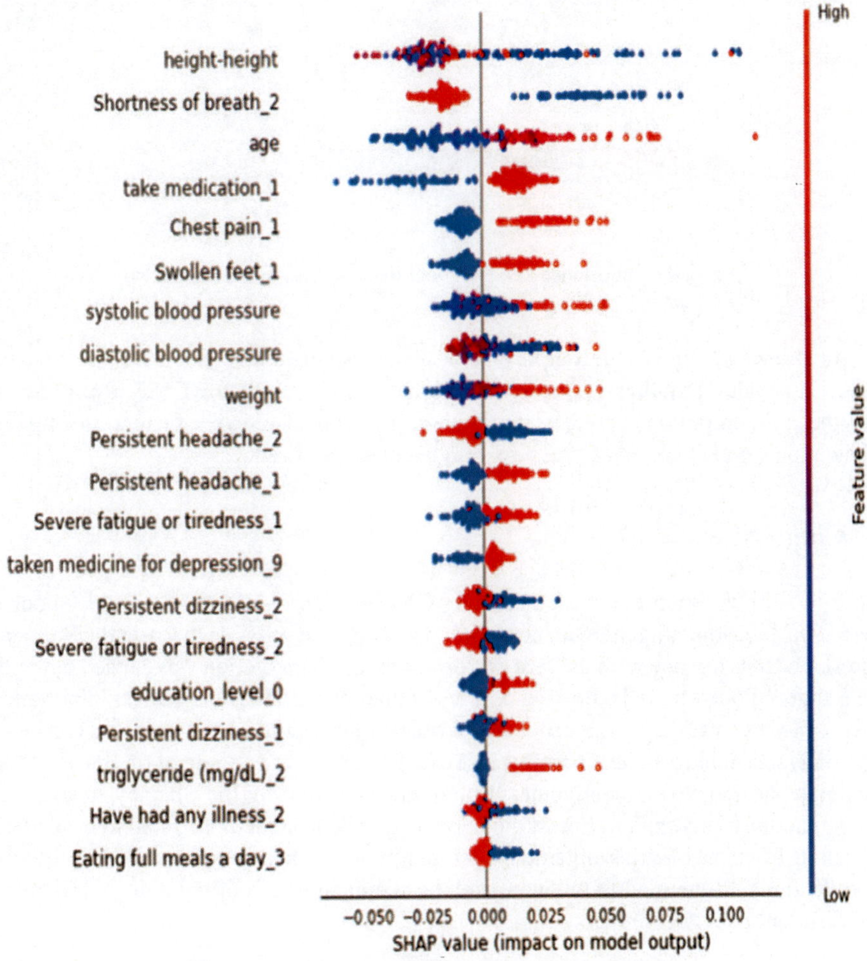

Fig. 3. Variables of greatest importance for the model.

As shown in Fig. 3, the variables with the highest value are age, taking medication, chest pain, swelling in the feet, systolic blood pressure, diastolic blood pressure, weight, Persistent headache, Severe fatigue or tiredness, taken medicine for depression, triglycine (mg/dL) and so on, the remaining ones can be analyzed in descending order, thus decreasing their importance value. As the value of the variable is shown, it is valid to clarify that the red color indicates a high impact of characteristics and the blue color indicates a low impact. In such a way that it can be concluded from these data that the most important variables on the model are: age, taking medicine, chest pain, swelling in the feet, severe fatigue or tiredness, taking medicine for depression, triglycine.

Another way to represent the results more precisely is observed using a force graph, which allows showing the importance and weight of the variables, which is shown in Fig. 4.

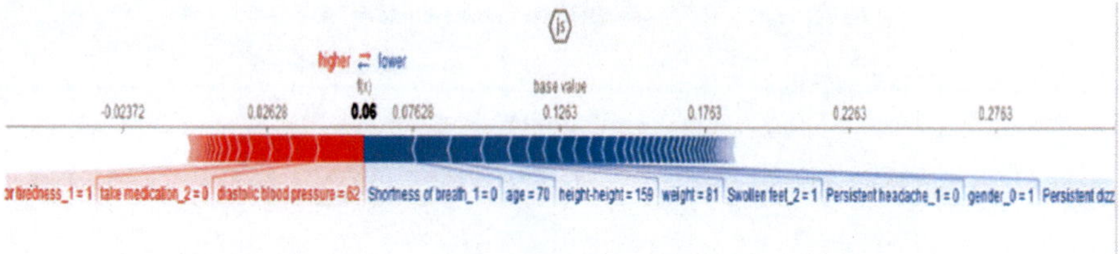

Fig. 4. Importance and weight of the variables in the model.

As shown in Fig. 4, the output value of the model (0.06) and the base-se value (0.1263) are the variables of greatest importance and weight for CVD. Resulting in that the most important variables for this disease considered risk factors are: taking medications, diasólico blood pressure, fatigue, or severe tiredness.

4 Conclusions

From sources of variables of risk factors for CVD in the province of Manabí, 13 models were evaluated adjusting them to these data, which made it possible to select the Random Forest model as the one with the best performance. CVD prediction was carried out with the Random Forest model, showing as a result that the variables of greatest importance or risk factors in CVD in the province of Manabí according to the data analyzed are: age, taking medication, chest pain, swelling in the feet, severe fatigue or tiredness, taken medicine for depression, triglycine. Which was validated by the Shapley index. The performance of the Random Forest model was validated and demonstrated to have good precision from the metrics obtained, such as the AUC. It is recommended for future work to apply all the models and compare the results in search of obtaining similarities and differences between them.

References

1. INEC: Encuesta de salud, bienestar del adulto mayor. https://www.ecuadorencifras.gob.ec/encuesta-de-salud-bienestar-del-adulto-mayor/. Accessed 10 Mar 2021
2. Góngora Alonso, S.: Análisis de fuentes y técnicas de Big Data en el sector sanitario (2017)
3. Zhang, L., Wang, Y., Niu, M., Wang, C., Wang, Z.: Machine learning for characterizing risk of type 2 diabetes mellitus in a rural Chinese population: the Henan Rural Cohort Study. Sci. Rep. **10**(1), 1–10 (2020)
4. Martínez-Toro, G., Rico-Bautista, D., Romero-Riaño, E.: Análisis comparativo de predicción dentro de bases de datos de cáncer: una aplicación de aprendizaje automático. Revista Ibérica de Sistemas e Tecnologias de Informação **E17**, 113–122 (2019)

5. Martínez, G., Bermúdez, Y.: Modelo híbrido para el diagnóstico de enfermedades cardiovasculares basado en inteligencia artificial. Tecnura: Tecnología y Cultura Afirmando el Conocimiento **16**(33), 35–52 (2012)
6. Aranda Núñez, A.: Estudio de la relación del esfuerzo de corte con la presión en aneurismas cerebrales y la predicción del riesgo de ruptura usando herramientas de inteligencia artificial basado en datos morfológicos, fluidodinámicos y estructurales. http://repositorio.uchile.cl/handle/2250/174776. Accessed 10 May 2021
7. Ambale-Venkatesh, B., et al.: Cardiovascular event prediction by machine learning: the multi-ethnic study of atherosclerosis. Circ. Res. **121**(9), 1092–1101 (2017)
8. Portugal, R., Carrasco, M.: Ensamble de Algoritmos Bayesianos con Árboles de decisión: una alternativa de clasificación. In: XVII Congreso Chileno de Control Automático ACCA, Universidad de la Frontera, Chile (2007)
9. Molina, C.: Detección temprana de riesgo cardiovascular usando text mining en los campos de texto no estructurado del registro clínico electrónico. http://repositorio.uchile.cl/handle/2250/130804. Accessed 15 May 2021
10. Jiménez, Y., Vivanco, O., Castillo, D., Torres, P., Jiménez, M.: Artificial intelligence in neuroeducation: the influence of emotions in the learning science. In: Botto-Tobar, M., Zambrano Vizuete, M., Díaz Cadena, A. (eds.) CI3 2020. AISC, vol. 1277, pp. 67–77. Springer, Cham (2021). https://doi.org/10.1007/978-3-030-60467-7_6
11. González, F.A.: Modelos de aprendizaje computacional en reumatología. Revista Colombiana de Reumatología **22**(2), 77–78 (2015)
12. Lagani, V., Koumakis, L., Chiarugi, F., Lakasing, E., Tsamardinos, I.: A systematic review of predictive risk models for diabetes complications based on large scale clinical studies. J. Diabetes Complications **27**(4), 407–413 (2013)
13. Sa, S.: Intelligent heart disease prediction system using data mining techniques. Int. J. Healthcare Biomed. Res. **1**, 94–101 (2013)
14. Rodriguez, C., Gallego, J., Mora, I., Duque, A., Bustamante, J.: Clasificación de latidos de contracción ventricular prematura basados en métodos de aprendizaje no supervisado. Revista Ingeniería Biomédica **8**(15), 51–58 (2014)
15. Palaniappan, S., Awang, R.: Intelligent heart disease prediction system using data mining techniques. In: 2008 IEEE/ACS International Conference on Computer Systems and Applications, pp. 108–115. IEEE (2008)
16. Moine, J., Haedo, A., Gordillo, S.: Estudio comparativo de metodologías para minería de datos. In: XIII Workshop de Investigadores en Ciencias de la Computación, pp. 278–281 (2011)
17. Dorado Díaz, P.I.: Contribuciones de las técnicas machine learning a la cardiología. Predicción de reestenosis tras implante de stent coronario (2019)

Organizational Management Analysis in Microenterprises: Colombian Case

Cliden Pereira(✉), Omar Borda, and Manuel Rey Borda

Corporación Universitaria Minuto de Dios, Bogotá D.C., Colombia
cliden.pereira@uniminuto.edu

Abstract. Companies have been evolving in their processes and adapting to the demands of the market, it can be seen during the last decade that especially SMEs have been applying different techniques and strategies in their management, however microenterprises are a little behind and even more so by the current situation of the economy. This research was carried out in order to analyze the main deficiencies in the organizational management of micro enterprises as specifically in Bogotá Colombia. This study is descriptive with a qualitative approach, since it seeks to know the causes of the difficulties in growth and sustenance in the market, a direct observation work is carried out with entrepreneurs through 12 focus group, the type of sampling is aleatorior, the area of greater business concentration of the city is taken as a reference, to know more precisely the activities that entrepreneurs perform in their businesses, a semi-structured survey of 35 questions oriented to know the functions related to the administration is applied to 120 microentrepreneurs. The results show the lack of projection and therefore the management of the same is focused on the daily activities of the establishment, also there is no awareness in the administrative, accounting and financial management, specifically in legalization and formalization, cost management, inventories, tax evasion among others. A training plan is proposed as a continuation of the project as a strategy to strengthen the organizational management of microenterprises.

Keywords: Management · Organizational management · Microenterprises

1 Introduction

The administration is one of the fundamental axes of the companies for their organizational management, since it allows the strengthening of the same, through the administrative process oriented to the fulfillment of the institutional objectives, through means and appropriate resources for the development of the activities. Microenterprises in Colombia present various problems in their management, according to the Chamber of Commerce of Bogotá 91.5% of the companies that close in the country are microenterprises (CCB, 2020), which do not reach the five-year lifecycle, usually caused by lack of credit and an adequate structure and administrative management [1].

The research is oriented to the characterization of the microenterprises of the south of Bogotá - Colombia in order to know the current situation of the administrative management in them, and to identify opportunities for improvement in these processes and generate a proposal that contributes to the strengthening, growth and positioning in the market.

M. Zambrano Vizuete et al. (Eds.): CI3 2021, LNNS 511, pp. 612–619, 2022.
https://doi.org/10.1007/978-3-031-11438-0_49

Therefore, the first phase of the research aims to identify the main needs in the administrative, accounting and financial management of microenterprises in Bogotá taking as a reference the area of greatest business concentration; the second phase shows the proposal for the strengthening of both administrative, accounting and financial management based on the diagnosis made in the business sector under study.

The structure of the article is composed in its first part by the theoretical framework where the concepts of organizational management are addressed, diagnosis of the processes in the administrative, accounting and financial management of microenterprises, also identification of strengths and weaknesses of the same; then, the methodology applied to carry out the study is described, continuing with the analysis of the data and results obtained. Finally, it describes the proposal for improvement to strengthen the management in microenterprises of the city and the conclusions of the article.

2 Theoretical Framework

Organizations and management are becoming increasingly important in today's society, due to the new and multiple challenges and opportunities determined by uncertainty, competitiveness, globalization and in general by the permanent interaction with the economic, cultural, social, political, environmental, and technological macroenvironment; likewise the microenvironment with its customers, suppliers, competition, allies, availability of resources and capacities for the development of the organization's own activities; given these conditions of the current world of societies where people are part of them, in different aspects of life determined by the actions carried out in and by organizations it is important to analyze their management and organizational development [2].

An organization is a group of people who direct all their efforts to achieve common goals, arise from solving the needs of people, individual or group; these in turn plan, organize, develop, produce, distribute different products and/or services that people require to satisfy a need or a desire [3], in this sense, "institutional objectives are the basic elements that guide organizational action. They reflect a future state desired by the organization" [4].

On the other hand, it is important to analyze the organization from two areas, one as an administrative function and the other as an entity. The first refers to the administrative process itself, related to the functions of the administration: planning, direction, coordination and control; hence the term organization refers to the act of organizing, integrating, and structuring resources, establishing relationships between the organizational structure and delegating responsibilities according to positions and functions to the collaborators of the process. The second, the organization as an entity, represents the formality in the organizations is visualized through the functional organizational structure that can be visualized in the organizational chart, in the same way the degree of authority and responsibility assigned.

There are administrative units or units, positions that are a set of similar tasks grouped under the same name, the positions or workplaces of people, the joint functions of activities related to each other, assigned to the positions that people perform to contribute to the achievement of the objectives of the organization, the activities that are a set of tasks for the fulfillment of a certain function and finally the rules, rules, procedures, methods and operations necessary for the fulfillment of the activities [5]. In this sense, the two meanings have enough correlation, therefore, all the processes are carried out according to a certain structure, they are taken together for the analysis of the administration and the relationship with the management, since both are oriented in the achievement of the objectives.

As for the concept of administration, it relates to the resources necessary for the normal development of the administrative process; However, nowadays the continuous changes and demands of the market on organizations to adapt and be more competitive arises the term management as a broader concept. That is, management is oriented to the integration of processes through actions over time that allow the organization to guide towards the achievement of the established goals; on the other hand, organizational theory focuses on the design of the organization and the organizational structure [6], for the functioning of the work teams according to the achievement of the proposed objectives. It can be said that, the administrator is the one who manages and manages an organization, also the decisions he makes can determine the success of the same in the short, medium and long term; since it is the one who decides what should and can be done, hence all organizations need to be managed and managed for the development of their plans and programs.

The management in the organization is based on the strategies that the leaders or managers of these develop for the achievement of their goals and objectives, the term management is related in the administrative field [7], in the various statements it is related to the form of organization and direction of the companies. Similarly, it can be said that, in organizational management, it is measured according to the performance of managers or directors through administrative functions, that is, when the four elements of administration are considered: plan, organize, direct and control to achieve the objectives [8].

In other words, management is understood to mean all those actions related to activities that are aimed at the realization and development of previously established objectives [9]. Also, to know the organizational management it is relevant to evaluate the process in decision-making; diagnostics is a tool of the management to obtain relevant information, analyze it and identify a set of variables of internal and external type. The external variables correspond to market evolution, demand, customers, competition, technological evolution, economic and social policies; internal variables such as use of material, human and financial resources, degree of technical knowledge, and organization [10]. Together, this will allow the owners and managers of the companies to establish the strategies to be followed to guide the actions for decision-making.

On the other hand, each company is responsible for establishing a correct direction of its processes towards the achievement of its goals, for this they must organize plans that allow to align the processes of their organization. The term process can be understood as "continuous sequence or transformation of an idea to reach precise purposes; it is also used to refer to the transformation of materials into products, or data into information with a specific objective" [11]. Thus, a process can be understood as a grouping of activities and operations that are carried out to generate value, seek to meet the needs of external or internal customers of the company, through the transformation of the raw material or providing a service [12].

Finally, it should be noted that to achieve an adequate integral management in small and medium-sized companies it is important to combine the components that are part of the organizational management such as: administrative, technical, legal, international, accounting and financial [13], these components are related as follows: the first considers the level of use of the resources, strategies and decisions that the company projects; the second the processes and other aspects of production to be competitive; the third component is associated with the formalization and accreditation of the organization; the fourth associated with foreign trade operations and finally the analysis of the accounting and financial system for decision-making being this component the basis for generating policies and practices in management within companies.

3 Theoretical Framework

3.1 Studio Design

A descriptive research is carried out with a qualitative approach that seeks to identify the main causes that occurs deficiencies in the organizational management of microentrepreneurs of Bogotá [14], the technique used was observation and field work, for his is taken as a population reference the area of the south of the city where the highest microenterprise concentration is located in the last year.

The study is carried out in three phases: the first phase of analysis of the management of microenterprises in the south of the city, with the purpose of knowing the perception of their administrative management [15] through 12 focus group in four localities or areas of the south of the city, with an average of 10 owners of establishments in each focus, made up of family groups and similar economic activities.

The second phase, corresponds to the planning of the design of strategy to strengthen the management of microentrepreneurs, for this, the information of the diagnosis previously made through the focus group is taken and semistructure survey is applied in order to analyze in depth the activities carried out by the mycormepressants for the administration of the businesses, the sample is calculated by means of random sampling for convenience [16] to 35 establishments that meet the characteristics of the population under study, finally the fieldwork information is crossed and the training plan structure is designed.

The third phase is oriented to the validation of the training plan by experts in the field and finally the socialization of the plan to microentrepreneurs through a seminar in order to adjust the final proposal and continue with the second phase of the Project.

3.2 Findings and Contributions

This research as mentioned above was carried out in three phases: The first phase begins with the documentary review of sources of information that allow to know the economic and business characterization, location of microentrepreneurs, through statistics, reports and business bulletins, presented by national and local entities, during the last five years, such as reports of the DANE, Chamber of Commerce of Bogotá, Secretariats of the Mayor's Office of Bogotá among other entities, and investigations currently carried out. According to this, it allows to identify that the concentration of microenterprises is located in the south of the city of Bogota, where 12 focus groups are carried out in different locations in the south of Bogotá during part of 2019 and the first quarter of 2020, as a result of the study of perception of the current situation of administrative and accounting management as well as its internal structure, it can be analyzed that most companies present difficulties in their management in the different functional areas. In industry, the main difficulties arise in the administrative, accounting and financial areas; in the commercial sector, management problems are concentrated in administrative, marketing and financial; while in the services sector the shortcomings are cost-oriented and financial as shown in Fig. 1. Other common aspects are also identified, most of the companies are family businesses, constituted mostly as natural persons, 47% of microentrepreneurs studied develop activities of a commercial nature, 44% in service activities, a minority of 7% industrial activities, and 2% any other type of commercial activity.

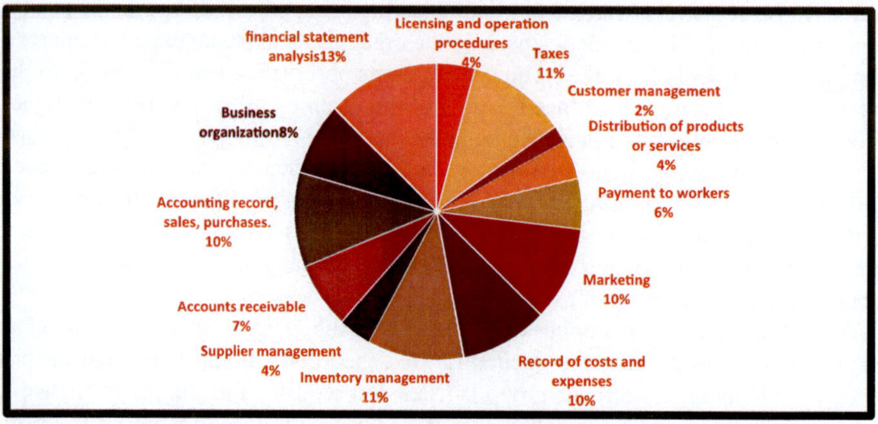

Fig. 1. Problematic organizational management of microenterprises. Research results.

Similarly, in phase two, based on the result of the perception study through the focus group of entrepreneurs, a diagnosis is made on specific needs in the administrative, accounting and financial area with the aim of examining in depth the organizational structure and way of managing the businesses to clearly identify the strengths and weaknesses in their management in each of its functional areas. For the study, the area with the highest business concentration is taken as a reference, in this case, the town of Kennedy located in the south of Bogotá [17]. 35 surveys are applied randomly to microentrepreneurs in the area, resulting in the following information:

As can be seen in Table 1, microenterprises present great difficulties in each of their functional areas, which allows to identify the ethical issue to be developed taking into account that currently the situation is even more complex due to the health emergency in the world.

Table 1. Problem of administrative, accounting and financial management in microenterprises

Areas	Description
Administrative	Lack of planning and business vision Informal organizational structure by family. Shared functions Verbal hiring, by referrals, remuneration for days Little training and welfare activities for workers Absence of marketing strategies High staff turnover and low motivation
Accounting and financial	Manual daily transaction logging Use of very poor, insecure and inadequate tools for the management of information Mixing personal expenses with the company Lack of knowledge in management of income, costs, expenses and inventories Lack of knowledge in tax obligations Difficulty in accessing sources of funding Capital comes from family savings Lack of knowledge and/or disinterest in handling accounting and financial information Lack of advice and importance for up-to-date business information

Therefore, once the main shortcomings of the companies in their management have been identified, it is proposed to structure a training plan in the areas of administrative, accounting and financial knowledge to strengthen the organizational management of microentrepreneurs in the south of Bogota to generate innovation and competitiveness strategies in the local, national and international market [18].

4 Results

According to the analysis of the documentary review, observation work and the results obtained from the surveys carried out, the main shortcomings of the microentrepreneurs of the south of Bogotá were identified, in their administrative, accounting and financial management; e can highlight that micro enterprises face great difficulty when seeking financing, due to the different requirements, which are mostly not easy to meet for this type of companies, due to their size, capital, opening time among others.

In addition, entrepreneurs do not have an administrative management itself there is no projection of their businesses and setting clear objectives that allow them to monitor and evaluate their goals over time and in general a clearly defined formal structure.

Therefore, taking into account the aforementioned results, it can be determined that there is an urgent need for awareness and training of the microentrepreneur, on the importance of carrying out a management oriented to the strengthening of their processes, hence, this leads to propose a training plan focused on the most important aspects of the administrative, accounting and finances area using tools of easy comprehension and learning in Table 2.

Table 2. Organizational management training plan.

Aspects	Objective	Thematic
Administrative management	Know the importance of management and organizational culture	Appropriate organizational models for business. Administration by objectives. Benchmarking, Outsourcing among others
Human talent management	Identify the process of human talent management in companies	Recruitment, selection, recruitment, training and evaluation of human talent. Organizational motivation
Marketing Management	Analyze marketing strategies and how to apply them to the business	Design of product design strategies, promotion, marketing distribution and customer service
Accounting management	Know the importance of accounting for decision-making in the company	Basics accounting Accounting record, accounting books, payroll and social security, income, costs, expenses
Financial management	Understand finances for microentrepreneurs	Projection of sales, expenses and profit of the business. Tax system and obligations according to economic activity. Sources of funding

Finally, it is validated with professionals of administrative sciences and the microentrepreneurs participating in the process by means of a sensitization seminar to know the importance of the management of their businesses for strengthening and competitiveness in the market, an application workshop is held identifying recommendations for each of the related topics, generating participation of questions that are satisfactorily resolved, it was carried out virtual in the month of December 2020.

5 Conclusions

As a result of the research it can be concluded that microentrepreneurs for the most part present great difficulties in the management of their businesses, mainly because they do not develop clear and efficient processes, or often lack them due to lack of knowledge and/or lack of interest; you can identify weaknesses in administrative, accounting areas in the microenterprises of southern Bogotá since they do not know the real situation of their organization, and this prevents them from taking both preventive and corrective actions, and thus viand losses and premature closures of their businesses. Microenterprises have a weak organizational culture, it is necessary to propose strategies that allow them to improve their organization or through innovation processes, in such a way that allows them to be more competitive and meet the needs of their customers.

References

1. Baquero, A.: Human resources management. Costa Rica (2005)
2. Bermudez, C.: Training, a tool to strengthen SMEs. Intersedes **38**(33) (2015)
3. Bernal, C., Sierra, D.: Administrative process for twenty-first century organizations (2015)
4. Confecámaras: Dynamics of business creation in Colombia. Bogotá (2020)
5. Creswell, J.: Educational Research. Planning, Conducting and evaluating Quantitative and Qualitative Research, Planning, Conducting and Evaluation in Quantitative and Qualitative Research, 4 edn. (2012)
6. Champoux, J.: Organizational Behavior: Integrating Individuals, Groups, and Organizations. New York, NY (2011)
7. Chiavenato, I.: The Administrative Process, 7 th edn. McGraw Hill (2011)
8. Flick, U.: Introduction to qualitative research, Ediciones Morata, Madrid, Spain (2011)
9. Garcia, J.: Marketing of products and services in small businesses or microenterprises. ADGD0210. Ic Editorial (2015)
10. García Pérez De Lema, D., Gálvez Albarracín, E.: Impact of innovation on MSME performance: an empirical study in Colombia. Manag. Stud. **28** (2012)
11. Gibson, J., Ivancevich, M., Donnelly, J.: Organizations: Behavior, Structure, and Processes. McGraw-Hill Inter-American (2020)
12. Góngora, N., Cicatelli, M.: Diagnosis and organizational evaluation: rethinking its approach (2012)
13. Hernández, R., Fernández, C., Baptista, P.: Research Methodology, 5th edn. McGraw-Hill, Mexico City (2013)
14. Lopez, P., Arevalo, M.: Towards an evolution in the field of knowledge of administrative discipline: from business administration to the management of organizations. Fac. Ciencias Econ. Univ. Militar Nueva Granada **18**(2), 75–93 (2010)
15. Mesa Arias, D., Marín Vélez, J.: Accounting tool for shopkeepers: case management of expenses. Doctoral dissertation, Universidad Libre Seccional Pereira (2018)
16. Salgueiro, A.: Management Indicators and Scorecard. Madrid (2001)
17. Portillo, R., et al.: Financial structure of the Colombian microenterprise. Opt.: J. Hum. Soc. Sci. **86**, 757–794 (2018)
18. Rios, M.: Marketing and business plan of the microenterprise: strategies and studies for the start-up of small businesses. Ideas propias Editorial SL (2015)

Evolution of Logistics in Small Businesses and Its Impact on Their Financial Performance in Times of Pandemic

Ximena Elizabeth Cayambe Badillo(✉), Marco Patricio Verdezoto Carrillo, and Luis Alberto Carrera Toro

Instituto Tecnológico Universitario Rumiñahui, Av. Atahualpa 1701 y 8 de Febrero, Sangolquí, Ecuador

ximena.cayambe@ister.edu.ec

Abstract. The research is structured on the basis of a review of the information provided by a group of small businesses using a non-probabilistic sampling. The information allows to determine, to what extent the implementation of logistic processes affected the financial results of the small businesses. In relation to the data obtained, a scenario of economic activities affected by a world crisis caused by a COVID 19 pandemic was considered. In accordance with the information provided for the year 2020, the behavior of sales data and growth percentages was reviewed, as well as the qualitative information, in order to conclude on the evolution of the logistics systems used by these companies in the period of study.

In conclusion, more than 65.7% of the micro-entrepreneurs and small businesses who implemented a logistics system, were able to maintain or increase their sales levels, and also staying in business on its markets.

Keywords: Small business · Logistics · Finance · Operation · Sales · COVID 19

1 Introduction

On December 2019 began a battle against an invisible enemy at worldwide level, which had a big economic impact on society, and also meant the disintegration of families worldwide. In the economic aspect, it is necessary to review the possible alternatives for small businesses to stay in business, since in 2019 they were 91% [1] of the total number of companies in Ecuador. In addition to above, in the last 10 years, they have had an almost sustained growth, as shown in the following (see Fig. 1).

In the development of the operations of small businesses during 2020, the importance of the implementation of logistics processes has been evidenced, since their implementation has allowed maintaining and even increasing the number of products or services distributed to customers and users. According to Mora García [2], "Logistics is an interdisciplinary activity that links the different areas of the company, from purchase scheduling to after sales service";

The financial results of small business are a direct consequence of the behavior of their operations, the level of sales and profitability are considered the most important

M. Zambrano Vizuete et al. (Eds.): CI3 2021, LNNS 511, pp. 620–633, 2022.
https://doi.org/10.1007/978-3-031-11438-0_50

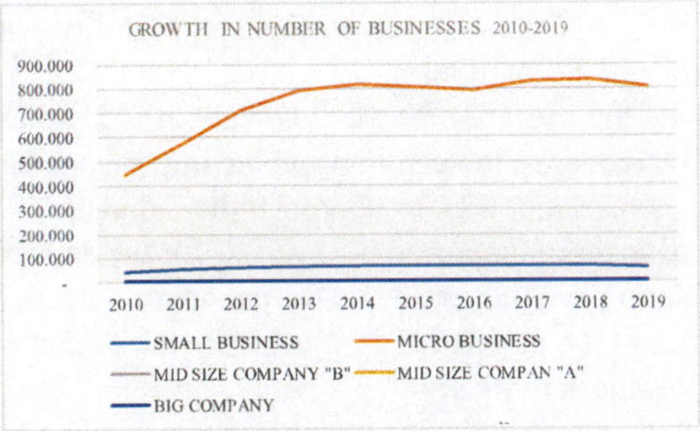

Fig. 1. Growth in number of businesses 2010–2019 [1]

business objective. According to Gonzales Urbina [3], "a successful company, usually achieves a significant growth rate in sales".

The financial crisis of several small businesses due pandemic, caused their closure. According to the report published by CEPAL [4], 21% of small businesses in Latin America and the Caribbean would close their operations in 2020, being these the most affected by the crisis due the pandemic declared by the World Health Organization in 2020. In Ecuador, as stated in the Regulation to the Structure of Productive Development of Investment issued according to Executive Decree 757 Official Record Supplement 450 of May 17, 2011 and reformed in 2017 and 2018, establishes the ranking of the companies as stated below (see Table 1).

Table 1. Ranking of companies according to their size

Variable	Micro business	Small business	Mid size company	Big company
Number of employees	01-sep	oct-49	50–199	>200
Sales	*<- $300.000*	$300.001-$ 1'000.000	$ 1'000.001-$ 5'000.000	> $ 5'000.000

According to Boero [5], the logistics system has as purpose, the efficient use of: investments in materials and equipment, materials, labor, etc. Additionally, it mentions the need for companies and small businesses to develop and control the processes of: purchasing, supplier management, inventory management, internal and external transportation, warehousing, packaging and distribution of finished products.

Within the advantages of logistics integration to achieve high performance in companies, Fernandez [6], refers to the connection of jobs and processes of the companies, which directly addressing the areas of: manufacturing, processes and activities, sales, product design and innovation, finance, IT.

Macías [7], indicates that the manager must have the ability to interpret and use the financial results to allow an optimal allocation of the company's resources, as well as, to facilitate the generation of the maximum possible profitability, without sacrificing the continuity of operations in the short term.

In order to evaluate the operation of the company, it is necessary for managers of companies and small businesses, to keep in mind the importance of sales [8].

In this context, this research, tries to approach the behavior of small businesses in times of COVID-19 pandemic, supported by a survey for the entrepreneurs. The survey was developed considering two things: 1) Did the pandemic and the quarantine have any impact on the activities of the company? 2) What were the decisions taken in the small businesses about to logistic activities?

This research is intended for researchers, public and private institutions interested in knowing about logistics and their financial impact.

This document is divided into 5 parts: 1. Introduction, 2. State of the Art, 3. Methodology and Methods, 4. Results, 5. Conclusions, and References.

2 State of the Art

Among the studies that have been conducted are the following (see Table 2).

Table 2. Evolution of logistics in small businesses and its impact on their financial performance in times of pandemic.

Research topic	Objectives	Conclusions
Small Businesses in pandemic: an approach from the entrepreneur's view	To find out about the behavior and expectations of micro and small enterprises considering a pandemic environment by COVID-19, from the perspective of the entrepreneur [1, p. 3]	The entrepreneur mentions that the COVID-19 pandemic and the quarantine in the country have negatively impacted the operations of micro and small businesses, especially in terms of sales volume and income [1, p. 1]
Administrative management and competitiveness of small businesses during and after the Covid-19 emergency	To compare the key indicators of economic competitiveness in some small businesses in Babahoyo Ecuador [2]	Statistical analysis shows that the Babahoyo region of Ecuador was affected due to salary adjustments and the drop of external demand [2]
The building values as a sum of capabilities, in the uncertainty generated by COVID 19	Review the value equation as the relationship between customer requirements, customer expectations and the company's offering [3]	The research shows that a new society has emerged, generating new tastes and preferences in consumers, determining factors for the growing of a new kind company that considers new social and individual requirements [3]

(continued)

Table 2. (*continued*)

Research topic	Objectives	Conclusions
The integration of finance into the logistics flow. Application: food processing	Propose a model to integrate logistics flow in supply chains in order to propose a tool that presents the integration between logistics and financial accounting [4]	Obtaining the MIFLCS as a tool for integrating logistics and accounting, facilitates the coordination between the institution's operational and logistics elements, with the statement of financial position and income statement [4]
Incidence of the implementation of logistics systems on the growing of mini-markets in zones five and nine of the Municipality of Envigado	To investigate the logistics systems implemented in some mini-markets in zones five and nine of the municipality of Envigado in order to identify the contribution of logistics to business growth [5]	Improving logistics systems has a strategic impact not only within the company, but also in the business environment, which will have a positive impact to support growth, and increasing sales, improving the product and quality at an affordable price, which generates a competitive advantage [5]
Sales levels during Covid-19, a systematic review of small business in Latin America	To know the level of sales in small business in Latin America during the period affected by Covid 19, using databases from Argentina, Colombia, México Perú Uruguay	Small business in Latin America played an important role in the economies, due to their ability to reinvent themselves, work from home, home delivery. Social networks are and will be allies in the new way of continuing to do business [6]

The continuity or not of a small business depends fundamentally on the entrepreneur who manages it, since decision-making and the choice of strategies become determining aspects for the continuity of the business, according to Ramírez [9], the results obtained show that behind the continuity of small business, an important factor is the role of the entrepreneur (owner, manager or director of the company).

The permanence of the small business depends largely on its competitive capacity, Checa [10], defines "competitive capacity, as the potential of the company to positioning itself in the market and stay in the market", in the same way, Galarza Villalva, Cruz Piza, Castro Pataron, & Marcial Coello, [11], mentions that the growth of small business, depends on the skills of entrepreneurs to assimilate knowledge and technology to compete in the markets, together with the ability to efficiently use their resources, to achieve high levels of productivity and competitiveness.

Changes in people's tastes and preferences worldwide were a determining factor for business continuity in 2020, and will be in the coming years, according to Amigo,

Arcieri, & Perozzi [12], it is necessary to consider a specific external environment composed of customers, suppliers, competitors and pressure groups; and a general external environment comprising economic, legal, political, technological, sociocultural, demographic and world conditions.

On the other hand, it is indicated that the infrastructure of information systems and logistics became the focus of attention in companies, as a result of the change in consumer behavior, as well as the psychological impact due to the concern about the vulnerability of the individual to the possible infection by Coronavirus. As a consequence of this, the circle of social influence changed to a circle of social isolation.

Martínez & Acevedo Suárez [13], mention that sales, costs, time of logistics cycles, idle and slow-moving inventories, collections and payments, determine the financial situation of the company.

One of the growth strategies of small business in times of pandemic will be the implementation of logistics systems, this will allow them to serve their customers; Bolivar, Taborda, & Quintero [14], mention that there is a lack of knowledge about logistics in business models. Logistics is performed empirically, without considering the profitability and effectiveness of their actions, when making decisions aimed at better performance and control of their operations.

3 Methodology and Methods

The research was addressed with the descriptive method which according to Caballero [15], "focuses on answering a question about how a specific part of the reality is".

Non-probabilistic sampling was carried out, with the aim of approximating the results that would be obtained, with probability samples [16].

Within this, self-chosen sampling is considered, which according to Eduardo [16] is mainly used to reduce costs, in this type of sampling, the subjects of the study are not selected, but people are asked to participate voluntarily.

Eduardo [16], mentions that in this type of sampling, the choice does not depend on chance but on someone's will.

In order to know the information regarding the implementation of logistics processes in small businesses during the year 2020, in which economic activities have been carried out in the midst of an environment of uncertainty and affected by the COVID-19 pandemic. Virtual surveys were sent to people related to MSMEs and the data obtained from the surveys were analyzed using the SPSS software. The survey addressed financial and operational aspects of the MSMEs, among these (see Table 3).

The information was collected between January and February 2021 with the support of the students of the Instituto Tecnológico Universitario Rumiñahui, 162 surveys were conducted of which 143 surveys are considered valid due to: repeated information, incomplete information, information from other countries (Colombia, Perú).

Table 3. Small business survey

Questions	Questions	Questions
Small business activity Objective: To know the economic activity of the company	Sales in 2019 Objective: To know what were the sales before the pandemic	Sales in 2020 Objective: To know what were the sales during the pandemic
Places were the companies are located Objective. To know location of the companies	Assets value in 2019 Objective: To know one of the parameters to rank the company	Assets value in 2020 Objective: To know one of the parameters to rank the company
Did you have logistics pro-cess before 2019? Objective: To know if be-fore pandemic, the company had logistics process implemented	Did you implement a logistics process in 2020? Objective: To know if it was nec-essary to modify the customer service process	The logistics process im-plemented in 2020, allowed to maintain or increasing the sales. Objective: To know if the changes in the customer service processes, allowed to maintain or increasing sales
What was the rate of rising sales in 2020? Objective: To know the contribution of the logistics system to increas-ing sales		

4 Results

First, the classification of small businesses was reviewed, considering the value of reported sales, as shown below (see Fig. 2).

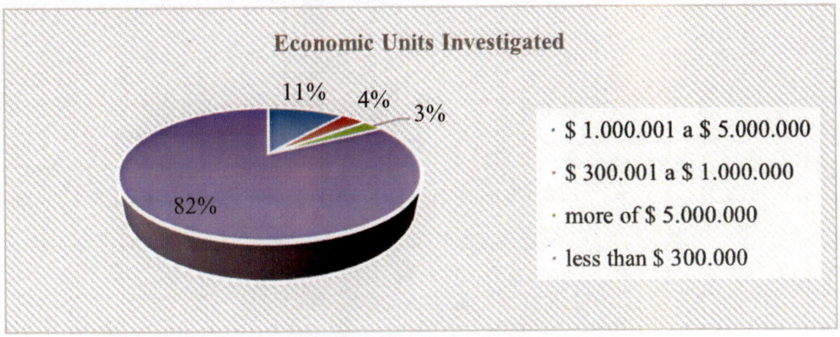

Fig. 2. Economic units investigated

Of the researched companies, 82% are small businesses, 11% are medium-sized companies, 4% are small companies, and 3% are large companies.

Secondly, it is important to know where the companies are located, hence they were classified as: Costa, Sierra, Oriente, and Galapagos regions. The analysis of the data shows that 70.4% of the companies are located in Sierra region, 22.5% of the companies are located in the Costa region, 3.5% of the companies are located in the Oriente region, and 3% of the companies are located in the Galapagos, as shown in the following graph (see Fig. 3).

Fig. 3. Location of the researched companies

If we review the sales variable, with the location, it was found that 70% of the researched companies were located in Sierra, 23% in Costa, 4% in Galapagos and 3% in Oriente, (see Fig. 4).

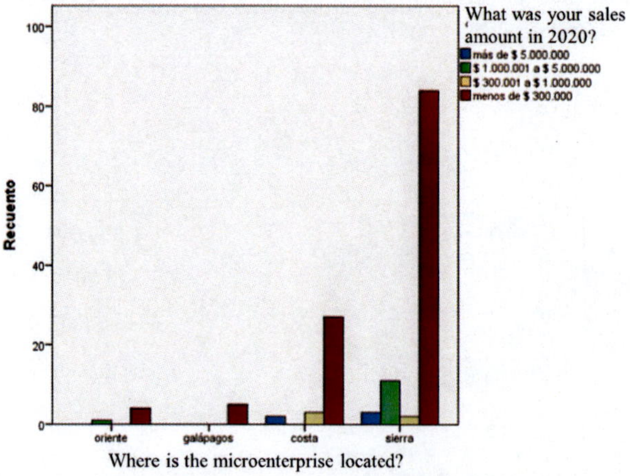

Fig. 4. Location by types of companies

To continue, the activities of the small businesses were reviewed, considering classifying them into: trading activities, manufacturing, service and construction, and grouped according to their line of business. The review showed that the majority of the companies were engaged in trade activities (41.5%), followed by service activities (29.6%), followed by manufacturing activities (26.8%) and finally construction (2.1%), as shown in Fig. 5:

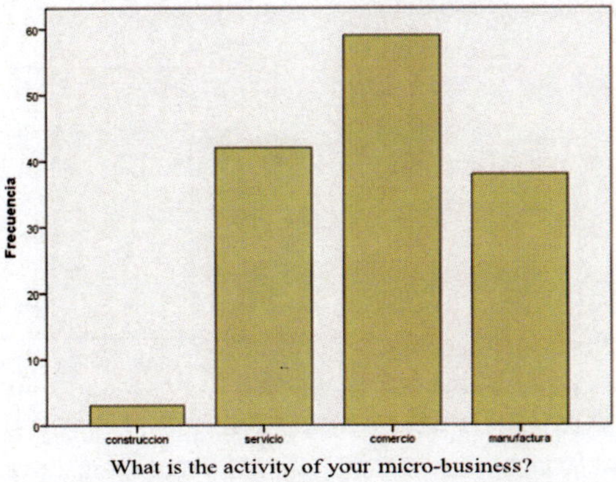

Fig. 5. Business line of the researched companies

Next, we reviewed whether the companies had a logistics system implemented before the 2020, in order to show the evolution of the implementation of logistics systems before the year in which the pandemic was declared worldwide. As a result, 57% did not have one, while 43% had implemented one, as can be seen in the graph (see Fig. 6).

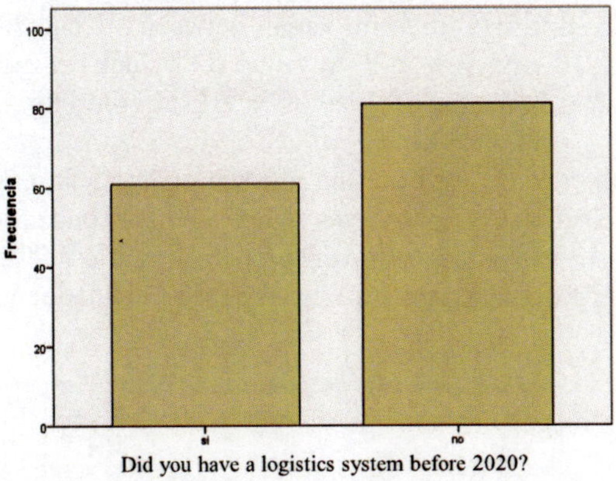

Fig. 6. Small businesses with logistics systems before 2020

Once reviewed whether the companies had logistics systems implemented before 2020, it was essential to determine how many implemented a logistics system during 2020, and among the results, it is observed that 59% had one implemented and 41% responded that they did not, which is shown in Fig. 7:

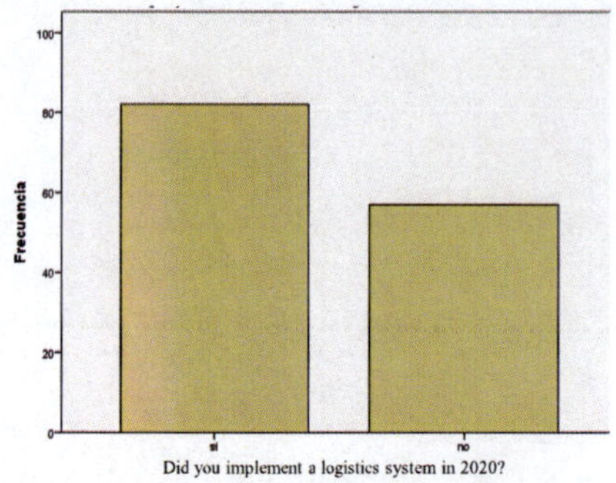

Fig. 7. Small businesses with logistics systems implemented in 2020

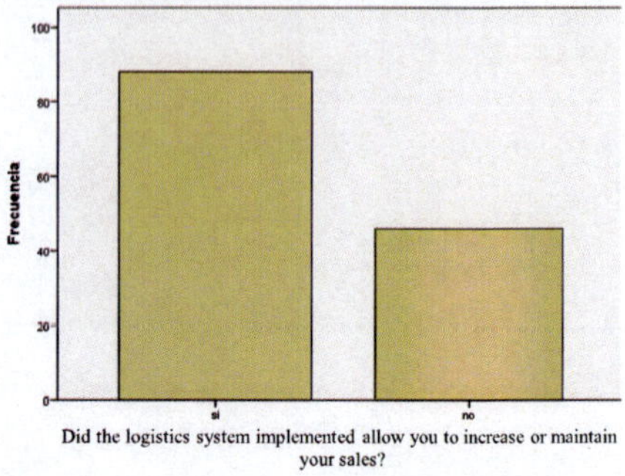

Fig. 8. Small businesses that maintaining their sales level or rising it with a logistics system implemented

At the same time, in Fig. 8, it is necessary to review whether the logistics system implemented in the companies during the year 2020, allowed them to maintain their sales level or rising it, and as a result of the data provided, a positive response of 65.7% and a negative response of 34.3% was obtained.

To clarify the impact on the company's sales, due the logistics systems implemented in 2020, the percentages of increase were researched; and, for the analysis, 5 groups of values were considered: a first group for values between 0% and 20%, a second group for values between 21% and 40%, a third group for values between 41% and 60%, a fourth group for values between 61% and 80%, a fifth group for values of more than 80%.

As a result of the review, we have the following participation percentages; In the first group the participation percentage was 76.8%, in the second group the participation percentage was 14.4%, in the third group the participation was 3.2%, in the fourth group the participation was 1.6%, and in the fifth group the participation was 4% (see Fig. 9).

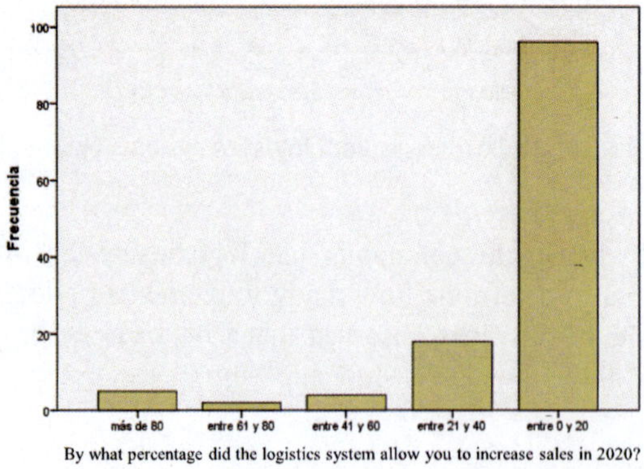

Fig. 9. Percentages of rising sales of the companies researched

Complementary to the analysis of individual questions and answers, it is necessary to cross-check the information, considering some variables that may be related to generate a better understanding of the phenomenon studied.

When reviewing the companies that had not implemented logistics systems before 2020 considering their economic activity, it was possible to identify that of the 100% of the companies did not have a logistics system before 2020, 43% were trading activities companies, followed by 30% of companies that performed services activities and 26% manufacturing activities (see Fig. 10).

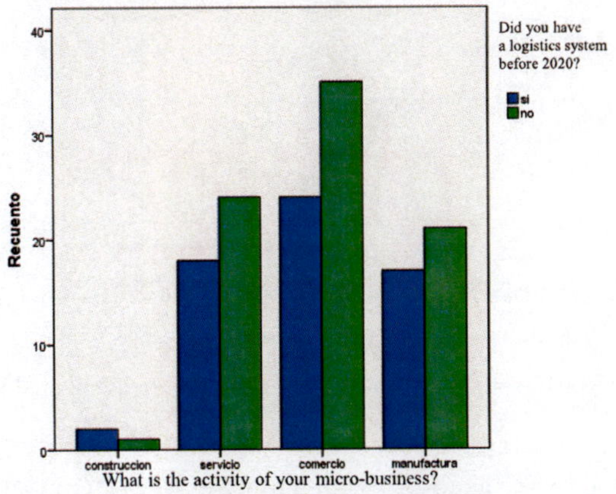

Fig. 10. Small businesses according their business line with logistics systems before 2020

Similarly, the type of company was analyzed to determine which did not have logistics systems, and the results showed that 96% of those who answered no, were small businesses (see Fig. 11).

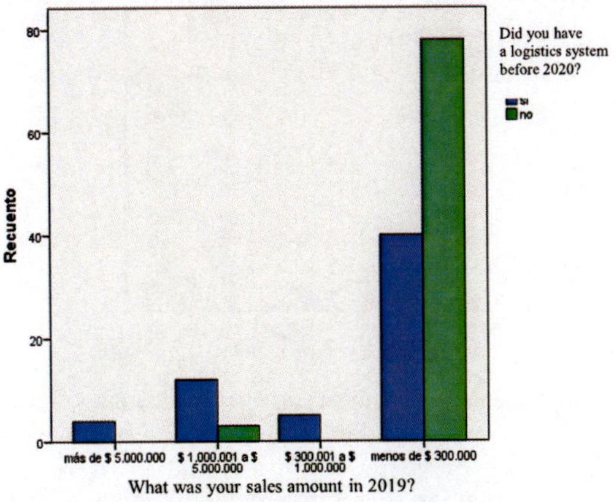

Fig. 11. Small businesses with logistics systems before 2020

Equally important was the review of the companies by economic activity, which implemented logistics systems in 2020; and, as a result, 51% of those who answered yes, were trading companies, followed by 26% of manufacturing companies and very closely with 22% of service companies (see Fig. 12).

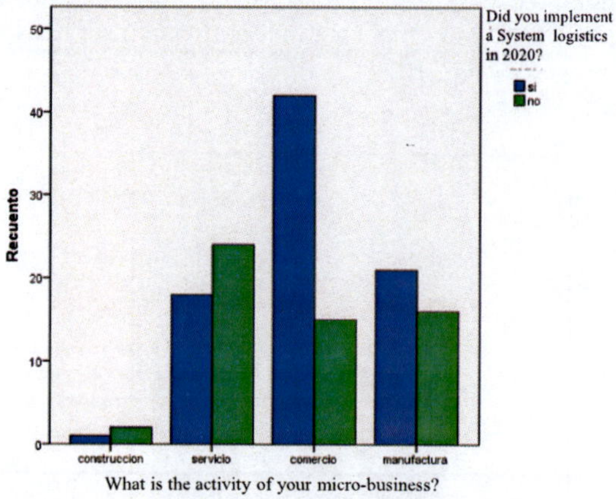

Fig. 12. Small businesses with logistics systems implemented in 2020

With respect to the same analysis, but relating the type of company, it was obtained that of the 100% of the companies that responded that they implemented logistics systems in 2020, 88% were companies ranked as small businesses (see Fig. 13).

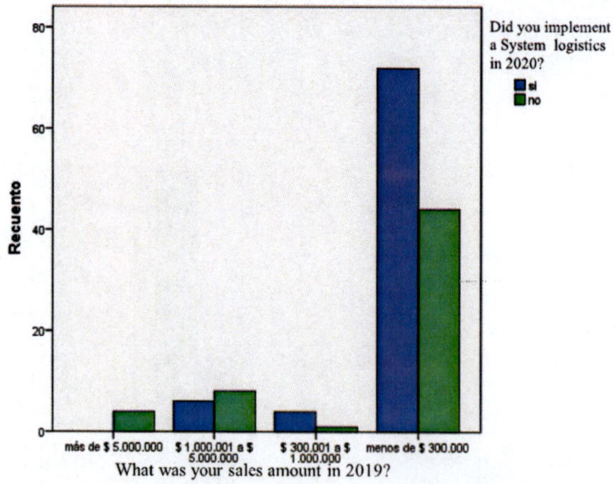

Fig. 13. Implementation of logistics systems according to sales levels

In addition to the companies that implemented logistics systems, we reviewed whether the logistics system allowed them to increase or maintain their sales, considering their economic activity and the type of companies, and the result showed that 48%

of the companies that answered yes, were trading companies, and 27% manufacturing companies. On the other hand, 89% of the companies that responded yes, (allowing them to increasing or maintain their sales), were small businesses (see Fig. 14).

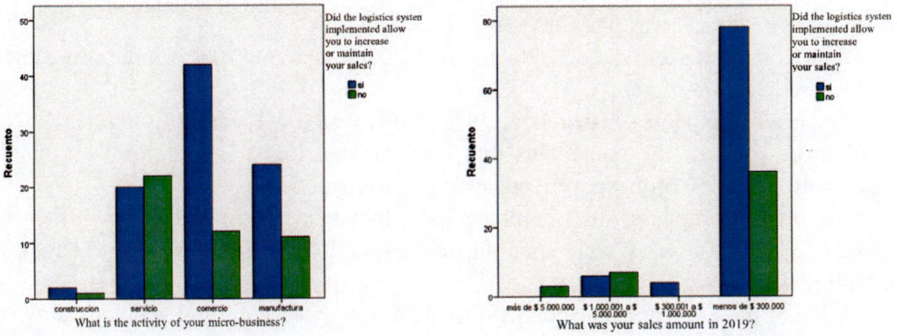

Fig. 14. Companies according to their business line and sales levels that maintained or increased their sales

After reviewing whether the logistics system implemented contributed to maintain or increase sales, the percentages of increasing sales mentioned by the companies were reviewed, and for the analysis, the data were grouped considering the following ranges; between 0% y 20%, between 21% y 40%, between 41% y 60%, between 61% y 80%, and more than 80%.

Based on the data processed, 82% of the companies that responded that they had had increases between 0% and 20% were small businesses, and of those, 38% were involved in trading activities, followed by 33% of services companies and 27% manufacturing companies (see Fig. 15).

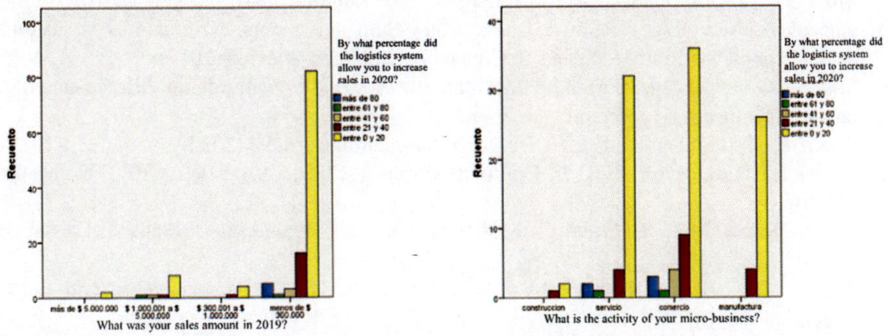

Fig. 15. Percentage of rising sales in companies ranked by business line and sales levels

5 Conclusions

In conclusion, from the non-probabilistic sampling, by which small businesses were asked to participate voluntarily, in order to establish whether the implementation of logistics systems during the pandemic in the year 2020 allowed them to maintain or increase their sales, it was possible to establish that:

Of the companies researched, 82% are small businesses; and 70.4% of the companies are located in Sierra region.

Of the all companies researched, 59% implemented a logistics system in 2020; in addition, 88% of the companies that implemented logistics systems were small businesses; and 51% of those were engaged in trading activities.

89% of the companies, that implemented a logistics system that allowed them to maintain or increase sales, were small businesses, and 48% of the total, were engaged in trading activities.

65.7% of the companies responded that the logistics system implemented, allowed them to maintain or increase their sales; and of those companies, 76,8% are between 0% and 20%; In addition, 82% of the companies that responded that they had had increases between 0% and 20% were small businesses, and of these, 38% were engaged in trading activities, followed by 33% of companies whose activities were services and 27% were in manufacturing.

Finally, with the data obtained, it can be concluded that the best strategy for small businesses, to stay in business in crisis times, and thus maintain and even increase their sales, will be with the implementation of a logistics systems. In a following research we will continue with the change of business models from the pandemic.

References

1. Instituto Nacional de Estadística y Censos, INEC: Directorio de Empresas y Establecimientos 2018. INEC, Quito (2019)
2. Mora García, L.A.: Gstión Logistica Integral. Ecoe Ediciones 2016, Bogotá (2016)
3. Gonzales Urbina, P.A.: Gestión de la inversión y el financiamiento: herramientas para la toma de decisiones. Instituto Mexicano de Contadores Públicos, México (2017)
4. Stumpo, G.: cepal.org, 13 March 2021. https://www.cepal.org/sites/default/files/events/files/ppt_covid_empresas_y_sectores_gs_v3.pdf
5. Boero, C.: Introducción a la logística. Jorge Sarmiento, Córdova (2020)
6. Fernandez Diez de los Ríos, J.: Optimización de la Cadena Logística. CEP S.L., Madrid (2016)
7. Macías Joven, M.T.: Guía del Gerente Financiero. Universidad de la Sabana, La Sabana (2020)
8. Aching, G.C.: Ratios Financieros y Matemáticas de la Mercadotecnica. El Cid Editor, Argentina (2015)
9. Ramírez Charca, E.D., Campos Lizarzaburu, W.B.: Microempresas en pandemia: una aproximación desde el discurso del emprendedor. Econ. Negocios **2**, 22–31 (2020)
10. Checa Hinojo, E.J.: Dirección de la actividad empresarial de pequeños negocios. IC Editorial, Málaga (2018)
11. Galarza Villalva, M.F., Cruz Piza, A., Castro Pataron, E.K., Marcial Coello, C.R.: La gestión administrativa y la competitividad de las microempresas durante y post la emergencia por Covid-19. Universidad y Sociedad (2020)

12. Amigo, A., Arcieri, H.C., Perozzi, A.R.: La construcción de valor como sumatoria de capacidades, en la incertidumbre que genera el Covid 19. Informes de Investigación (2020)
13. Martínez, D.A., Acevedo Suárez, J.A.: La integración de las finanzas al flujo logístico. Aplicación: proceso de alimentación. SCIELO (2019)
14. Bolívar, G., Taborda, J., Quintero, S.: Incidencia de la implementación de sistemas logísticos en el crecimiento de los minimercados de las zonas cinco y nueve del municipio de envigado. ESUMER Institución Universitaria (2020)
15. Caballero, R.A.: MEtodología Integral Innovadora para Planes y Tesis. Cengaje Elearning, México (2014)
16. Eduardo, B.: Métodos Estadísticos de Investigación. Editorial Brujas, Córdova (2018)
17. Instituto Nacional de Estadística y Censos, INEC: Visualizador de Estadisticas Empresariales, 13 March 2021. https://public.tableau.com/profile/instituto.nacional.de.estad.stica.y.censos.inec.#!/vizhome/VisualizadordeEstadisticasEmpresariales/Dportada
18. Lira, B.P.: Apuntes de Finanzas Operativas. Editor del proyecto editorial, Lima (2016)
19. P. S. y. G. A. d. l. C., Polanco, Y.J.: Análisis financiero para la toma de decisiones. Lennys Tejada Betancourt, República Dominicana (2020)
20. Tuesta-Panduro, J.A., Coronado-Chang, L.V., Pardo-HidalgoIII, C., SobeidaSalirrosas-Navarro, L.: Niveles de ventas durante el Covid-19 una revisión sistemática enmicroempresas en Latinoamérica. Anuario Fac. Ciencias Econ. Empresariales **12**, 280–286 (2021)

Author Index

M. Zambrano Vizuete et al. (Eds.): CI3 2021, LNNS 511, pp. 635–637, 2022.
https://doi.org/10.1007/978-3-031-11438-0